Advances in Damage Mechanics:
Metals and Metal Matrix Composites

Elsevier Titles of Related Intererest

Books

ABE & TSUTA
AEPA '96: Proceedings of the 3rd Asia-Pacific Symposium on Advances in Engineering Plasticity and its Applications (Hiroshima, August 1996). ISBN 008 042824 X

CARPINTERI
Handbook of Fatigue Crack Propagation in Metallic Structures. ISBN 0 444 81645 3

CARPINTERI
Minimum Reinforcement in Concrete Members. ISBN 008 043022 8

JONES
Failure Analysis Case Studies. ISBN 008 043338 3

KARIHALOO *ET AL.*
Advances in Fracture Research: Proceedings of the 9th International Conference on Fracture (Sydney, April 1997). ISBN 008 042820 7

LÜTJERING & NOWACK
Fatigue '96: Proceedings of the 6th International Fatigue Congress (Berlin, May 1996). ISBN 008 042268 3

MACHA ET AL.
Multiaxial Fatigue and Fracture. ISBN 008 043336 7

MARQUIS & SOLIN
Fatigue Design of Components. ISBN 008 043318 9

MARQUIS & SOLIN
Fatigue Design and Reliability. ISBN 008 043329 4

RAVICHANDRAN *ET AL.*
Small Faitigue Cracks: Mechanics, Mechanisms & Applications. ISBN 008 043011 2

RIE & PORTELLA
Low Cycle Fatigue and Elasto-Plastic Behaviour of Materials. ISBN 008 043326 X

VOYIADJIS *ET AL.*
Damage Mechanics in Engineering Materials. ISBN 008 043322 7

Journals

Acta Metallurgica et Materialia
Composite Structures
Computers and Structures
Corrosion Science
Engineering Failure Analysis
Engineering Fracture Mechanics
International Journal of Fatigue
International Journal of Impact Engineering
International Journal of Mechanical Sciences
International Journal of Non-Linear Mechanics
International Journal of Pressure Vessels & Piping
International Journal of Solids and Structures
Journal of Applied Mathematics and Mechanics
Journal of the Mechanics and Physics of Solids
Materials Research Bulletin
Mechanics of Materials
Mechanics Research Communications
NDT&E International
Scripta Metallurgica et Materialia
Theoretical and Applied Fracture Mechanics
Tribology International
Wear

For more information Elsevier's catalogue can be accessed via the internet on http://www.elsevier.com

Advances in Damage Mechanics: Metals and Metal Matrix Composites

G.Z. Voyiadjis

Department of Civil Engineering
Louisiana State University Batton Rouge, LA 70803-3808, U.S.A.

P.I. Kattan

Department of Civil Engineering
Applied Science University, Amman, Jordan 11931

1999
ELSEVIER
Amsterdam - Lausanne - New York - Oxford - Shannon - Singapore - Tokyo

ELSEVIER SCIENCE Ltd
The Boulevard, Langford Lane
Kidlington, Oxford OX5 1GB, UK

First edition 1999

Library of Congress Cataloging in Publication Data
Voyiadjis, G. Z.
        Advances in damage mechanics: metals and metal matrix composites
        / G.Z. Voyiadjis, P.I. Kattan. -- 1st ed.
        p.   cm.
        Includes bibliographical references (p.
        ISBN 0-08-043601-3 (alk. paper)
        1. Metallic composites--Mechanical properties. 2. Continuum damage mechanics. 3. Metals--Mechanical
        properties. I. Kattan, Peter Issa, 1961-   . II. Title.
        TA481.V69 1999
        620.1'692--dc21   99-38171 CIP

British Library Cataloguing in Publication Data
A catalogue record from the British Library has been applied for.

ISBN: 0 08 043601 3

∞ The paper used in this publication meets the requirements of ANSI/NISO Z39.48-1992 (Permanence of Paper).
Printed in The Netherlands.

Dedicated, in loving memory, to my parents

Zenon and Eleni (Lela) Voyiadjis

Advances in Damage Mechanics
Metals and Metal Matrix Composites

George Z. Voyiadjis
and
Peter I. Kattan

PREFACE

This book is intended to provide researchers and graduate students with a clear and thorough presentation of the recent advances in continuum damage mechanics for metals and metal matrix composites. Emphasis is placed on the theoretical formulation of the different constitutive models in this area. However, sections are added in the book to demonstrate the applications of the theory. A chapter is also included on experimental investigations and comparisons with theoretical predictions. In addition, some sections contain new material that does not appear before in the literature. Although the book covers damage mechanics in metals, the presentation is very minimal and many approaches to this topic are only briefly presented. This is mainly because there are other books that give a more thorough and adequate exposition to this subject. However the book goes in more details into the subject of damage mechanics in metal matrix composites. The primary reason for this is that no such book currently exists on this topic. Most of the available books that were published recently on this topic are edited books. This is the first book that attempts to bring together continuum damage mechanics and metal matrix composites in a single and unified volume. Therefore, a substantial part of the material reflects the authors' own work in this area, but other approaches are presented when they are clearly relevant to the topic under discussion.

The major goal of this text is to introduce many of the different constitutive models that recently appeared in different research publications. Another goal is to clearly present the different approaches to this topic in a single complete volume that will be easily accessible to researchers and graduate students in civil engineering, mechanical engineering, engineering mechanics, and materials science. Most of the available books on this subject are edited books; they are just collections of research papers with no clear relation between them. This book presents the material in well-organized chapters that start with the preliminaries and proceed to advanced topics. Furthermore, the book is divided into three major parts: Part I deals with the scalar formulation and is limited to the analysis of isotropic damage in materials. Thus this part can be read by a wide variety of readers; the only mathematical requirement is a knowledge of simple algebra. However, Parts II and III deal with the tensor formulation and is applied to general states of deformation and damage. The reader of these parts is assumed to have an advanced mathematical training in tensor algebra in order to fully grasp the intricacies and detailed mathematical derivations that appear in these parts.

The material appearing in this text is limited to plastic deformation and damage in ductile materials (e.g. metals and metal matrix composites). The authors elect to exclude many of the recent advances made in creep, brittle fracture, and temperature effects. The authors feel that these topics require a separate volume for this presentation. Furthermore, the applications presented in the book are the simplest possible ones and are mainly based on the uniaxial tension test. The presentation of more challenging examples is left to the researchers in this field.

This book does not claim to revolutionize the way in which research is done in this area, but it does advance a few new ideas and it does have several noteworthy features:

1. There is a complete separation of scalar and tensorial formulations.
2. New approaches to the analysis of damage in composite materials are presented in a mathematically consistent manner.
3. Much effort went into the design of the structure of the book into well organized chapters with specific topics that gradually increase with difficulty.
4. The book places heavy emphasis on the incorporation of the computer into the research process. The source code and binary files that may be requested from the first author contain many of the different models presented in this book. This is in an effort to bridge the gap between the theoretical development and engineering applications.
5. There is a direct link between each constitutive model and its associated computer subroutines. These links are shown very clearly at several locations throughout the book.

The book consists of seventeen chapters. The contents of each chapter are arranged in sections with specific topics and increasing difficulty. Chapter 1 is an introductory chapter that reviews the basic assumptions and outlines the scope of the book. Chapters 2-5 deal with damage in uniaxial tension of metals and metal matrix composites. Chapters 6 and 7 deal with general states of damage and plasticity in metals. This is followed in Chapters 8-11 by an extension of the theory to metal matrix composites. These chapters conclude the theoretical presentation for metals and metal matrix composites. However, the authors select to add two additional chapters on related topics. Chapter 12 deals with the problem of symmetrization of the effective stress tensor. This chapter involves highly complex algebraic manipulations and may be excluded from a preliminary reading of the book. Chapter 13, however, is very relevant to the main material and presents the recent experimental investigations and comparisons with theoretical predictions. Anisotropic cyclic damage with anisotropic plasticity is presented in Chapter 14. In Chapter 15, the generalized cells model, is applied to damage models. This is an alternate approach to the homogenization procedures that use the averaging scheme. The kinematic description of damage is presented in Chapter 16. Finally, the coupled theory of damage with inelastic behavior is presented in Chapter 17 for both room and elevated temperatures. This accomplished for both rate dependent and rate independent plasticity and damage. The lengthy equations are listed in Appendix A so as not to clutter the main body of the book. Selected computer subroutines of some of the important models are available in FORTRAN language for the interested readers. The source code and binary files may be requested from the first author. They may also be obtained from the web site www.rsip.lsu.edu/csmlab. These subroutines must be linked to a main program, preferably a general purpose finite element program.

This book can also serve as a textbook for an advanced course on damage mechanics. As far as the authors know, such a course is not offered at the universities worldwide. It is hoped that this book will open the way for teaching such a course in the near future. The reader is assumed to have a solid background in the theory of plasticity and mechanics of composite materials. Plasticity is required for a thorough understanding of Chapters 7-17. Mechanics of composite materials is needed for Chapters 4-5 and 8-17. No knowledge is assumed in the finite element method; however, such knowledge is helpful for a complete appreciation of the material in sections 6.3, and 7.5, and Chapter 17. The material appearing in section 2.4 and Chapter 3 is new and does not appear before in the literature.

The authors wish to express their sincere appreciation and thanks to many individuals and friends for their assistance, advice and encouragement throughout the writing of this book. In addition, this work would not have been completed without the help, support and persistent encouragement of our family members. Finally, we would like to thank the editor of the Elsevier for providing us the opportunity to bring this book in its present form.

George Z. Voyiadjis
Baton Rouge, Louisiana

Peter I. Kattan
Amman, Jordan

May 1999

CONTENTS

CHAPTER 1

INTRODUCTION

In this introductory chapter, several issues concerning history, problems and approaches to various topics are discussed. The three topics of continuum damage mechanics, finite-strain plasticity and mechanics of composite materials are introduced. First, a brief history of continuum damage mechanics is given. This is followed by outlining some recent problems in finite-strain plasticity. Then the different approaches in the mechanics of composite materials are described. The chapter is concluded with an outline of the scope of the book and the notation used.

1.1 Brief History of Continuum Damage Mechanics

Continuum damage mechanics was introduced by Kachanov [1] in 1958 and has now reached a stage which allows practical engineering applications. In contrast to fracture mechanics which considers the process of initiation and growth of micro-cracks as a discontinuous phenomenon, continuum damage mechanics uses a continuous variable, φ, which is related to the density of these defects as shown in Figures 1.1 - 1.3 (Wang et al. [2], Bettge et al. [3], Voyiadjis and Venson [4]) in order to describe the deterioration of the material before the initiation of macro-cracks.

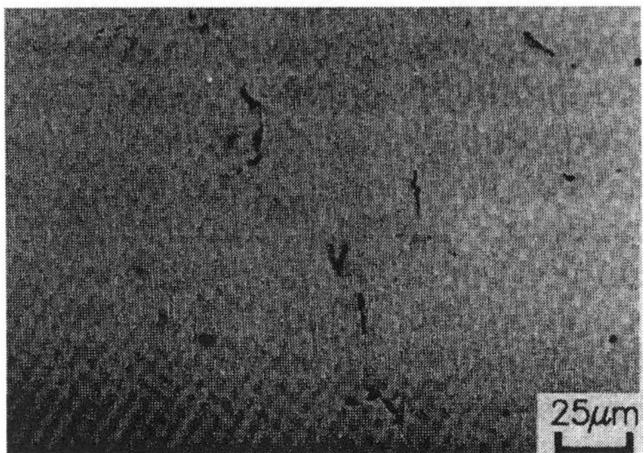

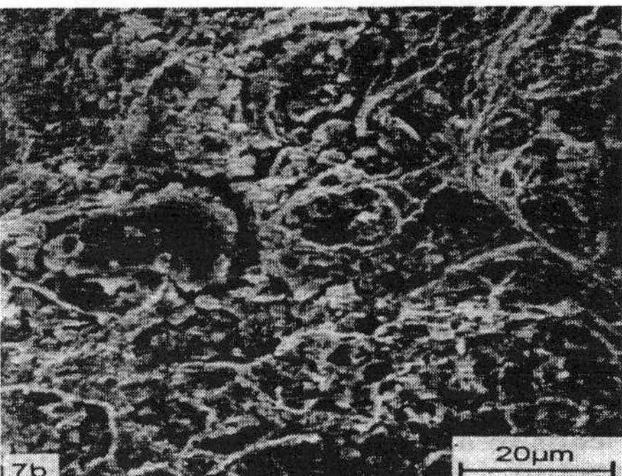

Fig. 1.1, Cavities and Micro-cracks in Grain Boundaries (Wang et al., [2])

Fig. 1.2, Cracks at Inclusion Surface (Bettge et al., [3])

2

Based on the damage variable φ, constitutive equations of evolution are developed to predict the initiation of macro-cracks for different types of phenomena. Lemaitre [5] and Chaboche [6] used it to solve different types of fatigue problems. Leckie and Hayhurst [7], Hult [8], and Lemaitre and Chaboche [9] used it to solve creep and creep-fatigue interaction problems. Also, it was used by Lemaitre for ductile plastic fracture [10,11] and for a number of other applications [12].

The damage variable, based on the effective stress concept, represents average material degradation which reflects the various types of damage at the micro-scale level like nucleation and growth of voids, cavities, micro-cracks, and other microscopic defects as shown in Figures 1.1 - 1.2.

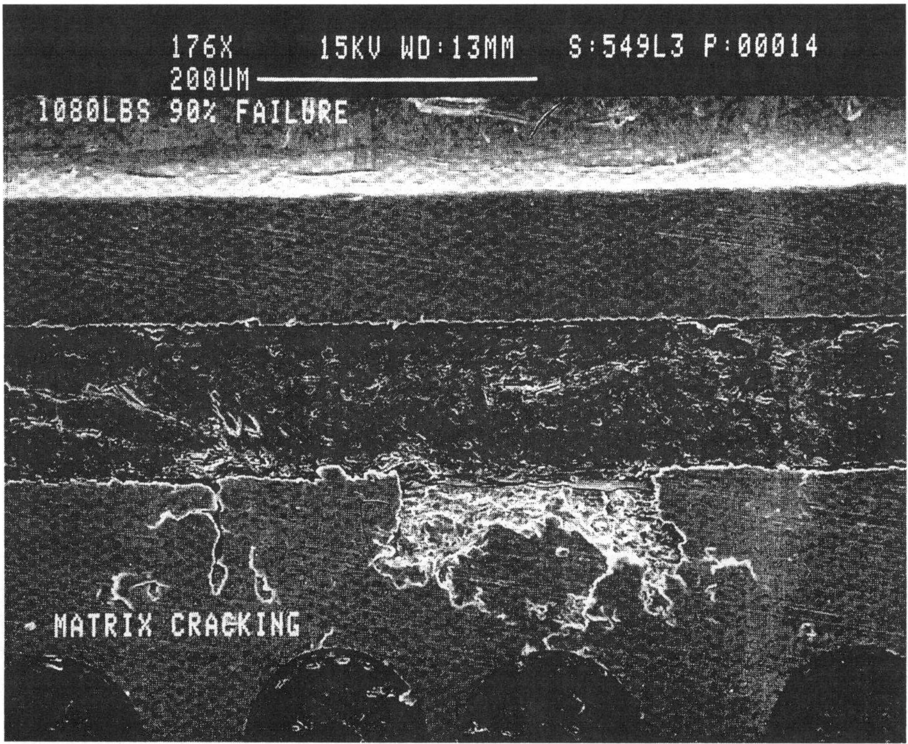

Fig. 1.3, Damage in Metal Matrix Composites (Voyiadjis and Venson [4])

For the case of isotropic damage, the damage variable is scalar and the evolution equations are easy to handle. It has been argued [12] that the assumption of isotropic damage is sufficient to give good predictions of the load carrying capacity, the number of cycles or the time to local failure in structural components. However, development of anisotropic damage and plasticity has been confirmed experimentally [13-15] even if the virgin material is isotropic. This has prompted several researchers to investigate the general case of anisotropic damage.

The theory of anisotropic damage mechanics was developed by Sidoroff and Cordebois [16-18], and later used by Lee, et al [15], and Chow and Wang [19, 20] to solve simple ductile fracture problems. Prior to this latest development, Krajcinovic and Foneska [21], Murakami and Ohno [22], Murakami [23], and Krajcinovic [24] investigated brittle and creep fracture using appropriate anisotropic damage models. Although these models are based on a sound physical background, they lack vigorous mathematical justification and mechanical consistency. Consequently, more work needs to be done to develop a more involved theory capable of producing results that can be used for practical applications [21,25].

In the general case of anisotropic damage, the damage variable has been shown to be tensorial in nature [22,26]. This damage tensor was shown to be an irreducible even-rank tensor [27, 28]. Several other basic properties of the damage tensor have been outlined by Betten [29, 30] in a rigorous mathematical treatment using the theory of tensor functions.

Lemaitre [31] summarized the work done in the last fifteen years to describe crack behavior using the theory of continuum damage mechanics. Also Lemaitre and Dufailly [32] described eight different experimental methods (direct and indirect) to measure damage according to the effective stress concept [33].

Chaboche [34-36] described different definitions of the damage variable based on indirect measurement procedures. Examples of these are damage variables based on the remaining life, the micro-structure and several physical parameters like density change, resistivity change, acoustic emissions, the change in fatigue limit, and the change in mechanical behavior through the concept of effective stress.

1.2 Finite-strain Plasticity

The widely used tools of classical fracture mechanics employ global concepts in analyzing

4

ductile rupture. These include strain energy release rate, contour integrals, and even stress intensity factors which are based on an overall global analysis of the cracked structure using energy considerations. These concepts have been very successful in predicting crack behavior in two-dimensional elasticity or small strain plasticity that involves only proportional loading paths. However, these concepts suffer from the following disadvantages:

1.   The hypotheses involved are too restrictive thus leading to large safety factors for their implementation.

2.   It is difficult to use the concepts of classical fracture mechanics for more sophisticated problems involving finite strain plasticity, ductile fracture due to large deformation, time-dependent behavior, three-dimensional effects (nonproportional loading paths), and delamination of composites.

In order to develop a model for a coupled theory of continuum damage mechanics and finite strain plasticity, a suitable stress corotational rate is needed. The Jaumann stress rate has been studied extensively in the past, but this rate will limit the theory to plasticity models which do not exhibit kinematic hardening (Lee, et al [37] and Dafalias [38]). According to these investigators, a monotonic simple shear loading causes oscillating shear stress response when use is made of the Jaumann stress rate for a kinematic hardening plasticity model.

A number of plausible explanations of the phenomenon have been presented. Lee, et al. [37] proposed a modified corotational rate using the spin of the principal direction of $\boldsymbol{\alpha}$ with the largest absolute eigenvalue, where $\boldsymbol{\alpha}$ is the deviatoric component of the shift stress tensor. An alternate approach by Onat [39, 40] defines the spatial spin equal to the anti-symmetric part of $d_{ij}^{\prime\prime}\,\alpha_{jk}$ multiplied by a constant, where $\boldsymbol{d}^{\prime\prime}$ is the plastic part of the spatial strain rate $\boldsymbol{d}$. The non-oscillatory solution for simple shear is obtained by the proper choice of the constant.

Dafalias [38, 41] and Loret [42] obtained similar relations by associating the corotational rate with the material substructure as defined by Mandel [43, 44]. Mandel [43] used the triad of director vectors attached to the material substructure and developed the theory of plasticity such that the substructure corotational rate is defined in terms of the spin of the director vectors. He postulated that the constitutive relations require not only the plastic component of the spatial strain rate tensor but also the plastic component of the spatial spin tensor. However, Onat and Leckie [28] have shown that it is advantageous to consider the internal structure and its orientations as a single entity and to use tensorial state variables for the representation of this entity [22, 26, 45-47].

Dafalias [41] and Loret [42] discussed the macroscopic constitutive relations for the plastic spin using the representation theorem for isotropic, second-rank, anti-symmetric, tensor-valued functions. The importance of the material substructure in defining objective corotational rates is also argued by Pecherski [48]. In inelastic finite deformations of polycrystalline metals, the material moves with respect to the underlying crystal lattice. The lattice itself undergoes elastic deformation and relative rigid-body rotations due to the lattice mis-orientation [48].

The work outlined above [37-42] imposes a retardation of the material spin $\mathbf{W}$ in order to obtain a non-oscillatory solution for the simple shear problem. The analysis of the solution of the simple shear test problem in [39-42] results in an unbounded non-oscillatory solution for the shear stress that increases montonically with increased deformation. Concurrently, the normal stress approaches an asymptotic upper bound. In the case of reference [37], both the shear and the normal stresses are unbounded and increase monotonically with increased deformation. We also note that in [39-42], the principal directions of $\boldsymbol{\alpha}$ tend toward the bisector direction of the plane coordinate axes while in [37] the maximum principal direction of $\boldsymbol{\alpha}$ inclines towards the horizontal axis. The above proposed solutions fail in the proper prediction of the shear stress-shear strain characteristic and the Swift effect in torsion of thin-walled tubes [49]. In Chapter 16, a damage spin associated with damage is introduced similar to the plastic spin concept.

Other authors have followed different approaches for the proper choice of the corotational objective stress rate. Atluri [50], based on the idea of a complete hypo-elastic law, modifies the rate of the back stress equation for the case of a rigid-kinematic hardening plastic model. Johnson and Bammann [51], Fressengeas and Molinari [52], Moss [53], Simo and Pister [54], Voyiadjis [55], and Voyiadjis and Kiousis [56], have also discussed different aspects of the proper choice of the objective stress rate in finite deformation analysis. An ASME publication by Willam [57] summarizes the debate on this subject.

Recently, Murakami [58] formulated a general theory of anisotropic damage mechanics based on a consistent mathematical and mechanical basis using the principles of continuum mechanics. He argued that since the material undergoes both damage and deformation at the same time, the damage tensor φ also depends on the current state of deformation and thus cannot describe properly the internal state of damage in the case of large deformation. Consequently, he introduced a new damage tensor $\bar{\varphi}$ that is derived with respect to the elastically unloaded damaged state.

6

1.3 Mechanics of Composite Materials

Fiber-reinforced composite materials play an important role in the industry today through the design and manufacture of advanced materials capable of attaining higher stiffness/density and strength/density ratios. Of particular importance is the problem of damage initiation and evolution in fiber-reinforced metal matrix composite plates. The analysis of damage mechanisms in two-phase composites is a rather complex problem that has challenged researchers during the past two decades. Although the literature is rich in new developments in the composite materials technology, it lacks tremendously a consistent and systematic approach to the study of damage in composite materials.

In reviewing the available literature concerning fiber-reinforced composites, it is clear that two different approaches are employed. In the first approach, the composite material is treated as a transversely isotropic medium and continuum theories are used in its analysis [59 - 64]. In this approach, the fiber direction is taken as the direction of anisotropy and the classical equations of orthotopic elasticity are used in the analysis [59 - 62]. The disadvantages of this approach are that no distinction is made regarding the different phases (matrix and fibers) in the analysis of stresses and strains and no consideration is given to the local effects of deformation and damage, especially the effects of the matrix-fiber interaction. Other researchers [65] used fracture mechanics techniques to analyze cracks in multi-layered plates.

In the second approach, micro mechanical models are used where the matrix and fibers are treated separately in a local analysis and this, in turn, is linked with the overall composite behavior. The advantages of using this approach are that local effects can be accounted for and different damage mechanisms can be identified. Different micro-mechanical models employ different methods of achieving the local-overall relations. Hill [66,67] employed volume averages of stress and strain increments in the different phases and introduced certain concentration factors to relate these volume averages of the local fields to the overall uniform increments. Dvorak and Bahei-El-Din [68-70] used Hill's technique to analyze the elasto-plastic behavior of fiber-reinforced composites. They considered elastic fibers embedded in an elasto-plastic matrix. In their micro mechanical analysis of elasto-plastic composites, Dvorak and Bahei-El-Din [68-70] identified two distinct deformation modes. One is matrix dominated and the other is fiber dominated. The first mode is prevalent in the case of stiff elastic fibers while the second mode is more general where the elastic fibers are more compliant and the mode is treated as a general case of plastic deformation of a heterogeneous medium. Aboudi [71] used an averaging technique in order to relate the local stresses to the overall composite stress.

A thermo-mechanical constitutive theory has recently been proposed by Allen and Harris [72] and Allen et al. [73] to analyze distributed damage in elastic composites. In particular, the problem of matrix cracking has been extensively studied in the literature [74-78].

A number of damage theories have been proposed with limited experimental investigation. These investigations are primarily confined to damage as a result of fatigue of fracture [79-82]. Each of these investigations does not present damage evolution as a function of the measured physical damage over a load history. A more recent work by Majumdar et al [83] provides a thorough examination and explanation of the microstructural evolution of damage. However, this work has not been extended to a constitutive theory for the quantification and evolution of physical damage. Recently, new experimental procedures have been introduced to quantify damage due to micro-cracks and micro-voids through X-ray diffraction, tomography, etc. [84-86]. Nevertheless, these procedures need to be refined in order to differentiate between the different types of damages such as voids and cracks as shown in Figure 1.3 (radial, debonding, z-type). Additional experiments need to be performed in order to quantify the damage parameters as well as evaluate the proposed damage theory. Much of the work in this area has been done using a continuum approach with various schemes of measuring the damage. In each of the schemes, damage is a measure of the ratio between an effective quantity and its respective damaged value. Lemaitre and Dufailly [32] listed several methods of obtaining ratios for the damage parameters based on area of resistance, material density, and elasto-plastic modulus. Obtaining the damage parameter as a ratio of the elastic-plastic modulus is most widely used because of the ease in evaluating the damaged and undamaged elasto-plastic moduli. As previously mentioned, methods such as this cannot capture or predict the effect of local components on the overall damage evolution. In Chapter 12, a method will be outlined to experimentally evaluate different types of damages in a metal matrix composite material that can be used in conjunction with a micro-mechanical damage theory. This is outlined through an overall damage quantification as well as a local damage quantification differentiating between damage in the matrix and in the fibers. Major topics covered are specimen design and preparation, mechanical testing (macro-analysis), Scanning Electron Microscope (SEM) analysis (micro-analysis), and evaluation of damage parameters based on the results of the micro-analysis.

1.4 Scope of The Book

The book is divided into two major parts: Part I (Chapters 2-4) deals with the scalar formulation and is limited to the analysis of isotropic damage in materials (both metals and metal matrix composites). Thus, this part can be read by a wide variety of readers; the only mathematical

8

requirement is a knowledge of simple algebra. However, Part II (Chapters 5-13) deals with the tensor formulation and is applicable to general states of deformation and damage. The reader of this part is assumed to have an advanced mathematical training in tensor algebra in order to fully grasp the intricacies and detailed mathematical derivations that appear in this part.

Chapters 2, 3 and 4 deal with damage in uniaxial tension of metals and metal matrix composites. Chapters 5 and 6 deal with general states of damages and plasticity in metals. This is followed, in Chapters 7, 8, 9 and 10, by an extension of the theory to metal matrix composites. Finally, three additional chapters are added on related topics. Chapter 11 deals with the problem of symmetrization of the effective stress tensor. This chapter involves highly complex algebraic manipulations and may be excluded from a preliminary reading of the book. Chapter 12, however, is very relevant to the main material and presents the recent experimental investigations and comparisons with theoretical predictions. Finally, an introduction to Fatigue Damage is given in Chapter 13. Several appendixes are included to supplement the material in the text. The lengthy equations are listed in Appendix A so as not to clutter the main body of the book. In Appendix B, selected computer subroutines of some of the important models are presented in the FORTRAN language for the interested readers. These subroutines must be linked to a main program, preferably a general purpose finite element program. Finally Appendix C contains a thorough list of all the references cited on this subject.

The material appearing in this text is limited to plastic deformation and damage in ductile materials (e.g. metals and metal matrix composites). The authors elect to exclude many of the recent advances made in creep, brittle fracture, viscoplasticity, fatigue and temperature effects. The authors feel that these topics require a separate volume for their presentation. Furthermore, the applications contained in this book are the simplest possible ones and are mainly based on the uniaxial tension test. The presentation of more challenging problems is left to the researchers in this field.

1.5 Notation

In Part I of the book (Chapters 2-4), ordinary symbols are used to represent scalar quantities. However, a problem in notation arises in Part II of the book (Chapters 5-17) when the tensor formulation is used. In general, tensional quantities are represented in the following three ways:

1.   Direct tensor notation where tensors are typed in boldface.

2.   Indicial notation where subscripts are used along with the Einstein summation convention.

3.   Matrix representation of tensors where tensors are represented by matrices. In this case, tensors are typed in boldface and enclosed between brackets.

The notation for tensor operations that is followed throughout the book is defined by the following: For second-rank tenors $\mathbf{A}$ and $\mathbf{B}$ and fourth-rank tenors $\mathbf{C}$ and $\mathbf{D}$, we have

$$\mathbf{A}{:}\mathbf{B} \equiv A_{ij}\, B_{ij}$$
$$\mathbf{A}{\otimes}\mathbf{B} \equiv A_{ij}\, B_{k\ell}$$
$$\mathbf{C}{:}\mathbf{A} \equiv C_{ijk\ell}\, A_{k\ell}$$
$$\mathbf{A}{:}\mathbf{C} \equiv A_{ij}\, C_{ijk\ell}$$
$$\mathbf{C}{\cdot}\mathbf{D} \equiv C_{ijmn}\, D_{mnk\ell}$$
$$\mathbf{C}{:}\mathbf{D} \equiv C_{mnij}\, D_{mnk\ell}$$

The trace "tr" of a second-rank tensor is defined by $\mathrm{tr}(\mathbf{A}) \equiv \mathbf{A}{:}\mathbf{I}_2 \equiv A_{ii}$. The symbols $\mathbf{I}_2$ and $\mathbf{I}_4$ are reserved for the second-rank and fourth-rank identity tenors $\mathbf{I}_2 \equiv \delta_{ij}$ and $\mathbf{I}_4 \equiv \frac{1}{2}(\delta_{ik}\delta_{jl} + \delta_{il}\delta_{jk})$, where δ_{ij} is known as the Kronecker delta. Brackets [ ] are used to denote 3 x 3 or 6 x 6 matrices, where braces { } are used to denote 3 x 1 or 6 x 1 vectors. Finally, the transpose and inverse of a tensor or matrix are denoted by the superscripts "T" and "-1", respectively, while the superscript "$-T$" stands for the inverse transpose of a tensor or matrix. This notation is used mainly in Chapters 5-17.

PART I

ISOTROPIC DAMAGE MECHANICS

SCALAR FORMULATION

CHAPTER 2

UNIAXIAL TENSION IN METALS

In this chapter the principles of continuum damage mechanics are introduced. The various assumptions and the equivalence principle are outlined clearly. This is followed by the derivation of the damage evolution equations. Finally, a new section is added on the separation of damage due to cracks and voids in metals. All the theory and derivations in this chapter are based on the uniaxial tension test. Therefore, isotropic damage is assumed and all the equations employ scalar variables. The extension of the theory to the general case of anisotropic damage is presented later in Part II of this book starting with Chapter 5.

2.1 Principles of Continuum Damage Mechanics

The limitation of classical fracture mechanics have been outlined recently by Lemaitre [31]. Parameters like the J-Integral and COD are difficult to use in cases of large strain plasticity, time-dependent behavior, crack evolution for non-proportional loading, and delamination of composites.

Murakami [58] indicated that proper understanding and the mechanical description of the damage process of materials brought about by the internal defects are of vital importance in discussing the mechanical effects of the material deterioration on the macroscopic behavior of materials, as well as in elucidating the process leading from these defects to the final fracture. A systematic approach to these problems of distributed defects can be provided by continuum damage mechanics (Chaboche [87], Hult [88], Kachanov [33], Krajcinovic [89], Lemaitre and Chaboche [90, 91], Murakami [23]). The fundamental notion of this theory, attributable originally to Kachanov [1] and modified somewhat by Rabotnov [92], is to represent the damage state of materials characterized by distributed cavities in terms of appropriate mechanical variables (internal state variables), and then to establish mechanical equations to describe their evolution and the mechanical behavior of damaged materials.

Lemaitre [12] indicated that damage in metals is mainly the process of the initiation and growth of micro-cracks and cavities. At that scale, the phenomenon is discontinuous. Kachanov in 1958 [1] was the first to introduce a continuous variable related to the density of such defects. This variable has constitutive equations for evolution, written in terms of stress or strain, which may be used in structural calculations in order to predict the initiation of macro-cracks. These constitutive equations have been formulated in the framework of thermodynamics and identified for many

14

phenomena: dissipation and low-cycle fatigue in metals [93], coupling between damage and creep [7,8], high-cycle fatigue [6], creep-fatigue interaction [9], and ductile plastic damage [10].

In continuum damage mechanics, a crack is considered to be a zone (process zone) of high gradients of rigidity and strength that has reached critical damage conditions. Thus, a major advantage of continuum damage mechanics is that it utilizes a local approach and introduces a continuous damage variable in the process zone, while classical fracture mechanics uses more global concepts like the J-Integral and COD.

The assumption of isotropic damage is often sufficient to give a good prediction of the carrying capacity, the number of cycles or the time to local failure in structural components. The calculations are not too difficult because of the scalar nature of the damage variable in this case. For anisotropic damage the variable is of tensorial nature [21,22,47] and the work to be done for identification of the models and for applications is much more complicated [12, 21 25]. Nevertheless, according to Lemaitre [12], damage mechanics has been applied since 1975 with success in several fields to evaluate the integrity of structural components and it will become one of the main tools for analyzing the strength of materials as a complement to fracture mechanics.

Kachanov [1] introduced the idea of damage in the framework of continuum mechanics. In a damaged body, consider a volume element at the macro-scale, that is of a size large enough to contain many defects, and small enough to be considered as a material point of the mechanics of continuum. For the case of isotropic damage and using the concept of effective stress (because of its suitability for continuum mechanics), the damage variable φ is defined as a scalar in the following manner:

$$\varphi = \frac{A - \overline{A}}{A} \tag{2.1}$$

where $\overline{A}$ is the effective (net) resisting area corresponding to the damaged area A. The effective area $\overline{A}$ is obtained from A by removing the surface intersections of the micro-cracks and cavities and correcting for the micro-stress concentrations in the vicinity of discontinuities and for the interactions between closed defects.

The expression given in equation (2.1) implies that $\varphi = 0$ corresponds to the undamaged state, and $\varphi = \varphi_{cr}$ is a critical value which corresponds to the rupture of the element in two parts.

According to Lemaitre [12], the critical value of the damage variable lies in the range $0.2 \leq \varphi_{cr} \leq 0.8$ for metals. In general, the theoretical value of φ should be between $0 \leq \varphi < 1$. Equation (2.1) can be re-written in a more suitable form as follows:

$$\overline{A} = (1-\varphi)\,A \qquad\qquad (2.2)$$

The cross-sectional areas A and $\overline{A}$ are shown in Figure 2.1 on cylindrical material elements in the damaged and effective states, respectively.

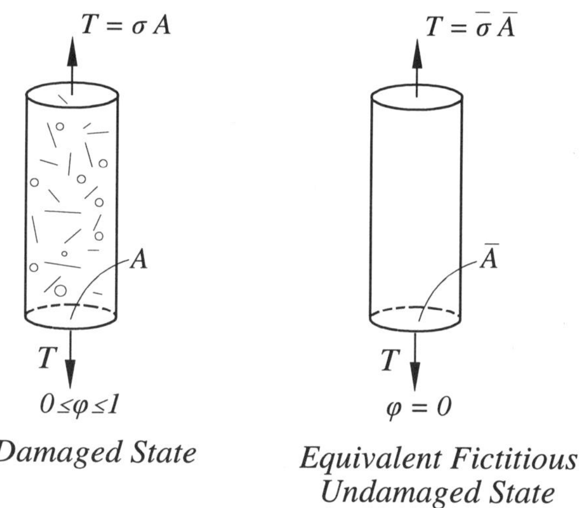

Figure 2.1. Isotropic Damage in Uniaxial Tension (Concept of Effective Stress)

2.2 Assumptions and the Equivalence Hypothesis

The assumption of isotropic damage has been stated previously and is used in this part of the book. Furthermore, the hypothesis of strain equivalence (Lemaitre [11, 12]) is assumed. The effective resisting area $\overline{A}$ can be calculated through mathematical homogenization techniques [94] but the shape and size of the defects must be known, which is somewhat difficult, even with a good electron microscope. To avoid this difficulty, the hypothesis of strain equivalence is made [90]. This hypothesis states that "every strain behavior of a damaged material is represented by constitutive equations of the undamaged material in the potential of which the stress is simply replaced by the effective stress." The effective stress $\bar{\sigma}$ is defined as the stress in the effective (undamaged) state. Considering Figure 2.1, the effective stress $\bar{\sigma}$ can be obtained from equation (2.2) by equating the force $T = \sigma A$ acting on the damaged area A with the force $T = \bar{\sigma}\,\bar{A}$ acting on the hypothetical undamaged area $\bar{A}$, i.e.

$$\sigma A = \bar{\sigma}\,\bar{A} \tag{2.3}$$

where σ is the Cauchy stress acting on the damaged area A. From equations (2.2) and (2.3), we can obtain the following expression for the effective Cauchy stress $\bar{\sigma}$:

$$\bar{\sigma} = \frac{\sigma}{1 - \varphi} \tag{2.4}$$

It should be noted that the effective stress $\bar{\sigma}$ can be considered as a fictitious stress acting on an undamaged equivalent (fictitious) area $\bar{A}$ (net resisting area).

For the uniaxial tension case shown in Figure 2.1, the constitutive relation is Hooke's law of linear elasticity given by

$$\sigma = E\varepsilon \tag{2.5}$$

where ε is the strain and E is the modulus of elasticity (Young's modulus). The same linear elastic constitutive relation applies to the effective (undamaged) state, i.e.

$$\bar{\sigma} = \bar{E}\bar{\varepsilon} \tag{2.6}$$

where $\bar{\varepsilon}$ and $\bar{E}$ are the effective counterparts of ε and E, respectively. Next, we will derive the necessary transformation equations between the damaged and the hypothetical undamaged states of the material. In the derivation, the following assumptions are incorporated: (1) the elastic deformations are small (infinitesimal) compared with the plastic deformations (finite), and (2) there exists an elastic strain energy scalar function U. This function is assumed based on the linear relation between the Cauchy stress σ and the engineering elastic strain ε given by equation (2.5). The elastic strain energy function U is defined by

$$U = \frac{1}{2}\sigma\varepsilon \tag{2.7}$$

It is clear from equations (2.5) and (2.7) that $\sigma = dU/d\varepsilon$ and $\varepsilon = dU/d\sigma$. Sidoroff [16] proposed the hypothesis of elastic energy equivalence. This latter hypothesis assumes that "the elastic energy for a damaged material is equivalent in form to that of the undamaged (effective) material except that the stress is replaced by the effective stress in the energy formulation." Thus, according to this hypothesis, the elastic strain energy $U = \frac{1}{2}\sigma\varepsilon$ is equated to the effective elastic strain energy $\overline{U} = \frac{1}{2}\overline{\sigma}\overline{\varepsilon}$ as follows:

$$\frac{1}{2}\sigma\varepsilon = \frac{1}{2}\overline{\sigma}\overline{\varepsilon} \tag{2.8}$$

Substituting equation (2.4) into equation (2.8) and simplifying, we obtain the following relation between the strain ε and the effective strain $\overline{\varepsilon}$:

$$\overline{\varepsilon} = (1 - \varphi)\,\varepsilon \tag{2.9}$$

Continuing further, we substitute equations (2.4) and (2.9) into equation (2.6), simplify the result and compare it with equation (2.5) to obtain:

$$E = \overline{E}\,(1 - \varphi)^2 \tag{2.10}$$

Equation (2.10) represents the transformation law for the modulus of elasticity. It is clear now that Young's modulus for the damaged material depends on the value of the damage variable φ. Further remarks on this relation and its experimental investigation are discussed in detail later in Chapter 12. Solving equation (2.10) for φ, one obtains:

$$\varphi = 1 - \sqrt{\frac{E}{\overline{E}}} \tag{2.11}$$

Once the values of $\overline{E}$ are measured experimentally, one can use equation (2.11) to obtain values of the damage variable φ. It should be noted that the value of $\overline{E}$ is constant for the effective (undamaged) material.

2.3 Damage Evolution

There are several approaches in the literature on the topic of evolution of damage and the proper form of the kinetic equation of the damage variable. Kachanov [33] proposed an evolution

18

equation of damage based on a power law with two independent material constants. However, the resulting kinetic equation for the damage variable evolution is complicated and difficult to solve. Therefore, a more rational approach based on energy considerations will be adopted in this book.

The approach followed will depend on the introduction of a damage strengthening criterion in terms of a scalar function g, and a generalized thermodynamic force that corresponds to the damage variable φ (Lemaitre [11], Lee et al [15]). Substituting equations (2.6) and (2.9) into the right-hand-side of equation (2.8), we obtain the elastic strain energy U in the damaged state of the material as follows:

$$U = \frac{1}{2}\,\overline{E}\,(1-\varphi)^2\,\varepsilon^2 \tag{2.12}$$

in which $\overline{E}$ is constant, therefore, the incremental elastic strain energy dU is obtained by differentiating equation (2.12):

$$dU = \overline{E}\,(1-\varphi)^2\,\varepsilon\,d\varepsilon \;-\; \overline{E}\,(1-\varphi)\,\varepsilon^2\,d\varphi \tag{2.13}$$

The generalized thermodynamic force y associated with the damage variable φ is thus defined by:

$$y \equiv \frac{\partial U}{\partial \varphi} = -\,\overline{E}\,(1-\varphi)\,\varepsilon^2 \tag{2.14}$$

Let $g(y, L)$ be the damage function (criterion) as proposed by Lee et al [15], where $L \equiv L(\ell)$ is a damage strengthening parameter which is a function of the "overall" damage parameter ℓ. For this problem, the scalar function g takes the following form:

$$g(y, L) = \frac{1}{2}\,y^2 - L(\ell) \equiv 0 \tag{2.15}$$

The damage strengthening criterion defined by equation (2.15) is similar to the von Mises yield criterion in the theory of plasticity. In order to derive a normality rule for the evolution of damage, we first start with the power of dissipation $\prod$ which is given by:

$$\prod = -y\,d\varphi - L\,d\ell \tag{2.16}$$

where the "d" in front of a variable indicates the incremental quantity of the variable. The problem is to extremize $\prod$ subject to the condition $g = 0$. Using the mathematical theory of functions of several variables, we introduce the Lagrange multiplier $d\lambda$ and form the objective function $\Psi(y, L)$ such that:

$$\Psi = \Pi - d\lambda \cdot g \tag{2.17}$$

The problem now reduces to extremizing the function Ψ. For this purpose, the two necessary conditions are $\partial\Psi/\partial y = 0$ and $\partial\Psi/\partial L = 0$. Using these conditions, along with equations (2.16) and (2.17), one obtains:

$$d\varphi = -d\lambda \frac{\partial g}{\partial y} \tag{2.18a}$$

$$d\ell = -d\lambda \frac{\partial g}{\partial L} \tag{2.18b}$$

Substituting for g from equation (2.15) into equation (2.18b), one concludes directly that $d\lambda = d\ell$. Substituting this into equation (2.18a), along with equation (2.15), we obtain:

$$d\varphi = -d\lambda \cdot y \tag{2.19}$$

In order to solve the differential equation (2.19), we must first find an expression for the Lagrange multiplier $d\lambda$. This can be done by invoking the consistency condition $dg = 0$. Applying this condition to equation (2.15), we obtain:

$$\frac{\partial g}{\partial y} \, dy + \frac{\partial g}{\partial L} \, dL = 0 \tag{2.20}$$

Substituting for $\partial g/\partial y$ and $\partial g/\partial L$ from equation (2.15) and for $dL = d\ell \, (\partial L/\partial \ell)$, from the chain rule of differentiation, and solving for $d\ell$, we obtain:

$$d\ell = d\lambda = \frac{y \, dy}{\partial L/\partial \ell} \tag{2.21}$$

Substituting the above expression of $d\lambda$ into equation (2.19), we obtain the kinetic (evolution)

20

equation of damage:

$$\left(\frac{\partial L}{\partial \ell}\right) d\varphi \ = \ -y^2 \, dy \tag{2.22}$$

with the initial condition that $\varphi = 0$ when $y = 0$. The solution of equation (2.22) depends on the form of the function $L(\ell)$. For simplicity, we may consider a linear function of the form $L(\ell) = c\ell + d$, where c and d are constants. This is motivated by the hardening parameter defined for isotropic hardening in the theory of plasticity as $\sqrt{d\varepsilon_{ij}'' \, d\varepsilon_{ij}''}$ where $d\varepsilon_{ij}''$ is the plastic component of the strain rate tensor[*]. The equivalent damage strengthening parameter can be analogously expressed as $\sqrt{d\ell \cdot d\ell}$ or simply $d\ell$ whereby giving a linear function in ℓ as discussed above. Substituting this into equation (2.22) and integrating, we obtain the following relation between the damage variable φ and its associated generalized thermodynamic force y:

$$\varphi = \ -\frac{y^3}{3c} \tag{2.23}$$

The above relation is shown graphically in Figure 2.2 where it is clear that φ is a monotonically increasing function of y. Next, we investigate the strain-damage relationship. Differentiating the expression of y in equation (2.14), we obtain:

$$dy \ = \ \overline{E}\,\varepsilon\,[\varepsilon\, d\varphi \ - \ 2\, d\varepsilon\,(1-\varphi)] \tag{2.24}$$

Substituting the expressions of y and dy of equations (2.14) and (2.24), respectively, into equation (2.22), we obtain the strain-damage differential equation:

$$\left(\frac{\partial L}{\partial \ell}\right) d\varphi \ = \ \overline{E}^3\,\varepsilon^5\,(1-\varphi)^2\,[2\, d\varepsilon\,(1-\varphi) - \varepsilon\, d\varphi] \tag{2.25}$$

The above differential equation can be solved easily by the simple change of variables $x = \varepsilon^2\,(1-\varphi)$ and noting that the expression on the right-hand-side of equation (2.25) is nothing but $\overline{E}^3\,x^2\,dx$. Performing the integration with the initial condition that $\varphi = 0$ when $\varepsilon = 0$ along with the linear expression of $L(\ell)$, we obtain:

---

[*] The notation used in this formula is defined later in Part II of the book.

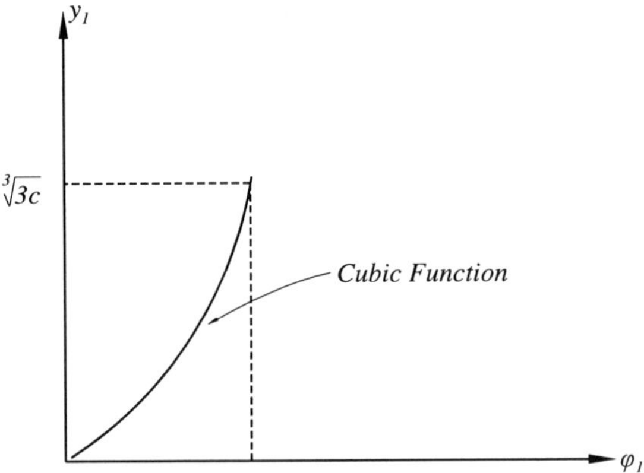

Figure 2.2. Relation Between the Overall Damage Variable φ_1 and its Associated Generalized Force y_1

$$\frac{\varphi}{(1-\varphi)^3} = \frac{\overline{E}^3}{3c}\,\varepsilon^6 \tag{2.26}$$

One should note that an initial condition involving an initial damage variable φ^o could have been used, i.e., $\varphi = \varphi^o$ when $\varepsilon = 0$. In addition, the strain-damage relation of equation (2.26) could easily have been obtained by substituting the expression of y of equation (2.14) directly into equation (2.23). However, it is preferable to derive it directly from the strain-damage differential equation (2.25) without the use of the generalized thermodynamic force y.

2.4 Separation of Damage Due to Cracks and Voids

In this section[**] the damage in the cylindrical bar of Figure 2.1 is assumed to consist of voids and cracks only. We will then proceed to separate the total damage variable φ into two damage variable, φ^v and φ^c, representing damage due to voids and cracks, respectively. A consistent

---

[**]The work appearing in this section has been recently done by the authors [95].

mathematical formulation is presented for this separation using the principles of continuum damage mechanics outlined in section 2.1. Throughout the formulation, isotropic damage is assumed for the uniaxial tension case. The cross-sectional area A of the damaged bar can be decomposed as follows:

$$A = \overline{A} + A^{v} + A^{c} \tag{2.27}$$

where A^{v} is the total area of voids in the cross-section and A^{c} is the total area of cracks (measured lengthwise) in the cross-section[***]. In addition to the total damage variable φ, two additional damage variables φ^{v} and φ^{c} are introduced to represent the damage due to voids and cracks respectively. Our goal is to find a representation of the total damage variable φ in terms of φ^{v} and φ^{c}. In order to do this, we need to separate the damage due to voids and cracks when constructing the effective damaged configuration. This separation can be performed in two different methods. We can start by removing the voids only then we can remove the cracks separately, or we can start by removing the cracks only then we can remove the voids separately. The detailed formulation based on each of these two methods is discussed below and is shown schematically in Figures 2.3 and 2.4.

In the first method, we first remove the voids only from the damaged configuration shown in Figure 2.3a. In this way we obtain the damaged configuration shown in Figure 2.3b which contains damage due to cracks only. This is termed the undamaged configuration with respect to voids. The cross-sectional area of the bar in this configuration is clearly $\overline{A} + A^{c}$ while the uniaxial stress is denoted by $\overline{\sigma}^{v}$. The total tensile force T in this configuration is then given by $T = \overline{\sigma}^{v} (\overline{A} + A^{c})$. This expression is equated to the total tensile force $T = \sigma A$ in the damaged configuration from which we obtain:

$$\overline{\sigma}^{v} = \frac{A}{\overline{A} + A^{c}}\, \sigma \tag{2.28}$$

The damage variable φ^{v} due to voids is defined by the ratio A^{v}/A. Substituting for A^{v} from equation (2.27), we obtain:

$$\varphi^{v} = 1 - \frac{\overline{A} + A^{c}}{A} \tag{2.29}$$

Substituting equation (2.29) into equation (2.28), we obtain the following relation between $\overline{\sigma}^{v}$ and σ:

[***]The superscripts "v" and "c" used in the notation denote voids and cracks, respectively.

$$\overline{\sigma}^{v} = \frac{\sigma}{1 - \varphi^{v}} \tag{2.30}$$

The similarity between equations (2.30) and (2.4) is very clear. The next step involves removing the cracks from the intermediate configuration in order to obtain the effective undamaged configuration shown in Figure 2.3c. Equating the previous expression for the tensile force $T = \overline{\sigma}^{v} (\overline{A} + A^{c})$ with the tensile force $T = \overline{\sigma}\,\overline{A}$ in the effective undamaged configuration, we obtain:

$$\overline{\sigma} = \overline{\sigma}^{v} \left(1 + \frac{A^{c}}{\overline{A}} \right) \tag{2.31}$$

The damage variable φ^{c} due to cracks is now defined by the ratio

$$\varphi^{c} = \frac{A^{c}}{\overline{A} + A^{c}} \tag{2.32}$$

Substituting equation (2.32) into equation (2.31) and simplifying, we obtain the following relation between $\overline{\sigma}$ and $\overline{\sigma}^{v}$:

$$\overline{\sigma} = \frac{\overline{\sigma}^{v}}{(1 - \varphi^{c})} \tag{2.33}$$

Finally, we substitute equation (2.30) into equation (2.33) to obtain the sought relationship between σ and $\overline{\sigma}$:

$$\overline{\sigma} = \frac{\sigma}{(1 - \varphi^{v})(1 - \varphi^{c})} \tag{2.34}$$

The above relation represents a formula for the effective stress in terms of the separate damage variables due to voids and cracks. The same result can be obtained by reversing the order of removal of voids and cracks. In the second method, we first remove the cracks only from the damaged configuration shown in Figure 2.4a. In this way, we obtain the damaged configuration shown in Figure 2.4b which contains damage due to voids only. This is termed the undamaged configuration with respect to cracks. The cross-sectional area of the bar in this configuration is clearly $\overline{A} + A^{v}$

24

while the uniaxial stress is denoted by $\bar{\sigma}^c$. The total tensile force T in this configuration is then given by $T = \bar{\sigma}^c\,(\bar{A} + A^v)$. This expression is equated to the total tensile force $T = \sigma A$ in the

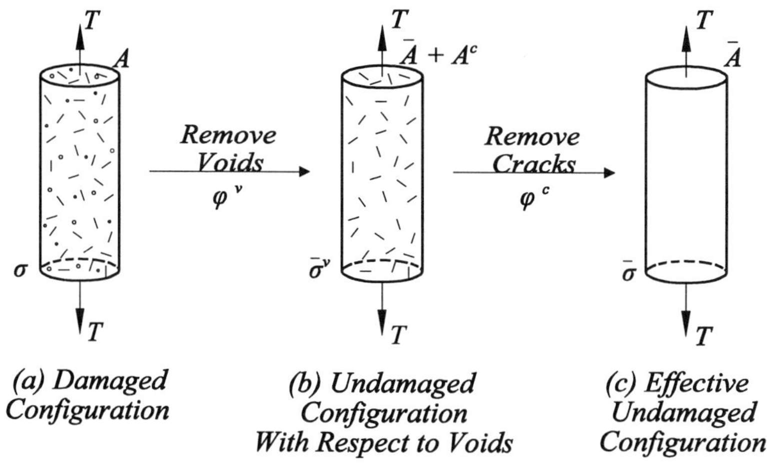

Figure 2.3. A Cylindrical Bar Subjected to Uniaxial Tension: Voids are Removed
First then Followed by Cracks

damaged configuration from which we obtain:

$$\bar{\sigma}^c = \frac{A}{\bar{A} + A^v}\,\sigma \tag{2.35}$$

The damage variable φ^c due to cracks is defined by the ratio A^c/A. Substituting for A^c from equation (2.27), we obtain:

$$\varphi^c = 1 - \frac{\bar{A} + A^v}{A} \tag{2.36}$$

Substituting equation (2.36) into equation (2.35), we obtain the following relation between $\overline{\sigma}^c$ and σ:

$$\overline{\sigma}^c = \frac{\sigma}{1 - \varphi^c} \tag{2.37}$$

The similarity between equations (2.37), (2.30), and (2.4) is very clear. The next step involves removing the voids from the intermediate configuration in order to obtain the effective undamaged configuration shown in Figure 2.4c. Equating the previous expression for the tensile force $T = \overline{\sigma}^c (\overline{A} + A^{\,v})$ with the tensile force $T = \overline{\sigma}\,\overline{A}$ in the effective undamaged configuration, we obtain:

$$\overline{\sigma} = \overline{\sigma}^c \left(1 + \frac{A^{\,v}}{\overline{A}} \right) \tag{2.38}$$

The damage variable φ^v due to voids is now defined by the ratio:

$$\varphi^v = \frac{A^{\,v}}{\overline{A} + A^{\,v}} \tag{2.39}$$

Substituting equation (2.39) into equation (2.38) and simplifying, we obtain the following relation between $\overline{\sigma}$ and $\overline{\sigma}^c$:

$$\overline{\sigma} = \frac{\overline{\sigma}^c}{1 - \varphi^v} \tag{2.40}$$

Finally, we substitute equation (2.37) into equation (2.40) to obtain the sought relationship between σ and $\overline{\sigma}$:

$$\overline{\sigma} = \frac{\sigma}{(1 - \varphi^c)\,(1 - \varphi^v)} \tag{2.41}$$

It is clear that the above relation between the two stresses in the damaged and the effective configurations is exactly the same relation obtained using the first method, i.e. equation (2.34). Thus both methods of constructing the effective undamaged configuration give the same relation between the stresses in the respective configurations. In this way, the separation of damage due to voids and cracks has been completed. In order to derive the final result, we compare either equation (2.34) or (2.41) with the total damage appearing in equation (2.4). Equating the denominators on the right-

26

hand-side of these equations, we can easily obtain the formula:

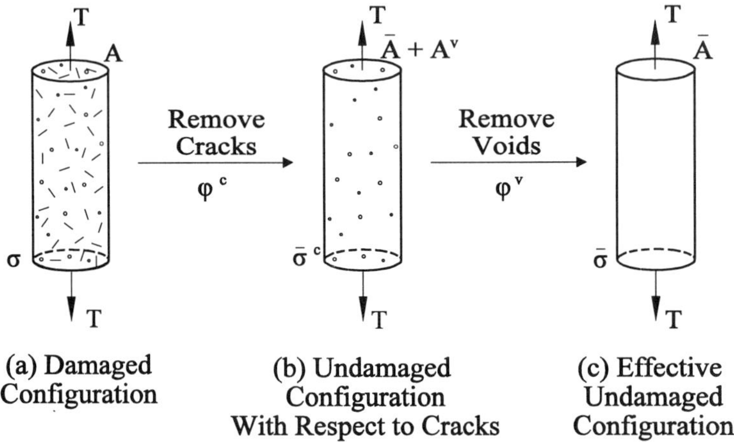

Figure 2.4. A Cylindrical Bar Subjected to Uniaxial Tension: Cracks are Removed
First then Followed by Voids

$$1 - \varphi = (1 - \varphi^v)\,(1 - \varphi^c) \tag{2.42}$$

Equation (2.42) represents the general form for the decomposition of the damage variable into its two respective components, φ^v and φ^c. The result can be further simplified by expanding equation (2.42) and simplifying to obtain:

$$\varphi = \varphi^v + \varphi^c - \varphi^v\,\varphi^c \tag{2.43}$$

Equation (2.42) gives a very clear picture of how the total damage variable φ can be decomposed into a damage variable φ^v due to voids and a damage variable φ^c due to cracks. It is also clear that equation (2.43) satisfies the constraint $0 \leq \varphi \leq 1$ whenever each of the other two damage variables satisfies it. It is also clear that when damage in the material is produced by voids only ($\varphi^c = 0$), then $\varphi = \varphi^v$. Alteratively, $\varphi = \varphi^c$ when damage in the material is produced by cracks only ($\varphi^v = 0$).

The above result given by equation (2.42) can be generalized to include other types of damage. For example, if we consider damage in the material to be due to void initiation, void growth and void coalescence, then the total damage variable φ can be decomposed as follows (based on equation (2.42)):

$$1 - \varphi = (1 - \varphi^{vi})\,(1 - \varphi^{vg})\,(1 - \varphi^{vc}) \qquad (2.44)$$

where φ^{vi} is the damage variable due to void initiation, φ^{vg} is the damage variable due to void growth, and φ^{vc} is the damage variable due to void coalescence. Once the result given in equation (2.44) is established, we can expand the right-hand-side and simplify to obtain the following decomposition:

$$\varphi = \varphi^{vi} + \varphi^{vg} + \varphi^{vc} - \varphi^{vi}\varphi^{vg} - \varphi^{vi}\varphi^{vc} - \varphi^{vc}\varphi^{vg} + \varphi^{vi}\varphi^{vg}\varphi^{vc} \qquad (2.45)$$

Equation (2.45) gives the explicit decomposition of the total damage variable in terms of the three other separate damage variables, where each of the three damage variables may represent a separate damage mechanism contributing to the total damage in the material.

Generalization of the result given in equation (2.42) to three-dimensional states of deformation and damage is possible using tensional variables, but this is still the subject of ongoing research and is beyond the scope of this book.

CHAPTER 3

UNIAXIAL TENSION IN ELASTIC METAL MATRIX COMPOSITES

A micromechemical composite model is used to study damage in a uniaxially-loaded, unidirectional fiber-reinforced composite thin lamina. The matrix and fibers are assumed to be elastic with the fibers continuous and aligned. An overall damage variable is introduced based on the concept of effective stress. The local damage effects are modeled through two additional separate damage variables which represent matrix and fiber damage. In addition, a local-overall relation for the damage variables is derived. Stress and strain concentration factors are derived for the damaged composite. Finally, damage evolution is also considered using both local and overall analysis based on an extremum principle.

Kachanov [1] introduced the idea of effective stress in order to characterize damage initiation and evolution within the framework of the mechanics of continuous media. In this approach, a damage variable is defined and used to represent degradation of the material which reflects various types of damage at the micro-scale level like nucleation and growth of voids, cavities, micro-cracks and other microscopic defects.

In the case of composite materials, the damage variable will also reflect the additional types of damage that occur in these materials like fracture of fibers, debonding and delamination, etc. In this chapter, an overall damage variable is introduced for the whole composite system. This damage variable is found to be decomposable into two local damage variables that are directly related to the matrix and fibers. The discussion is limited to damage due to uniaxial tension in a unidirectional fiber-reinforced composite thin lamina. This is done deliberately in order to keep the mathematical formulation simple and accessible to the general reader. Analysis of general states of damage and deformation in composite materials will require the use of tensor analysis and is left to Part II of this book.

3.1 Stresses

Consider a unidirectional fiber-reinforced composite thin lamina that is subjected to an uniaxial tensile force T along the x-direction as shown in Figure 3.1a. Both the matrix and fibers are assumed to be linearly elastic with the fibers being continuous, aligned, and symmetrically distributed along the x_1-axis. Let dA be the cross-sectional area of the lamina with dA^M and dA^F

30

being the cross-sectional areas of the matrix and fibers, respectively.[1] Since the composite lamina is assumed to consist of two phases only, it is clear that $dA = dA^M + dA^F$. The overall uniaxial stress increment, $d\sigma$, is clearly T/dA and the local uniaxial stress increments, $d\sigma^M$ and $d\sigma^F$, are related to the overall uniaxial stress increment, $d\sigma$, by:

$$d\sigma = c^M d\sigma^M + c^F d\sigma^F \tag{3.1}$$

where c^M and c^F are the matrix and fiber volume fractions (or area fractions here) given by dA^M/dA and dA^F/dA, respectively. It should be clear to the reader that $c^M + c^F = 1$. The local transverse stress increments $d\sigma_2^M$, $d\sigma_2^F$, $d\sigma_3^M$, and $d\sigma_3^F$, although non-zero, are not considered in this formulation.

Using the concept of effective stress, we now consider a fictitious lamina (see Figure 3.1b) made of the same composite material described above and subjected to the same uniaxial tensile force T. This lamina is assumed to undergo deformation with no damage. In other words, it can be hypothetically obtained from the lamina in

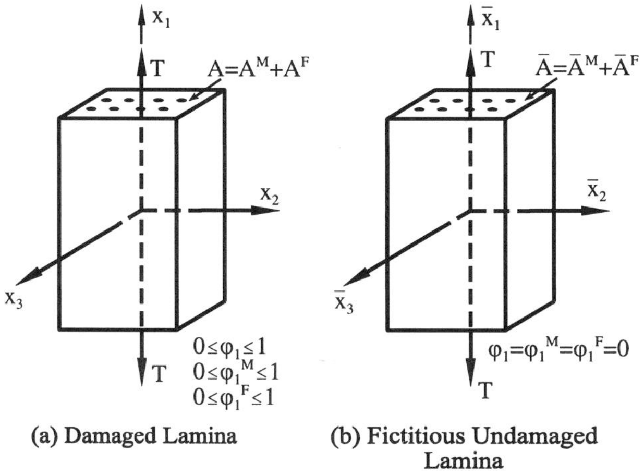

(a) Damaged Lamina   (b) Fictitious Undamaged Lamina

Figure 3.1   Damage Due to Uniaxial Tension

---

[1] Superscripts "M" and "F" are used throughout the book to denote matrix- and fiber- related quantities, respectively.

Figure 3.1 by removing all the damage that the lamina has experienced. Let $d\overline{A}$ denote the cross-sectional area of the effective undamaged lamina with $d\overline{A}^M$ and $d\overline{A}^F$ denoting the cross-sectional areas of the effective undamaged matrix and fibers, respectively. These quantities represent net or effective areas that include no damage. Also, let $\overline{c}^M$ and $\overline{c}^F$ denote the effective volume (or area) fractions for the undamaged matrix and fibers, respectively. The following relations should be clear:

$$d\overline{A}^M + d\overline{A}^F = d\overline{A} \, , \; d\overline{A} \leq dA \, , \; d\overline{A}^M \leq dA^M \, , \text{and} \; d\overline{A}^F \leq dA^F$$

The overall uniaxial effective stress increment $d\overline{\sigma}$ is taken to be the uniaxial stress in the fictitious lamina and it is clear that $d\overline{\sigma} = T/d\overline{A}$. We also consider the two local effective uniaxial stress increments, $d\overline{\sigma}^M$ and $d\overline{\sigma}^F$, and as before, it can be shown that they are related to $d\overline{\sigma}$ by:

$$d\overline{\sigma} = \overline{c}^M d\overline{\sigma}^M + \overline{c}^F d\overline{\sigma}^F \tag{3.2}$$

Since the two laminae are assumed to be mechanically equivalent (in terms of the uniaxial tensile force T that is applied to each one), it follows directly that $d\overline{\sigma} = d\sigma \, (dA/d\overline{A})$. The ratio of the damaged area, $dA - d\overline{A}$, to the original area, dA, is now used to define an overall damage variable φ_1 in the x_1-direction as follows:

$$\varphi_1 = \frac{dA - d\overline{A}}{dA} \tag{3.3}$$

Equation (3.3) is the same as equation (2.1) for metals. It is clear that the values of φ_1 range from 0, for undamaged material, to 1, for (theoretical) complete rupture. The effective uniaxial stress increment can now be written in terms of the damage variable:

$$d\overline{\sigma} = \frac{d\sigma}{1 - \varphi} \tag{3.4}$$

The above expression has been used extensively in the literature (Kachanov [1], Lemaitre [11, 31], Chaboche [35, 36]) to model various types of phenomena like ductile failure, brittle fracture, creep, etc.

In order to represent local damage effects in the matrix and fibers, we define two additional (local) damage variables, φ_1^M and φ_1^F. The first one, φ_1^M, is used to model damage in the matrix like nucleation, growth, and coalescence of voids and micro-cracks, etc., while the second one, φ_1^F, is

32

used to model damage in the fibers and that due to fiber-matrix interaction such as fiber fracture, debonding, etc. These two variables are defined, as before, based on the ratios of the relevant cross-sectional areas of the matrix and fibers as follows:

$$\varphi_1^M = \frac{dA^M - d\overline{A}^M}{dA^M} \tag{3.5a}$$

$$\varphi_1^F = \frac{dA^F - d\overline{A}^F}{dA^F} \tag{3.5b}$$

It is clear from equations (3.5) that the local damage variables satisfy the inequalities $0 \le \varphi_1^M \le 1$ and $0 \le \varphi_1^F \le 1$.

We can now derive equations for the effective matrix and fiber volume fractions, $\overline{c}^M$ and $\overline{c}^F$, in terms of c^M and c^F and the damage variables, φ_1^M and φ_1^F. Starting with $\overline{c}^M = d\overline{A}^M/d\overline{A}$ and $\overline{c}^F = d\overline{A}^F/d\overline{A}$, along with equations (3.3) and (3.5), we can show that:

$$\overline{c}^M = c^M \frac{1 - \varphi_1^M}{1 - \varphi_1} \tag{3.6a}$$

$$\overline{c}^F = c^F \frac{1 - \varphi_1^F}{1 - \varphi_1} \tag{3.6b}$$

Also, using equations (3.3) along with equations (3.5), we derive:

$$\varphi_1 \, dA = \varphi_1^M \, dA^M + \varphi_1^F \, dA^F \tag{3.7}$$

Dividing equation (3.7) through by dA, we derive the relationship between the local damage variables, φ_1^M and φ_1^F, and the overall damage variable φ_1 as follows:

$$\varphi_1 = c^M \varphi_1^M + c^F \varphi_1^F \tag{3.8}$$

The relationship between the matrix damage ratio φ_1^M/φ_1 and the fiber damage ratio φ_1^F/φ_1 is shown in Figure 3.2 for different values of the matrix volume fraction c^M. It is clear from the figure that these ratios are always greater than or equal to one, implying that $\varphi_1^M \geq \varphi_1$ and $\varphi_1^F \geq \varphi_1$. This remark does not contradict the fact that the matrix and fiber damage should be a part of the composite damage since the damage variables are defined as ratios of areas and do not reflect the absolute amount of damage in the material.

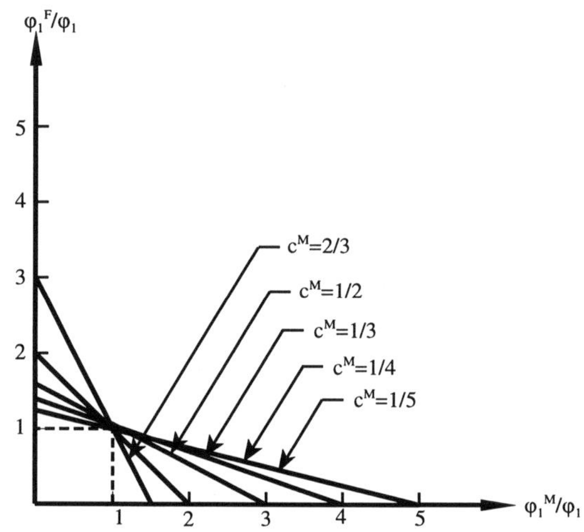

Figure 3.2   Relationship Between Local Damage Parameters
φ_1^M/φ_1 and φ_1^F/φ_1 for Different Matrix Volume Fractions

Adding equations (3.6a) and (3.6b), and utilizing equation (3.8) and the previous relation between c^M and c^F, we conclude that $\bar{c}^M + \bar{c}^F = 1$, that is, the phase volume fractions of the damaged material satisfy the same relation as that of the effective undamaged material indicating no significant (or large) changes in the geometry of the composite system.   Some authors use the "continuity" variable Ψ_1, defined by $\Psi_1 = 1 - \varphi_1$ (e.g. Kachanov [33]). In this case, we can easily show that Ψ_1 satisfies a relation similar to that of equation (3.8), namely, $\Psi_1 = c^M \Psi_1^M + c^F \Psi_1^F$ where $\Psi_1^M = 1 - \varphi_1^M$ and $\Psi_1^F = 1 - \varphi_1^F$.

Substituting equations (3.4) and (3.6) into equation (3.2) and simplifying, we obtain the following relation for the effective local stress increments, $\bar{d\sigma}^M$ and $\bar{d\sigma}^F$:

34

$$d\sigma = c^M(1-\varphi_1^M)\,d\bar{\sigma}^M + c^F(1-\varphi_1^F)\,d\bar{\sigma}^F \tag{3.9}$$

In the derivation of equation (3.9), it is assumed that $\varphi_1 \neq 0$. Therefore, the case of complete rupture is excluded from the discussion that follows. In view of the effective stress equation (3.4), we can assume similar expressions for the effective local stresses as follows [96]:

$$d\bar{\sigma}^M = \frac{d\sigma^M}{1-\varphi_1^M} \tag{3.10a}$$

$$d\bar{\sigma}^F = \frac{d\sigma^F}{1-\varphi_1^F} \tag{3.10b}$$

It is clear that equations (3.10) satisfy the requirement given by equation (3.9). However, the constraint given in equation (3.9) is a necessary condition that must be satisfied by any alternative expressions for the effective local stresses other than equations (3.10).

Next, we consider the relations between the local and overall uniaxial stresses in the composite system. Following the work of Dvorak and Bahei-El-Din [68, 69] and Bahei-El-Din and Dvorak [97], we consider a micromechemically based approach and introduce the effective matrix and fiber stress concentration factors $\bar{B}^M$ and $\bar{B}^F$ in the effective undamaged lamina shown in Figure 3.1b. Therefore, we can write the following local-overall relations for the effective uniaxial stress increments:

$$d\bar{\sigma}^M = \bar{B}^M\,d\bar{\sigma} \tag{3.11a}$$

$$d\bar{\sigma}^F = \bar{B}^F\,d\bar{\sigma} \tag{3.11b}$$

The effective stress concentration factors $\bar{B}^M$ and $\bar{B}^F$ can be derived from the solution of an inclusion problem in the effective undamaged configuration of the material. However, certain models have been proposed by Dvorak and Bahei-EL-Din [68, 69] in order to derive simple expressions for $\bar{B}^M$ and $\bar{B}^F$. Two of these methods will be discussed at the end of section 3.3 as they relate to the problem at hand. Substituting equations (3.11) into equation (3.2), we obtain the relation between the stress concentration factors and the effective volume fractions:

$$\overline{c}^M \, \overline{B}^M + \overline{c}^F \, \overline{B}^F = 1 \tag{3.12}$$

Substituting further for $\overline{c}^M$ and $\overline{c}^F$ from equations (3.6) into equation (3.12), we obtain:

$$c^M \, (1 - \varphi_1^M) \, \overline{B}^M + c^F \, (1 - \varphi_1^F) \, \overline{B}^F = 1 - \varphi_1 \tag{3.13}$$

Assuming that stress concentration factors B^M and B^F exist in the actual damaged lamina, we can write the following local-overall relations for the corresponding stress increments:

$$d\sigma^M = B^M d\sigma \tag{3.14a}$$

$$d\sigma^F = B^F d\sigma \tag{3.14b}$$

Substituting equations (3.14) into equation (3.1), we obtain the relation between the volume fractions and the damaged concentration factors (see equation (3.12) for comparison):

$$c^M B^M + c^F B^F = 1 \tag{3.15}$$

Finally, we substitute equations (3.11) into equations (3.10) along with equation (3.4). Comparing the resulting two equations with equation (3.14), we conclude that the damaged stress concentration factors are given by:

$$B^M = \overline{B}^M \, \frac{1 - \varphi_1^M}{1 - \varphi_1} \tag{3.16a}$$

$$B^F = \overline{B}^F \, \frac{1 - \varphi_1^F}{1 - \varphi_1} \tag{3.16b}$$

Therefore, once appropriate expressions are derived for the effective undamaged stress concentration factors $\overline{B}^M$ and $\overline{B}^F$, we can use equations (3.16) to derive the corresponding expressions for the damaged stress concentration factors B^M and B^F.

The relation given in equation (3.16a) is now investigated in Figures 3.3 and 3.4. In Figure 3.3, the relation between the matrix damage variable φ_1^M and the ratio $B^M/\overline{B}^M$ is shown for different values of the overall damage variable φ_1. It is noticed that the damaged matrix stress concentration factor becomes larger (i.e. the ratio $B^M/\overline{B}^M$ grows) with the decrease in the matrix damage variable φ_1^M.

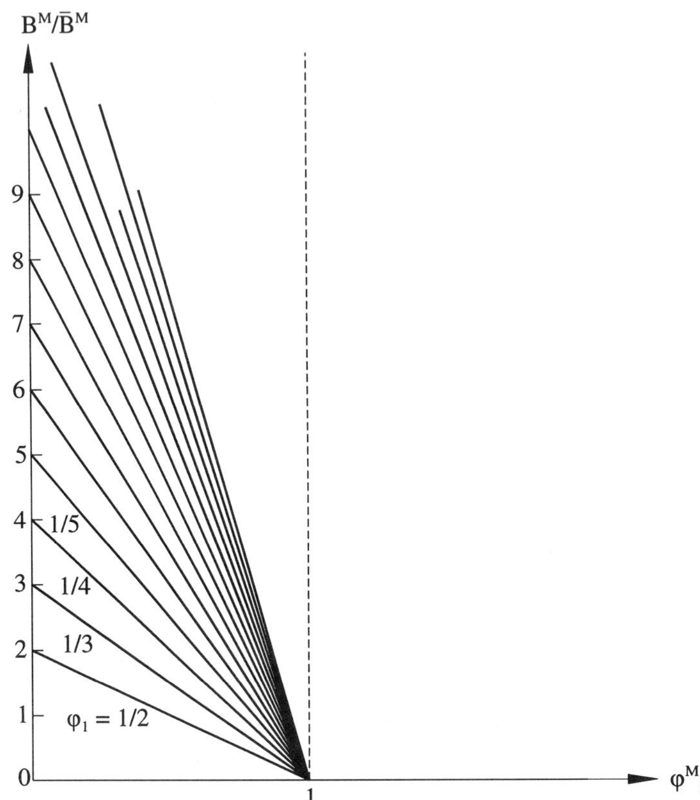

Figure 3.3   Effect of Matrix Damage φ^M_1 on the Stress Concentration for
Different Overall Damage Parameters φ_1

This is also clear in Figure 3.4. However, Figure 3.4 also shows that $B^M/\overline{B}^M$ increases with the increase in the overall damage variable φ_1 . Similar remarks apply to the fiber stress concentration ratio $B^F/\overline{B}^F$ of equation (3.16b).

3.2 Strains

In this section, the appropriate expressions for the effective strain increments $d\bar{\varepsilon}_1$, $d\bar{\varepsilon}_2$ and $d\bar{\varepsilon}_3$ will be developed in terms of the strain increments $d\varepsilon_1$, $d\varepsilon_2$ and $d\varepsilon_3$, and the damage variables

φ_1, φ_2 and φ_3 (φ_2 and φ_3 are overall transverse damage variables along the x_2- and x_3-directions, respectively). In addition, the local-overall strain equations will be derived for both the damaged and the effective undamaged configurations of the material.

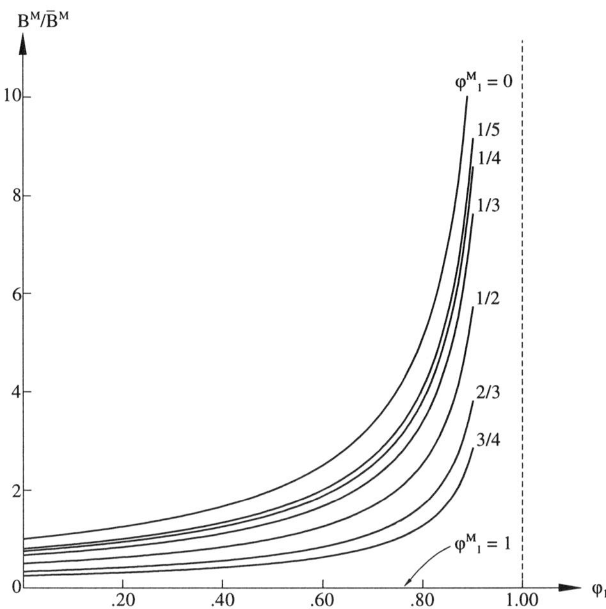

Figure 3.4   Effect of Overall Damage φ_1 on the Stress Concentration Factor for Different Matrix Damage Parameters φ^M_1

In order to derive the required relations, the hypothesis of elastic energy equivalence [16] is used (see section 2.2). In this hypothesis, it is assumed that the elastic energy for the damaged material is equivalent in form to that of the effective undamaged material except that the stress is replaced by the effective stress in the energy formulation. Applying this to the overall composite system considered here, this hypothesis takes the form:

$$\frac{1}{2} \, d\sigma \, d\varepsilon_1 = \frac{1}{2} \, d\bar{\sigma} \, d\bar{\varepsilon}_1 \qquad (3.17)$$

where $d\varepsilon_1$ is the overall axial strain increment in the x_1-direction and $d\bar{\varepsilon}_1$, is its effective counterpart.

Substituting for $d\bar{\sigma}$ from equation (3.4) into equation (3.17), we obtain the following expression for the effective overall axial strain increment $d\bar{\varepsilon}_1$:

$$d\bar{\varepsilon}_1 = (1 - \varphi_1)d\varepsilon_1 \tag{3.18}$$

The above relation is very similar to equation (2.9), which was derived previously for metals. In view of the above relation, we can assume similar relations for the transverse overall strain increments $d\varepsilon_2$ and $d\varepsilon_3$:

$$d\bar{\varepsilon}_2 = (1 - \varphi_2)d\varepsilon_2 \tag{3.19a}$$

$$d\bar{\varepsilon}_3 = (1 - \varphi_3)d\varepsilon_3 \tag{3.19b}$$

where φ_2 and φ_3 are the overall transverse damage variables. The reader should note that definitions for φ_2 and φ_3 similar to the definition of φ_1 in equation (3.3) are not possible. A more convenient way to define these two variables is suggested in the next section.

Next, the local-overall relations are discussed. The matrix and fiber axial strain increments are related to the overall axial strain increment in the fictitious undamaged state by the following equations:

$$d\bar{\varepsilon}_1^M = \bar{C}_{11}^M d\bar{\varepsilon}_1 + \bar{C}_{12}^M d\bar{\varepsilon}_2 + \bar{C}_{13}^M d\bar{\varepsilon}_3 \tag{3.20a}$$

$$d\bar{\varepsilon}_1^F = \bar{C}_{11}^F d\bar{\varepsilon}_1 + \bar{C}_{12}^F d\bar{\varepsilon}_2 + \bar{C}_{13}^F d\bar{\varepsilon}_3 \tag{3.20b}$$

where $\bar{C}_{11}^M, \bar{C}_{12}^M, \ldots, \bar{C}_{13}^F$ are the appropriate matrix and fiber strain concentration factors. Using the definition of Poisson's ratios, $\bar{\nu}_{21} = -d\bar{\varepsilon}_2/d\bar{\varepsilon}_1$ and $\bar{\nu}_{31} = -d\bar{\varepsilon}_3/d\bar{\varepsilon}_1$, equation (3.20) can be rewritten in the simplified form:

$$d\bar{\varepsilon}_1^M = \bar{C}_1^M \, d\bar{\varepsilon}_1 \tag{3.21a}$$

$$d\bar{\varepsilon}_1^F = \bar{C}_1^F \, d\bar{\varepsilon}_1 \tag{3.21b}$$

where the modified strain concentration factors $\overline{C}_1^M$ and $\overline{C}_1^F$ are given by:

$$\overline{C}_1^M = \overline{C}_{11}^M - \overline{\nu}_{21}\overline{C}_{12}^M - \overline{\nu}_{31}\overline{C}_{13}^M \tag{3.22a}$$

$$\overline{C}_1^F = \overline{C}_{11}^F - \overline{\nu}_{21}\overline{C}_{12}^F - \overline{\nu}_{31}\overline{C}_{13}^F \tag{3.22b}$$

Similarly, we can write the following relations for the transverse strains:

$$d\overline{\varepsilon}_2^M = \overline{C}_2^M \, d\overline{\varepsilon}_2 \tag{3.23a}$$

$$d\overline{\varepsilon}_3^M = \overline{C}_3^M \, d\overline{\varepsilon}_3 \tag{3.23b}$$

$$d\overline{\varepsilon}_2^F = \overline{C}_2^F \, d\overline{\varepsilon}_2 \tag{3.23c}$$

$$d\overline{\varepsilon}_3^F = \overline{C}_3^F \, d\overline{\varepsilon}_3 \tag{3.23d}$$

where $\overline{C}_2^M$, $\overline{C}_3^M$, $\overline{C}_2^F$ and $\overline{C}_3^F$ are modified strain concentration factors having expressions similar to those of equations (3.22).

The strain concentration factors can be obtained from the solution of an appropriate inclusion problem. However, in this chapter, a much simpler approach is followed. This approach is based on deriving a relation between the strain and stress concentration factors as follows. Starting with the expression $d\overline{\sigma} \, d\overline{\varepsilon}_1$ and expanding it in terms of local axial stresses and strains using equation (3.2) and a similar equation for the effective overall axial strain increment, we obtain:

$$d\overline{\sigma} d\overline{\varepsilon}_1 = (\overline{c}^M d\overline{\sigma}^M + \overline{c}^F d\overline{\sigma}^F)\,(\overline{c}^M d\overline{\varepsilon}_1^M + \overline{c}^F d\overline{\varepsilon}_1^F) \tag{3.24}$$

Substituting for the effective local stresses and strains from equations (3.11) and (3.21) into equation (3.24), and simplifying the result, we obtain the following constraint equation regarding the concentration factors for stresses and strains:

$$(\overline{c}^M \overline{B}^M + \overline{c}^F \overline{B}^F)\,(\overline{c}^M \overline{C}_1^M + \overline{c}^F \overline{C}_1^F) = 1 \tag{3.25a}$$

In view of the constraint relation (3.12), the above constraint relation can be further simplified to:

$$(\overline{c}^{M} \ \overline{C}_{1}^{M} + \overline{c}^{F} \ \overline{C}_{1}^{F}) = 1 \tag{3.25b}$$

Therefore, once the stress concentration factors, $\overline{B}^{M}$ and $\overline{B}^{F}$, are determined, we can use equations (3.25) to derive suitable expressions for the strain concentration factors, $\overline{C}_{1}^{M}$ and $\overline{C}_{1}^{F}$.

In order to formulate the transformation equations for the local axial strain increments, $d\varepsilon_{1}^{M}$ and $d\varepsilon_{1}^{F}$, we use the hypothesis of elastic energy equivalence using local quantities. Therefore, equation (3.17) is rewritten in the form:

$$\frac{1}{2} d\sigma^{M} d\varepsilon_{1}^{M} + \frac{1}{2} d\sigma^{F} d\varepsilon_{1}^{F} = \frac{1}{2} d\overline{\sigma}^{M} d\overline{\varepsilon}_{1}^{M} + \frac{1}{2} d\overline{\sigma}^{F} d\overline{\varepsilon}_{1}^{F} \tag{3.26}$$

Substituting for $d\overline{\sigma}^{M}$ and $d\overline{\sigma}^{F}$ from equations (3.10) into equation (3.26), we obtain the following relation between the local axial strain increments and their effective counterparts:

$$d\sigma^{M} d\varepsilon_{1}^{M} + d\sigma^{F} d\varepsilon_{1}^{F} = \frac{d\sigma^{M} d\overline{\varepsilon}_{1}^{M}}{1 - \varphi_{1}^{M}} + \frac{d\sigma^{F} d\overline{\varepsilon}_{1}^{F}}{1 - \varphi_{1}^{F}} \tag{3.27}$$

After studying equation (3.27), it is noticed that it is difficult to derive explicit formulae for $d\overline{\varepsilon}_{1}^{M}$ and $d\overline{\varepsilon}_{1}^{F}$ without making a simplifying assumption. We are led directly to assume local axial strain relations similar to the overall axial strain relation given by equation (3.18). Assuming that:

$$d\overline{\varepsilon}_{1}^{M} = (1 - \varphi_{1}^{M}) \ d\varepsilon_{1}^{M} \tag{3.28a}$$

$$d\overline{\varepsilon}_{1}^{F} = (1 - \varphi_{1}^{F}) \ d\varepsilon_{1}^{F} \tag{3.28b}$$

we conclude directly that these relations satisfy equation (3.27). Similar relations can be assumed for the local-overall transverse strains, as those of equations (3.19), by replacing φ_{2} by φ_{2}^{M} or φ_{2}^{F} and replacing φ_{3} by φ_{3}^{M} or φ_{3}^{F}.

Substituting for $d\overline{\varepsilon}_{1}^{M}$ and $d\overline{\varepsilon}_{1}^{F}$ from equations (3.28) and for $d\overline{\varepsilon}_{1}$, from equation (3.18) into equations (3.21), we obtain the following equations for the local axial strain increments in the damaged state:

$$d\varepsilon_1^M = C_1^M d\varepsilon_1 \qquad (3.29a)$$

$$d\varepsilon_1^F = C_1^F d\varepsilon_1 \qquad (3.29b)$$

where the strain concentration factors C_1^M and C_1^F are now defined in the damaged lamina (i.e. these are damaged strain concentration factors) and are given by:

$$C_1^M = \overline{C}_1^M \frac{1-\varphi_1}{1-\varphi_1^M} \qquad (3.30a)$$

$$C_1^F = \overline{C}_1^F \frac{1-\varphi_1}{1-\varphi_1^F} \qquad (3.30b)$$

Equations (3.30) can be investigated in a similar way to those of equations (3.16) and some figures can be similarly obtained. However, this is not shown here since the resulting figures will be somewhat similar to Figures 3.3 and 3.4 and there is no need to repeat them here.

Similarly, using equations (3.23) and the appropriate transformation equations for the transverse strains, we obtain:

$$d\varepsilon_2^M = C_2^M d\varepsilon_2 \qquad (3.31a)$$

$$d\varepsilon_3^M = C_3^M d\varepsilon_3 \qquad (3.31b)$$

$$d\varepsilon_2^F = C_2^F d\varepsilon_2 \qquad (3.31c)$$

$$d\varepsilon_3^F = C_3^F d\varepsilon_3 \qquad (3.31d)$$

where C_2^M, C_3^M, C_2^F and C_3^F are related to $\overline{C}_2^M$, $\overline{C}_3^M$, $\overline{C}_2^F$ and $\overline{C}_3^F$ by the local damage variables. Using generalized forms for equations (3.22), we can show that:

$$C_{ij}^M = \overline{C}_{ij}^M \frac{1-\varphi_j}{1-\varphi_i^M} , \qquad i,j = 1, 2, 3 \qquad \text{(no sum over i or j)} \qquad (3.32a)$$

42

$$C_{ij}^{F} = \overline{C}_{ij}^{F} \, \frac{1 - \varphi_j}{1 - \varphi_i^{F}} \, , \qquad i, j = 1, 2, 3 \qquad \text{(no sum over i or j)} \qquad (3.32a)$$

Substituting equations (3.16) and (3.30) into equation (3.22) and using equation (3.15), we obtain the following constraint relation for the damaged stress and strain concentration factors:

$$c^{M} C_{1}^{M} + c^{F} C_{1}^{F} = 1 \qquad (3.33)$$

Equations (3.29) provide the required local-overall strain relations that are needed in the next section in order to formulate the damage constitutive equations.

In general, we can show that the constraint relations for the strain concentration factors, appearing partially in equations (3.20), take the following form:

$$\overline{c}^{M} \, \overline{C}_{ij}^{M} + \overline{c}^{F} \, \overline{C}_{ij}^{F} = \delta_{ij} \, , \qquad i, j = 1, 2, 3 \qquad (3.34a)$$

$$c^{M} \, C_{ij}^{M} + c^{F} \, C_{ij}^{F} = \delta_{ij} \, , \qquad i, j = 1, 2, 3 \qquad (3.34b)$$

Where δ_{ij} is equal to 1 when $i = j$ and 0 when $i \neq j$.

3.3 Constitutive Relations

The elastic constitutive relations are now developed in both the damaged and the effective undamaged configurations. In addition, the local-overall constitutive relations are also discussed. In the fictitious undamaged lamina, the overall strain increments are given by:

$$d\overline{\varepsilon}_{1} = \frac{d\overline{\sigma}}{\overline{E}} \qquad (3.35a)$$

$$d\overline{\varepsilon}_{2} = - \frac{\overline{v}_{21} \, d\overline{\sigma}}{\overline{E}} \qquad (3.35b)$$

$$d\bar{\varepsilon}_3 = - \frac{\bar{v}_{31}\, d\bar{\sigma}}{\bar{E}} \tag{3.35c}$$

where the material constants $\bar{E}$, $\bar{v}_{21}$, and $\bar{v}_{31}$ are the effective overall Young's modulus of elasticity, and effective overall Poisson's ratios, respectively. Based on equations (3.35), we can write a similar set of overall constitutive relations in the damaged lamina as follows:

$$d\varepsilon_1 = \frac{d\sigma}{E} \tag{3.36a}$$

$$d\varepsilon_2 = - \frac{v_{21}\, d\sigma}{E} \tag{3.36b}$$

$$d\varepsilon_3 = - \frac{v_{31}\, d\sigma}{E} \tag{3.36c}$$

where E, v_{21} and v_{31} are the damaged overall Young's modulus of elasticity and damaged overall Poisson's ratios respectively.

It is noted that E, v_{21} and v_{31} are no longer constants but depend on the damage variables. In order to demonstrate this, we substitute for $d\bar{\varepsilon}_1$ and $d\bar{\sigma}$ from equations (3.18) and (3.4), respectively, into equation (3.35a) and compare the result with equation (3.36a). It follows that:

$$E = \bar{E}\, (1 - \varphi_1)^2 \tag{3.37}$$

Similarly, substituting for $d\bar{\varepsilon}_2$ and $d\bar{\varepsilon}_3$ from equations (3.19) and for $d\bar{\sigma}$ from equation (3.4) into equations (3.35b) and (3.35c), and comparing the results with equations (3.36b) and (3.36c), we then obtain:

$$v_{21} = \bar{v}_{21}\, \frac{1 - \varphi_1}{1 - \varphi_2} \tag{3.38a}$$

$$v_{31} = \bar{v}_{31}\, \frac{1 - \varphi_1}{1 - \varphi_3} \tag{3.38b}$$

Alternatively, solving equations (3.37) and (3.38) for the three damage variables, φ_1, φ_2 and φ_3, we

44

obtain:

$$\varphi_1 = 1 - \sqrt{\frac{E}{\overline{E}}} \tag{3.39a}$$

$$\varphi_2 = 1 - \frac{\overline{v}_{21}}{v_{21}} \sqrt{\frac{E}{\overline{E}}} \tag{3.39b}$$

$$\varphi_3 = 1 - \frac{\overline{v}_{31}}{v_{31}} \sqrt{\frac{E}{\overline{E}}} \tag{3.39c}$$

Equations (3.39b) and (3.39c) may be viewed as suitable definitions for the transverse damage variables φ_2 and φ_3 for this problem. However, generalization of these definitions to other states of deformation and damage is not possible. In general, a fourth-rank damage effect tensor should be considered as shown later in Part II of this book. For more details, see references [58, 98-99].

It should be mentioned that equations (3.37) - (3.39) are available in the literature [19, 100]. Next, the more difficult task of developing similar relations on the local level as well as the local-overall constitutive relations is considered.

The local elastic stress-strain relations for the fibers and matrix along the fiber direction are given now in the fictitious undamaged configuration:

$$d\overline{\sigma}^M = \overline{E}^M \, d\overline{\varepsilon}_1^M \tag{3.40a}$$

$$d\overline{\sigma}^F = \overline{E}^F \, d\overline{\varepsilon}_1^F \tag{3.40b}$$

where $\overline{E}^M$ and $\overline{E}^F$ are the constant moduli of elasticity for the matrix and fiber materials, respectively. Substituting for $d\overline{\sigma}^M$ and $d\overline{\sigma}^F$ from equations (3.10) and for $d\overline{\varepsilon}_1^M$ and $d\overline{\varepsilon}_1^F$ from equations (3.28) into equations (3.40), we obtain:

$$d\sigma^M = E^M \, d\varepsilon_1^M \tag{3.41a}$$

$$d\sigma^F = E^F \, d\varepsilon_1^F \tag{3.41b}$$

where E^M and E^F are the damaged moduli of elasticity given by:

$$E^M = \overline{E}^M \, (1 - \varphi_1^M)^2 \tag{3.42a}$$

$$E^F = \overline{E}^F \, (1 - \varphi_1^F)^2 \tag{3.42b}$$

Equations (3.41) represent the local elastic stress-strain relations for the matrix and fibers in the damaged configuration of the lamina.

Finally, the local-overall relations for the modulus of elasticity are now presented. Substituting $d\sigma^M$ and $d\sigma^F$, from equations (3.41), for $d\varepsilon_1^M$ and $d\varepsilon_1^F$, from equations (3.29), and for $d\sigma$ from equation (3.36a) into equation (3.1), we obtain:

$$E = c^M E^M C_1^M + c^F E^F C_1^F \tag{3.43}$$

Performing similar substitutions using equations (3.40), (3.21), and (3.35a) along with equation (3.2), we obtain:

$$\overline{E} = \overline{c}^M \overline{E}^M \overline{C}_1^M + \overline{c}^F \overline{E}^F \overline{C}_1^F \tag{3.44}$$

Equations (3.43) and (3.44) are equivalent when we consider the transformation relations for $\overline{E}$, $\overline{E}^M$, $\overline{E}^F$, $\overline{C}_1^M$, $\overline{C}_1^F$, $\overline{c}_1^M$ and $\overline{c}_1^F$ given by equations (3.37), (3.42), (3.30) and (3.6). Using equations (3.6) and substituting them into equation (3.44), we obtain the following expression for the effective overall elasticity modulus, $\overline{E}$, in terms of the local parameters and the overall damage variable, φ_1:

$$\overline{E} = \frac{c^M (1 - \varphi_1^M) \overline{E}^M \overline{C}_1^M + c^F (1 - \varphi_1^F) \overline{E}^F \overline{C}_1^F}{1 - \varphi_1} \tag{3.45}$$

Alternatively, substituting for $\overline{E}$ from equation (3.37) into equation (3.45), we obtain the following expression for E:

$$E = c^M \overline{E}^M \overline{C}_1^M (1 - \varphi_1^M)(1 - \varphi_1) + c^F \overline{E}^F \overline{C}_1^F (1 - \varphi_1^F)(1 - \varphi_1) \qquad (3.46)$$

The above expression for E can also be derived from equation (3.43). Equations (3.45) and (3.46) represent local-overall relations for the modulus of elasticity.

Using similar relations for the local transverse strains as those of equations (3.35b), (3.35c), (3.36b) and (3.36c), we can easily prove the following:

$$v_{21}^M = \overline{v}_{21}^M \frac{1 - \varphi_1^M}{1 - \varphi_2^M} \qquad (3.47a)$$

$$v_{21}^F = \overline{v}_{21}^F \frac{1 - \varphi_1^F}{1 - \varphi_2^F} \qquad (3.47b)$$

where $\overline{v}_{21}^M$ and $\overline{v}_{21}^F$ are the effective Poisson's ratios for the matrix and fiber material, respectively. Similar expressions exist for v_{31}^M and v_{31}^F. In addition, we can derive relations for the local damage variables φ_1^M, φ_2^M, φ_3^M, φ_1^F, φ_2^F and φ_3^F similar to those of equations (3.39) with all overall quantities replaced by their local counterparts. Finally, we can derive the following overall-local relations for Poisson's ratios by using equations (3.1) and (3.2) and substituting the transverse strain increments for the stress increments:

$$\frac{c^M \overline{E}^M \overline{C}_1^M (1 - \varphi_1^M) + c^F \overline{E}^F \overline{C}_1^F (1 - \varphi_1^F)}{v_{21}} = \frac{c^M \overline{E}^M \overline{C}_2^M (1 - \varphi_1^M)}{v_{21}^M} + \frac{c^F \overline{E}^F \overline{C}_2^F (1 - \varphi_1^F)}{v_{21}^F}$$

$$(3.48a)$$

$$\frac{E}{v_{21}} = \frac{c^M E^M C_2^M}{v_{21}^M} + \frac{c^F E^F C_2^F}{v_{21}^F} \qquad (3.48b)$$

Equations (3.48) are the transverse local-overall relations for Poisson's ratio v_{21} in both the damaged and effective undamaged configurations. In view of the definition of the matrix Poisson's ratio $v_{21}^M = -d\varepsilon_2^M/d\varepsilon_1^M$ and equations (3.21a) and (3.23a), we can show that $\overline{C}_1^M v_{21}^M = \overline{C}_2^M v_{21}$. Similarly, we can show that $C_1^M \overline{v}_{21}^M = C_2^M \overline{v}_{21}$. These two relations can be substituted into equations (3.48) appropriately to show that the two equation (3.48a) and (3.48b) are equivalent. It should also be noted that similar relations can be shown to exist for Poisson's ratio v_{31}. The rest of this section is left for a brief discussion of the stress and strain concentration factors $\overline{B}^M$, $\overline{B}^F$, $\overline{C}_1^M$ and $\overline{C}_1^F$.

In order to determine the concentration factors, we may use the Voigt model [68, 97]. In this model, it is assumed that the phase strain increments are equal to the overall strain increment. This assumption will be applied here to the effective undamaged configuration, that is $d\overline{\varepsilon}_1^M = d\overline{\varepsilon}_1^F = d\overline{\varepsilon}_1$. Incorporating this assumption into the presented theory by comparing with equations (3.21), we directly conclude that $d\overline{C}_1^M = d\overline{C}_1^F = 1$. Upon further using equations (3.40), we have $d\overline{\sigma}^M = \overline{E}^M d\overline{\varepsilon}_1 = \overline{E}^M d\overline{\sigma}/\overline{E}$. Comparing this with equation (3.11a), we conclude that $\overline{B}^M = \overline{E}^M/\overline{E}$. A similar argument shows that $\overline{B}^F = \overline{E}^F/\overline{E}$.

The reader should be cautious, however, in using the Voigt model. Although the expressions obtained for the stress and strain concentration factors are very simple, there are certain inconsistencies that arise as a result of adopting this model. For example, using a local relation for the matrix similar to that of equation (3.35b), we have $d\overline{\sigma}^M = (-\overline{E}^M/\overline{v}_{21}^M)d\overline{\varepsilon}_2^M = (\overline{E}^M \overline{v}_{21}/\overline{E} \overline{v}_{21}^M) d\overline{\sigma}$. Comparing this with equation (3.11a) and the above result for $\overline{B}^M$, we conclude that $\overline{v}_{21}^M = \overline{v}_{21}$. This is obviously a contradiction since the matrix and overall Poisson's ratios are generally different. This contradiction arises directly from the simple assumption of the Voigt model. In addition, the derived expressions for the concentration factors using this model validates the constraint equation, (3.12) and (3.25b). Other more realistic models for determining concentration factors are available, however they are far from being simple.

The above contradiction can be corrected by employing the Vanishing Fiber Diameter (VFD) model [68, 97]. In this model, it is assumed that each of the cylindrical fibers has a vanishing diameter and that the fibers occupy a finite volume fraction of the composite (in order to provide axial constraint of the phase, [68, 69]). For the problem considered here, these assumptions reduce to:

$$d\overline{\sigma} = c^M d\overline{\sigma}^M + c^F d\overline{\sigma}^F \tag{3.49a}$$

$$d\overline{\varepsilon}_1 = d\overline{\varepsilon}_1^M = d\overline{\varepsilon}_1^F \tag{3.49b}$$

$$d\overline{\varepsilon}_2 = c^M d\overline{\varepsilon}_2^M + c^F d\overline{\varepsilon}_2^F \tag{3.49c}$$

$$d\bar{\varepsilon}_3 = c^M d\bar{\varepsilon}_3^M + c^F d\bar{\varepsilon}_3^F \tag{3.49d}$$

It is clear that the axial strain increment assumption (3.49b) conforms with that of the Voigt model. However, a more realistic assumption is provided for the transverse strain increments (3.49c) and (3.49d) which is compatible with the physics of the problem. Considering the argument of the previous paragraph, it can be seen that the contradiction concerning Poisson's ratio no longer exists in the VFD model and, therefore, this model is appropriate to use for this problem. Nevertheless, more sophisticated models for determining the concentration factors will be discussed in Part II of this book.

3.4 Damage Evolution

The problem of damage evolution has been studied previously for metals in section 2.3. The same equations presented before can be used for evolution of the overall damage variable φ_1. In addition, we an easily incorporate local damage evolution based on the same principles outlined in section 2.3. We assume that there exist two local damage strengthening criteria, $g^M(y_1^M, L^M)$ and $g^F(y_1^F, L^F)$, having the same form as that of equation (2.15), where y_1^M and y_1^F are the generalized thermodynamic forces associated with φ_1^M and φ_1^F, respectively, and L^M and L^F are the local counterparts of L. Linear expressions are also used for L^M and L^F such that $L^M = c_1 \ell^M + d_1$ and $L^F = c_2 \ell^F + d_2$, where ℓ^M and ℓ^F are local counterparts of ℓ_1 and c_1, c_2, d_1, d_2 are constants.

Assuming matrix and fiber damage evolution laws similar to that of equations (2.22) and (2.23), we can write:

$$\varphi_1^M = -\frac{(y_1^M)^3}{3 c_1} \tag{3.50a}$$

$$\varphi_1^F = -\frac{(y_1^F)^3}{3 c_2} \tag{3.50b}$$

Substituting equations (2.23) and (3.50) into equation (3.8) and simplifying the result, we obtain the local-overall relation for the generalized thermodynamic force associated with the damage variable:

$$y_1^3 = c\left[\frac{c^M}{c_1}(y_1^M)^3 + \frac{c^F}{c_2}(y_1^F)^3\right]$$

(3.51)

Finally, using the above equation along with the fact that $y_1 = \partial g/\partial y_1$, and similar expressions for y_1^M, and y_1^F, we obtain:

$$\left(\frac{\partial g}{\partial y_1}\right)^3 = c\left[\frac{c^M}{c_1}\left(\frac{\partial g^M}{\partial y_1^M}\right)^3 + \frac{c^F}{c_2}\left(\frac{\partial g^F}{\partial y_1^F}\right)^3\right]$$

(3.52)

Equation (3.52) is a nonlinear partial differential equation that represents the local-overall relation for the damage strengthening criteria for the matrix, fibers and the overall composite system. The generalization of this damage evolution model to general states of deformation and damage in metal matrix composites is presented in Part II of this book.

A micromechemical damage analysis was presented in this chapter for a unidirectional fiber-reinforced composite thin lamina subjected to uniaxial tension. The analysis was based on a combination of the micromechanical composite model coupled with continuum damage mechanics. The mathematical formulation appearing in this chapter was taken from the authors' own work in this area [100]. The theory presented in this chapter can be generalized for general states of deformation and damage in metal matrix composites. However, tensor analysis is needed for the mathematical formulation. Therefore, the generalization of this theory is left to Part II of this book.

CHAPTER 4

**UNIAXIAL TENSION IN ELASTO - PLASTIC METAL MATRIX COMPOSITES:
VECTOR FORMULATION OF THE OVERALL APPROACH**

The initiation and evolution of damage and elasto-plastic deformation in metal matrix laminae is studied in this chapter using an overall approach. The recent work of the authors [101, 102] is applied to the problem of damage initiation and growth in a uniaxially loaded unidirectional fiber-reinforced composite lamina. Damage is modeled according to the overall approach in which one damage variable is used to describe damage in the lamina including the initiation, growth and coalescence of voids and cracks in the matrix, fiber fracture, and debonding. A governing system of nine simultaneous ordinary differential equations is established for this problem. The system is solved numerically and the results are discussed. This problem is selected because it can be solved numerically without the use of the finite element method. A subsequent chapter demonstrates the implementation of the model using finite elements.

4.1 Preliminaries

An overall approach to the characterization of damage in elasto-plastic fiber-reinforced metal matrix composites [101, 102] is formulated using simple mathematical techniques. A vector formulation of the model is presented in a simple form without the use of tensors or advanced mathematics. However, the reader should view the mathematical formulation of this chapter as a transitional device to the use of tensor algebra that starts with Chapter 5. In the formulation, the notation described in section 1.5 is used. In particular, brackets [ ] are used to denote 3 x 3 matrices, while braces { } are used to denote 3 x 1 vectors. A superscript T indicates the transpose of a vector or matrix. The formulation is general , except that the only restriction is the formula for the derivative of the damage effect matrix $[M]$ which is valid only for problems involving principal damage variables (e.g. uniaxial tension). The composite system consists of an elasto-plastic matrix reinforced with continuous, perfectly aligned, cylindrical elastic fibers.

Let $\{\bar{\sigma}^M\}$ and $\{\bar{\sigma}^F\}$ be the matrix and fiber effective stress vectors, respectively. In the formulation given in this chapter, the vectors $\{\bar{\sigma}^M\}$ and $\{\bar{\sigma}^F\}$ take the form:

$$\{\bar{\sigma}^M\} = \begin{bmatrix} \bar{\sigma}_1^M & \bar{\sigma}_2^M & \bar{\sigma}_3^M \end{bmatrix}^T \tag{4.1a}$$

52

$$\{\overline{\sigma}^F\} = \begin{bmatrix} \overline{\sigma}_1^F & \overline{\sigma}_2^F & \overline{\sigma}_3^F \end{bmatrix}^T \tag{4.1b}$$

Similarly, the overall effective stress vector $\{\overline{\sigma}\}$ takes the form:

$$\{\overline{\sigma}\} = \begin{bmatrix} \overline{\sigma}_1 & \overline{\sigma}_2 & \overline{\sigma}_3 \end{bmatrix}^T \tag{4.1c}$$

The elastic stress concentration matrices $[\overline{B}^M]$ and $[\overline{B}^F]$ for the matrix and fibers, respectively, are defined as follows:

$$\{\overline{\sigma}^M\} = [\overline{B}^M]\{\overline{\sigma}\} \tag{4.2a}$$

$$\{\overline{\sigma}^F\} = [\overline{B}^F]\{\overline{\sigma}\} \tag{4.2b}$$

where $[\overline{B}^M]$ and $[\overline{B}^F]$ are 3 x 3 constant matrices. For the case of plastic loading or elastic unloading, equation (4.2a) is rewritten in incremental form as follows:

$$\{d\overline{\sigma}^M\} = [\overline{B}^{MP}]\{d\overline{\sigma}\} \tag{4.2c}$$

where $[\overline{B}^{MP}]$ is a 3 x 3 elasto-plastic stress concentration matrix for the matrix material.[1] Several models are available in the literature for the determination of the three matrices $[\overline{B}^M]$, $[\overline{B}^F]$, and $[\overline{B}^{MP}]$ (see references [68 - 70]). Some of these models are discussed briefly at the end of this chapter and in more detail in subsequent chapters.

Let $\{\overline{\tau}\}$ and $\{\overline{\tau}^M\}$ be the overall and matrix deviatoric stress vectors, respectively. Then, they are related to the total overall and matrix stress vectors $\{\overline{\sigma}\}$ and $\{\overline{\sigma}^M\}$, respectively, as follows:

$$\{\overline{\tau}\} = [a]\{\overline{\sigma}\} \tag{4.3a}$$

$$\{\overline{\tau}^M\} = [a]\{\overline{\sigma}^M\} \tag{4.3b}$$

where the constant 3 x 3 matrix $[a]$ is given by:

---

[1] A superscript "p" indicates a plastic quantity

$$[a] = \frac{1}{3} \begin{bmatrix} 2 & -1 & -1 \\ -1 & 2 & -1 \\ -1 & -1 & 2 \end{bmatrix} \tag{4.3c}$$

The matrix $[a]$ is idempotent, i.e. $[a]^n = [a]$, where n is a positive integer. In particular, the relation $[a]^2 = [a]$ will be used in the present formulation.

4.2 Effective Stresses and the Yield Function

The relations governing the overall and matrix backstress vectors are assumed identical to those of the corresponding stress vectors. They are listed here as follows:

$$\{\bar{\boldsymbol{\beta}}^M\} = \left[\bar{\boldsymbol{B}}^M\right]\{\bar{\boldsymbol{\beta}}\} \tag{4.4a}$$

$$\{d\bar{\boldsymbol{\beta}}^M\} = \left[\bar{\boldsymbol{B}}^{MP}\right]\{d\bar{\boldsymbol{\beta}}\} \tag{4.4b}$$

The deviatoric backstress vectors are also given by:

$$\{\bar{\boldsymbol{\alpha}}\} = [a]\{\bar{\boldsymbol{\beta}}\} \tag{4.4c}$$

$$\{\bar{\boldsymbol{\alpha}}^M\} = [a]\{\bar{\boldsymbol{\beta}}^M\} \tag{4.4d}$$

Substituting equation (4.2a) into (4.3b), we obtain the following expression for the matrix deviatoric effective stress vector:

$$\{\bar{\boldsymbol{\tau}}^M\} = [a]\left[\bar{\boldsymbol{B}}^M\right]\{\bar{\boldsymbol{\sigma}}\} \tag{4.5a}$$

Similarly, substituting equation (4.4a) into (4.4d), we obtain the following expression for the matrix deviatoric effective backstress vector:

$$\{\bar{\boldsymbol{\alpha}}^M\} = [a]\left[\bar{\boldsymbol{B}}^M\right]\{\bar{\boldsymbol{\beta}}\} \tag{4.5b}$$

The effective yield function $\bar{f}^M$ for the matrix material is given here as a von Mises type with kinematic hardening:

$$\bar{f}^M = \frac{3}{2}\{\bar{\boldsymbol{\tau}}^M - \bar{\boldsymbol{\alpha}}^M\}^T\{\bar{\boldsymbol{\tau}}^M - \bar{\boldsymbol{\alpha}}^M\} - \bar{\sigma}_o^{M2} \equiv 0 \tag{4.6}$$

54

where $\overline{\sigma}_o^M$ is the yield strength of the matrix material. Substituting equations (4.5a) and (4.5b) into equation (4.6) and simplifying, we obtain the following expression for the effective yield function $\overline{f}$ for the overall composite system:

$$\overline{f}^M = \frac{3}{2}\{\overline{\sigma} - \overline{\beta}\}^T [\overline{B}^M]^T [a] [\overline{B}^M] \{\overline{\sigma} - \overline{\beta}\} - \overline{\sigma}_o^{M2} \equiv 0 \tag{4.7}$$

Using the yield function expressions of equations (4.6) and (4.7) and simplifying, we obtain the following formulae for the yield function partial derivative vectors $\{\partial \overline{f}^M / \partial \overline{\sigma}^M\}$ and $\{\partial \overline{f} / \partial \overline{\sigma}\}$.

$$\left\{\frac{\partial \overline{f}^M}{\partial \overline{\sigma}^M}\right\} = 3[a]\{\overline{\sigma}^M - \overline{\beta}^M\} \tag{4.8a}$$

$$\left\{\frac{\partial \overline{f}}{\partial \overline{\sigma}}\right\} = 3[\overline{B}^M]^T [a][\overline{B}^M]\{\overline{\sigma} - \overline{\beta}\} \tag{4.8b}$$

In fact, we can show (using only equation (4.2a)) that the two derivative vectors given above are related by the following equation which is independent of the yield function:

$$\left\{\frac{\partial \overline{f}}{\partial \overline{\sigma}}\right\} = [\overline{B}^M]^T \left\{\frac{\partial \overline{f}^M}{\partial \overline{\sigma}^M}\right\} \tag{4.9}$$

4.3 Effective Strains and the Flow Rule

We now introduce the effective strain vectors $\{\overline{\varepsilon}^M\}$ and $\{\overline{\varepsilon}^F\}$ for the matrix and fibers, respectively, as follows:

$$\{\overline{\varepsilon}^M\} = \begin{bmatrix} \overline{\varepsilon}_1^M & \overline{\varepsilon}_2^M & \overline{\varepsilon}_3^M \end{bmatrix}^T \tag{4.10a}$$

$$\{\overline{\varepsilon}^F\} = \begin{bmatrix} \overline{\varepsilon}_1^F & \overline{\varepsilon}_2^F & \overline{\varepsilon}_3^F \end{bmatrix}^T \tag{4.10b}$$

Similarly, the effective strain vector $\{\overline{\varepsilon}\}$ for the overall composite system is given by:

$$\{\overline{\varepsilon}\} = \begin{bmatrix} \overline{\varepsilon}_1 & \overline{\varepsilon}_2 & \overline{\varepsilon}_3 \end{bmatrix}^T \tag{4.10c}$$

The elastic constant strain concentration matrices $[\overline{C}^M]$ and $[\overline{C}^F]$ for the matrix and fibers, respectively, are defined by:

$$\left\{\bar{\varepsilon}^M\right\}' = \left[\bar{C}^M\right]\left\{\bar{\varepsilon}\right\}' \tag{4.11a}$$

$$\left\{\bar{\varepsilon}^F\right\} = \left[\bar{C}^F\right]\left\{\bar{\varepsilon}\right\} \tag{4.11b}$$

where the prime $'$ indicates elastic strains. For the case of plastic loading or elastic unloading, the elasto-plastic strain concentration matrix $\left[\bar{C}^{MP}\right]$ is defined by the following incremental relation:

$$\left\{d\bar{\varepsilon}^M\right\}'' = \left[\bar{C}^{MP}\right]\left\{d\bar{\varepsilon}\right\}'' \tag{4.11c}$$

where the double prime $''$ indicates plastic strains. It is noticed that in equation (4.11b), the total effective fiber strain vector $\left\{\bar{\varepsilon}^F\right\}$ is used because the fibers undergo only elastic deformation. The elastic and plastic parts of the effective strain vectors are given by the additive decomposition:

$$\left\{d\bar{\varepsilon}\right\} = \left\{d\bar{\varepsilon}\right\}' + \left\{d\bar{\varepsilon}\right\}'' \tag{4.12a}$$

$$\left\{d\bar{\varepsilon}^M\right\} = \left\{d\bar{\varepsilon}^M\right\}' + \left\{d\bar{\varepsilon}^M\right\}'' \tag{4.12b}$$

An effective associated flow rule is used for the "undamaged" matrix material as follows:

$$\left\{d\bar{\varepsilon}^M\right\}'' = d\bar{\lambda}^M\left\{\frac{\partial\bar{f}^M}{\partial\bar{\sigma}^M}\right\} \tag{4.13}$$

where $d\bar{\lambda}^M$ is a scalar multiplier to be determined. Substituting equations (4.11c) and (4.9) into equation (4.13) and simplifying, we obtain:

$$\left\{d\bar{\varepsilon}\right\}'' = \left[d\bar{\lambda}\right]\left\{\frac{\partial\bar{f}}{\partial\bar{\sigma}}\right\} \tag{4.14}$$

where the multiplier matrix $\left[d\bar{\lambda}\right]$ is given by:

$$\left[d\bar{\lambda}\right] = d\bar{\lambda}^M\left[\bar{C}^{MP}\right]^{-1}\left[\bar{B}^M\right]^{-T} \tag{4.15}$$

Equation (4.14) clearly indicates a non-associated flow rule due to the presence of the 3 x 3 matrix $\left[d\bar{\lambda}\right]$.

56

In the formulation, the "undamaged" matrix material undergoes kinematic hardening of the Prager - Ziegler type. This is represented by the evolution equation for the effective matrix backstress vector $\{\bar{\alpha}^M\}$ as follows:

$$\{d\,\bar{\alpha}^M\} = d\bar{\mu}^M \{\bar{\tau}^M - \bar{\alpha}^M\} \tag{4.16}$$

where $d\bar{\mu}^M$ is a scalar multiplier to be determined. In order to obtain a relation between the two scalar multipliers $d\bar{\mu}^M$ and $d\bar{\lambda}^M$, we equate the projection of the effective matrix incremental backstress vector on the yield surface $\bar{f}^M$ to $b\{d\bar{\varepsilon}^M\}''$:

$$b\{d\bar{\varepsilon}^M\}'' = \frac{\{d\,\bar{\alpha}^M\}^T \left\{\dfrac{\partial \bar{f}^M}{\partial \bar{\sigma}^M}\right\}}{\left\{\dfrac{\partial \bar{f}^M}{\partial \bar{\sigma}^M}\right\}^T \left\{\dfrac{\partial \bar{f}^M}{\partial \bar{\sigma}^M}\right\}} \left\{\frac{\partial \bar{f}^M}{\partial \bar{\sigma}^M}\right\} \tag{4.17}$$

where b is a constant material parameter determined from experiments [55, 56]. Post-multiplying equation (4.17) by $\{\partial \bar{f}^M/\partial \bar{\sigma}^M\}^T$, simplifying and using equation (4.8a), we obtain the desired relation as follows:

$$d\bar{\mu}^M = 3\,b\,d\bar{\lambda}^M \tag{4.18}$$

It is noted that equation (4.18) is valid only for the von Mises yield function $\bar{f}^M$ since it is used in the derivation. Next, we determine an expression for the scalar multiplier $d\bar{\lambda}^M$ using the consistency condition $d\bar{f}^M = 0$ as follows:

$$\left\{\frac{\partial \bar{f}^M}{\partial \bar{\sigma}^M}\right\}^T \{d\bar{\sigma}^M\} + \left\{\frac{\partial \bar{f}^M}{\partial \bar{\alpha}^M}\right\}^T \{d\bar{\alpha}^M\} = 0 \tag{4.19}$$

4.4 Effective Constitutive Relation

Consider the elastic matrix relation:

$$\left\{ d\,\overline{\sigma}^M \right\} = \left[\overline{E}^M \right] \left\{ d\,\overline{\varepsilon}^M \right\}'$$

(4.20a)

where $\left[\overline{E}^M \right]$ is the effective elasticity matrix for the matrix material. Substituting for $\left\{ d\,\overline{\varepsilon}^M \right\}'$ from equation (4.12b), and for $\left\{ d\,\overline{\varepsilon}^M \right\}''$ from equation (4.13), we obtain:

$$\left\{ d\,\overline{\sigma}^M \right\} = \left[\overline{E}^M \right] \left(\left\{ d\,\overline{\varepsilon}^M \right\} - d\overline{\lambda}^M \left\{ \frac{\partial \overline{f}^M}{\partial \overline{\sigma}^M} \right\} \right)$$

(4.20b)

Substituting equations (4.16), (4.18) and (4.20b) into equation (4.19) and simplifying, we obtain the following expression for $d\overline{\lambda}^M$:

$$d\overline{\lambda}^M = \frac{1}{\overline{Q}^M} \left\{ \frac{\partial \overline{f}^M}{\partial \overline{\sigma}^M} \right\}^T \left[\overline{E}^M \right] \left\{ d\,\overline{\varepsilon}^M \right\}$$

(4.21a)

where the scalar quantity $\overline{Q}^M$ is given by:

$$\overline{Q}^M = 9 \left\{ \overline{\tau}^M - \overline{\alpha}^M \right\}^T \left(\left[\overline{E}^M \right] + b[I] \right) \left\{ \overline{\tau}^M - \overline{\alpha}^M \right\}$$

(4.21b)

where $[I]$ is the 3 x 3 identity matrix. The expression of $\overline{Q}^M$ given in equation (4.21b) is valid only when using the von Mises yield function $\overline{f}^M$ given by equation (4.6).

Next, we derive the effective matrix elasto-plastic stiffness matrix $\left[\overline{D}^M \right]$. This is performed by substituting equation (4.21a) into (4.20b) and simplifying. Therefore, we obtain:

$$\left\{ d\,\overline{\sigma}^M \right\} = \left[\overline{D}^M \right] \left\{ d\,\overline{\varepsilon}^M \right\}$$

(4.22a)

where $\left[\overline{D}^M \right]$ is given by:

$$\left[\overline{D}^M \right] = \left[\overline{E}^M \right] - \frac{3}{\overline{Q}^M} \left[\overline{E}^M \right] \left\{ \frac{\partial \overline{f}^M}{\partial \overline{\sigma}^M} \right\} \left\{ \overline{\sigma} - \overline{\beta} \right\}^T \left[\overline{B}^M \right]^T \left[a \right] \left[\overline{E}^M \right]$$

(4.22b)

The above equation can be used with any yield function $\bar{f}^M$, except when using the specific expression of $\overline{Q}^M$ given in equation (4.21b).

Next, we derive an expression for the evolution of the effective overall backstress vector $\{\bar{\beta}\}$ based on equation (4.16). Subtracting equation (4.4a) from equation (4.2a) and rewriting the resulting equation in incremental form, we obtain:

$$\{d\bar{\sigma}^M - d\bar{\beta}^M\} = [\overline{B}^M]\{d\bar{\sigma} - d\bar{\beta}\} \tag{4.23}$$

Upon plastic loading, we substitute equation (4.2c) into equation (4.23) and solve for $\{d\bar{\beta}\}$ to obtain:

$$\{d\bar{\beta}\} = \left([I] - [\overline{B}^M]^{-1}[\overline{B}^{MP}]\right)\{d\bar{\sigma}\} - [\overline{B}^M]^{-1}\{d\bar{\beta}^M\} \tag{4.24}$$

To find an expression for $\{d\bar{\beta}^M\}$ based on the Prager-Ziegler evolution law of equation

(4.16), we substitute equations (4.3b) and (4.4d) into equation (4.16) and simplify. The resulting equation is:

$$\{d\bar{\beta}^M\} = d\bar{\mu}^M[\overline{B}^M]\{\bar{\sigma} - \bar{\beta}\} \tag{4.25}$$

Finally, substituting equation (4.25) into equation (4.24), we obtain the following evolution law for the effective overall backstress vector $\{\bar{\beta}\}$:

$$\{d\bar{\beta}\} = \left([\bar{I}] - [\overline{B}^M]^{-1}[\overline{B}^{MP}]\right)\{d\bar{\sigma}\} - d\bar{\mu}^M\{\bar{\sigma} - \bar{\beta}\} \tag{4.26}$$

It is clear from equation (4.26) that kinematic hardening of the composite material consists of two types. The first type is due to the kinematic hardening of the matrix material and is represented by the second term on the right-hand-side of equation (4.26). The second type is represented by the first term on the right-hand-side of equation (4.26) due to the interaction of the matrix and fibers. Therefore, the composite material will still undergo kinematic hardening (of the second type) even if the matrix does not.

Equation (4.20a) introduced the effective elastic constitutive relation for the matrix. Similarly, we can introduce an effective elastic constitutive relation for the fibers in the form:

$$\{d\bar{\boldsymbol{\sigma}}^F\} = \left[\bar{\boldsymbol{E}}^F\right]\{d\bar{\boldsymbol{\varepsilon}}^F\} \tag{4.27}$$

where $\left[\bar{\boldsymbol{E}}^F\right]$ is the effective elasticity matrix for the fiber material and $\{d\bar{\boldsymbol{\varepsilon}}^F\}$ consists entirely of elastic strain. The effective overall elastic constitutive relation for the composite system can now be written in the form:

$$\{d\bar{\boldsymbol{\sigma}}\} = \left[\bar{\boldsymbol{E}}\right]\{d\bar{\boldsymbol{\varepsilon}}\}' \tag{4.28}$$

where $\left[\bar{\boldsymbol{E}}\right]$ is the effective overall elasticity matrix for the composite system. The matrix $\left[\bar{\boldsymbol{E}}\right]$ is obtained from the matrices $\left[\bar{\boldsymbol{E}}^M\right]$ and $\left[\bar{\boldsymbol{E}}^F\right]$ as is shown shortly. We now introduce the following relation between the effective incremental overall and local stresses [68 - 70]:

$$\{d\bar{\boldsymbol{\sigma}}\} = \bar{c}^M\{d\bar{\boldsymbol{\sigma}}^M\} + \bar{c}^F\{d\bar{\boldsymbol{\sigma}}^F\} \tag{4.29}$$

where $\bar{c}^M$ and $\bar{c}^F$ are the matrix and fiber volume fractions, respectively. We substitute equations (4.20a), (4.27) and (4.28) into equation (4.29), and simplify to obtain:

$$\left[\bar{\boldsymbol{E}}\right] = \bar{c}^M\left[\bar{\boldsymbol{E}}^M\right]\left[\bar{\boldsymbol{C}}^M\right] + \bar{c}^F\left[\bar{\boldsymbol{E}}^F\right]\left[\bar{\boldsymbol{C}}^F\right] \tag{4.30}$$

The above equation is the generalization of equation (3.44) which was derived previously for metals.

In order to derive the effective overall elasto-plastic constitutive relation for the composite system, we need first to find an expression for the multiplier matrix $\left[d\bar{\lambda}\right]$ of equation (4.14) in terms of the effective overall quantities. Therefore, we first invoke the consistency condition $d\bar{f} = 0$:

$$\left\{\frac{\partial\bar{f}}{\partial\bar{\boldsymbol{\sigma}}}\right\}^T\{d\bar{\boldsymbol{\sigma}}\} + \left\{\frac{\partial\bar{f}}{\partial\bar{\boldsymbol{\beta}}}\right\}^T\{d\bar{\boldsymbol{\beta}}\} = 0 \tag{4.31}$$

Substituting for $\{d\bar{\boldsymbol{\beta}}\}$ from equation (4.26), for $\{d\bar{\boldsymbol{\sigma}}\}$ from equation (4.28), for $\{d\bar{\boldsymbol{\varepsilon}}\}'$ from equation (4.12a), for $\{d\bar{\boldsymbol{\varepsilon}}\}''$ from equation (4.14), and for $\left[d\bar{\lambda}\right]$ from equation (4.15), we obtain (after simplifying and solving for $d\bar{\lambda}^M$):

60

$$d\bar{\lambda}^M = \{\bar{T}\}^T [\bar{E}] \{d\bar{\varepsilon}\} \qquad (4.32)$$

where the 3 x 1 vector $\{\bar{T}\}$ is given by:

$$\{\bar{T}\} = \frac{\left\{\dfrac{\partial \bar{f}}{\partial \bar{\sigma}}\right\} + \left([I] - [\bar{B}^M]^{-1}[\bar{B}^{MP}]\right)^T \left\{\dfrac{\partial \bar{f}}{\partial \bar{\beta}}\right\}}{\left\{\dfrac{\partial \bar{f}}{\partial \bar{\sigma}}\right\}^T + \left\{\dfrac{\partial \bar{f}}{\partial \bar{\beta}}\right\}^T \left([I] - [\bar{B}^M]^{-1}[\bar{B}^{MP}]\right) [\bar{E}][\bar{C}^{MP}]^{-1}[\bar{B}^M]^{-T} \left\{\dfrac{\partial \bar{f}}{\partial \bar{\sigma}}\right\} + 3b \left\{\dfrac{\partial \bar{f}}{\partial \bar{\beta}}\right\}^T \{\bar{\sigma} - \bar{\beta}\}}$$

$$(4.33)$$

We now start with equation (4.28) and substitute for $\{d\bar{\varepsilon}\}'$ from equation (4.12a), for $\{d\bar{\varepsilon}\}''$ from equation (4.14), for $[d\bar{\lambda}]$ from equation (4.15), and for $d\bar{\lambda}^M$ from equation (4.32), to obtain:

$$\{d\bar{\sigma}\} = [\bar{D}]\{d\bar{\varepsilon}\} \qquad (4.34a)$$

where the effective overall elasto-plastic stiffness matrix $[\bar{D}]$ is given by:

$$[\bar{D}] = [\bar{E}] - [\bar{E}][\bar{C}^{MP}]^{-1}[\bar{B}^M]^{-1}\left\{\dfrac{\partial \bar{f}}{\partial \bar{\sigma}}\right\}\{\bar{T}\}^T[\bar{E}] \qquad (4.34b)$$

Equation (4.34a) represents the effective elasto-plastic constitutive relation for the overall composite material.

4.5 Stresses in the Damaged Composite System

The second step of the formulation involves the incorporation of damage in the constitutive equations. This is performed by using the effective overall constitutive relation given in the equation (4.34a) and transforming it into a constitutive equation for the whole composite system. Therefore, all the quantities appearing in equations (4.34a) and (4.34b) need to be transformed using the damage variable.

We first start by using the linear transformation $[M]$ between the effective stress vector $\{\bar{\sigma}\}$ and the stress vector $\{\sigma\}$ as follows:

$$\{\overline{\sigma}\} = [M]\{\sigma\} \tag{4.35}$$

where $[M]$ is a 3 x 3 matrix of the damage variables ϕ_1, ϕ_2, and ϕ_3. The matrix $[M]$ is represented in principal form as follows:

$$[M] = \begin{bmatrix} \dfrac{1}{1-\varphi_1} & 0 & 0 \\[2ex] 0 & \dfrac{1}{1-\varphi_2} & 0 \\[2ex] 0 & 0 & \dfrac{1}{1-\varphi_3} \end{bmatrix} \tag{4.36}$$

and the stress vector is given as $\{\sigma\} = \begin{bmatrix} \sigma_1 & \sigma_2 & \sigma_2 \end{bmatrix}^T$. The damage transformation equations (4.35) and (4.36) should be compared with equation (2.4) for the case of uniaxial tension in metals.

It is clear from equation (4.36) that the matrix $[M]$ reduces to the identity matrix $[I]$ when there is no damage in the material, i.e. when $\phi_1 = \phi_2 = \phi_3 = 0$. On the other hand, the elements of the matrix $[M]$ become very large when the material approaches complete rupture, i.e. when the values of ϕ_1, ϕ_2, and ϕ_3 approach 1. Actually, the values of ϕ_1, ϕ_2, and ϕ_3 do not need to approach 1 separately for rupture to occur. A representative scalar parameter (e.g. $\varphi_{cr} = \sqrt{\varphi_1^2 + \varphi_2^2 + \varphi_3^2}$) could be defined to characterize rupture. In the following formulation, the derivative matrix $d[M]$ is needed and is calculated using the chain rule as follows:

$$d[M] = \left[\dfrac{\partial M}{\partial \varphi_1}\right] d\varphi_1 + \left[\dfrac{\partial M}{\partial \varphi_2}\right] d\varphi_2 + \left[\dfrac{\partial M}{\partial \varphi_3}\right] d\varphi_3 \tag{4.37}$$

Substituting equation (4.36) into equation (4.37) and simplifying, we obtain:

62

$$d[M] = \begin{bmatrix} \dfrac{d\varphi_1}{(1-\varphi_1)^2} & 0 & 0 \\[2em] 0 & \dfrac{d\varphi_2}{(1-\varphi_2)^2} & 0 \\[2em] 0 & 0 & \dfrac{d\varphi_3}{(1-\varphi_3)^2} \end{bmatrix} \tag{4.38}$$

Taking the derivative of equation (4.35) and utilizing equations (4.36) and (4.38), and simplifying, we obtain:

$$\{d\overline{\sigma}\} = \begin{Bmatrix} \dfrac{d\sigma_1}{1-\varphi_1} + \dfrac{\sigma_1\,d\varphi_1}{(1-\varphi_1)^2} \\[2em] \dfrac{d\sigma_2}{1-\varphi_2} + \dfrac{\sigma_2\,d\varphi_2}{(1-\varphi_2)^2} \\[2em] \dfrac{d\sigma_3}{1-\varphi_3} + \dfrac{\sigma_3\,d\varphi_3}{(1-\varphi_3)^2} \end{Bmatrix} \tag{4.39}$$

Using equation (4.38), we can obtain the following expression for the quantity $d[M]\cdot\{\sigma\}$ which is used extensively in the derivations that follow:

$$d[M]\cdot\{\sigma\} = [K^\sigma]\{d\varphi\} \tag{4.40a}$$

where the matrix $[K^\sigma]$ is given by:

$$[K^\sigma] = \begin{bmatrix} \dfrac{\sigma_1}{(1-\varphi_1)^2} & 0 & 0 \\[2em] 0 & \dfrac{\sigma_2}{(1-\varphi_2)^2} & 0 \\[2em] 0 & 0 & \dfrac{\sigma_3}{(1-\varphi_3)^2} \end{bmatrix} \tag{4.40b}$$

and the damage vector is $\{d\varphi\} = \begin{bmatrix} d\varphi_1 & d\varphi_2 & d\varphi_3 \end{bmatrix}^T$. Similarly, we can derive following equation for the quantity $d[M] \cdot \{\beta\}$:

$$d[M] \cdot \{\beta\} = [\mathbf{K}^\beta]\{d\varphi\} \tag{4.41a}$$

where the matrix $[\mathbf{K}^\beta]$ is given by:

$$[\mathbf{K}^\beta] = \begin{bmatrix} \dfrac{\beta_1}{(1-\varphi_1)^2} & 0 & 0 \\ 0 & \dfrac{\beta_2}{(1-\varphi_2)^2} & 0 \\ 0 & 0 & \dfrac{\beta_3}{(1-\varphi_3)^2} \end{bmatrix} \tag{4.41b}$$

The expressions in equations (4.40a) and (4.41a) are used extensively in the derivations below. However, the reader must keep in mind that these expressions are valid only when using principal values and the representation of $[M]$ given in equation (4.36). In fact, these expressions cannot be easily generalized.

Substituting equation (4.35) into equation (4.3a) and simplifying, we obtain the following relation for the effective overall deviatoric stress vector $\{\bar{\tau}\}$:

$$\{\bar{\tau}\} = [N]\{\sigma\} \tag{4.42a}$$

where the 3 x 3 matrix $[N]$ is given by:

$$[N] = [a][M] \tag{4.42b}$$

and $[a]$ is the 3 x 3 constant matrix given in equation (4.3c). The effective overall backstress vector $\{\beta\}$ is assumed to transform in a similar way to $\{\sigma\}$. Therefore, the following damage transformation equation is used (compare with equation (4.35)):

$$\{\bar{\beta}\} = [M]\{\beta\} \tag{4.43a}$$

64

Substituting equation (4.43c) into equation (4.4c) and simplifying, we obtain the following equation which is analogous to equation (4.42a):

$$\{\overline{\alpha}\} = [N]\{\beta\} \tag{4.43b}$$

Equations (4.42a) and (4.43b) represent the damage transformation equations for the effective overall stress and backstress vectors, respectively. They will be used in the transformation of the yield function, the flow rule, the kinematic hardening rule, and the constitutive relations. Starting with the effective yield function $\overline{f}$ given in equation (4.7) and substituting for $\{\overline{\sigma}\}$ from equation (4.35) and for $\{\overline{\beta}\}$ from equation (4.43a) and simplifying, we obtain:

$$f = \frac{3}{2}\{\sigma - \beta\}^T [H] \{\sigma - \beta\} - \overline{\sigma}_o^{M\,2} \equiv 0 \tag{4.44a}$$

where the 3 x 3 matrix $[H]$ is given by:

$$[H] = [M]^T \left[\overline{B}^M\right]^T [a] \left[\overline{B}^M\right] [M] \tag{4.44b}$$

Equation (4.44a) represents the yield function for the damaged composite system. The partial derivative $\{\partial f / \partial \sigma\}$ is now readily obtained from equation (4.44a) as follows:

$$\left\{\frac{\partial f}{\partial \sigma}\right\} = 3[H]\{\sigma - \beta\} \tag{4.45}$$

Using equation (4.35), we can show, using the chain rule, that the following general relation exists between the partial derivatives $\{\partial \overline{f} / \partial \overline{\sigma}\}$ and $\{\partial f / \partial \sigma\}$.

$$\left\{\frac{\partial f}{\partial \sigma}\right\} = [M]^T \left\{\frac{\partial \overline{f}}{\partial \overline{\sigma}}\right\} \tag{4.46}$$

The above relation is independent of the yield function.

4.6 Damage Evolution

Several criteria are available in the literature for the description of damage evolution. The one chosen here is that proposed by Lee et al. [15] for its simplicity and ease of integration in the constitutive model. This criterion has been used previously in section 2.3 to describe damage evolution in uniaxial tension of metals. However, it should be emphasized that the constitutive model is so general that any viable damage criterion can be used. This point is further elaborated on in Part II of this book.

Let g be the scalar damage function given by Lee et al. [15]:

$$g = \frac{1}{2}\{\bar{\sigma}\}^{T}[J]\{\bar{\sigma}\} - \ell_{o}^{2} - L(\ell) \equiv 0 \tag{4.47a}$$

Where ℓ is a scalar "overall" damage parameter, and $[J]$ is a constant 3 x 3 matrix given by:

$$[J] = \begin{bmatrix} 1 & \mu & \mu \\ \mu & 1 & \mu \\ \mu & \mu & 1 \end{bmatrix} \tag{4.47b}$$

and μ is a constant damage parameter, $-0.5 \leq \mu \leq 1.0$. The matrix representative of $[J]$ given in equation (4.47b) applies only for the problem considered in this chapter. A more general representation of $[J]$ is given later in Part II of the book. Substituting equation (4.35) into equation (4.47a) and simplifying, we obtain:

$$g = \frac{1}{2}\{\sigma\}^{T}[M]^{T}[J][M]\{\sigma\} - \ell_{o}^{2} - L(\ell) \equiv 0 \tag{4.48a}$$

It should be noted that the form of the scalar damage function g considered in this section is different from that used in section 2.3 for the case of damage evolution in uniaxial tension in metals. Using equation (4.48a), we can readily determine the following partial derivatives of g:

$$\frac{\partial g}{\partial L} = -1 \tag{4.48b}$$

$$\left\{\frac{\partial g}{\partial \sigma}\right\} = [M]^{T}[J][M]\{\sigma\} \tag{4.48c}$$

In order to determine the evolution equation for the damage vector $\{\varphi\}$, we start with the power of dissipation Π given by:

$$\Pi = \{\sigma\}^T \{d\varepsilon\}'' + \{\sigma\}^T \{d\varphi\} - L\,d\ell \qquad (4.49)$$

The problem of damage evolution now reduces to the problem of extermination of Π subject to the constraints $f = 0$ and $g = 0$. We, therefore, introduce the objective function[2] Ψ given by:

$$\Psi = \Pi - d\lambda_1 \cdot f - d\lambda_2 \cdot g \qquad (4.50)$$

where $d\lambda_1$ and $d\lambda_2$ are scalar Lagrange multipliers.   Using the two conditions $\{\partial\Psi / \partial\sigma\} = \{0\}$ and $\partial\Psi / \partial L = 0$ and simplifying, we obtain:

$$\{d\varphi\} = d\lambda_2 \left\{\frac{\partial g}{\partial\sigma}\right\} \qquad (4.51a)$$

$$d\ell = -d\lambda_2 \frac{\partial g}{\partial L} \qquad (4.51b)$$

Substitution of equation (4.48b) into equation (4.51b) results in $d\lambda_2 = d\ell$. Substituting this result into equation (4.51a), we obtain:

$$\{d\varphi\} = d\ell \left\{\frac{\partial g}{\partial\sigma}\right\} \qquad (4.52)$$

In order to determine the scalar damage multiplier $d\ell$, we need to apply the consistency condition $dg = 0$:

$$\left\{\frac{\partial g}{\partial\sigma}\right\}^T \{d\sigma\} + \left\{\frac{\partial g}{\partial\varphi}\right\}^T \{d\varphi\} + \frac{\partial g}{\partial\ell}\,d\ell = 0 \qquad (4.53)$$

---

[2] This is the same as the function H used in section 2.3

Substituting for $\{d\varphi\}$ from equation (4.52), using
$\partial g/\ell = (\partial g/\partial L)(\partial L/\partial \ell) = -dL/d\ell$, and solving for $d\ell$, we obtain:

$$d\ell = \frac{1}{r}\left\{\frac{\partial g}{\partial \sigma}\right\}^{T}\{d\sigma\} \tag{4.54a}$$

where the scalar quantity r is given by:

$$r = \frac{dL}{d\ell} - \left\{\frac{\partial g}{\partial \varphi}\right\}^{T}\left\{\frac{\partial g}{\partial \sigma}\right\} \tag{4.54b}$$

Finally, substituting equation (4.54a) into equation (4.52), we obtain the required evolution equation for the damage vector $\{\varphi\}$:

$$\{d\varphi\} = \left(\frac{1}{r}\left\{\frac{\partial g}{\partial \sigma}\right\}^{T}\left\{\frac{\partial g}{\partial \sigma}\right\}\right)\{d\sigma\} \tag{4.55}$$

Equation (4.55) can be rewritten simply as $\{d\varphi\} = r^{*}\{d\sigma\}$, where r^{*} is the scalar quantity shown in parentheses in equation (4.55). It should be noted that equation (4.55) represents a set of three simultaneous ordinary differential equations in the variables ϕ_{1}, ϕ_{2}, and ϕ_{3}. This set of differential equations will be used in section 4.9 in the numerical solution of the problem.

4.7 Elastic Constitutive Relation in the Damaged Composite System

The next step is the derivation of the elastic constitutive relation. This is accomplished by first determining the damage transformation equation for the elastic strain rate vector $\{d\varepsilon\}'$. Starting with the effective elastic strain energy $\overline{U}$ given by:

$$\overline{U} = \frac{1}{2}\{\overline{\sigma}\}^{T}\{\overline{\varepsilon}\}' \tag{4.56}$$

and using the hypothesis of elastic strain energy equivalence $\left(\overline{U} = U\right)$, we obtain:

68

$$\{\bar{\sigma}\}^T \{\bar{\varepsilon}\}' = \{\sigma\}^T \{\varepsilon\}' \tag{4.57}$$

Substituting equation (4.35) into equation (4.57) and solving for $\{\bar{\varepsilon}\}'$, we obtain:

$$\{\bar{\varepsilon}\}' = [M]^{-T} \{\varepsilon\}' \tag{4.58}$$

The above transformation equation for the elastic strain vector should be compared with equation (2.9) for the case of uniaxial tension in metals. Using the method outlined in section 4.5 for the derivation of equations (4.40a) and (4.41a), we can show that:

$$d[M]^{-T} \cdot \{\varepsilon\}' = [\mathbf{K}^{\varepsilon'}] \{d\varphi\} \tag{4.59a}$$

where the matrix $[\mathbf{K}^{\varepsilon'}]$ is obtained from equation (4.40b) by replacing σ by ε'. Taking the derivative of equation (4.58) and substituting equation (4.59a) into the resulting expression, we obtain the damage transformation equation for $\{d\varepsilon\}'$:

$$\{d\bar{\varepsilon}\}' = [M]^{-T} \{\varepsilon\}' + [\mathbf{K}^{\varepsilon'}] \{d\varphi\} \tag{4.59b}$$

In order to find a relation between $\{d\bar{\sigma}\}$ and $\{d\sigma\}$, we take the derivative of equation (4.35) and substitute equations (4.40a) and (4.55) into the resulting expression. After simplification, we obtain:

$$\{d\bar{\sigma}\} = [M^*] \{d\sigma\} \tag{4.60a}$$

where the 3 x 3 matrix $[M^*]$ is given by:

$$[M^*] = [M] + \frac{1}{r} [\mathbf{K}^{\sigma}] \left(\left\{ \frac{\partial g}{\partial \sigma} \right\}^T \left\{ \frac{\partial g}{\partial \sigma} \right\} \right) \tag{4.60b}$$

Finally, substituting equations (4.55), (4.59b) and (4.60a) into the effective elastic constitutive relation given in equation (4.28) and simplifying, we obtain the elastic constitutive relation for the damaged composite system as follows:

$$\{d\sigma\} = [E] \{d\varepsilon\}' \tag{4.61a}$$

where the 3 x 3 damage-elasticity matrix $[E]$ is given by:

$$[E] = \left([M^*] - \frac{1}{r} \left(\left\{ \frac{\partial g}{\partial \sigma} \right\}^T \left\{ \frac{\partial g}{\partial \sigma} \right\} \right) [\bar{E}] [K^{\varepsilon'}] \right)^{-1} [\bar{E}] [M]^{-T} \quad (4.61b)$$

4.8 Elasto-Plastic Constitutive Relation in the Damaged Composite System

The kinematic hardening rule given in equation (4.26) can now be transformed to the damaged composite system. Substituting equations (4.35) and (4.43a) into equation (4.26), simplifying and solving for $\{d\beta\}$, we obtain:

$$\{d\beta\} = \left([M]^{-1} - [M]^{-1} [\bar{B}^M]^{-1} [\bar{B}^{MP}] \right) [M] \{d\sigma\}$$

$$- \left(\left([M]^{-1} - [M]^{-1} [\bar{B}^M]^{-1} [\bar{B}^{MP}] \right) [dM] + d\bar{\mu}^M [I] \right) \{\sigma\}$$

$$- \left([M]^{-1} [dM] - d\bar{\mu}^M [I] \right) \{\beta\} \quad (4.62)$$

The additive decomposition of the strain rate vector is taken in the form:

$$\{d\varepsilon\} = \{d\varepsilon\}' + \{d\varepsilon\}'' \quad (4.63)$$

It can be shown that the above decomposition is compatible with the decomposition given in equation (4.12a).

The flow rule for the damaged composite system is taken in the form:

$$\{d\varepsilon\}'' = [d\lambda] \left\{ \frac{\partial f}{\partial \sigma} \right\} \quad (4.64)$$

The above equation clearly provides for a non-associated flow rule. This is in agreement with the recent results of Stolz [103] where it was shown that an associated flow rule may not be derivable for damaged materials. The multiplier matrix $[d\lambda]$ is determined from the consistency condition $df = 0$ as follows:

$$\left\{\frac{\partial f}{\partial \sigma}\right\}^T \{d\sigma\} + \left\{\frac{\partial f}{\partial \beta}\right\}^T \{d\beta\} = 0 \tag{4.65}$$

Substituting for $\{d\beta\}$ from equation (4.62), for $\{d\sigma\}$ from equation (4.61a), for $\{d\varepsilon\}'$ from equation (4.63), for $\{d\varepsilon\}''$ from equation (4.64), for $\{d\varphi\}$ from equation (4.55), and using equations (4.40a) and (4.41a), and simplifying, we obtain:

$$\{\gamma\}^T \{d\sigma\} = d\overline{\mu}^M \left\{\frac{\partial f}{\partial \beta}\right\}^T \{\sigma - \beta\} \tag{4.66a}$$

where the 3 x 1 vector $\{\gamma\}$ is given by:

$$\{\gamma\} = \left\{\frac{\partial f}{\partial \sigma}\right\} + [M]^T \left([I] - [\overline{B}^M]^{-1} [\overline{B}^{MP}]\right)^T [M]^{-T} \left\{\frac{\partial f}{\partial \beta}\right\}$$

$$- \frac{1}{r} \left(\left\{\frac{\partial g}{\partial \sigma}\right\}^T \left\{\frac{\partial g}{\partial \sigma}\right\}\right) \left([K^\sigma]^T \left([I] - [\overline{B}^M]^{-1} [\overline{B}^{MP}]\right)^T + [K^\beta]^T\right) [M]^{-T} \left\{\frac{\partial f}{\partial \beta}\right\}$$

$$\tag{4.66b}$$

The solution of equation (4.66a) for $d\overline{\mu}^M$ yields:

$$d\overline{\mu}^M = \frac{\{\gamma\}^T \{d\sigma\}}{\left\{\dfrac{\partial f}{\partial \beta}\right\}^T \{\sigma - \beta\}} \tag{4.67}$$

Substituting equation (4.67) into equation (4.18), solving for $d\overline{\lambda}^M$ and substituting the result into equation (4.15), we obtain the following expression for the multiplier matrix $[d\overline{\lambda}]$:

$$[d\overline{\lambda}] = \frac{\{\gamma\}^T [E] \{d\varepsilon\}}{3b\left\{\dfrac{\partial f}{\partial \beta}\right\}^T \{\sigma - \beta\}} [\overline{C}^{MP}]^{-1} [\overline{B}^M]^{-T} - \frac{\{\gamma\}^T [E] \{d\varepsilon\}''}{3b\left\{\dfrac{\partial f}{\partial \beta}\right\}^T \{\sigma - \beta\}} [\overline{C}^{MP}]^{-1} [\overline{B}^M]^{-T}$$

$$\tag{4.68}$$

Equating the plastic energy of dissipation $\{\sigma\}^T\{d\varepsilon\}''/2$ in the damaged configuration with the plastic energy of dissipation $\{\overline{\sigma}\}^T\{d\overline{\varepsilon}\}''/2$ in the effective undamaged configuration, and using equation (4.35), we obtain:

$$\{d\overline{\varepsilon}\}'' = [M]^{-T}\{d\varepsilon\}'' \tag{4.69}$$

Equation (4.69) is the damage transformation equation for the plastic strain rate vector. Finally, in order to derive the elasto-plastic constitutive relation for the damaged composite system, we substitute equations (4.60a), (4.12a), (4.59b), (4.69), (4.55), (4.61a) (for $\{d\varepsilon\}'$) into equation (4.34a) and simplify to obtain:

$$\{d\sigma\} = [D]\{d\varepsilon\} \tag{4.70a}$$

where the damage-elasto-plastic stiffness matrix is given by:

$$[D] = \left([M] + \frac{1}{r}\left(\left\{\frac{\partial g}{\partial\sigma}\right\}^T \left\{\frac{\partial g}{\partial\sigma}\right\}\right) [K^\sigma] \right)^{-1} [\overline{D}]\,[M]^{-T} \tag{4.70b}$$

Finally, one needs to rewrite equation (4.62) in a form suitable for numerical implementation. In order to rewrite it in the required incremental form, we substitute equations (4.40a) and (4.41a) into equation (4.62) to obtain:

$$\{d\beta\} = [X^*]\{d\sigma\} + [Y^*]\{d\varphi\} \tag{4.71a}$$

where the 3 x 3 matrices $[X^*]$ and $[Y^*]$ are given by:

$$[X^*] = [I] - [M]^{-1}[\overline{B}^M]^{-1}[\overline{B}^{MP}][M]$$

$$- \frac{\left\{\dfrac{\partial f}{\partial\sigma}\right\}^T [\overline{B}^M][\overline{C}^{MP}]^{-T}[E]^T\{\gamma\}}{3b\left\{\dfrac{\partial f}{\partial\beta}\right\}^T\{\sigma-\beta\}} \{\sigma-\beta\}\{\gamma\}^T \tag{4.71b}$$

$$[Y^*] = -[M]^{-1}\left([K^\sigma] - [\overline{B}^M]^{-1}[\overline{B}^{MP}][K^\sigma] + [K^\beta]\right) \tag{4.71c}$$

It is noticed that equation (4.71a) represents a set of three simultaneous ordinary differential equations in β_1, β_2, and β_3. This set will be used in the next section for the numerical solution of the problem.

4.9 Numerical Implementation - Example

Consider a unidirectional fiber-reinforced thin composite lamina that is subjected to a uniaxial tensile force T along the x_1-direction as shown in Figure 4.1. The matrix is assumed to be elasto-plastic and the fibers elastic and cylindrical in shape. The fibers are also assumed to be continuous, perfectly aligned and symmetrically distributed along the x_1-axis. For this problem, the stress vector $\{\sigma\}$ is given by

$$\{\sigma\} = \begin{bmatrix} \sigma & 0 & 0 \end{bmatrix}^T \tag{4.72}$$

where σ is the uniaxial stress in the lamina, obtained by dividing T by the cross-sectional area of the lamina. Substituting $\{\sigma\}$ into equation (4.70a), we can write the constitutive equation for this problem as follows:

$$\begin{Bmatrix} d\varepsilon_1 \\ d\varepsilon_2 \\ d\varepsilon_3 \end{Bmatrix} = \begin{bmatrix} D_{11} & D_{12} & D_{13} \\ D_{21} & D_{22} & D_{23} \\ D_{31} & D_{32} & D_{33} \end{bmatrix}^{-1} \begin{Bmatrix} d\sigma \\ 0 \\ 0 \end{Bmatrix} \tag{4.73}$$

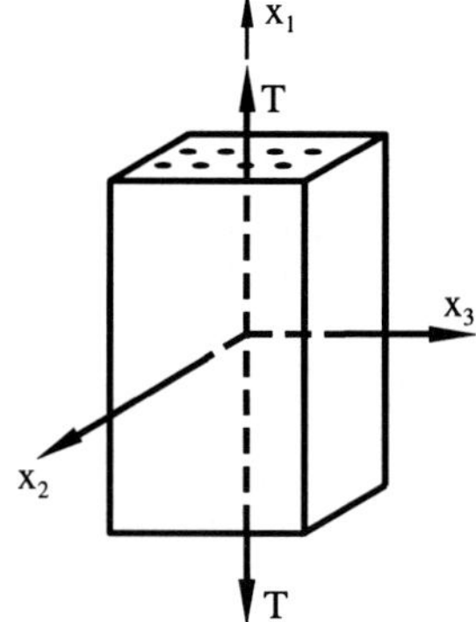

Figure 4.1   Unidirectional Thin Lamina Under Uniaxial Tension

where $d\sigma$ is the increment of uniaxial stress. Since we have only one independent component; namely σ, in the stress vector, a system of simultaneous ordinary differential equations can be written for this problem with σ as the independent variable. In this way, this problem can be solved numerically using a suitable differential equation solver without the use of finite elements. Let the matrix $[S]$ denote the inverse of $[D]$, and rewrite equation (4.73) as follows:

$$\left\{ \begin{array}{c} d\varepsilon_1 \\ d\varepsilon_2 \\ d\varepsilon_3 \end{array} \right\} = \left\{ \begin{array}{c} S_{11} \\ S_{21} \\ S_{31} \end{array} \right\} d\sigma \tag{4.74}$$

Equation (4.74) represents the first set of the governing system of differential equations for this problem. It should be mentioned that the expressions of S_{11}, S_{21}, and S_{31} are obtained by numerically inverting the elasto-plastic matrix $[D]$. The matrix $[D]$ is obtained using equation (4.70b) with the condition that the stress vector is given by equation (4.72).

The second set of differential equations is obtained from the evolution of the backstress vector $\{\beta\}$ given by equation (4.71a). However, equation (4.71a) must be rewritten in the required format to be used in the system of differential equations. In other words, the right-hand-side should be a function of the independent variable $d\sigma$. Therefore, the second term of the right-hand-side of equation (4.71a) will be rewritten in terms of the vector $\{d\sigma\}$. Substituting for $\{d\varphi\}$ from equation (4.55) into equation (4.71a) and simplifying, we obtain:

$$\{d\beta\} = [Z^*]\{d\sigma\} \tag{4.75a}$$

where the matrix $[Z^*]$ is given by:

$$[Z^*] = [X^*] + \frac{1}{r}\left(\left\{\frac{\partial g}{\partial \sigma}\right\}^T \left\{\frac{\partial g}{\partial \sigma}\right\}\right)[Y^*] \tag{4.75b}$$

Substituting equation (4.72) into equation (4.75a), we can rewrite the resulting equation as follows:

74

$$\begin{Bmatrix} d\beta_1 \\ d\beta_2 \\ d\beta_3 \end{Bmatrix} = \begin{Bmatrix} Z_{11}^* \\ Z_{21}^* \\ Z_{31}^* \end{Bmatrix} d\sigma \qquad (4.76)$$

Equation (4.76) represents the second set of differential equations required for the solution of this problem. Finally, the last set of differential equations uses the evolution of the damage vector $\{\varphi\}$ as given by equation (4.55). Equation (4.55) is rewritten in the required format as follows:

$$\begin{Bmatrix} d\varphi_1 \\ d\varphi_2 \\ d\varphi_3 \end{Bmatrix} = \frac{1}{r}\left(\left\{ \frac{\partial g}{\partial \boldsymbol{\sigma}} \right\}^T \left\{ \frac{\partial g}{\partial \boldsymbol{\sigma}} \right\} \right) \begin{Bmatrix} 1 \\ 0 \\ 0 \end{Bmatrix} d\sigma \qquad (4.77)$$

The three sets of equations: (4.74), (4.76) and (4.77) represent the governing system of ordinary differential equations for this problem. Taking the independent variable σ as the time t, the governing system of differential equations is given by :

$$d\varepsilon_1/dt = S_{11} \qquad (4.78\text{a})$$

$$d\varepsilon_2/dt = S_{21} \qquad (4.78\text{b})$$

$$d\varepsilon_3/dt = S_{31} \qquad (4.78\text{c})$$

$$d\beta_1/dt = Z_{11}^* \qquad (4.78\text{d})$$

$$d\beta_2/dt = Z_{21}^* \qquad (4.78\text{e})$$

$$d\beta_3/dt = Z_{31}^* \qquad (4.78\text{f})$$

$$d\varphi_1/dt = \frac{1}{r}\left\{ \frac{\partial g}{\partial \boldsymbol{\sigma}} \right\}^T \left\{ \frac{\partial g}{\partial \boldsymbol{\sigma}} \right\} \qquad (4.78\text{g})$$

$$d\varphi_2/dt = 0 \qquad (4.78\text{h})$$

$$d\varphi_3/dt = 0 \tag{4.78i}$$

Equations (4.78) form a system of nine simultaneous ordinary differential equations that can be solved numerically using a Runge-Kutta type method. In the numerical solution, it is assumed that the elastic strains are infinitesimal; therefore, they are neglected. Consequently, the solution scheme starts at the initiation of yielding. This means that the initial conditions for this problem are zero strains, backstresses, and damage variables. Therefore, in this problem, damage is initiated at the same time yielding starts; though this may not be the case in a general problem where the amount of elastic strain may be significant. As initial conditions to the boundary value problem, all nine dependent variables $(\varepsilon_1,\ \varepsilon_2,\ \varepsilon_3,\ \beta_1,\ \beta_2,\ \beta_3,\ \phi_1,\ \phi_2,\ \phi_3)$ are taken to be zero while the initial value of the independent variable $\sigma = \sigma_o \neq 0$. The initial value σ_o is needed for the solution of the differential equations. It is computed by setting the yield function f of equation (4.44a) to be equal to zero. It should also be pointed out that at the start of plasticity and damage, the backstresses and damage variables are also zero. Substituting zero for $\{\beta\}$ in equation (4.44a), we obtain the following condition at yielding:

$$\frac{3}{2} \{\sigma\}^T [H] \{\sigma\} - \overline{\sigma}_o^{M^2} = 0 \tag{4.79}$$

where $[H]$ is given by equation (4.44b). However, since $\phi_1 = \phi_2 = \phi_3 = 0$ at yielding, the matrix $[M]$ becomes the identity matrix $[I]$. Substituting $[I]$ for $[M]$ in equation (4.44b), we obtain:

$$[H] = \left[\overline{B}^{\,M}\right]^T [a] \left[\overline{B}^{\,M}\right] \tag{4.80}$$

The stress concentration matrix $\left[\overline{B}^{\,M}\right]$ is obtained using either the Voigt model or the Mori-Tanaka model as discussed shortly, while $[a]$ is the constant matrix given in equation (4.3c).

Substituting $\begin{bmatrix}\sigma_o & 0 & 0\end{bmatrix}^T$ for $\{\sigma\}$ in equation (4.79) and solving for σ_o , we obtain:

$$\sigma_o = \sqrt{\frac{2}{3\,H_{11}}}\ \overline{\sigma}_o^{\,M} \tag{4.81}$$

where H_{11} is the first term in the matrix $[\boldsymbol{H}]$ of equation (4.80), and $\bar{\sigma}_o^M$ is the yield strength of the matrix material. Equation (4.81) represents the initial condition for the uniaxial stress σ to be used in the solution of the differential equations.

Equations (4.78) are solved simultaneously using the IMSL routine DIVPRK. This routine uses a Runge-Kutta-Verner fifth-order and sixth-order method for the solution of the differential equations. Figure 4.2 shows a schematic diagram of the numerical computations. In the determination of the stress and strain concentration matrices, two different models are used. The first is the Voigt model which is based on the assumption that the strains in the matrix, fibers and composite are equal. The Voigt model was discussed previously in section 3.3. The second model is the Mori-Tanaka model which uses the Eshelby tensor and theory of inclusions and inhomogenities. The Mori-Tanaka model is more sophisticated than the Voigt model but the latter is considered here for comparison. This model is discussed in detail in Part II of the book. Details about the two models can be found in the papers of Dvorak and Bahei-El-Din [68-70], Voyiadjis and Kattan [99, 104, 105], and Mori and Tanaka [106]. Details about the numerical scheme used in calculating the Eshelby tensor are found in the papers of Gavazzi and Lagoudas [107] and Lagoudas et al. [108].

The lamina consists of matrix and fibers with volume fractions 55% and 45% respectively. The material properties used are: $\bar{E}^M = 84.1$ GPa, $\bar{v}^M = 0.3$, $\bar{E}^F = 414$ GPa, $\bar{v}^F = 0.22$, $\bar{E}_{11} = 200$ GPa, $\bar{E}_{22} = 137$ GPa, $\bar{v}_{12} = 0.27$, $\bar{v}_{23} = 0.31$, and $\bar{G}_{12} = 52.6$ GPa. The yield strength of the matrix material is 0.35 GPa. The damage parameters are $\mu = 0.5$ and $\partial L / \partial \ell = 1.0 \times 10^{12}$. Using equation (4.81), we find that the stress at which yielding occurs is $\sigma_o = 1$ GPa for the Voigt model, and $\sigma_o = 0.4$ GPa for the Mori-Tanaka model. It is noted that the material yields at a higher yield stress when using the Voigt model because of the assumption of equal strains in the material thus making it stiffer. In the numerical calculations, the stress is increased monotonically starting from the yield stress σ_o in increments of 1 GPa for a total of 100 increments. The tolerance factor for convergence of the iterative scheme is taken as 0.005. The results are shown in Figures 4.3 - 4.5.

In Figure 4.3, the variation of the damage variable ϕ_1 is shown vs. the strain ε_1. It is clear that the value of ϕ_1 is monotonically increasing for both the Voigt and Mori-Tanaka models although the rate of increase of damage is higher when the Voigt model is used. This may be attributed to the use of constant concentration factors when the Voigt model is used.

In the Mori-Tanaka model, the concentration matrices change depending on the stiffness of the material. The values of ϕ_2 and ϕ_3 are identically zero, therefore, no plots for these damage variables are shown. The variations of the backstresses β_1 and β_2 are shown vs. the strain ε_1 in Figures 4.4 and 4.5, respectively. It is noticed that the Voigt model gives higher values of the backstress β_1. However, the backstress β_2 vanishes when using the Voigt model. It is apparent that the values of μ and $\partial L/\partial \ell$ provide for very small values of the damage variable ϕ_1. It should be emphasized that the solution of practical problems in this area requires the use of the finite element method. An example using this technique is provided later in Part II of the book.

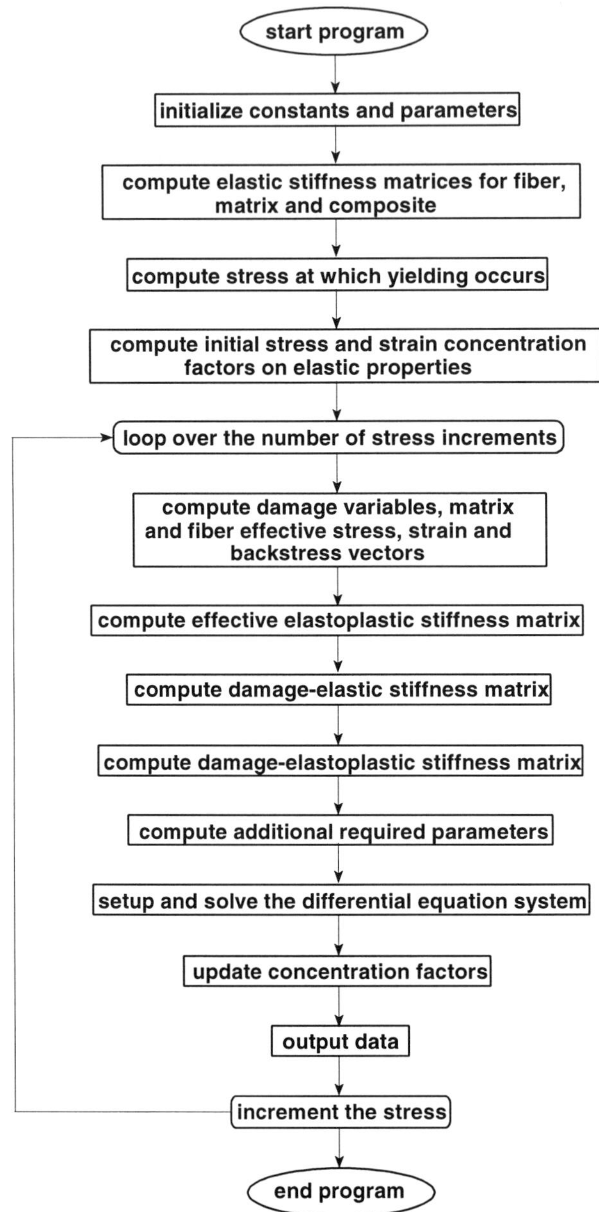

Figure 4.2 Schematic Diagram of the Numerical Computations

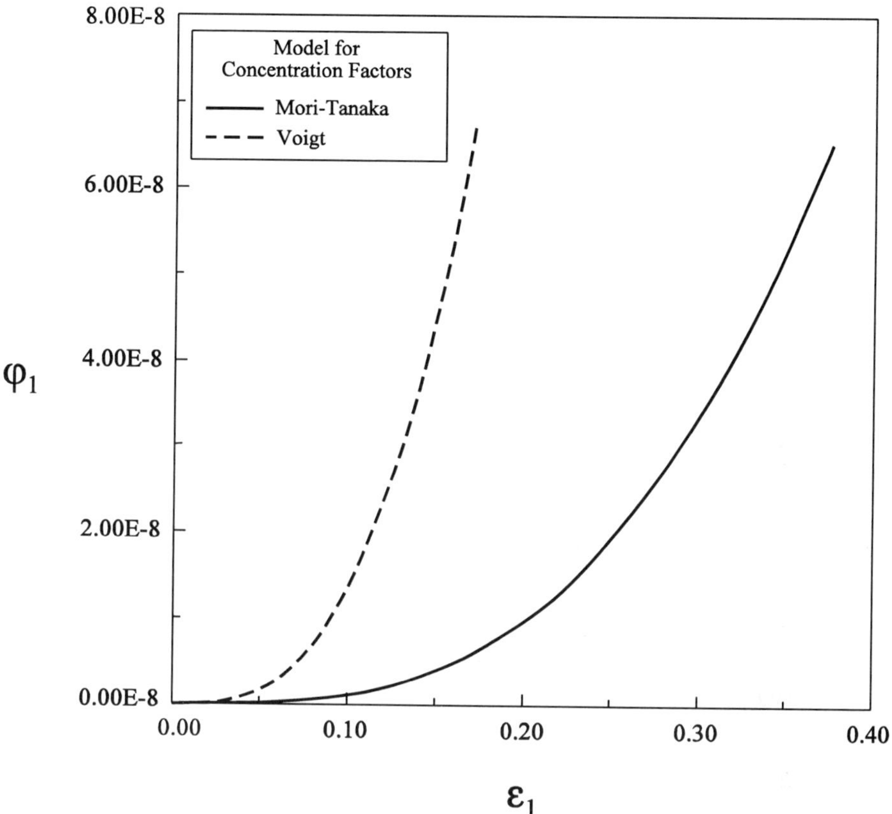

Figure 4.3   Damage Variable ϕ_1 vs. ε_1.

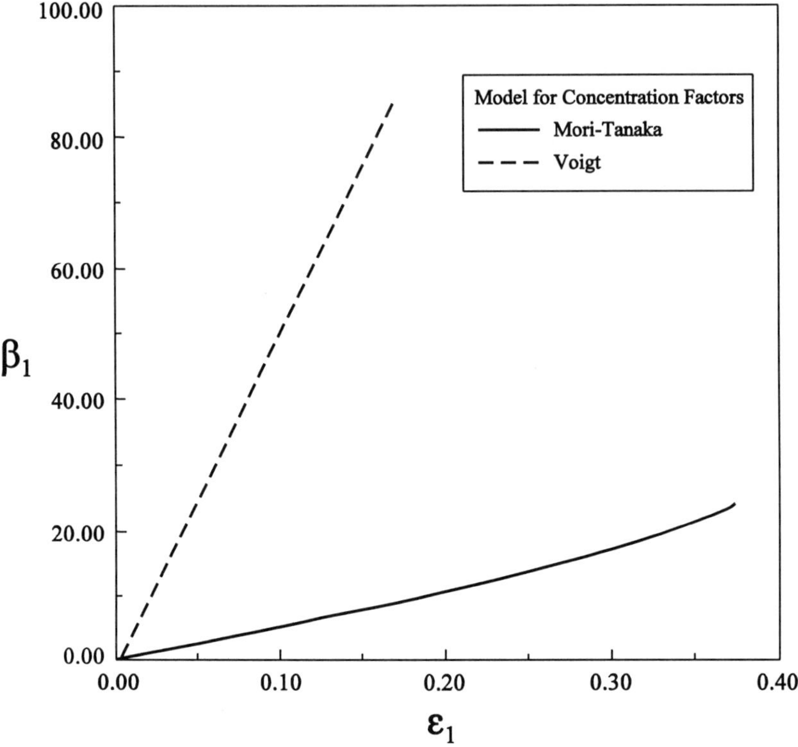

Figure 4.4   Backstress β_1 vs. ε_1.

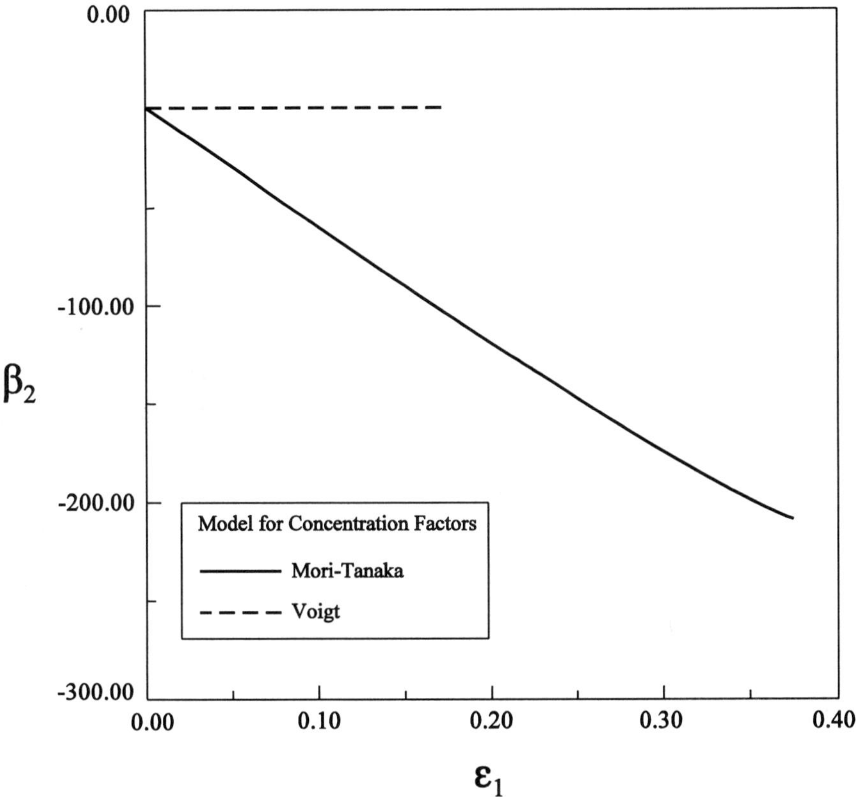

Figure 4.5   Backstress β_1 vs. ε_1.

<u>PART II</u>

ANISOTROPIC DAMAGE MECHANICS

TENSOR FORMULATION

CHAPTER 5

DAMAGE AND ELASTICITY IN METALS

For the case of isotropic damage, the damage variable is scalar and the evolution equations are easy to handle, as was shown in Part I of this book. It has been argued [12] that the assumption of isotopic damage is sufficient to give good predictions of the carrying capacity, the number of cycles, or the time to local failure in structural components. However, the development of anisotropic damage and plasticity has been confirmed experimentally [13 - 15] even if the virgin material is isotropic. This has prompted several researchers to investigate the general case of anisotropic damage.

The theory of anisotropic damage mechanics was developed by Sidoroff and Cordebois [16 - 18], and later used by Lee et al. [15] and Chow and Wang [19, 20] to solve simple ductile fracture problems. Prior to this development, Krajcinovic and Foneska [21], Murakami and Ohno [22], Murakami [23] and Krajcinovic [24] investigated brittle and creep fracture using appropriate anisotropic damage models. Although these models are based on a sound physical background, they lack rigorous mathematical justification and mechanical consistency. Consequently, more work was recently done to develop a more involved theory capable of producing results that can be used for practical applications [21-25, 98, 99, 109, 110].

In the general case of anisotropic damage, the damage variable has been shown to be tensorial in nature [22, 26]. This damage tensor was shown to be an irreducible even-rank tensor [27, 28]. Several other properties of the damage tensor have been outlined by Betten [29, 30] in a rigorous mathematical treatment using the theory of tensor functions.

A coupled theory of elasticity and continuum damage mechanics is formulated in this chapter for metals. It is assumed that the material undergoes damage with small elastic strains. The hypothesis of elastic energy equivalence is used in order to produce the proposed coupling. The damage variable used represents average material degradation which reflects the various types of damage at the micro-scale level like nucleation and growth of voids, cavities, micro-cracks and other microscopic defects.

The constitutive model is numerically implemented using finite elements with an updated Lagrangian description. It is also shown how the model can be applied to problems of ductile fracture. The problem of crack initiation in a thin plate with a center crack that is subjected to uniaxial tension is analyzed using the constitutive model.

5.1 General States of Damage

The principles of the continuum mechanics theory for the general case of anisotropic damage was recently [98, 99, 109-111] cast in a consistent mathematical and mechanical framework. Equation (2.4) is generalized for the anisotropic case, in indicial notation, as follows [58]:

$$\overline{\sigma}_{ij} = M_{ijk\ell}\, \sigma_{k\ell} \tag{5.1}$$

Where M is a symmetric fourth-rank tensor called the damage effect tensor, σ is the Cauchy stress tensor, and $\overline{\sigma}$ is the corresponding effective stress tensor. A special vector form of equation (5.1) was presented previously in Chapter 4 in equation (4.35). The damage effect tensor M was shown by Murakami [58] to be given by (in terms of its matrix representation):

$$M = (I - \varphi)^{-1} = \det (G)^{-1} G^T \tag{5.2}$$

where I is the second-rank identity tensor, φ is the second-rank damage tensor, "det" is the determinant function, $(\ )^{-1}$ is the generalized inverse of a tensor, and G is a fictitious deformation gradient give by:

$$G_{ij} = \frac{\partial \overline{x}_i}{\partial x_j} \tag{5.3}$$

where x and $\overline{x}$ are the coordinates in the damaged state and the fictitious effective undamaged state, respectively (see Figure 5.1).

In this section, we derive the necessary transformation equations between the damaged and the hypothetical undamaged states of the material. In the derivation, the following two assumptions are incorporated: (1) the elastic deformations are small (infinitesimal) compared with the plastic deformations (which are finite), and (2) there exists an elastic strain energy function $U(\varepsilon', \varphi)$. This

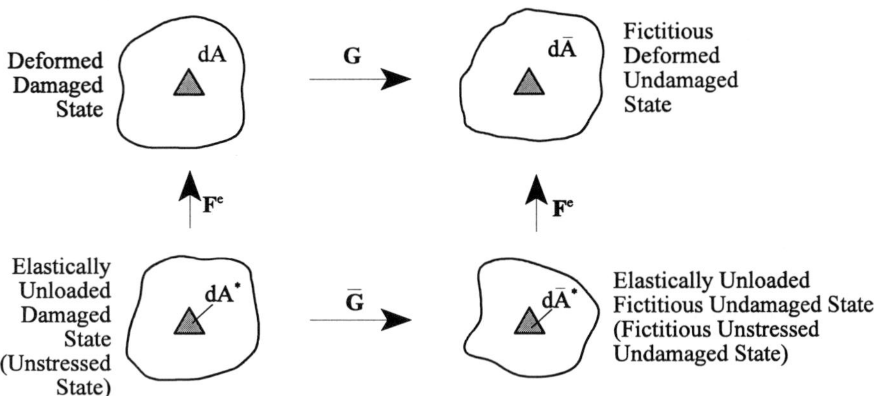

Figure 5.1   States of Deformation and Damage

function is assumed based on a linear relation between the Cauchy stress $\boldsymbol{\sigma}$ and the engineering elastic strain ε'. The tensor ε' constitutes the linear part of the elastic component of the spatial strain tensor (second order terms are neglected). This relation can be written in the effective undamaged state of the material as follows:

$$\bar{\sigma}_{ij} = \bar{E}_{ijk\ell} \, \bar{\varepsilon}'_{k\ell} \tag{5.4}$$

where $\bar{E}$ is the effective constant elasticity tensor. Using equation (5.4) and the second assumption given above, the elastic strain energy function $U(\varepsilon', \varphi)$ can be expressed in the effective undamaged state $(\varepsilon' \equiv \bar{\varepsilon}', \varphi \equiv 0)$ as follows:

$$U(\bar{\varepsilon}', \mathbf{0}) = \frac{1}{2} \, \bar{E}_{ijk\ell} \, \bar{\varepsilon}'_{ij} \, \bar{\varepsilon}'_{k\ell} \tag{5.5}$$

Using the Legendre transformation, the complementary elastic strain energy function $V(\boldsymbol{\sigma}, \varphi)$ is defined by:

$$V(\boldsymbol{\sigma}, \varphi) = \sigma_{ij} \, \varepsilon'_{ij} - U(\varepsilon', \varphi) \tag{5.6}$$

It follows that the engineering elastic strain tensor ε' is derived from the complementary energy $V(\sigma, \varphi)$, i.e.

$$\varepsilon'_{ij} = \frac{\partial V(\sigma, \varphi)}{\partial \sigma_{ij}} \tag{5.7}$$

Substituting equation (5.5) for $U(\varepsilon', \varphi)$ into equation (5.6) (where σ and ε' are replaced by their effective counterparts), we obtain the following expression for $V(\sigma, \varphi)$ in the effective undamaged state of the material:

$$V(\overline{\sigma}, 0) = \frac{1}{2} \overline{E}^{-1}_{ijk\ell} \, \overline{\sigma}_{ij} \, \overline{\sigma}_{k\ell} \tag{5.8}$$

When the material is deformed and damaged, the complementary elastic strain energy function takes the form:

$$V(\sigma, \varphi) = \frac{1}{2} E^{-1}_{ijk\ell} (\varphi) \, \sigma_{ij} \, \sigma_{k\ell} \tag{5.9}$$

where $E(\varphi)$ is the elasticity tensor for the damaged state, i.e. $E(\varphi)$ includes the effects of damage. In order to determine $E(\varphi)$, the hypothesis of elastic energy equivalence is used in the form:

$$V(\sigma, \varphi) = V(\overline{\sigma}, 0) \tag{5.10}$$

Equating the two expressions given in equations (5.8) and (5.9) and substituting for $\overline{\sigma}$ from equation (5.1), we obtain the following transformation relation between $\overline{E}$ and $E(\varphi)$:

$$E_{k\ell mn} (\varphi) = M^{-1}_{ijk\ell} (\varphi) \, \overline{E}_{ijpq} \, M^{-T}_{pqmn} (\varphi) \tag{5.11}$$

Similarly, starting with equation (5.7) and utilizing equations (5.8), (5.10) and (5.11), we obtain the following relation between the elastic strain tensor and its effective counterpart:

$$\overline{\varepsilon}'_{mn} = M^{-T}_{mnpq} \, \varepsilon'_{pq} \tag{5.12}$$

The nature and matrix representation of the damage effect tensor M is discussed in detail in Chapter 6 of the book. We will only mention here the following proposed expression for $M(\varphi)$:

$$M_{ijk\ell}(\varphi) = \frac{\delta_{ijk\ell}}{\sqrt{(1 - \varphi_i)(1 - \varphi_j)}} \qquad \text{(no sum over } i, j) \qquad (5.13)$$

where $\delta_{ijk\ell}$ represents the fourth-rank identity tensor. The fourth-rank tensorial expression given in equation (5.13) applies only when principal values are used. No tensorial generalization of M is available, but a generalized matrix representation is given in Chapter 6.

5.2 Damage Evolution

The damage evolution criterion used is proposed by Lee et al. [15] and is given by the scalar damage function $g(\overline{\sigma}, L)$:

$$g(\overline{\sigma}, L) = \frac{1}{2} J_{ijk\ell} \, \overline{\sigma}_{ij} \, \overline{\sigma}_{k\ell} - \ell_o^2 - L(\ell) \equiv 0 \qquad (5.14)$$

where ℓ_o is the initial damage threshold, $L(\ell)$ is the increment of damage threshold, and ℓ is a scalar variable that represents "overall" damage. In equation (5.14), J is a fourth-rank symmetric tensor that is represented by the following matrix [15]:

$$[J] = \begin{bmatrix} 1 & \mu & \mu & 0 & 0 & 0 \\ \mu & 1 & \mu & 0 & 0 & 0 \\ \mu & \mu & 1 & 0 & 0 & 0 \\ 0 & 0 & 0 & 2(1-\mu) & 0 & 0 \\ 0 & 0 & 0 & 0 & 2(1-\mu) & 0 \\ 0 & 0 & 0 & 0 & 0 & 2(1-\mu) \end{bmatrix} \qquad (5.15)$$

where μ is a material constant satisfying $-1/2 \leq \mu \leq 1$. Equations (5.14) and (5.15) are the tensorial generalizations of equations (4.47a) and (4.47b), respectively.

During the process of elastic deformation and damage, the power of dissipation Π is defined by:

$$\Pi = \sigma_{k\ell}\, d\varphi_{k\ell} - L\, d\ell \qquad (5.16)$$

The actual values of the variables $\boldsymbol{\sigma}$, $\boldsymbol{\varphi}$, and L will extremize the power of dissipation Π subject to certain constraints. The problem is to determine a stationary value for Π as given in equation (5.16) subject to the constraint $g\,(\overline{\boldsymbol{\sigma}}, L) = 0$. Using the theory of functions of several variables, we introduce the Lagrange multiplier $d\lambda$ and construct the objective function $\boldsymbol{\Psi}$ such that:

$$\boldsymbol{\Psi} = \boldsymbol{\Pi} - d\lambda \cdot \boldsymbol{g} \qquad (5.17)$$

The problem now reduces to that of determining the extremum of the objective function $\boldsymbol{\Psi}$. This is done by satisfying the following two necessary conditions:

$$\frac{\partial \boldsymbol{\Psi}}{\partial \boldsymbol{\sigma}} = \boldsymbol{0} \qquad (5.18a)$$

$$\frac{\partial \boldsymbol{\Psi}}{\partial L} = 0 \qquad (5.18b)$$

Substituting equation (5.17) into equations (5.18), we obtain:

$$d\varphi - d\lambda\, \frac{\partial g}{\partial \boldsymbol{\sigma}} = 0 \qquad (5.19a)$$

$$-d\ell - d\lambda\, \frac{\partial g}{\partial L} = 0 \qquad (5.19b)$$

It is clear from equation (5.14) that $\partial g/\partial L = -1$. Using this in equation (5.19b) leads to $d\lambda = d\ell$. Consequently, equation (5.19a) reduces to the following evolution equation for φ :

$$d\varphi = d\ell\, \frac{\partial g}{\partial \boldsymbol{\sigma}} \qquad (5.20)$$

In order to determine $d\ell$, we consider the damage strengthening criterion given by equation (5.14) and invoke the consistency condition $dg = 0$. This leads to:

$$\frac{\partial g}{\partial \sigma_{ij}}\, d\sigma_{ij} \;+\; \frac{\partial g}{\partial \varphi_{k\ell}}\, d\varphi_{k\ell} \;+\; \frac{\partial g}{\partial L}\, dL \;=\; 0 \tag{5.21}$$

Substituting for $d\varphi$ from equation (5.20), $\partial g/\partial L = -1$, and $dL = d\ell\,(\partial L/\partial \ell)$ into equation (5.21), and then solving for the "overall" damage parameter $d\ell$, we obtain:

$$d\ell \;=\; \frac{\dfrac{\partial g}{\partial \sigma_{ij}}\, d\sigma_{ij}}{\dfrac{\partial L}{\partial \ell} \;-\; \dfrac{\partial g}{\partial \varphi_{pq}}\dfrac{\partial g}{\partial \sigma_{pq}}} \tag{5.22}$$

The evolution equation for the damage tensor φ is now obtained by substituting the expression of $d\ell$ given in equation (5.22) into equation (5.20):

$$d\varphi_{k\ell} \;=\; \frac{\dfrac{\partial g}{\partial \sigma_{ij}}\, d\sigma_{ij}}{\dfrac{\partial L}{\partial \ell} \;-\; \dfrac{\partial g}{\partial \varphi_{pq}}\dfrac{\partial g}{\partial \sigma_{pq}}} \;\frac{\partial g}{\partial \sigma_{k\ell}} \tag{5.23}$$

The above relation represents damage evolution for a general case of elastic deformation and damage.

5.3 Finite Element Formulation

In this section we present the necessary equations that enable the numerical analyst to implement the elastic-damage constitutive model on high-speed computers using the finite element method. An Updated Lagrangian description is used in the numerical implementation of each load increment. This is done in such a way that use is made of the current deformed configuration of the body (assumed known) in order to determine the required quantities in the neighboring incremented configuration. Linearization of the nonlinear equations is performed and the Newton-Raphson method is used in the solution of the resulting equations. The necessary discretized equations will be derived here based on the principle of virtual work although other methods may be used (Zienkiewicz and Morgan [112]).

92

The use of an Updated Langrangian description in finite element analysis is well documented in the literature (Bathe [113], Cescotto et al. [114]). In the following derivation, emphasis is placed on the important aspect of this method rather than on the details of the finite element equations. For the details, the reader is referred to the references by Zienkiewicz [115] and Oden [116].

Consider the motion of a body in three successive configurations (see Figure 5.2): the initial configuration $\Omega(t_o)$, the current deformed configuration $\Omega(t)$ and the incremented configuration $\overline{\Omega}(t)$ with total volumes V_o, V, and $\overline{V}$, respectively. Let $\Gamma(t_o)$, $\Gamma(t)$, and $\overline{\Gamma}(t)$ denote the boundaries of the above configurations with total surface areas A_o, A, and $\overline{A}$, respectively. It is assumed that the initial configuration $\Omega(t_o)$ is both unstressed and undeformed. Using the principle of virtual work, we can write the equilibrium equations of the body in $\Omega(t)$ as follows:

$$\iiint_{\Omega(t)} \sigma_{ij}\, \delta\varepsilon_{ij}\, dV = \iiint_{\Omega(t)} \rho\, p_i\, \delta u_i\, dV + \iint_{\Gamma(t)} t_i\, \delta u_i\, dA \tag{5.24}$$

where δu_i is a field of virtual displacements that is compatible with the applied forces and $\delta\varepsilon_{ij}$ is the corresponding field of compatible virtual strains given by:

$$\delta\varepsilon_{ij} = \frac{1}{2}\left[\frac{\partial(\delta u_i)}{\partial x_j} + \frac{\partial(\delta u_j)}{\partial x_i}\right] \tag{5.25}$$

The displacement field u_i from $\Omega(t)$ to $\overline{\Omega}(t)$ is discretized as follows:

$$u_i = h_i(v_a, x_j) \tag{5.26}$$

where v_a ($a = 1, 2, \ldots, n$) are the unknown nodal displacements. The discretization functions h_i are defined with respect to the configuration $\Omega(t)$. Taking the variation of equation (5.26), we obtain the following expression for the virtual displacement field δu_i:

$$\delta u_i = \frac{\partial h_i}{\partial v_a}(v_b, x_j)\, \delta v_a \tag{5.27}$$

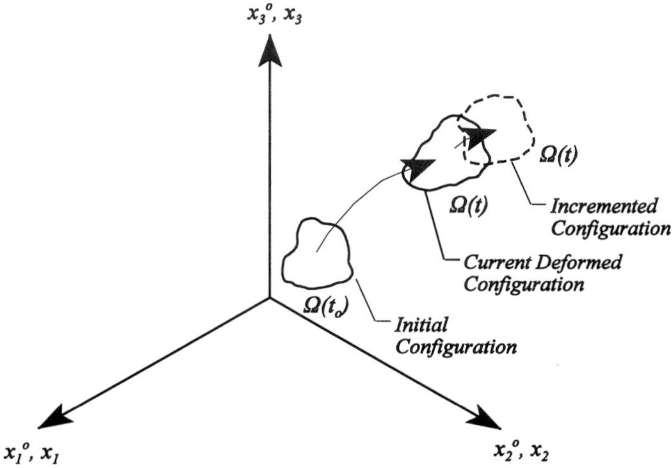

Figure 5.2   Different Configurations of the Body

Denoting L_{ia} as the partial derivatives $\partial h_i/\partial v_a$, we can now express equation (5.27) as follows:

$$\delta u_i = L_{ia}\,\delta v_a \tag{5.28}$$

It is noted that equations (5.26) through (5.28) are valid for any large displacement field u_i. However, the displacement field u_i from $\Omega(t)$ to $\overline{\Omega}(t)$ is incremental (infinitesimal). This fact imposes certain restrictions on the discretization functions h_i. When the body is in $\Omega(t)$, the functions h_i and their material derivatives must vanish, thus:

$$h_i\,(v_a = 0) = \left.\left(\frac{\partial h_i}{\partial x_j}\right)\right|_{v_a = 0} = \left.\left(\frac{\partial^2 h_i}{\partial x_j\,\partial x_k}\right)\right|_{v_a = 0} = 0 \tag{5.29}$$

where it is emphasized that x_i denotes the current cartesian coordinates of material points in the deformed configuration $\Omega(t)$. Since the discretization involves small displacements, we can consider only the linear part of the Taylor series expansion of the functions h_i about $v_a = 0$ as follows:

$$u_i = h_i(v_a = 0) + \left.\frac{\partial h_i}{\partial x_j}\right|_{v_a = 0} dx_j + \left.\frac{\partial h_i}{\partial v_b}\right|_{v_a = 0} dv_b \qquad (5.30)$$

Recognizing that the first two terms on the right-hand side of the above expansion vanish (since they are in $\Omega(t)$), equation (5.30) may be expressed as follows:

$$u_i = L_{ib}(v_a = 0)\, dv_b \qquad (5.31)$$

Let $N_{ib} = L_{ib}(v_a = 0)$ and let $q_b = dv_b$, then the discretized displacement field is finally written as:

$$u_i = N_{ib}\, q_b \qquad (5.32)$$

where q_b are infinitesimal (incremental) nodal displacements and N_{ib} are the shape functions.

Substituting for δu_i from expressions (5.27) and (5.32) into equation (5.24) and utilizing equation (5.25) for the virtual strains, we obtain the discretized equilibrium equation as follows (note that the quantities q_b are arbitrary and thus are eliminated from the final equation):

$$\iiint_{\Omega(t)} \sigma_{ij} \frac{\partial N_{ia}}{\partial x_j}\, dV = \iiint_{\Omega(t)} \rho\, p_i\, N_{ia}\, dV + \iint_{\Gamma(t)} t_i\, N_{ia}\, dA \qquad (5.33)$$

We finally obtain the incremental equilibrium equations in the Updated Lagrangian description by differentiating both sides of equation (5.33) with respect to x_k:

$$\left([K] + [K]^{(\sigma)} + [K]^{(NC)}\right)\{dv\} = \{dP\} \qquad (5.34)$$

where $\{dv\}$ is the unknown incremental vector of nodal displacements and $\{dP\}$ is the corresponding incremental vector of nodal forces which is given by:

$$dP_a = \iiint\limits_{\Omega(t)} \rho \; (dp_i) \; N_{ia} \; dV + \iint\limits_{\Gamma(t)} (dt_i) \; N_{ia} \; dA \tag{5.35}$$

In equation (5.34), $[K]$ is the symmetric "large displacement" matrix, $[K]^{(\sigma)}$ is the symmetric "initial stress" matrix, and $[K]^{(NC)}$ is the non-symmetric "displacement dependent load" matrix. These matrices are given by:

$$K_{ab} = \iiint\limits_{\Omega(t)} \frac{\partial N_{ia}}{\partial x_j} \; E_{ijk\ell} \; \frac{\partial N_{kb}}{\partial x_\ell} \; dV \tag{5.36a}$$

$$K_{ab}^{(\sigma)} = \iiint\limits_{\Omega(t)} \frac{\partial N_{ka}}{\partial x_i} \; \sigma_{ij} \; \frac{\partial N_{kb}}{\partial x_j} \; dV \tag{5.36b}$$

$$K_{ab}^{(NC)} = \iiint\limits_{\Omega(t)} \rho \; \frac{\partial p_i}{\partial x_j} \; N_{jb} \; N_{ia} \; dV + \iint\limits_{\Gamma(t)} T_{ib} \; N_{ia} \; dA \tag{5.36c}$$

where T_{ib} is defined by the following relation:

$$\frac{\partial t_i}{\partial x_j} \; u_j = T_{ib} \; q_b \tag{5.37}$$

The discretized equilibrium equation (5.34) expresses the equilibrium between the internal forced $\{Q\}$ (on the left-hand side) and the external forces $\{P\}$ (on the right-hand-size). The residual force vector $\{R_e\}$ is defined by:

$$\{R_e\} = \{P\} - \{Q\} \tag{5.38}$$

Finally, it should be noted that a new independent variable, namely, the damage variable φ, appears in the finite element formulation. Therefore, the necessary modifications should be made so that the evolution equation (5.23) is incorporated in the finite element routines. Furthermore, it must be emphasized that the variable φ is based on the deformed configuration $\Omega(t)$ which makes the Updated Lagrangian description very convenient to use. If the Total Lagrangian description were to be used, then the damage variable φ must first be transformed into the initial configuration $\Omega(t_o)$

which makes the resulting equations more complicated.

5.4 Application to Ductile Fracture - Example

The elastic constitutive model derived in this chapter is now applied to solve problems in elastic ductile fracture. As an example, the problem of crack initiation in a center-cracked thin plate that is subjected to inplane tension is analyzed. The plate is symmetrical in geometry and loaded as shown in Figure 5.3a. It is made of aluminum alloy 2024-T3 ($\overline{E}$ = 73,087 MPa, $\overline{v}$ = 0.3) with a thickness of 3.175 mm. Since the thickness is small compared with the other dimensions, a state of plane stress is assumed. Due to symmetry, in Figure 5.3b, only one-quarter of the plate is discretized by finite elements.

An optimum finite element mesh around the crack tip is used as shown in Figure 5.4. This grid has been previously used [117] to analyze plane stress and plane strain conditions under mode I tensile fracture. The use of this grid has been proven to be successful [117] as it is especially designed to be used around crack tips of the type considered here.

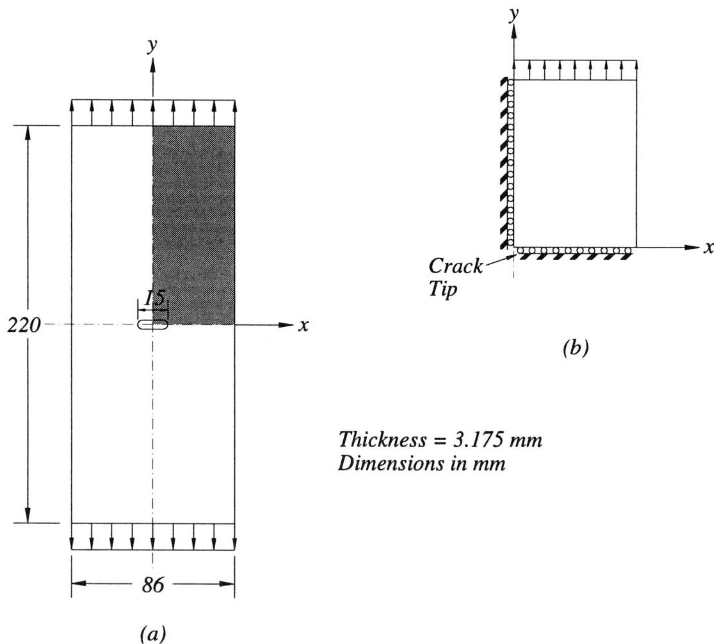

Figure 5.3   (a) Thin Plate with a Center Crack. (b) Quarter of Plate to be discretized by Finite Elements
(Aluminum Alloy 2024-T3)

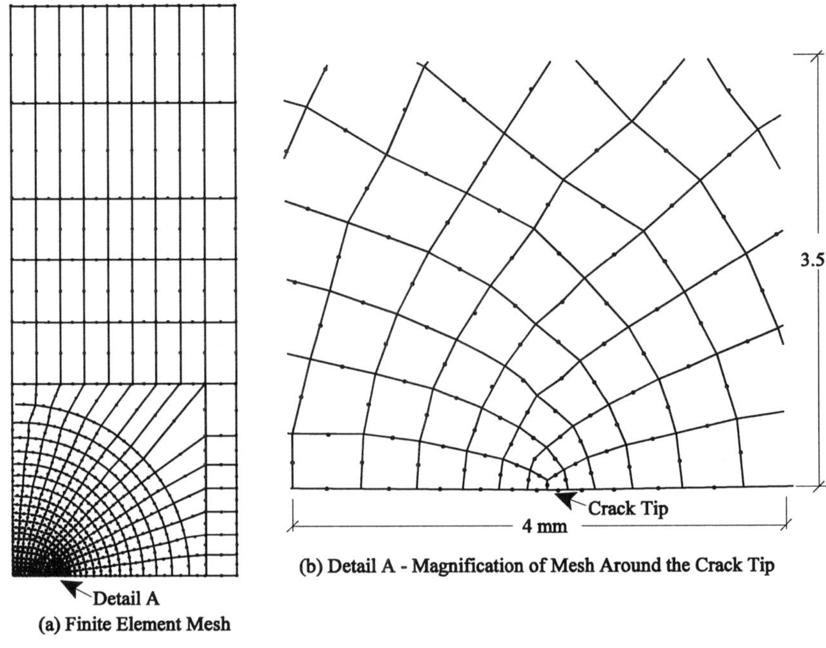

Figure 5.4   Optimum Finite Element Mesh Around the Crack Tip.

The eight-node quadrilateral isoparametric element is used in the finite element analysis. It is noticed that a large number of regular elements is used around the crack tip in order to avoid the use of special (singularity) elements at that point [118, 119]. Consequently, a total of 381 elements and 1228 nodes is used.

The problem is solved independently by first assuming an elastic material behavior and then using the proposed coupled theory. The load is incremented with uniform load increments of 10 MPa. This process is terminated when a final load of 300 MPa is reached, which is far beyond the anticipated crack initiation load. The "overall" damage parameter ℓ is monitored in the elements surrounding the crack tip at each load increment because it is this factor that is used to determine crack initiation. The results of this analysis are examined after the 9th, 18th, and 27th load increments. However, the results are shown here after the 27th load increment is completed for a value $\mu = 0.4$.

The critical value of the "overall" damage parameter ℓ is taken to be 0.115. This value is obtained experimentally from the uniaxial test performed by Chow and Wang [120]. The corresponding value for the load causing crack initiation is determined here to be 243 MPa. It should be noted that this value is dependent upon the appropriate choice of the constant μ. It is noticed that convergence is obtained in less than 20 iterations for each increment of load. The results are shown in Figures 5.5 - 5.8.

In Figure 5.5, the distribution of the axial strain ε_{xx} is shown around the crack tip for the two cases of elasticity and elasticity with damage. It is noticed that the values of ε_{xx} are highest at the crack tip and they decrease in magnitude as we move away from the crack tip. It is also noticed that the incorporation of damage in the analysis has reduced the axial strains although the order of the strains is the same for both constitutive models. The maximum value of ε_{xx} is 0.0212 for the coupled model compared with 0.0269 for the elastic solution.

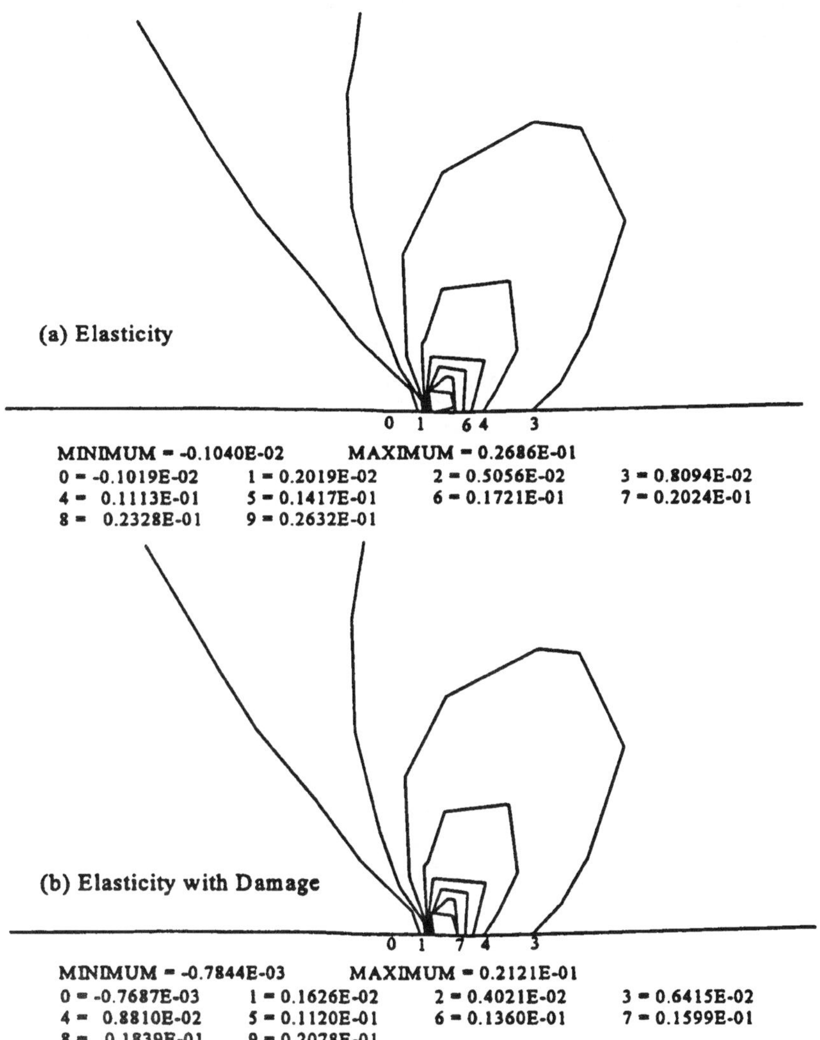

Figure 5.5   Distribution of ε_{xx} around the crack tip.

The distribution of the axial Cauchy stress σ_{xx} is shown in Figure 5.6. The stress contours are shown around the crack tip for both constitutive models. It is noticed that slightly higher stresses are obtained when the coupled theory is used. This is mainly attributed to the incorporation of the damage parameters in the equations. The concentration of high stresses is clearly displayed in the figure. The Cauchy stress σ_{xx} decreases in magnitude as we move away from the crack tip. A maximum stress of 2657 MPa was obtained using the coupled model compared with 2651 MPa for the elastic solution.

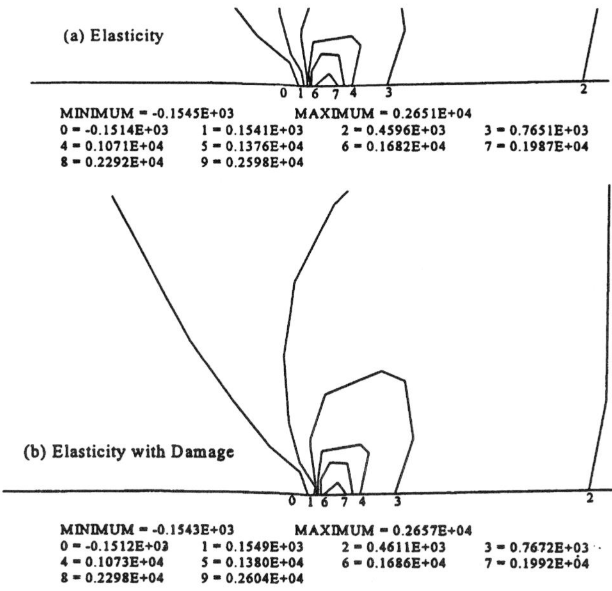

Figure 5.6   Distribution of σ_{xx} around the crack tip

The volumetric Cauchy stress, σ_v, where $\sigma_v = \sigma_{xx} + \sigma_{yy}$ is shown in Figure 5.7. It is noticed that generally a similar stress distribution is obtained as that of σ_x. Again, we notice that the coupled damage model gives slightly higher volumetric stresses than the elasticity theory. In this case also, the volumetric stress contours decrease in magnitude away from the crack tip. In Figure 5.8, the Cauchy stress, τ_{xy}, is shown around the crack tip. In this case, we notice that the damage model gives slightly lower stress values when compared with the elastic solution.

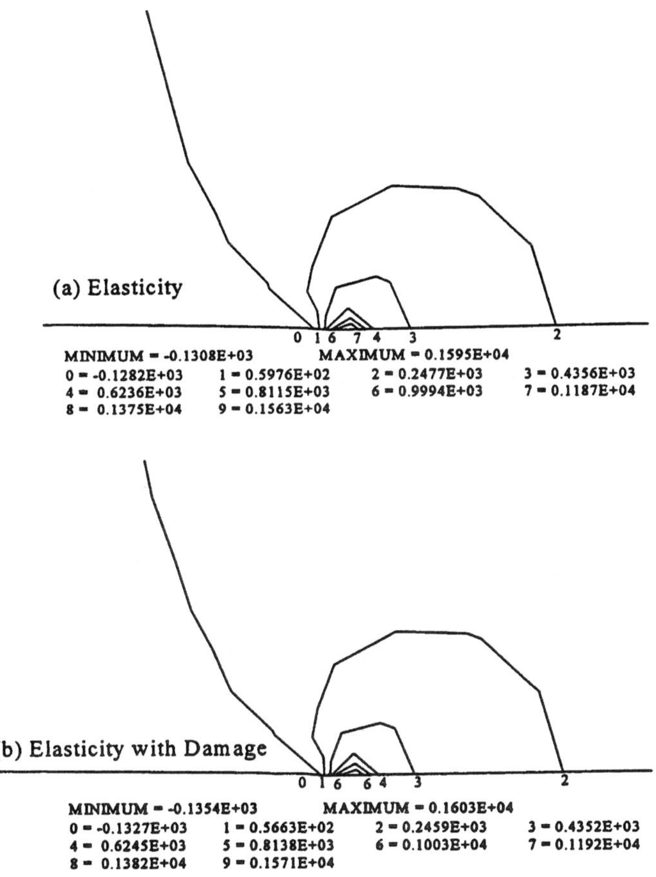

Figure 5.7   Distribution of σ_v around the crack tip.

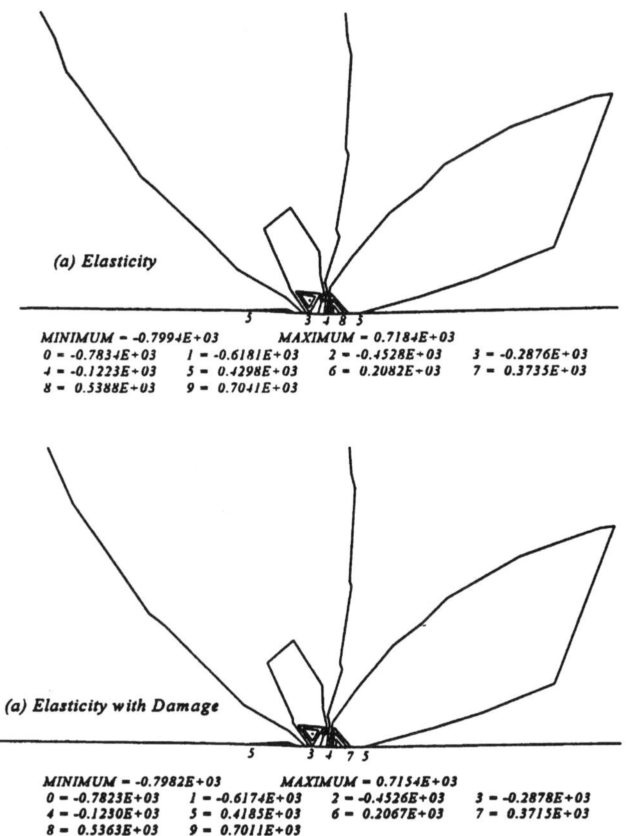

Figure 5.8 Distribution of σ_{xy} around the crack tip.

It is noticed that higher normal stresses and lower shear stresses are obtained when the damage model is incorporated in the elasticity solution. However, the magnitude of the strains is decreased for both normal and shear strains. It is thus demonstrated that the presented damage-elasticity coupled model provides a powerful tool to tackle problems involving stress concentrations that may arise from material defects. The presented model can now be used to solve more complicated engineering problems especially in elastic ductile fracture.

In order to ascertain the accuracy of the finite element solution, an alternate finite element mesh is used as shown in Figure 5.9. This mesh comprises of 836 eight-noded isoparametric quadrilateral elements with a total of 2629 nodes. The results obtained for this mesh are shown in Figures 5.10 to 5.13. The results obtained from the two meshes are identical and confirm the accuracy of the solution.

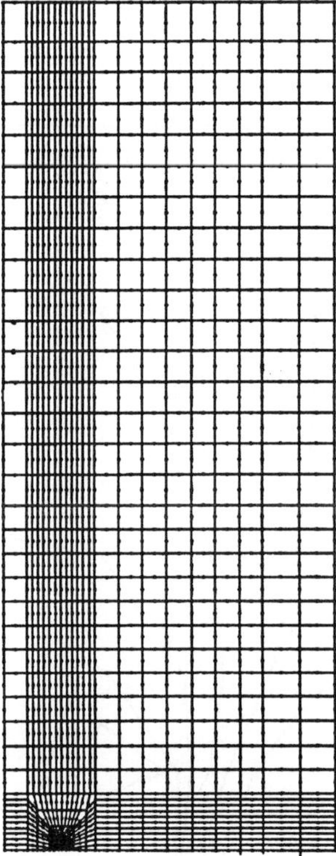

Figure 5.9   Alternate Finite Element Mesh

104

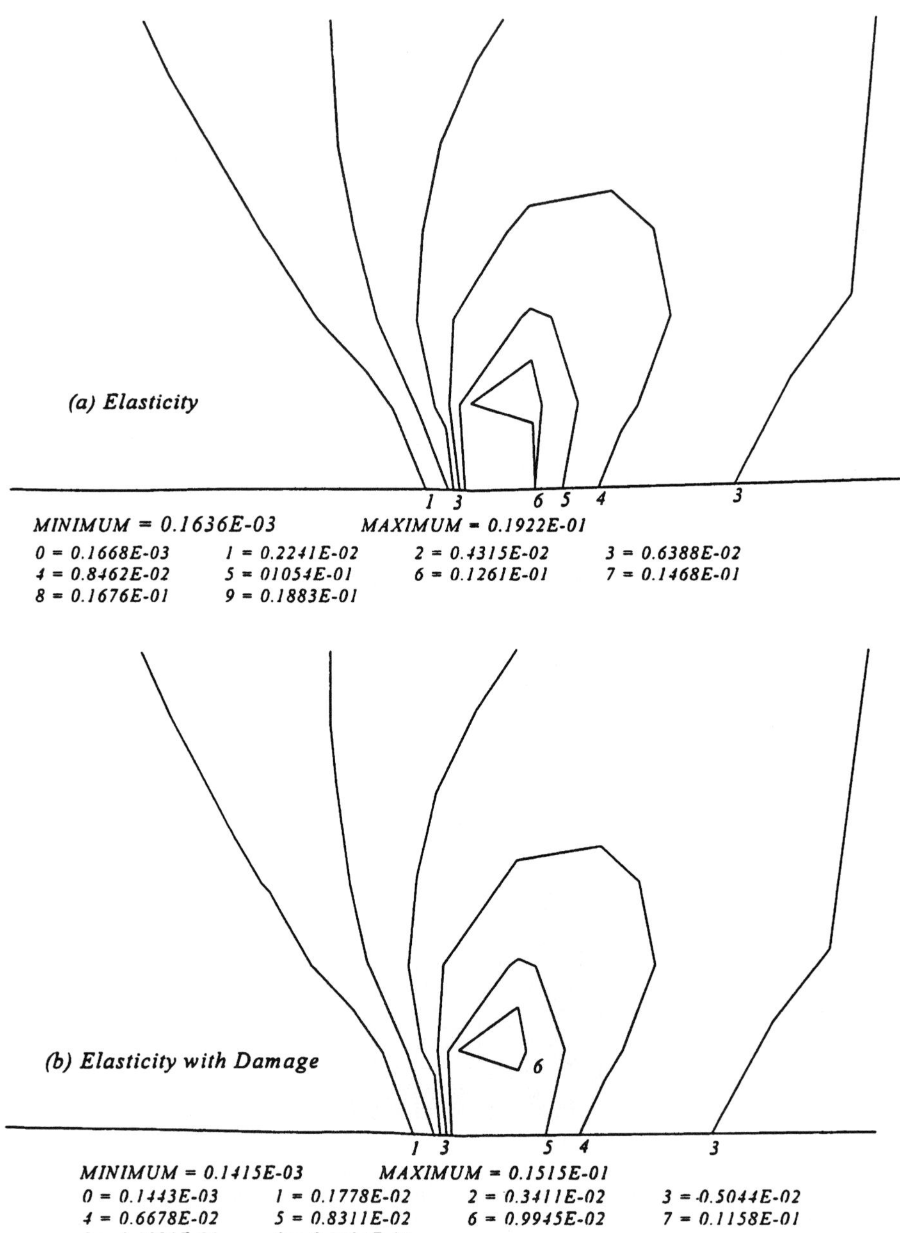

Figure 5.10   Distribution of ε_{yy} around the crack tip.

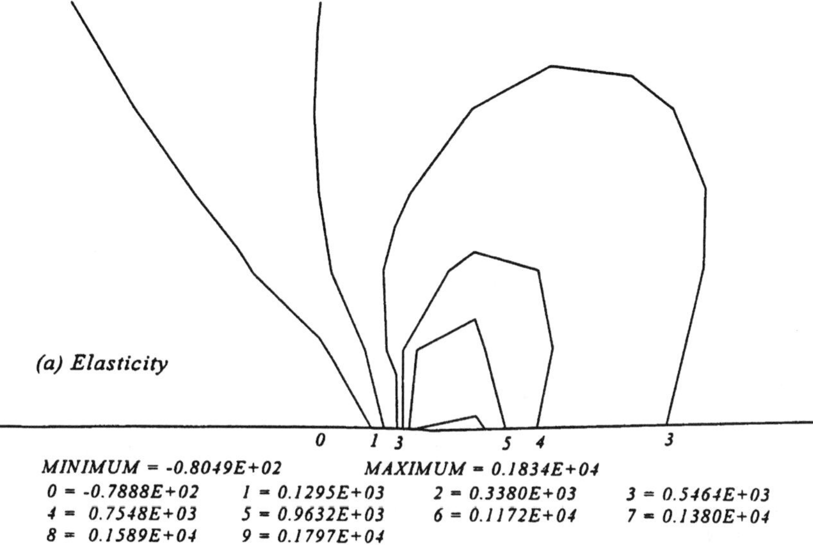

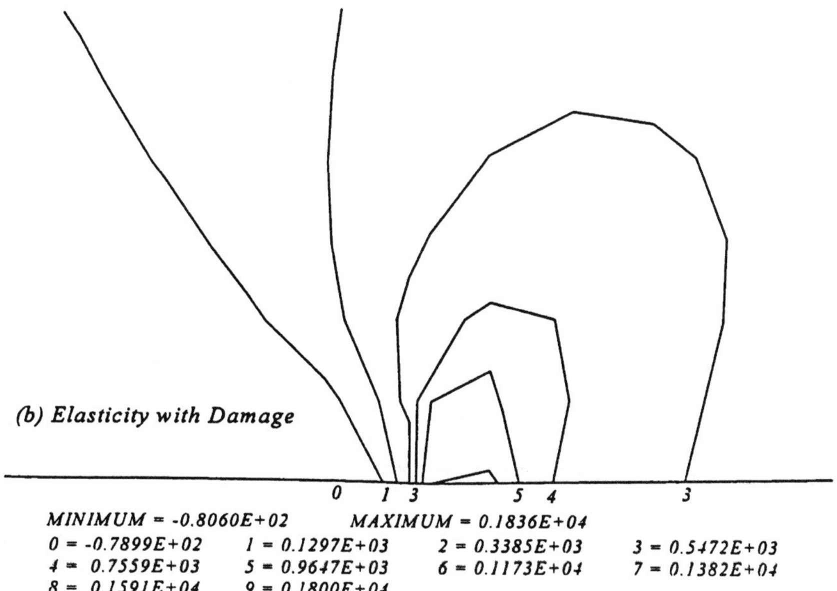

Figure 5.11   Distribution of σ_{yy} around the crack tip.

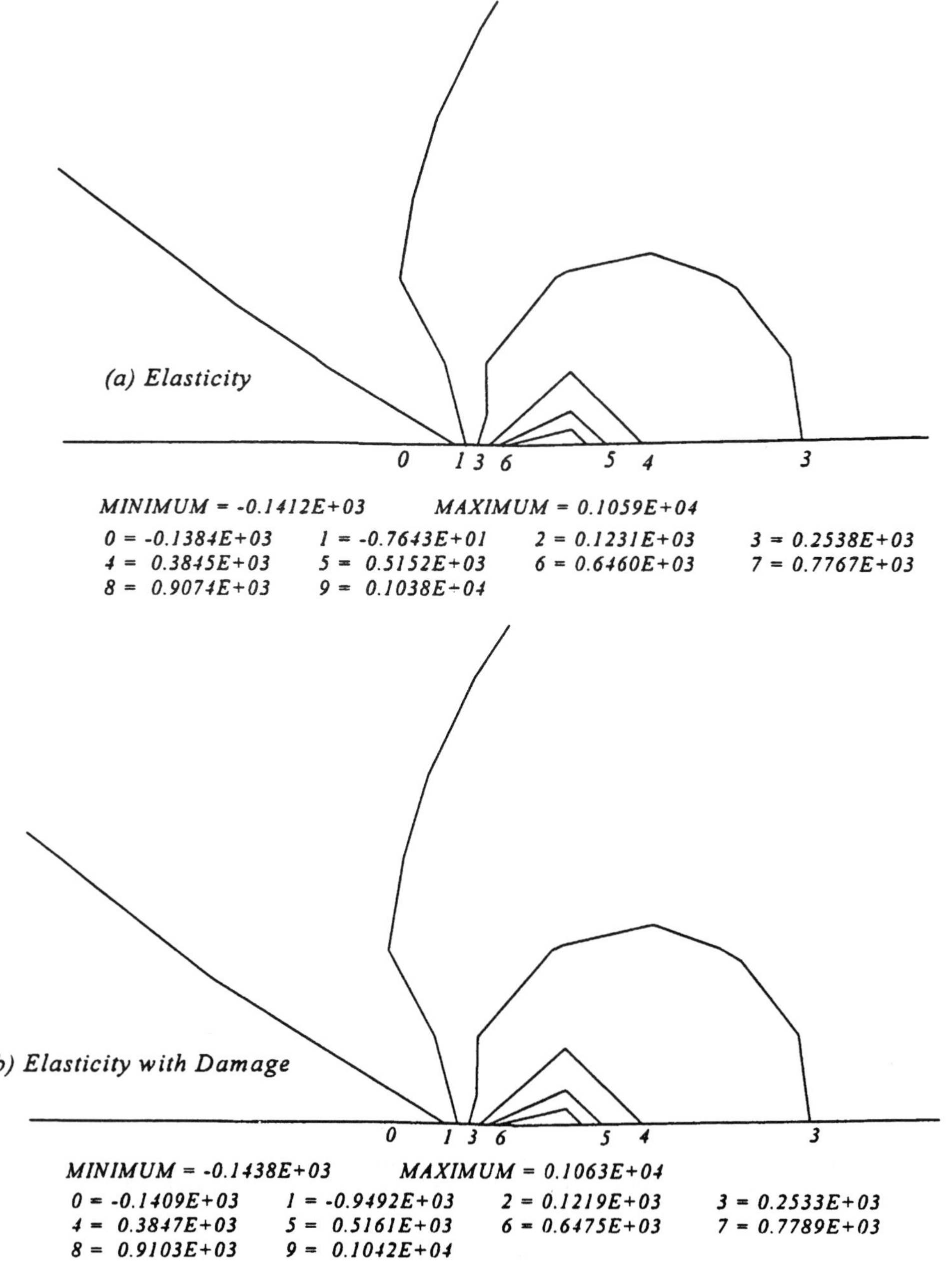

Figure 5.12   Distribution of σ_y around the crack tip.

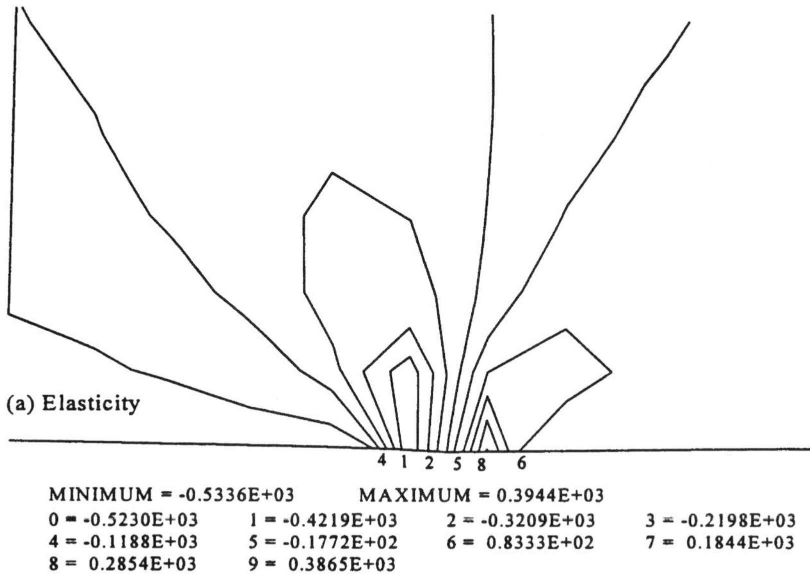

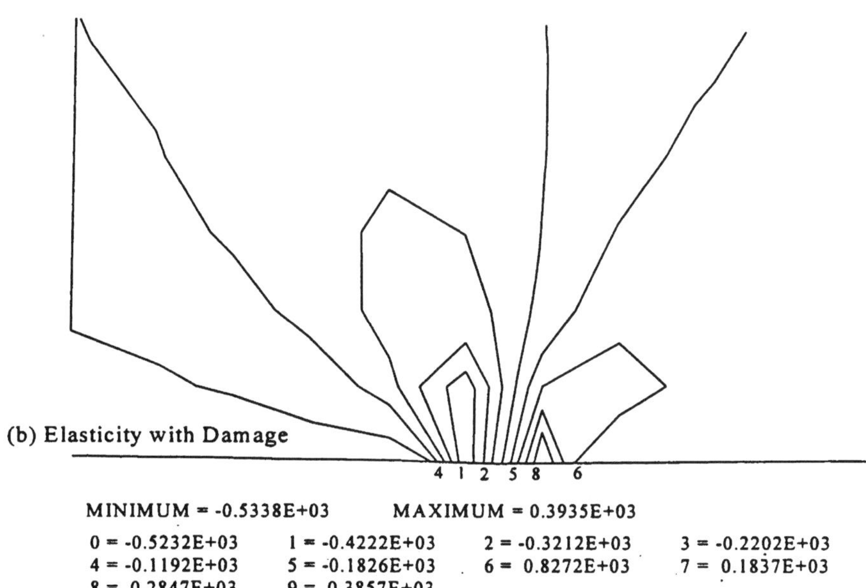

Figure 5.13   Distribution of σ_{xy} around the crack tip.

CHAPTER 6

DAMAGE AND PLASTICITY IN METALS

A constitutive model is developed in this chapter for anisotropic continuum damage mechanics using finite strain plasticity. The formulation is given in spatial coordinates (Eulerian reference frame) and incorporates both isotropic and kinematic hardening. The von Mises yield criterion is modified to include the effects of damage through the use of the hypothesis of elastic energy equivalence. A modified elasto-plastic stiffness tensor that includes the effects of damage is derived within the framework of the proposed model.

It is also shown how the model can be used in conjunction with other damage-related yield functions. In particular Gurson's yield function [121, 122] which was later modified by Tvergaard [123], and Tvergaard and Needleman [124] is incorporated in the proposed theory. This yield function is derived based on the presence of spherical voids in the material and an evolution law for the void growth is also incorporated. It is also shown how a modified Gurson yield function can be related to the proposed model. Some interesting results are obtained in this case.

Numerical implementation of the proposed model includes the finite element formulation where an updated Lagrangian description is used. The problem of crack initiation is solved for a thin elasto-plastic plate with a center crack that is subjected to inplane tension.

6.1 Stress Transformation Between Damaged and Undamaged States

Consider a body in the initial undeformed and undamaged configuration C_o. Let C be the configuration of the body that is both deformed and damaged after a set of external agencies act on it. Next consider a fictitious configuration of the body $\overline{C}$ obtained from C by removing all the damage that the body has undergone. In other words, $\overline{C}$ is the state of the body after it had only deformed without damage. Therefore, in defining a damage tensor φ, its components must vanish in the configuration $\overline{C}$ (see Figure 6.1).

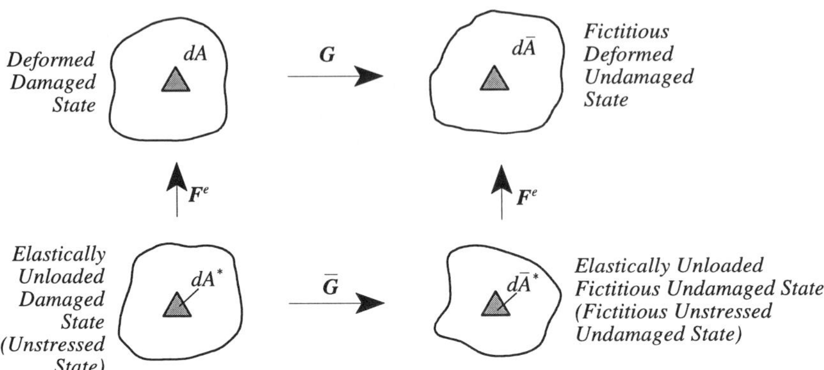

Figure 6.1   States of Deformation and Damage

6.1.1 Effective Stress Tensor

In the formulation that follows, the Eulerian reference system is used, i.e. all the actual quantities are referred to the configuration C while the effective quantities referred to $\overline{C}$. One first introduces a linear transformation between the Cauchy stress tensor $\boldsymbol{\sigma}$ and the effective Cauchy stress tensor $\overline{\boldsymbol{\sigma}}$ in the form:

$$\overline{\sigma}_{ij} = M_{ijk\ell}\, \sigma_{k\ell} \tag{6.1}$$

where $M_{ijk\ell}$ are the components of the fourth-rank linear operator called the damage effect tensor. As defined in equation (6.1), the effective Cauchy stress need not be symmetric or frame-invariant under the given transformation. Although the Cauchy stress $\boldsymbol{\sigma}$ is frame-invariant, the effective Cauchy stress $\overline{\boldsymbol{\sigma}}$ does not necessarily satisfy the frame-invariance principle. Proof of frame-invariance depends on the particular expression of M that is used. However, once the effective Cauchy stress is symmetrized as shown in Chapter 11, it can be easily shown that it satisfies the frame-invariance principle. It has been shown [58] that φ is symmetric and that M can be represented by a 6 x 6 matrix in the form:

$$[M] = [I - \varphi]^{-1} \tag{6.2}$$

where I is the second-rank identity tensor and φ is the second-rank damage tensor. Alternatively, one can use the tensorial equation $M_{ijk\ell} = (I - \varphi)_{ik}\, \delta_{j\ell}$, but the matrix equation is preferred since it is used in section 6.3. It has been shown [19] that for the one-dimensional case, the scalar counterparts of M and φ are M and φ and are related by $M = 1/(1-\varphi)$ which is a special case of equation (6.2). The nature of the tensor M is discussed in detail in section 6.3. Equation (6.1) can be used to describe anisotropic damage in general. However, if the specific matrix representation of equation (6.2) is used, then the formulation is restricted only to isotropy. This can be easily proved by showing that the effective stress is frame-invariant when expression (6.2) is used. This argument can be similarly applied to the matrix expression of M given later in section 6.3.

A transformation relation for the deviatoric Cauchy stress tensor is next derived. One first writes the deviatoric part τ in the configuration C:

$$\tau_{ij} = \sigma_{ij} - \frac{1}{3}\sigma_{mm}\,\delta_{ij} \tag{6.3}$$

where δ_{ij} are the components of the identity tensor I. A similar relation exists in the configuration $\overline{C}$ between $\overline{\sigma}$ and $\overline{\tau}$ in the form:

$$\overline{\tau}_{ij} = \overline{\sigma}_{ij} - \frac{1}{3}\overline{\sigma}_{mm}\,\delta_{ij} \tag{6.4}$$

where δ_{ij} is the same in both C and $\overline{C}$. Substituting for $\overline{\sigma}_{ij}$ from equation (6.1) into equation (6.4) while using equation (6.3), we obtain the following:

$$\overline{\tau}_{ij} = M_{ijk\ell}\,\tau_{k\ell} + \frac{1}{3}M_{ijmm}\,\sigma_{nn} - \frac{1}{3}M_{ppqr}\,\sigma_{qr}\,\delta_{ij} \tag{6.5}$$

It is clear from equation (6.5) that a linear relation does not exist between $\overline{\tau}$ and τ. On the other hand, one might suspect that the last two terms on the right-hand side of equation (6.5) cancel each other when they are written in expanded form. However, this possibility can be easily dismissed as follows: suppose one assumes $\overline{\tau}_{ij} = M_{ijk\ell}\,\tau_{k\ell}$. Using this with equation (6.5) one concludes that $M_{ijmm}\,\sigma_{nn} = M_{ppqr}\,\sigma_{qr}\,\delta_{ij}$. Now consider the case when $i \neq j$. One has $\delta_{ij} = 0$ and therefore $M_{ijmm}\,\sigma_{nn} = 0$. It is clear that this is a contradiction to the fact that generally $M_{ijmm} \neq 0$ and $\sigma_{nn} \neq 0$. Therefore, the additional terms in equation (6.5) are non-trivial and such a linear transformation cannot be assumed.

112

Upon examining equation (6.5) in more detail, eliminating $\tau_{k\ell}$ by using equation (6.3) and simplifying the resulting expression, one obtains the following:

$$\bar{\tau}_{ij} = N_{ijk\ell}\,\sigma_{k\ell} \tag{6.6}$$

where $N_{ijk\ell}$ are the components of a fourth-rank tensor given by:

$$N_{ijk\ell} = M_{ijk\ell} - \frac{1}{3}M_{rrk\ell}\,\delta_{ij} \tag{6.7}$$

Equation (6.6) represent a linear transformation between the effective deviatoric Cauchy stress tensor $\bar{\tau}$ and the Cauchy stress tensor $\boldsymbol{\sigma}$. However, in this case the linear operator N is not simply the damage effect tensor M but a linear function of M as seen by equation (6.7). The tensors M and N are mappings $S \to \bar{S}$ and $S \to \bar{S}_{dev}$, respectively, where S is the stress space in the current configuration C and $\bar{S}$ is the stress space in the fictitious undamaged state $\bar{C}$, with $\boldsymbol{\sigma} \in S$.

Next, we consider the effective stress invariants and their transformations in the configuration $\bar{C}$. It was earlier seen from equation (6.6) that the first effective deviatoric stress invariant $\bar{\tau}_{ii}$ is given by:

$$\bar{\tau}_{ii} = N_{iik\ell}\,\sigma_{k\ell} = 0 \tag{6.8}$$

since $N_{iik\ell} = 0$ by direct contraction in equation (6.7). Therefore, one obtains $\bar{\tau}_{ii} = \tau_{ii} = 0$.

The problem becomes more involved when considering the second effective stress invariant $\bar{\tau}_{ij}\bar{\tau}_{ij}$. Using equation (6.5) along with equation (6.3), one obtains:

$$\bar{\tau}_{ij}\bar{\tau}_{ij} = A_{k\ell mn}\,\tau_{k\ell}\tau_{mn} + B_{pq}\,\tau_{pq} + C \tag{6.9}$$

where

$$A_{k\ell mn} = M_{ijk\ell}M_{ijmn} - \frac{1}{3}M_{ttmn}M_{rrk\ell} \tag{6.10a}$$

$$B_{pq} = \frac{2}{3}\sigma_{mm}\left(M_{ijkk}M_{ijpq} - \frac{1}{3}M_{ttpq}M_{rrnn}\right) \tag{6.10b}$$

$$C = \frac{1}{2}\,\sigma_{mm}^2 \left(M_{ijpp}\,M_{ijqq} - \frac{1}{3}\,M_{ttmm}\,M_{rrnn} \right) \qquad (6.10c)$$

Substituting for τ from equation (6.3) into equation (6.9) [or more directly using equation (6.6) along with equation (6.7)], one obtains:

$$\bar{\tau}_{ij}\,\bar{\tau}_{ij} = H_{k\ell mn}\,\sigma_{k\ell}\,\sigma_{mn} \qquad (6.11)$$

where the fourth-rank tensor H is given by:

$$H_{k\ell mn} = N_{ijk\ell}\,N_{ijmn} \qquad (6.12)$$

and the tensor N is given by equation (6.7). The transformation equation (6.11) will be used in the next sections to transform the von Mises yield criterion into the configuration $\overline{C}$.

6.1.2 Effective Backstress Tensor

In the theory of plasticity, kinematic hardening is modeled by the motion of the yield surface in the stress space. This is implemented mathematically by the evolution of the shift or backstress tensor β. The backstress tensor β denotes the position of the center of the yield surface in the stress space. For this purpose, one studies now the transformation of this tensor in the configurations C and $\overline{C}$.

Let α be the deviatoric part of β. Therefore, one has

$$\alpha_{ij} = \beta_{ij} - \frac{1}{3}\,\beta_{mm}\,\delta_{ij} \qquad (6.13)$$

where both α and β are referred to the configuration C. Let their effective counterparts $\overline{\alpha}$ and $\overline{\beta}$ be referred to the configuration $\overline{C}$. Similarly to equation (6.13) we have:

$$\overline{\alpha}_{ij} = \overline{\beta}_{ij} - \frac{1}{3}\,\overline{\beta}_{mm}\,\delta_{ij} \qquad (6.14)$$

Assuming a linear transformation (based on the same argument used for the stresses) similar to equation (6.1) between the effective backstress tensor $\overline{\beta}$ and backstress tensor β:

$$\overline{\beta}_{ij} = M_{ijk\ell}\,\beta_{k\ell} \qquad (6.15)$$

114

and following the same procedure in the derivation of equation (6.6), we obtain the following linear transformation between $\boldsymbol{\beta}$ and $\overline{\boldsymbol{\alpha}}$:

$$\overline{\alpha}_{ij} = N_{ijk\ell}\,\beta_{k\ell} \qquad\qquad (6.16)$$

The effective backstress invariants have similar forms to those of the effective stress invariants, mainly, $\overline{\alpha}_{ii} = \alpha_{ii} = 0$ and

$$\overline{\alpha}_{ij}\,\overline{\alpha}_{ij} = H_{k\ell mn}\,\beta_{k\ell}\,\beta_{mn} \qquad\qquad (6.17)$$

In addition, one more transformation equation needs to be given before one proceeds to the constitutive model. By following the same procedure for the other invariants, the mixed invariant $\sigma_{ij}\,\beta_{ij}$ in the configuration C is transformed to $\overline{\tau}_{ij}\,\overline{\alpha}_{ij}$ as follows:

$$\overline{\tau}_{ij}\,\overline{\alpha}_{ij} = H_{k\ell mn}\,\sigma_{k\ell}\,\beta_{mn} \qquad\qquad (6.18)$$

and a similar relation holds for the invariant $\overline{\alpha}_{ij}\,\overline{\tau}_{ij}$. The stress and backstress transformation equations will be used later in the constitutive model.

6.2 Strain Rate Transformation Between Damaged and Undamaged States

In the general elasto-plastic analysis of deforming bodies, the spatial strain rate tensor $\boldsymbol{d}$ in the configuration C is decomposed additively (Nemat-Nasser [125, 126] and Lee [127]):

$$d_{k\ell} = d_{k\ell}' + d_{k\ell}'' \qquad\qquad (6.19)$$

where $\boldsymbol{d}'$ and $\boldsymbol{d}''$ denote the elastic and plastic parts of $\boldsymbol{d}$, respectively. In equation (6.19), the assumption of small elastic strains is made, however, finite plastic deformations are allowed. On the other hand, the decomposition in equation (6.19) will be true for any amount of elastic strain if the physics of elasto-plasticity is invoked, for example the case of single crystals. A thorough account of this is given by Asaro [128].

In the next two subsections the necessary transformation equations between the configuration C and $\overline{C}$ will be derived for the elastic strain and the plastic strain rate tensors. In this derivation, it is assumed that the elastic strains are small compared to the plastic strains and consequently the elastic strain tensor is taken to be the usual engineering elastic strain tensor ε'. In addition, it is assumed that an elastic strain energy function exists such that a linear relation can be used between the Cauchy stress tensor $\boldsymbol{\sigma}$ and the engineering elastic strain tensor ε'. The tensor ε' is defined here as the linear term of the elastic part of the spatial strain tensor where second order terms are neglected. For more details, see the work by Kattan and Voyiadjis [98].

6.2.1 Effective Elastic Strain

The elastic constitutive equation to be used is based on one of the assumptions outlined in the previous paragraph and is represented by the following linear relation in the configuration $\overline{C}$:

$$\overline{\sigma}_{ij} = \overline{E}_{ijk\ell}\,\overline{\varepsilon}'_{k\ell} \tag{6.20}$$

where $\overline{E}$ is the fourth-rank elasticity tensor given by:

$$\overline{E}_{ijk\ell} = \lambda\delta_{ij}\delta_{k\ell} + G\,(\,\delta_{ik}\delta_{j\ell} + \delta_{i\ell}\delta_{jk}\,) \tag{6.21}$$

and λ and G are Lame's constants. Based on the constitutive equation (6.20), the elastic strain energy function $U(\varepsilon',\,\varphi)$ in the configuration $\overline{C}$ is given by:

$$U(\overline{\varepsilon}',\mathbf{0}) = \frac{1}{2}\,\overline{E}_{ijk\ell}\,\overline{\varepsilon}'_{ij}\,\overline{\varepsilon}'_{k\ell} \tag{6.22}$$

One can now define the complementary elastic energy function $V(\boldsymbol{\sigma},\,\varphi)$, based on a Legendre transformation, as follows:

$$V(\boldsymbol{\sigma},\varphi) = \sigma_{ij}\varepsilon'_{ij} - U(\overline{\varepsilon}',\varphi) \tag{6.23}$$

By taking the partial derivative of equation (6.23) with respect to the stress tensor $\boldsymbol{\sigma}$, one obtains:

$$\varepsilon'_{ij} = \frac{\partial V(\boldsymbol{\sigma},\varphi)}{\partial\sigma_{ij}} \tag{6.24}$$

Substituting expression (6.22) into equation (6.23) in the configuration $\overline{C}$, one obtains the following expression for $V(\boldsymbol{\sigma},\,\varphi)$ in the configuration $\overline{C}$ as follows:

116

$$V(\overline{\sigma},0) \;=\; \frac{1}{2}\,\overline{E}^{-1}_{ijk\ell}\,\overline{\sigma}_{ij}\,\overline{\sigma}_{k\ell} \tag{6.25}$$

The hypothesis of elastic energy equivalence, which was initially proposed by Sidoroff [16], is now used to obtain the required relation between ε' and $\overline{\varepsilon}'$. In this hypothesis, one assumes that the elastic energy $V(\sigma,\varphi)$ in the configuration C is equivalent in form to $V(\overline{\sigma},0)$ in the configuration $\overline{C}$. Therefore, one has:

$$V(\sigma,\varphi) \;=\; V(\overline{\sigma},0) \tag{6.26}$$

where $V(\sigma,\varphi)$ is the complementary elastic energy in C and is given by:

$$V(\sigma,\varphi) \;=\; \frac{1}{2}\,E^{-1}_{ijk\ell}(\varphi)\,\sigma_{ij}\,\sigma_{k\ell} \tag{6.27}$$

where the superscript -1 indicates the inverse of the tensor.

In equation (6.27), the effective elasticity modulus $E(\varphi)$ is a function of the damage tensor φ and is no longer constant. Using equation (6.26) along with expressions (6.25) and (6.27), one obtains the following relation between $\overline{E}$ and $E(\varphi)$:

$$E_{k\ell mn}(\varphi) \;=\; M^{-1}_{ijk\ell}(\varphi)\,\overline{E}_{ijpq}\,M^{-T}_{pqmn}(\varphi) \tag{6.28}$$

where the superscript $-T$ indicates the transpose of the inverse of the tensor as defined by $M^{-1}_{ijk\ell}\,M_{mnk\ell} = \delta_{im}\,\delta_{jn}$. Finally, using equation (6.24) along with equations (6.25), (6.26) and (6.28), one obtains the desired linear relationship between the elastic strain tensor ε' and its effective counterpart $\overline{\varepsilon}'$:

$$\overline{\varepsilon}'_{k\ell} \;=\; M^{-T}_{k\ell mn}\,\varepsilon'_{mn} \tag{6.29}$$

The two transformation equations (6.28) and (6.29) will be incorporated in section 6.4 in the general inelastic constitutive model that will be developed later.

6.2.2 Effective Plastic Strain Rate

The constitutive model to be developed here is based on a von Mises type yield function $f(\tau,\,\alpha,\,\kappa,\,\varphi)$ in the configuration C that involves both isotropic and kinematic hardening through the evolution of the plastic work κ and the backstress tensor $\boldsymbol{\alpha}$, respectively. The corresponding yield function $f(\overline{\tau}, \overline{\boldsymbol{\alpha}}, \overline{\kappa}, \mathbf{0})$ in the configuration $\overline{C}$ is given by:

$$f = \frac{3}{2}(\overline{\tau}_{k\ell} - \overline{\alpha}_{k\ell})(\overline{\tau}_{k\ell} - \overline{\alpha}_{k\ell}) - \overline{\sigma}_o^2 - c\,\overline{\kappa} = 0 \tag{6.30}$$

where σ_o and c are material parameters denoting the uniaxial yield strength and isotropic hardening, respectively. The plastic work κ is a scalar function and its evolution in the configuration $\overline{C}$, is taken here to be in the form:

$$\overset{\circ}{\overline{\kappa}} = \sqrt{\overline{d}''_{ij}\,\overline{d}''_{ij}} \tag{6.31}$$

where $\overline{d}''_{ij}$ is the plastic part of the spatial strain rate tensor $\boldsymbol{d}$.

Isotropic hardening is described by the evolution of the plastic work κ as given above. In order to describe kinematic hardening, the Prager-Ziegler evolution law [132] is used here in the configuration $\overline{C}$, as follows:

$$\overset{\circ}{\overline{\alpha}}_{ij} = d\overline{\mu}\;(\overline{\tau}_{ij} - \overline{\alpha}_{ij}) \tag{6.32}$$

where $d\overline{\mu}$ is a scalar function to be determined shortly. The superposed "$\circ$" in equation (6.32) indicates a suitable corotational derivative which is defined later in section 6.4.

The plastic flow in the configuration $\overline{C}$ is described by the associated flow rule in the form:

$$\overline{d}''_{ij} = d\overline{\Lambda}\,\frac{\partial f}{\partial \overline{\sigma}_{ij}} \tag{6.33}$$

where $d\overline{\Lambda}$ is a scalar function introduced as a Lagrange multiplier in the constraint thermodynamic equations (see section 6.4), that is still to be determined. In the present formulation, it is assumed that the associated flow rule of plasticity will still hold in the configuration C, that is:

$$d''_{ij} = d\Lambda\,\frac{\partial f}{\partial \sigma_{ij}} \tag{6.34}$$

where $d\Lambda$ is another scalar function that is to be determined.

118

Substituting the yield function f of equation (6.30) into equation (6.33) and using the transformation equations (6.6) and (6.16), one obtains:

$$\bar{d}''_{ij} = 3\,d\bar{\Lambda}\,N_{ijk\ell}\,(\sigma_{k\ell} - \beta_{k\ell})$$ (6.35)

On the other hand, substituting the yield function f of equation (6.30) into equation (6.34) and noting the appropriate transformations (6.11) and (6.17), one obtains:

$$d''_{ij} = 3\,d\Lambda\,H_{ijk\ell}\,(\sigma_{k\ell} - \beta_{k\ell})$$ (6.36)

It is noticed that plastic incompressibility exists in the configuration $\overline{C}$ as seen from equation (6.35) where $\bar{d}''_{mm} = 0$ since $N_{mmk\ell} = 0$. However, this is not true in the configuration C since d''_{mm} does not vanish depending on $H_{mmk\ell}$ as shown in equation (6.36).

In order to derive the transformation equation between d'' and $\bar{d}''$, one first notices that:

$$\frac{\partial f}{\partial \sigma_{ij}} = \frac{\partial f}{\partial \bar{\sigma}_{pq}}\frac{\partial \bar{\sigma}_{pq}}{\partial \sigma_{ij}} = \frac{\partial f}{\partial \bar{\sigma}_{pq}}\,M_{pqij}$$ (6.37)

where M_{pqij} is defined in equation (6.1) as $\partial \bar{\sigma}_{pq}/\partial \sigma_{ij}$. Using the above relation along with equations (6.33) and (6.34), one obtains:

$$\bar{d}''_{ij} = \frac{d\bar{\Lambda}}{d\Lambda}\,M^{-1}_{ijmn}\,d''_{mn}$$ (6.38)

the above equation represents the desired relation, except that the expression $d\bar{\Lambda}/d\Lambda$ needs to be determined. This is done by finding explicit expressions for both $d\Lambda$ and $d\bar{\Lambda}$ using the consistency condition. The rest of this section is devoted to this task; but first one needs to determine an appropriate expression for $d\bar{\mu}$ that appears in equation (6.32), since it plays an essential role in the determination of $d\bar{\Lambda}$.

In order to determine an expression for $d\bar{\mu}$, one assumes that the projection of $\overset{\circ}{\boldsymbol{\alpha}}$ on the gradient of the yield surface f in the stress space is equal to $b\bar{d}''$ in the configuration $\overline{C}$, where b is a material parameter to be determined from the uniaxial tension test [55, 56]. This assumption is written as:

$$b\,\overline{d}''_{k\ell} = \overset{\circ}{\overline{\alpha}}_{mn}\,\frac{\dfrac{\partial f}{\partial \overline{\sigma}_{mn}}}{\dfrac{\partial f}{\partial \overline{\sigma}_{pq}}\dfrac{\partial f}{\partial \overline{\sigma}_{pq}}}\,\frac{\partial f}{\partial \overline{\sigma}_{k\ell}} \tag{6.39}$$

Substituting for $\overset{\circ}{\overline{\alpha}}$ and $\overline{d}''$ from equation (6.32) and (6.33), respectively, into equation (6.39) and post-multiplying the resulting equation by $\partial f/\partial \overline{\sigma}_{k\ell}$, one obtains the required expression for $d\overline{\mu}$:

$$d\overline{\mu} = b\,d\overline{\Lambda}\,\frac{\dfrac{\partial f}{\partial \overline{\sigma}_{mn}}\dfrac{\partial f}{\partial \overline{\sigma}_{mn}}}{(\overline{\tau}_{pq} - \overline{\alpha}_{pq})\dfrac{\partial f}{\partial \overline{\sigma}_{pq}}} \tag{6.40}$$

Using the elastic linear relationship in equation (6.20), and taking its corotational rate, one obtains:

$$\overset{\circ}{\overline{\sigma}}_{ij} = \overline{E}_{ijk\ell}\,\overline{d}'_{k\ell} \tag{6.41}$$

where $\overline{d}'_{k\ell}$ is assumed to be equal to $\overset{\circ}{\overline{\varepsilon}}{}'_{k\ell}$ based on the assumption of small elastic strains as discussed in section 6.2.1. Eliminating $\overline{d}'$ from equation (6.41) through the use of expressions (6.19) and (6.33), one obtains:

$$\overset{\circ}{\overline{\sigma}}_{ij} = \overline{E}_{ijk\ell}\left(\overline{d}_{k\ell} - d\overline{\Lambda}\,\frac{\partial f}{\partial \overline{\sigma}_{k\ell}}\right) \tag{6.42}$$

The scalar multiplier $d\overline{\Lambda}$ is obtained from the consistency condition $df(\overline{\tau}_{k\ell},\,\overline{\alpha}_{k\ell},\,0,\,\overline{\kappa}) = 0$, such that:

$$\frac{\partial f}{\partial \overline{\tau}_{k\ell}}\,\overset{\circ}{\overline{\tau}}_{k\ell} + \frac{\partial f}{\partial \overline{\alpha}_{k\ell}}\,\overset{\circ}{\overline{\alpha}}_{k\ell} + \frac{\partial f}{\partial \overline{\kappa}}\,d\overline{\kappa} = 0 \tag{6.43}$$

Using equations (6.19), (6.31), (6.33), (6.40), and (6.42) into equation (6.43), one obtains the following expression for $d\overline{\Lambda}$:

$$d\overline{\Lambda} = \frac{1}{\overline{Q}} \frac{\partial f}{\partial \overline{\tau}_{k\ell}} \overline{E}_{k\ell mn} \overline{d}_{mn} \qquad (6.44)$$

where $\overline{Q}$ is given by:

$$\overline{Q} = \frac{\partial f}{\partial \overline{\tau}_{ab}} \overline{E}_{abcd} \frac{\partial f}{\partial \overline{\tau}_{cd}} - \frac{\partial f}{\partial \overline{\kappa}} \sqrt{\frac{\partial f}{\partial \overline{\sigma}_{pq}} \frac{\partial f}{\partial \overline{\sigma}_{pq}}} - b \frac{\partial f}{\partial \overline{\alpha}_{ef}} (\overline{\tau}_{ef} - \overline{\alpha}_{ef}) \frac{\dfrac{\partial f}{\partial \overline{\sigma}_{ij}} \dfrac{\partial f}{\partial \overline{\sigma}_{ij}}}{(\overline{\tau}_{gh} - \overline{\alpha}_{gh}) \dfrac{\partial f}{\partial \overline{\sigma}_{gh}}} \qquad (6.45)$$

Assuming that the Prager-Ziegler kinematic hardening rule holds in the configuration C along with the projection assumption of equation (6.14), one can derive a similar equation to (6.44), in the form:

$$d\Lambda = \frac{1}{Q} \frac{\partial f}{\partial \tau_{k\ell}} E_{k\ell mn} d_{mn} \qquad (6.46)$$

where Q is given by:

$$Q = \frac{\partial f}{\partial \tau_{ab}} E_{abcd} \frac{\partial f}{\partial \tau_{cd}} - \frac{\partial f}{\partial \kappa} \sqrt{\frac{\partial f}{\partial \sigma_{pq}} \frac{\partial f}{\partial \sigma_{pq}}} - b \frac{\partial f}{\partial \alpha_{ef}} (\tau_{ef} - \alpha_{ef}) \frac{\dfrac{\partial f}{\partial \sigma_{ij}} \dfrac{\partial f}{\partial \sigma_{ij}}}{(\tau_{gh} - \alpha_{gh}) \dfrac{\partial f}{\partial \sigma_{gh}}} \qquad (6.47)$$

In contrast to the method used in reference [109] where the two yield functions in the configurations C and $\overline{C}$ are assumed to be equal, a more consistent approach is adopted here. This approach is based on the assumptions used to derive equation (4.46). It is clear that in this method, the two yield functions in the configurations C and $\overline{C}$ are treated separately and two separate consistency conditions are thus invoked. In the authors' opinion, this emphasizes a more consistent approach than the method used in reference [109].

One is now left with some tedious algebraic manipulations of equations (6.44) and (6.46) in order to derive an appropriate form for the ratio $d\overline{\Lambda} / d\Lambda$. First, equation (6.46) is re-written in the following form, where the appropriate transformations $\overline{E} \rightarrow E$ and $\sigma \rightarrow \overline{\sigma}$ are used:

$$Q\,d\Lambda \;=\; \frac{\partial f}{\partial \overline{\tau}_{ij}} \; \overline{E}_{ijpq} \; M^{-T}_{pqmn} \; d_{mn} \tag{6.48}$$

Then, one expands equation (6.44) by using appropriate transformation $\overline{d}' \rightarrow d'$ and $\overline{d}'' \rightarrow d''$ to obtain:

$$d\overline{\Lambda} \;=\; \frac{1}{Q} \, \frac{\partial f}{\partial \overline{\tau}_{k\ell}} \; \overline{E}_{k\ell mn} \left[\overset{\circ}{M}{}^{-T}_{mnpq} E^{-1}_{ijpq} \;+\; M^{-T}_{mnpq}\, d_{pq} \;-\; \left(M^{-T}_{mnpq} - \frac{d\overline{\Lambda}}{d\Lambda} M^{-1}_{mnpq} \right) d''_{pq} \right] \tag{6.49}$$

where the corotational derivative $\overset{\circ}{M}_{mnpq}$ is defined later in section 6.4.3. The last major step in the derivation is to substitute the term on the right-hand side of equation (6.48) for the last term of equation (6.49) and simplify the results using the transformations $\sigma \rightarrow \overline{\sigma}$, $\sigma \rightarrow \overline{\tau}$, $E \rightarrow \overline{E}$ and others. Once this is done, the following relation is obtained:

$$\left(\overline{Q} - \frac{\partial f}{\partial \overline{\tau}_{k\ell}} \overline{E}_{k\ell mn} \frac{\partial f}{\partial \overline{\sigma}_{mn}} \right) d\overline{\Lambda} \;=\; \left(Q - \frac{\partial f}{\partial \tau_{ij}} E_{ijpq} \frac{\partial f}{\partial \sigma_{pq}} \right) d\Lambda \;+\; \frac{\partial f}{\partial \overline{\tau}_{k\ell}} \overline{E}_{k\ell mn} \overset{\circ}{M}{}^{-T}_{mnpq} E^{-1}_{ijpq}\, \sigma_{ij} \tag{6.50}$$

The above equation is rewritten in the form:

$$a_1\, d\overline{\Lambda} \;=\; a_2\, d\Lambda \;+\; a_3 \tag{6.51}$$

where

$$a_1 \;=\; \overline{Q} \;-\; \frac{\partial f}{\partial \overline{\tau}_{k\ell}} \overline{E}_{k\ell mn} \frac{\partial f}{\partial \overline{\sigma}_{mn}} \tag{6.52a}$$

$$a_2 \;=\; Q \;-\; \frac{\partial f}{\partial \tau_{k\ell}} E_{k\ell mn} \frac{\partial f}{\partial \sigma_{mn}} \tag{6.52b}$$

$$a_3 \;=\; \frac{\partial f}{\partial \overline{\tau}_{k\ell}} \overline{E}_{k\ell mn} \overset{\circ}{M}{}^{-T}_{mnpq} E^{-1}_{ijpq}\, \sigma_{ij} \tag{6.52c}$$

It is noticed that a_1 and a_2 are the last two terms on the right-hand side of equations (6.45) and (6.47), respectively.

It should be noted that when the material undergoes only plastic deformation without damage, that is when the configurations C and $\overline{C}$ coincide, then $a_1 = a_2$ and, $a_3 = 0$ since M vanishes in this case, thus leading to $d\overline{\Lambda} \,/\, d\Lambda$.

The relation (6.51) is now substituted into equation (6.38) along with equation (6.36) to obtain the following nonlinear transformation equation for the plastic part of the spatial strain rate:

$$\overline{d}_{ij}'' = X_{ijk\ell} \, d_{k\ell}'' + Z_{ij} \tag{6.53}$$

where the tensors X and Z are given by:

$$X_{ijk\ell} = \frac{a_2}{a_1} \, M_{ijk\ell}^{-1} \tag{6.54}$$

$$Z_{ij} = 3 \, \frac{a_3}{a_1} \, M_{ijk\ell}^{-1} \, H_{k\ell mn} \, (\sigma_{mn} - \beta_{mn}) \tag{6.55}$$

The transformation equation (6.53) will be used later in the derivation of the constitutive equations.

6.3 The Damage Effect Tensor M

In this section, the nature of the damage effect tensor will be discussed for a general state of deformation and damage. Also, the explicit matrix representation of this tensor will be presented to be used in future applications. Considering the transformation in equation (6.1) along with equation (6.2), it is easily seen that the effective stress tensor $\overline{\sigma}$ is non-symmetric. Different methods of symmetrization are available and are presented later in detail in Chapter 11. The three proposed techniques [130] to symmetrize $\overline{\sigma}$ are shown below:

$$\overline{\sigma}_{ij} = \frac{\sigma_{ik} \, (\delta_{kj} - \varphi_{kj})^{-1} + (\delta_{i\ell} - \varphi_{i\ell})^{-1} \, \sigma_{\ell j}}{2} \tag{6.56a}$$

$$\overline{\sigma}_{ij}'' = (\delta_{ik} - \varphi_{ik})^{-1/2} \, \sigma_{k\ell} \, (\delta_{\ell j} - \varphi_{\ell j})^{-1/2} \tag{6.56b}$$

$$\sigma_{ij} = \frac{\overline{\sigma}_{ik}(\delta_{kj} - \varphi_{kj}) + (\delta_{i\ell} - \varphi_{i\ell})\overline{\sigma}_{\ell j}}{2} \tag{6.56c}$$

It can easily be shown that all the stresses given above are frame-indifferent. Upon examining equations (6.56), one can see that the first two symmetrization methods in equations (6.56a) and (6.56b) are explicit definitions for $\overline{\sigma}$. However, equation (6.56c) defines $\overline{\sigma}$ implicitly and one needs to solve a system of algebraic equations in this case to obtain the components of the effective stress tensor $\overline{\sigma}$.

In the following, we will adopt the first symmetrization technique appearing in equation (6.56a). The same procedures to be discussed below can be applied in a similar way to the other symmetrization methods. See Chapter 11 for more details. We will now use the symmetrization in equation (6.56a) to derive explicit matrix representations for the damage effect tensor M for a general state of deformation and damage.

In trying to carry out the procedure for a general state of deformation and damage, one is faced with many difficulties. First, the inverse of the matrix $[\delta_{ij} - \varphi_{ij}]$ is not readily derivable. In addition, when representing the stress tensor σ as a vector $[\sigma_{11}\ \sigma_{22}\ \sigma_{33}\ \sigma_{23}\ \sigma_{31}\ \sigma_{12}]^T$ it is difficult to write the matrix of M explicitly. However, with the aid of symbolic programming, this process was performed and the results are shown below.

For a general state of deformation and damage, the stress and damage tensors are represented by the matrices:

$$[\sigma_{ij}] = \begin{bmatrix} \sigma_{11} & \sigma_{12} & \sigma_{13} \\ \sigma_{12} & \sigma_{22} & \sigma_{23} \\ \sigma_{13} & \sigma_{23} & \sigma_{33} \end{bmatrix} \tag{6.57}$$

$$[\varphi_{ij}] = \begin{bmatrix} \varphi_{11} & \varphi_{12} & \varphi_{13} \\ \varphi_{12} & \varphi_{22} & \varphi_{23} \\ \varphi_{13} & \varphi_{23} & \varphi_{33} \end{bmatrix} \tag{6.58}$$

Using equation (6.58), one easily writes the matrix expression for the tensor $I - \varphi$ as:

124

$$[\delta_{ij} - \varphi_{ij}] = \begin{bmatrix} 1 - \varphi_{11} & -\varphi_{12} & -\varphi_{13} \\ -\varphi_{12} & 1 - \varphi_{22} & -\varphi_{23} \\ -\varphi_{13} & -\varphi_{23} & 1 - \varphi_{33} \end{bmatrix} \tag{6.59}$$

In order to find a closed form for the inverse of the matrix in equation (6.59), the authors used the symbolic manipulation program REDUCE. Therefore, one obtains:

$$[\delta_{ij} - \varphi_{ij}]^{-1} = \frac{1}{\nabla} \begin{bmatrix} \Psi_{22}\Psi_{33} - \varphi_{23}^2 & \varphi_{12}\varphi_{23} + \varphi_{12}\Psi_{33} & \varphi_{12}\varphi_{23} + \varphi_{13}\Psi_{22} \\ \varphi_{13}\varphi_{23} + \varphi_{12}\Psi_{33} & \Psi_{11}\Psi_{33} - \varphi_{13}^2 & \varphi_{12}\varphi_{13} + \varphi_{23}\Psi_{11} \\ \varphi_{12}\varphi_{23} + \varphi_{13}\Psi_{22} & \varphi_{12}\varphi_{13} + \varphi_{23}\Psi_{11} & \Psi_{11}\Psi_{22} - \varphi_{12}^2 \end{bmatrix} \tag{6.60}$$

where ∇ is given by:

$$\nabla = \Psi_{11}\Psi_{22}\Psi_{33} - \varphi_{23}^2\Psi_{11} - \varphi_{13}^2\Psi_{22} - \varphi_{12}^2\Psi_{33} - 2\varphi_{12}\varphi_{23}\varphi_{13} \tag{6.61}$$

where the notation Ψ_{ij} is used to denote $\delta_{ij} - \varphi_{ij}$.

Using the symmetrization procedure in equation (6.56a) along with equations (6.57) and (6.60), one obtains the matrix representation for the effective stress tensor $\overline{\sigma}$. Rewriting the resulting matrix of $\overline{\sigma}$ in vector form $[\sigma_{11} \quad \sigma_{22} \quad \sigma_{33} \quad \sigma_{12} \quad \sigma_{31} \quad \sigma_{23}]^T$ and re-arranging the terms, one finally obtains the explicit matrix representation for the damage effect tensor M as follows:

$$[M] = \frac{1}{2\nabla} \begin{bmatrix} 2\Psi_{22}\Psi_{33} - 2\varphi_{23}^2 & 0 & 0 \\ 0 & 2\Psi_{11}\Psi_{33} - 2\varphi_{13}^2 & 0 \\ 0 & 0 & 2\Psi_{11}\Psi_{22} - 2\varphi_{12}^2 \\ \varphi_{13}\varphi_{23} + \varphi_{12}\Psi_{33} & \varphi_{13}\varphi_{23} + \varphi_{12}\Psi_{33} & 0 \\ \varphi_{12}\varphi_{23} + \varphi_{13}\Psi_{22} & 0 & \varphi_{12}\varphi_{23} + \varphi_{13}\Psi_{22} \\ 0 & \varphi_{12}\varphi_{13} + \varphi_{23}\Psi_{11} & \varphi_{12}\varphi_{13} + \varphi_{23}\Psi_{11} \end{bmatrix}$$

$$
\left.
\begin{array}{ccc}
2\varphi_{13}\,\varphi_{23} + 2\varphi_{12}\,\psi_{33} & 2\varphi_{12}\,\varphi_{23} + 2\varphi_{13}\,\psi_{22} & 0 \\[4pt]
2\varphi_{13}\,\varphi_{23} + 2\varphi_{12}\,\psi_{33} & 0 & 2\varphi_{12}\,\varphi_{13} + 2\varphi_{23}\,\psi_{11} \\[4pt]
0 & 2\varphi_{12}\,\varphi_{23} + 2\varphi_{13}\,\psi_{22} & 2\varphi_{12}\,\varphi_{13} + 2\varphi_{23}\,\psi_{11} \\[4pt]
\psi_{22}\,\psi_{33} + \psi_{11}\,\psi_{33} - \varphi_{23}^{2} - \varphi_{13}^{2} & \varphi_{12}\,\varphi_{13} + \varphi_{23}\,\psi_{11} & \varphi_{12}\,\varphi_{23} + \varphi_{13}\,\psi_{22} \\[4pt]
\varphi_{12}\,\varphi_{13} + \varphi_{23}\,\psi_{11} & \psi_{22}\,\psi_{33} + \psi_{11}\,\psi_{22} - \varphi_{23}^{2} - \varphi_{12}^{2} & \varphi_{13}\,\varphi_{23} + \varphi_{12}\,\psi_{33} \\[4pt]
\varphi_{12}\,\varphi_{23} + \varphi_{13}\,\psi_{22} & \varphi_{13}\,\varphi_{23} + \varphi_{12}\,\psi_{33} & \psi_{11}\,\psi_{33} + \psi_{11}\,\psi_{22} - \varphi_{13}^{2} - \varphi_{12}^{2}
\end{array}
\right]
$$

$$\tag{6.62}$$

Using the principal damage variables φ_1, φ_2 and φ_3, one can easily see that the matrix in equation (6.62) reduces to the following diagonalized form:

$$
[M]_{diag.} =
\left[
\begin{array}{cccccc}
\dfrac{1}{1-\varphi_1} & 0 & 0 & 0 & 0 & 0 \\[10pt]
0 & \dfrac{1}{1-\varphi_2} & 0 & 0 & 0 & 0 \\[10pt]
0 & 0 & \dfrac{1}{1-\varphi_3} & 0 & 0 & 0 \\[10pt]
0 & 0 & 0 & \dfrac{(1-\varphi_3)+(1-\varphi_2)}{2\,(1-\varphi_3)\,(1-\varphi_2)} & 0 & 0 \\[14pt]
0 & 0 & 0 & 0 & \dfrac{(1-\varphi_3)+(1-\varphi_1)}{2\,(1-\varphi_3)\,(1-\varphi_1)} & 0 \\[14pt]
0 & 0 & 0 & 0 & 0 & \dfrac{(1-\varphi_2)+(1-\varphi_1)}{2\,(1-\varphi_2)\,(1-\varphi_1)}
\end{array}
\right]
$$

$$\tag{6.63}$$

126

The diagonalized matrix of M in equation (6.63) should be compared with the following diagonalized form proposed by Sidoroff [16] where the difference between the two forms is attributed to the symmetrization procedure used:

$$[M]_{diag.} = \begin{bmatrix} \dfrac{1}{1-\varphi_1} & 0 & 0 & 0 & 0 & 0 \\[2ex] 0 & \dfrac{1}{1-\varphi_2} & 0 & 0 & 0 & 0 \\[2ex] 0 & 0 & \dfrac{1}{1-\varphi_3} & 0 & 0 & 0 \\[2ex] 0 & 0 & 0 & \dfrac{1}{\sqrt{(1-\varphi_2)(1-\varphi_1)}} & 0 & 0 \\[2ex] 0 & 0 & 0 & 0 & \dfrac{1}{\sqrt{(1-\varphi_3)(1-\varphi_1)}} & 0 \\[2ex] 0 & 0 & 0 & 0 & 0 & \dfrac{1}{\sqrt{(1-\varphi_3)(1-\varphi_2)}} \end{bmatrix} \qquad (6.64)$$

where the stress and strain tensors are represented by the following notation, respectively:

$$[\sigma_1 \;\; \sigma_2 \;\; \sigma_3 \;\; \sigma_4 \;\; \sigma_5 \;\; \sigma_6]^T = [\sigma_{11} \;\; \sigma_{22} \;\; \sigma_{33} \;\; \sigma_{23} \;\; \sigma_{31} \;\; \sigma_{12}]^T$$

and

$$[\varepsilon_1 \;\; \varepsilon_2 \;\; \varepsilon_3 \;\; \varepsilon_4 \;\; \varepsilon_5 \;\; \varepsilon_6]^T = [\varepsilon_{11} \;\; \varepsilon_{22} \;\; \varepsilon_{33} \;\; \varepsilon_{23} \;\; \varepsilon_{31} \;\; \varepsilon_{12}]^T$$

The diagonalized matrix shown above is written in terms of the principal damage variables φ_1, φ_2 and φ_3. The appropriate transformation to any coordinate system can now be easily performed in order to obtain a more generalized matrix representation for M.

6.4 Constitutive Model

In this section, a coupled constitutive model will be derived incorporating both elasto-plasticity and damage. This section is divided into three subsections detailing the derivation starting with the equations of damage evolution then proceeding to the desired coupling.

6.4.1 Damage Evolution

In this section, an inelastic constitutive model is derived in conjunction with the damage transformation equations presented in the previous sections. An elasto-plastic stiffness tensor that involves damage effects is derived in the Eulerian reference system. In this formulation, rate-dependent effects are neglected and isothermal conditions are assumed. The damage evolution criterion to be used here is that proposed by Lee et al. [15] and is given by:

$$g\left(\overline{\sigma}, L\right) = \frac{1}{2} J_{ijk\ell} \, \overline{\sigma}_{ij} \, \overline{\sigma}_{k\ell} - \ell_o^2 - L(\ell) \equiv 0 \tag{6.65}$$

where $J_{ijk\ell}$ are the components of a constant fourth-rank tensor that is symmetric and isotropic. This tensor is represented by the following matrix [15]:

$$[J] = \begin{bmatrix} 1 & \mu & \mu & 0 & 0 & 0 \\ \mu & 1 & \mu & 0 & 0 & 0 \\ 0 & 0 & 0 & 2(1-\mu) & 0 & 0 \\ 0 & 0 & 0 & 0 & 2(1-\mu) & 0 \\ 0 & 0 & 0 & 0 & 0 & 2(1-\mu) \end{bmatrix} \tag{6.66}$$

where μ is a material constant satisfying $-1/2 \le \mu \le 1$. In equation (6.65), ℓ_o represents the initial damage threshold, $L(\ell)$ is the increment of damage threshold, and ℓ is a scalar variable that represents overall damage.

During the process of plastic deformation and damage, the power of dissipation Π is given by [109]:

128

$$\Pi = \sigma_{ij} d_{ij}'' + \sigma_{k\ell} \overset{\circ}{\varphi}_{k\ell} - L d\ell \tag{6.67}$$

In order to obtain the actual values of the parameters σ, φ, κ and ℓ, one needs to solve an extremization problem, i.e. the power of dissipation Π is to be extremized subject to two constraints, namely $f\ (\tau,\ \alpha,\ \kappa,\ \varphi) = 0$ and $g\ (\overline{\sigma},\ L) = 0$. Using the method of the calculus of functions of several variables, one introduces two Lagrange multipliers $d\lambda_1$ and $d\lambda_2$ and forms the function Ψ such that:

$$\Psi = \Pi - d\lambda_1 f - d\lambda_2 g \tag{6.68}$$

The problem now reduces to that of extremizing the function Ψ. Therefore, one uses the necessary conditions $\partial \Psi / \partial \sigma = 0$ and $\partial \Psi / \partial L = 0$ and obtains:

$$d_{ij}'' + \overset{\circ}{\varphi}_{ij} - d\lambda_1 \frac{\partial f}{\partial \sigma_{ij}} - d\lambda_2 \frac{\partial g}{\partial \sigma_{ij}} = 0 \tag{6.69a}$$

$$- d\ell - d\lambda_2 \frac{\partial g}{\partial L} = 0 \tag{6.69b}$$

Next, one obtains from equation (6.65) that $\partial g / \partial L = -1$. Substituting this into equation (6.69b), one obtains $d\lambda_2 = d\ell$. Thus $d\lambda_2$ describes the evolution of the overall damage parameter ℓ which is to be derived shortly. Using equation (6.69b) and assuming that damage and plastic deformation are two independent processes, one obtains the following two rate equations for the plastic strain and damage tensors:

$$d_{ij}'' = d\lambda_1 \frac{\partial f}{\partial \sigma_{ij}} \tag{6.70a}$$

$$\overset{\circ}{\varphi}_{ij} = d\ell \frac{\partial g}{\partial \sigma_{ij}} \tag{6.70b}$$

The first of equations (6.70) is the associated flow rule for the plastic strain introduced earlier in equation (6.34), while the second is the evolution of the damage tensor. It is to be noted that $d\lambda_1$ is exactly the same as the multiplier $d\Lambda$ used earlier. However, one needs to obtain explicit expressions for the multipliers $d\Lambda$ and $d\ell$. The derivation of an expression for $d\Lambda$ will be left for the next section when the inelastic constitutive model is discussed. Now one proceeds to derive an expression for $d\ell$. This is done by invoking the consistency condition $dg(\sigma, \varphi, L) = 0$. Therefore one obtains:

$$\frac{\partial g}{\partial \sigma_{ij}} \overset{\circ}{\sigma}_{ij} + \frac{\partial g}{\partial \varphi_{k\ell}} \overset{\circ}{\varphi}_{k\ell} + \frac{\partial g}{\partial L} dL = 0 \tag{6.71}$$

Substituting for $\overset{\circ}{\varphi}$ from equation (6.70b) along with $\partial g/\partial L = -1$ and $dL = d\ell \,(\partial L/\partial \ell)$, one obtains:

$$d\ell = \frac{\dfrac{\partial g}{\partial \sigma_{ij}} \overset{\circ}{\sigma}_{ij}}{\dfrac{\partial L}{\partial \ell} - \dfrac{\partial g}{\partial \varphi_{pq}} \dfrac{\partial g}{\partial \sigma_{pq}}} \tag{6.72}$$

Finally, by substituting equation (6.72) into equation (6.70b), one obtains the general evolution equation for the damage tensor φ as:

$$\overset{\circ}{\varphi}_{k\ell} = \frac{\dfrac{\partial g}{\partial \sigma_{ij}} \overset{\circ}{\sigma}_{ij}}{\dfrac{\partial L}{\partial \ell} - \dfrac{\partial g}{\partial \varphi_{pq}} \dfrac{\partial g}{\partial \sigma_{pq}}} \frac{\partial g}{\partial \sigma_{k\ell}} \tag{6.73}$$

The evolution equation (6.73) is to be incorporated in the constitutive model in the next two sections. It will also be used in the derivation of the effective elasto-plastic stiffness tensor. It should be noted that equation (6.73) is based on the damage criterion of equation (6.65) which is applicable to anisotropic damage. However, using the form for J given in equation (6.66) restricts the formulation to isotropy.

6.4.2 Plastic Deformation

In the analysis of finite strain plasticity one needs to define an appropriate corotational stress rate that is objective and frame-indifferent. Detailed discussions of these types of stress rates are available in the papers by Voyiadjis and Kattan [131], and Paulun and Pecherski [49]. The corotational stress rate to be adopted in this model is given for σ in the form:

$$\overset{\circ}{\sigma}_{ij} = d\sigma_{ij} - \Omega_{ip}\,\sigma_{pj} + \sigma_{iq}\,\Omega_{qj} \tag{6.74}$$

where the modified spin tensor Ω is given by [49,131]:

$$\Omega_{ij} = \omega\, W_{ij} \tag{6.75}$$

In equation (6.75), W is the material spin tensor (the antisymmetric part of the velocity gradient) and ω is an influence scalar function to be determined. The effect of ω on the evolution of the stress and backstress is discussed in detail in [131]. The corotational rate $\overline{\boldsymbol{\alpha}}$ has a similar expression as that in equation (6.74) keeping in mind that the modified spin tensor $\boldsymbol{\Omega}$ remains the same in both equations.

The yield function to be used in this model is the function f given by equation (6.30) with both isotropic and kinematic hardening. Isotropic hardening is described by the evolution of the plastic work as given earlier by equation (6.31), while kinematic hardening is given by equation (6.32). Most of the necessary plasticity equations were given in section 6.2.2 and the only thing remaining is the derivation of the constitutive equation.

By substituting for $d\overline{\Lambda}$ from equation (6.44) into equation (6.42), one derives the general inelastic constitutive equation in the configuration $\overline{C}$ as follows:

$$\overline{\sigma}_{ij} = \overline{D}_{ijk\ell}\, \overline{d}_{k\ell} \tag{6.76}$$

where the elasto-plastic stiffness tensor $\boldsymbol{D}$ is given by:

$$\overline{D}_{ijk\ell} = \overline{E}_{ijk\ell} - \frac{1}{\overline{Q}}\, \frac{\partial f}{\partial \overline{\tau}_{mn}}\, \overline{E}_{mnk\ell}\, \overline{E}_{ijpq}\, \frac{\partial f}{\partial \overline{\sigma}_{pq}} \tag{6.77}$$

The next step is to use the transformation equations developed in the previous sections in order to obtain a constitutive equation in the configuration C similar to that of equation (6.76).

6.4.3 Coupling of Damage and Plastic Deformation

In this section the transformation equations developed in sections 6.1 and 6.2 are used with the constitutive model of the previous section in order to transform the inelastic constitutive equation (6.76) in the configuration $\overline{C}$ to a general constitutive equation in the configuration C that accounts for both damage and plastic deformation.

Using equation (6.29) and taking its corotational derivative, one obtains the following transformation equation for $\overline{d}^{\,\prime}$:

$$\overline{d}^{\,\prime}_{ij} = \overset{\circ}{M}{}^{-T}_{ijmn}\,\varepsilon'_{mn} + M^{-T}_{ijk\ell}\,\overset{\circ}{d}{}^{\prime}_{k\ell} \qquad (6.78)$$

where $\overset{\circ}{M}{}^{-T}$ is obtained by taking the corotational derivative of the identity $M^{T}M^{-T} = I$ and noting that $\overset{\circ}{I} = 0$. Thus, one has:

$$\overset{\circ}{M}{}^{-T}_{k\ell mn} = -\,M^{-T}_{ijk\ell}\,\overset{\circ}{M}{}^{T}_{ijpq}\,M^{-T}_{pqmn} \qquad (6.79)$$

The corotational derivative $\overset{\circ}{M}$ is obtained by using the chain rule as follows:

$$\overset{\circ}{M}_{ijpq} = \frac{\partial M_{ijpq}}{\partial\varphi_{mn}}\,\overset{\circ}{\varphi}_{mn} \qquad (6.80a)$$

Alternatively, $\overset{\circ}{M}$ is given by the following Lie derivative derived by Oldroyd [132]:

$$\overset{\circ}{M}_{ijmn} = dM_{ijmn} - \Omega_{ip}\,M_{pjmn} - \Omega_{jq}\,M_{iqmn} - \Omega_{mr}\,M_{ijrn} - \Omega_{ns}\,M_{ijms} \qquad (6.80b)$$

The transformation equation (6.78) for the effective elastic strain rate tensor $\overline{d}^{\,\prime}$ represents a nonlinear relation unlike that of the effective stress tensor of equation (6.1). A similar nonlinear transformation equation (6.53) was previously derived for the effective plastic strain rate tensor d''. These two equations will now be used in the derivation of the constitutive model.

Now one is ready to derive the inelastic constitutive relation in the configuration C. Starting with the constitutive equation (6.76) and substituting for $\overline{\sigma}_{ij}$ and $\overline{d}_{k\ell}$ from equations (6.1) and (6.19), respectively, along with equations (6.78) and (6.53), one obtains:

$$\overset{\circ}{M}_{ijmn}\,\sigma_{mn} + M_{ijpq}\,\overset{\circ}{\sigma}_{pq} = D_{ijk\ell}\,(\overset{\circ}{M}{}^{-T}_{k\ell mn}\,d'_{mn} + M^{-T}_{k\ell pq}\,\varepsilon'_{pq} + X_{k\ell mn}\,d''_{mn} + Z_{k\ell}) \qquad (6.81)$$

Next, one substitutes for $\overset{\circ}{M}$ and $\overset{\circ}{M}{}^{-T}$ from equations (6.80) and (6.79), respectively, for d'' from equation (6.19) and for d' from a similar equation for (6.41), i.e., $d'_{ij} = E^{-1}_{ijk\ell}\,\overset{\circ}{\sigma}_{k\ell}$, into equation (6.81) the resulting expression is:

$$\frac{\partial M_{ijmn}}{\partial\varphi_{pq}}\,\overset{\circ}{\varphi}_{pq}\,\sigma_{mn} + M_{ijpq}\,\overset{\circ}{\sigma}_{pq} =$$

132

$$D_{ijk\ell} \left(M_{k\ell mn}^{-T} E_{pqmn}^{-1} \overset{\circ}{\sigma}_{pq} - M_{xyk\ell}^{-T} \frac{\partial M_{xyuv}^{T}}{\partial \varphi_{mn}} \overset{\circ}{\varphi}_{mn} M_{uvpq}^{-T} E_{cdpq}^{-1} \sigma_{cd} \right.$$

$$\left. + X_{k\ell mn} d_{mn} - X_{k\ell mn} E_{abmn}^{-1} \overset{\circ}{\sigma}_{ab} + Z_{k\ell} \right) \tag{6.82}$$

Finally, one substitutes for $\overset{\circ}{\varphi}$ from equation (6.73) into equation (6.82) and solves for $\overset{\circ}{\sigma}$ in terms of d. After several algebraic manipulations, one obtains the desired inelastic constitutive relation in the configuration C as:

$$\overset{\circ}{\sigma}_{ij} = D_{ijk\ell} d_{k\ell} + G_{ij} \tag{6.83}$$

where the effective elasto-plastic stiffness tensor D and the additional tensor G (comparable to the plastic relaxation stress introduced by Simo and Ju [133]) are given by:

$$D_{ijk\ell} = O_{pqij}^{-1} \overline{D}_{pqmn} X_{mnk\ell} \tag{6.84a}$$

$$G_{ij} = O_{pqij}^{-1} \overline{D}_{pqmn} Z_{mn} \tag{6.84b}$$

and the fourth-rank tensor O is given by:

$$O_{ijpq} = M_{ijpq} + \frac{\dfrac{\partial M_{ijmn}}{\partial \varphi_{uv}} \dfrac{\partial g}{\partial \sigma_{pq}} \dfrac{\partial g}{\partial \sigma_{uv}} \sigma_{mn}}{\dfrac{\partial L}{\partial \ell} - \dfrac{\partial g}{\partial \varphi_{xy}} \dfrac{\partial g}{\partial \sigma_{xy}}}$$

$$- \overline{D}_{ijk\ell} \left(M_{k\ell mn}^{-T} E_{pqmn}^{-1} - X_{k\ell mn} E_{pqmn}^{-1} \right.$$

$$\left. - M_{xyk\ell}^{-T} \frac{\partial M_{xyuv}^{T}}{\partial \varphi_{mn}} \frac{\dfrac{\partial g}{\partial \sigma_{pq}} \dfrac{\partial g}{\partial \sigma_{mn}}}{\dfrac{\partial L}{\partial \ell} - \dfrac{\partial g}{\partial \varphi_{ab}} \dfrac{\partial g}{\partial \sigma_{ab}}} M_{uvcd}^{-T} M_{rscd}^{-1} \sigma_{rs} \right) \tag{6.85}$$

The effective elasto-plastic stiffness tensor D in equation (6.84a) is the stiffness tensor including the effects of damage and plastic deformation. It is derived in the configuration C which is the actual current configuration of the deformed and damaged body. Equations (6.84) can now be used in finite element analysis. However, it should be noted that the constitutive relation in equation (6.83) represents a nonlinear transformation that makes the numerical implementation of this model impractical. This is due to the additional term G_{ij} which can be considered as some residual stress due to the damaging process. Nevertheless, the constitutive equation becomes linear provided that $G_{ij} = 0$. This is possible only when the term $(\sigma_{mn} - \beta_{mn}) N_{ijrr}$ vanishes as seen in equations (6.84b) and (6.55) and therefore

$$\overset{\circ}{\sigma}_{ij} = D_{ijk\ell}\, d_{k\ell} \tag{6.86}$$

Upon investigation of the nonlinear constitutive equation (6.83), it is seem that the extra term G_{ij} is due to the linear transformation of the effective stress $\boldsymbol{\sigma}$ and $\overline{\boldsymbol{\sigma}}$ in equation (6.1). It was shown in equation (6.5) that this transformation leads to a nonlinear relation between $\boldsymbol{\tau}$ and $\overline{\boldsymbol{\tau}}$. The authors have shown in a recent paper [98] that a linear constitutive equation similar to equation (6.86) can be obtained if a linear transformation is assumed between the deviatonic stresses $\boldsymbol{\tau}$ and $\overline{\boldsymbol{\tau}}$ in the form $\overline{\tau}_{ij} = M_{ijk\ell}\, \tau_{k\ell}$.

For completeness, one can obtain an identity that may be helpful in the numerical calculations. This is done by using the plastic volumetric incompressibility condition (which results directly from equation (6.35):

$$\overline{d}_{kk}^{\,\prime\prime} = 0 \tag{6.87}$$

in the configuration $\overline{C}$. Equation (6.87) is commonly used in metal plasticity without damage [49, 131]. Using equation (6.35) along with the condition (6.87), one obtains the useful identity

$$N_{rrk\ell}\,(\sigma_{k\ell} - \beta_{k\ell}) = 0 \tag{6.88}$$

Equation (6.88) is consistent with the previous conclusion of equation (6.8) since it was shown earlier that $N_{rrk\ell} = 0$.

In finite element calculations the critical state of damage is reached when the overall damage parameter ℓ reaches a critical value called ℓ_{cr} in at least one of the elements. This value determines the initiation of micro-cracks and other damaging defects. Alternatively, one can assign several critical values $\ell_{cr}^{(1)}$, $\ell_{cr}^{(2)}$, etc. for different damage effects. In order to determine these critical values, which may be considered as material parameters, a series of uniaxial extension tests are to be performed on tensile specimens and the stress-strain curves drawn.

In order to determine $\ell_{cr}^{(i)}$ (the value of ℓ at which damage initiation starts for a particular damage process "i"), the tensile specimen has to be sectioned at each load increment. The cross-section is to be examined for any cracks or cavities. The load step when cracks first appear in terms of the strain ε_1 is to recorded and compared with the graph of ℓ vs ε_1. The corresponding value of ℓ obtained in this way will be taken to be the critical value $\ell_{cr}^{(i)}$. This value is to be used in the finite element analysis of more complicated problems. For more details, see the papers by Chow and Wang [20] and Voyiadjis [55, 134].

6.5 Application to Void Growth: Gurson's Model

Gurson [121, 122] proposed a yield function $f\,(\boldsymbol{\sigma},\,v)$ for a porous solid with a randomly distributed volume fraction v of voids. This function was obtained based on an approximate analysis of spherical voids. Gurson's model was used later [123, 124] to study necking and failure of damaged solids. Tvergaard and Needleman [124] modified Gurson's yield function in order to account for rate sensitivity and necking instabilities in plastically deforming solids. The modified yield function is used here in the form (which includes kinematic hardening)

$$f = (\tau_{ij} - \alpha_{ij})(\tau_{ij} - \alpha_{ij}) + 2q_1\,\sigma_F^2\,v\,\cosh\left(\frac{\sigma_{kk}}{2\sigma_F}\right) - \sigma_F^2\,(1 + q_2\,v^2) = 0 \tag{6.89}$$

where σ_F is the yield strength of the matrix material and q_1 and q_2 are material parameters introduced by Tvergaard [123] to improve agreement between Gurson's model and other results. In equation (6.89), the variable v denotes the void volume fraction in the damaged material. In Gurson's model, damage is characterized by void growth only. The void growth is described by the rate of change of v given by [134]:

$$dv = (1 - v)\,d_{kk}^{\prime\prime} \tag{6.90}$$

In Gurson's model, it is assumed that the voids remain spherical in shape through the whole process of deformation and damage. The change of shape of voids, their coalescence and nucleation of new voids are ignored in the model. Equation (6.90) implies also that the plastic volumetric change, $d_{kk}^{\prime\prime}$ does not vanish for a material with voids.

In the following, it is shown how the proposed model outlined in the first sections of this chapter can be used to obtain the damage effect tensor M as applied to Gurson's yield function. It is also shown how certain expressions can be derived for the parameter q_1 and q_2 in a consistent manner. One first starts with the yield function f in the configuration $\overline{C}$ and transforms it to the configuration C. Therefore, using equation (6.30) in the form:

$$f = (\overline{\tau}_{mn} - \overline{\alpha}_{mn})(\overline{\tau}_{mn} - \overline{\alpha}_{mn}) - \frac{2}{3}\,\overline{\sigma}_o^2 \tag{6.91}$$

where the term $-\overline{c}\overline{\kappa}$ is dropped since isotropic hardening is not displayed by Gurson's function. Using the transformation equations (6.11), (6.17) and (6.18) and noting that $\sigma_F^2 = 2\overline{\sigma}_o^2/3$, equation (6.91) becomes:

$$f = H_{ijk\ell}(\sigma_{ij} - \beta_{ij})(\sigma_{k\ell} - \beta_{k\ell}) - \sigma_F^2 \tag{6.92}$$

It is noticed that equation (6.91) corresponds exactly to Gurson's function of equation (6.89) with $v = 0$. Using equation (6.3) and (6.13) to transform the total stresses in equation (6.92) into deviatoric stresses, one obtains:

$$f = H_{ijk\ell}(\tau_{ij} - \alpha_{ij})(\tau_{k\ell} - \alpha_{k\ell}) + \frac{1}{9}H_{mmnn}(\sigma_{pp} - \beta_{pp})^2 - \sigma_F^2 \tag{6.93}$$

Equation (6.93) represents the yield function f in the configuration C, which can now be compared with Gurson's yield function of equation (6.89). Thus, upon comparing equation (6.89) with equation (6.93), it is clear that the deviatoric parts of the two functions have to be equal. Therefore, one obtains:

$$H_{ijk\ell}(\tau_{ij} - \alpha_{ij})(\tau_{k\ell} - \alpha_{k\ell}) = (\tau_{rs} - \alpha_{rs})(\tau_{rs} - \alpha_{rs}) \tag{6.94}$$

On the other hand, upon equating the remaining parts of the two functions, one obtains:

$$\frac{1}{9} H_{mmnn} (\sigma_{pp} - \beta_{qq})^2 = 2 q_1 \sigma_F^2 \, v \, \cosh\left(\frac{\sigma_{kk}}{2\sigma_F} \right) - q_2 \sigma_F^2 v^2 \qquad (6.95)$$

The problem is now reduced to manipulating equations (6.94) and (6.95). Rewriting equation (6.94) in the form:

$$(H_{ijk\ell} - \delta_{ik} \delta_{j\ell})(\tau_{ij} - \alpha_{ij})(\tau_{k\ell} - \alpha_{k\ell}) = 0 \qquad (6.96)$$

One concludes that the tensor H is constant for Gurson's model and can be expressed by:

$$H_{ijk\ell} = \delta_{ik} \delta_{j\ell} \qquad (6.97)$$

It is clear that the deviatoric part of Gurson's yield function does not display any damage characteristics as given by equation (6.89). This is further supported by equation (6.97) where the damage effect tensor is independent of the damages variable φ. Upon considering equation (6.97), one obtains $H_{mmnn} = 3$. Substituting this into equation (6.95) yields:

$$\frac{1}{3}(\sigma_{pp} - \beta_{qq})^2 = \left[2 q_1 \cosh\left(\frac{\sigma_{kk}}{2\sigma_F} \right) - q_2 v \right] v \, \sigma_F^2 \qquad (6.98)$$

Equation (6.98) must be satisfied for a possible relationship between Gurson's model and the proposed model. Equation (6.98), as it stands, does not seem to merit an explicit relationship between the parameters q_1, q_2, and v. This is due to the presence of the "cosh" term on the right-hand side. Therefore, it is clear that one cannot proceed further without making some assumptions. In particular, two assumptions are to be employed. The first assumption is valid for small values of $\sigma_{kk}/2\sigma_F$, where the first two terms in the "cosh" series expansion are considered:

$$\cosh\left(\frac{\sigma_{kk}}{2\sigma_F} \right) = 1 + \frac{\sigma_{kk}^2}{8\sigma_F^2} \qquad (6.99)$$

The second assumption concerns the term β_{qq} which appears in equation (6.98). For the following to be valid, one needs to consider a modified Gurson yield function where the volumetric stress σ_{kk} is replaced by $(\sigma_{kk} - \beta_{kk})$. Therefore, upon incorporating the above two assumptions into equation

(6.98), one obtains:

$$\frac{1}{3}(\sigma_{kk} - \beta_{qq})^2 = \frac{1}{4} q_1 v (\sigma_{kk} - \beta_{qq})^2 + (2q_1 - q_2 v) v \sigma_F^2 \tag{6.100}$$

It is clear from equation (6.100) that the following two expressions for q_1 and q_2 in terms of v, need to be satisfied.:

$$q_1 = \frac{4}{3v} \qquad , \qquad q_2 = \frac{8}{3v^2} \tag{6.101}$$

The relations (6.101) represent variable expressions for the parameters q_1 and q_2 in terms of the void volume fraction, in contrast to the constant values that were suggested earlier by various authors. The relations (6.101) are consistently derived and although they are approximate, in the authors' opinion, they form a basis for further more sophisticated expressions. In addition, they are based on a solid derivation which cannot be said for the constant values that were used in the literature. Finally, one more important point that came up in the derivation needs to be considered. As it stands, Gurson's function of equation (6.89) cannot be related to the work presented here. It is a modified form of it containing the term $\cosh[(\sigma_{kk} - \beta_{qq})/2\sigma_F]$ instead of $\cosh(\sigma_{kk}/2\sigma_F)$ that is used in the derivation of the relations (6.101). The authors believe that this point should be pursued and the proposed modified Gurson function explored further. However, this may well be the subject of future research.

6.6 Effective Spin Tensor

In this section, a formal derivation is presented for the transformation equation of the modified spin tensor that is used in the corotational rate equations. In the configuration $\overline{C}$, the corotational derivative of the effective Cauchy stress tensor is given by:

$$\overset{\circ}{\overline{\sigma}}_{ij} = d\overline{\sigma}_{ij} - \overline{\Omega}_{ip} \overline{\sigma}_{pj} + \overline{\sigma}_{iq} \overline{\Omega}_{qj} \tag{6.102}$$

where $\overline{\Omega}$ is the effective modified spin tensor. The problem now reduces to finding a relation between Ω and $\overline{\Omega}$. One should keep in mind, however, that equation (6.102) is valid only when a Cartesian coordinate system is used. The same remark applies to equation (6.74) in the configuration C.

In order to derive the required relation, one first starts with the transformation equation (6.1). Taking the corotational derivative of this equation and rearranging the terms, one obtains:

$$\overset{\circ}{\sigma}_{k\ell} = M_{ijk\ell}^{-1} \, (\overset{\circ}{\overline{\sigma}}_{ij} - \overset{\circ}{M}_{ijmn} \, \sigma_{mn})$$

(6.103)

Substituting for $\overset{\circ}{\overline{\sigma}}_{ij}$ from equation (6.102) into equation (6.103) and using the material time derivative $d\sigma_{ij} = dM_{ijrs} \, \sigma_{rs} + M_{ijpq} \, d\sigma_{pq}$, one obtains:

$$\overset{\circ}{\sigma}_{k\ell} = M_{ijk\ell}^{-1} \, dM_{ijrs} \, \sigma_{rs} + d\sigma_{k\ell} - M_{ijk\ell}^{-1} \, \overline{\Omega}_{ip} \, M_{pjab} \, \sigma_{ab} + M_{ijk\ell}^{-1} \, M_{iqcd} \, \sigma_{cd} \, \overline{\Omega}_{qj} - M_{ijk\ell}^{-1} \, M_{ijmn} \, \sigma_{mn}$$

(6.104)

Comparing the two corotational derivatives appearing in equations (6.74) and (6.104), and after some tedious algebraic manipulations, one can finally obtain a relation between Ω and $\overline{\Omega}$ in the form:

$$\overline{\Omega}_{mn} = A_{mnpq} \, \Omega_{pq} + B_{mn}$$

(6.105)

where the tensors A and B are given by:

$$A_{mnpq} = - C_{mnk\ell}^{-1} \, (\delta_{kp} \, \sigma_{q\ell} - \delta_{\ell q} \, \sigma_{kp})$$

(6.106a)

$$B_{mn} = - C_{mnk\ell}^{-1} \, M_{pqk\ell}^{-1} \, (dM_{pqmn} - \overset{\circ}{M}_{pqmn}) \, \sigma_{mn}$$

(6.106b)

and the tensor C is given by:

$$C_{k\ell ip} = (M_{qpk\ell}^{-1} \, M_{qicd} - M_{ijk\ell}^{-1} \, M_{pqcd}) \, \sigma_{cd}$$

(6.106c)

With the availability of the transformation equation of the spin tensor, the theory presented in this chapter is now complete. Next, one investigates the applicability of the proposed theory to ductile fracture.

6.7 Application to Ductile Fracture - Example

A center-cracked thin plate as shown in Figure 6.2 in analyzed. The plate is subjected to uniaxial tension in the y-direction. The material used is aluminum alloy 2024 T3 (E = 73,087 MPa,

$\nu = 0.3$) with both the kinematic and isotropic hardening parameters of b = 275.8 MPa and c = 792.9 MPa, respectively. Initial yielding is characterized by $\sigma_y/\sqrt{3}$ = 226.8 MPa. Since the thickness of the plate (t = 3.175 mm) is small compared with other dimensions, a state of plane stress is assumed.

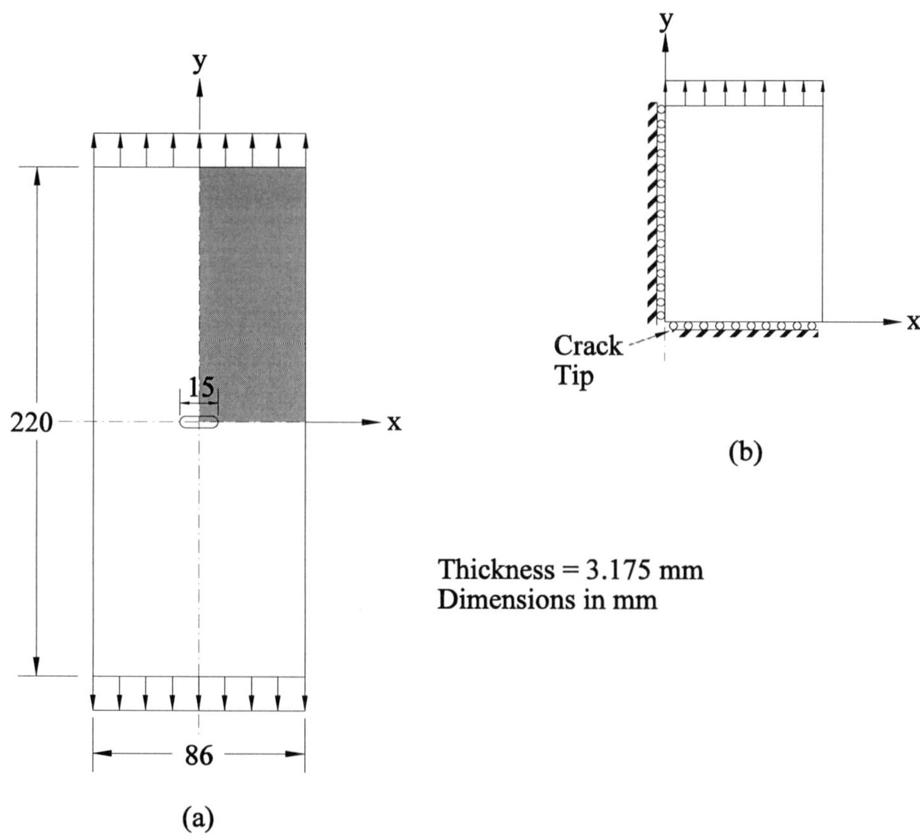

Figure 6.2   (a) Thin Plate with a Center Crack, (b) Quarter of Plate to be Discretized by Finite Elements. (Aluminum Alloy 2024-T3)

140

Since the plate geometry and loading are symmetrical, only one-quarter of the plate is discretized by finite elements as shown in Figure 6.3. The same optimum finite element mesh used in Chapter 5 is utilized again here. Eight node isoparametric quadrilateral elements are used in the finite element grid. Again, we avoid the use of singularity elements around the crack tip by using a large number of regular elements at that point. The total number of elements used is 381 with 1228 nodes.

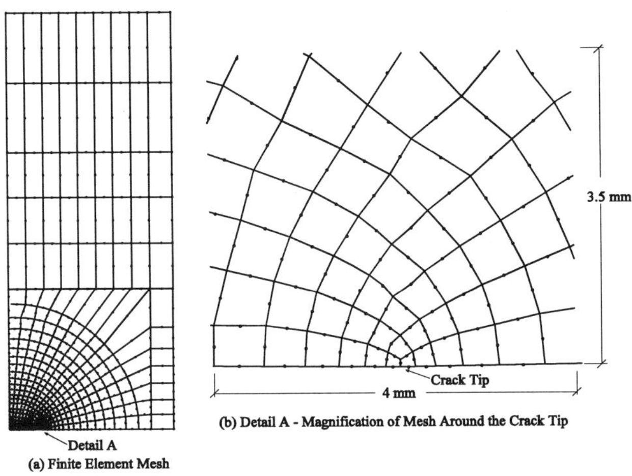

Figure 6.3   Optimum Finite Element Mesh Around the Crack Tip.

A load increment of 10 MPa is used until a total load of 300 MPa is reached. The results of the finite element analysis are shown here for a value of $\mu = 0.4$ afer the 27th load increment. It is noticed that convergence is achieved in less than 20 iterations for each load increment.

In Figure 6.4, the development of the plastic zone is shown around the crack tip. The results are shown after the 8, 14, 22 and 30th increments of load in Figure 6.4(a) - 6.4(d), respectively. The distribution of the axial strain ε_{yy} is shown after the 27th load increment in Figure 6.5. A band of very close contours around the crack tip indicates high axial strain gradients in that area. It is noticed that we obtain smaller values for ε_{yy} when using the coupled damage model. The maximum axial strain ε_{yy} obtained when using the coupled theory is 0.04355 compared to 0.05489 when the plasticity model is used.

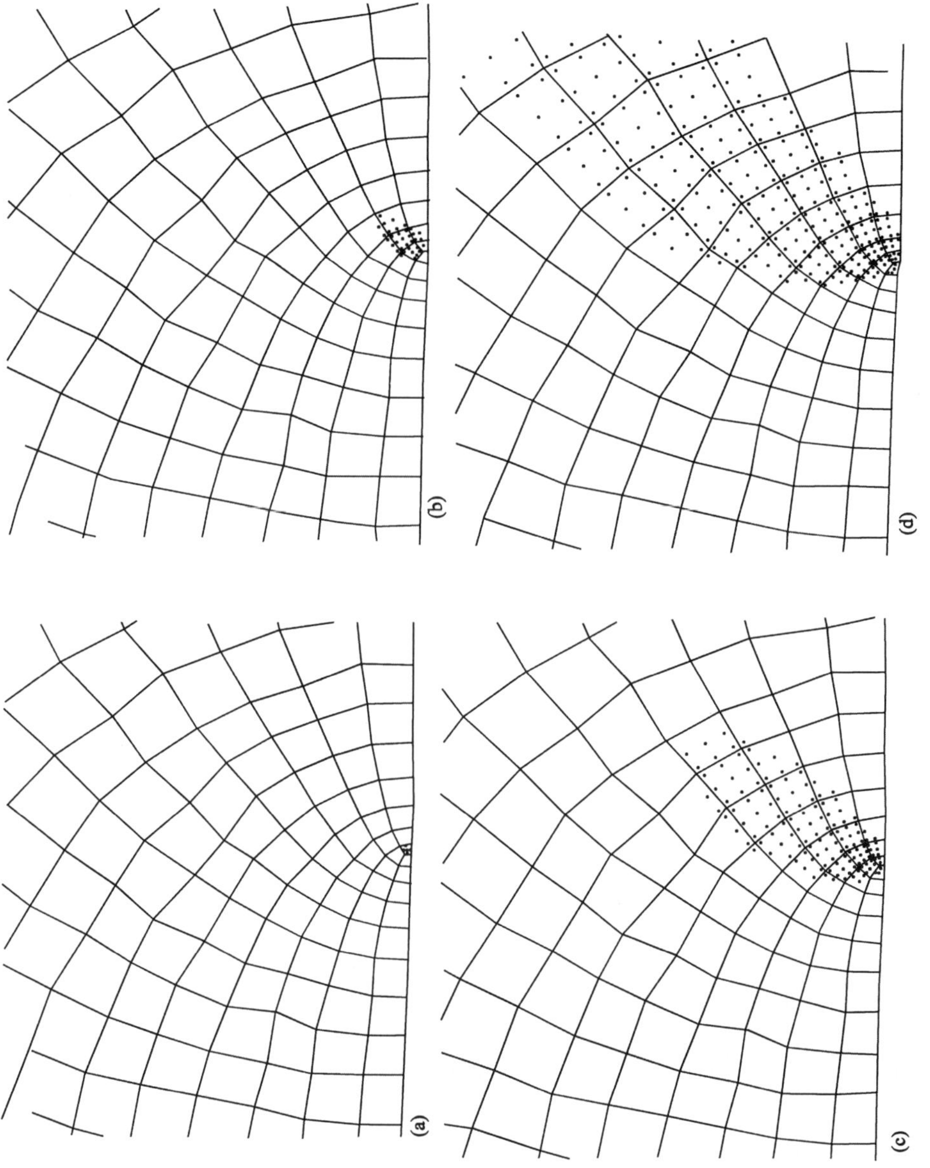

Figure 6.4 Development of Plastic Zone for the Damaged Model

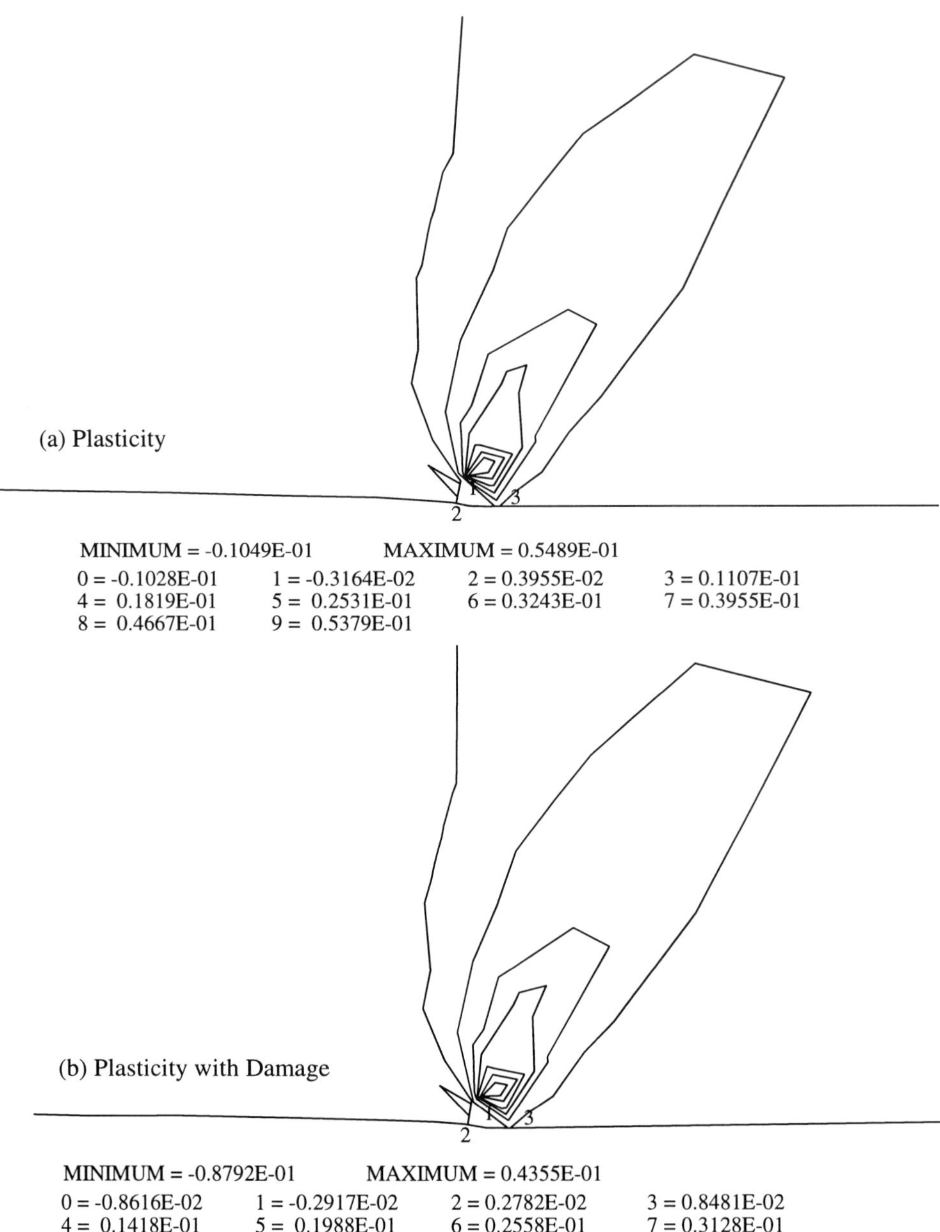

Figure 6.5   Distribution of ε_{yy} Around the Crack Tip

The distributions of the normal Cauchy stress σ_{yy} and the volumetric Cauchy stress σ_v are shown in Figures 6.6 and 6.7, respectively, where $\sigma_v = \sigma_{xx} + \sigma_{yy}$. Again it is noticed that smaller values for the stresses are obtained when using the damage theory. The maximum and minimum values for σ_{xx} are 1835 MPa and -626 MPa, respectively, compared to 1830 MPa and -572 MPa when the plasticity model is used.

The shear stress contours are shown in Figure 6.8. It is noticed that the damage model has a slight increase in σ_{xy}. However, the order of the stress σ_{xy} remains the same when using the two models.

Finally, the contours of plastic work κ are shown in Figure 6.9. These contours clearly show the development of the plastic zone around the crack tip. This result is shown after the 9, 18 and 27th increments of load. The maximum amount of plastic work is reduced due to the inclusion of the damage parameters in the constitutive equations. We notice also how the plastic work decreases in magnitude as one moves away from the crack tip.

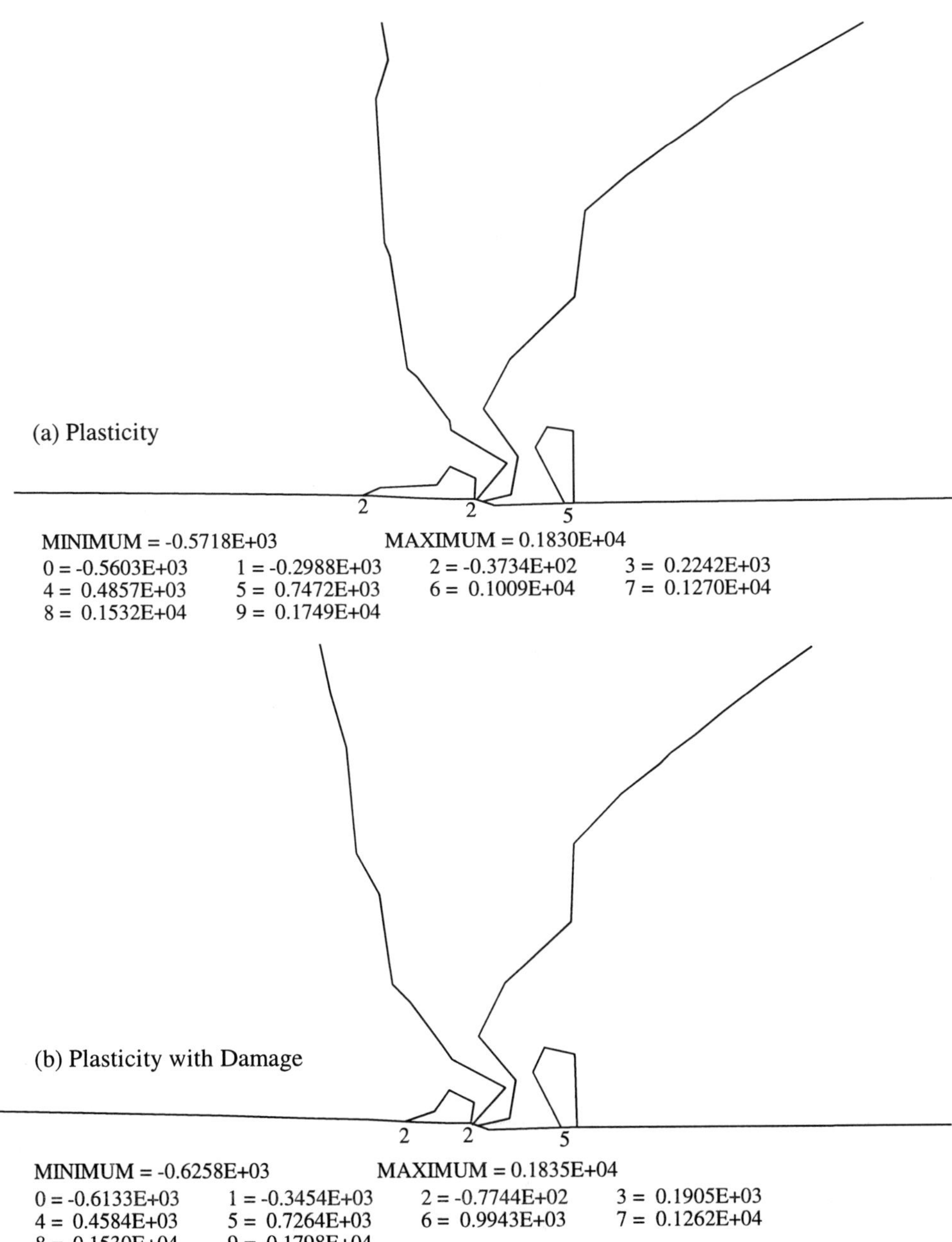

Figure 6.6   Distribution of σ_{yy} Around the Crack Tip

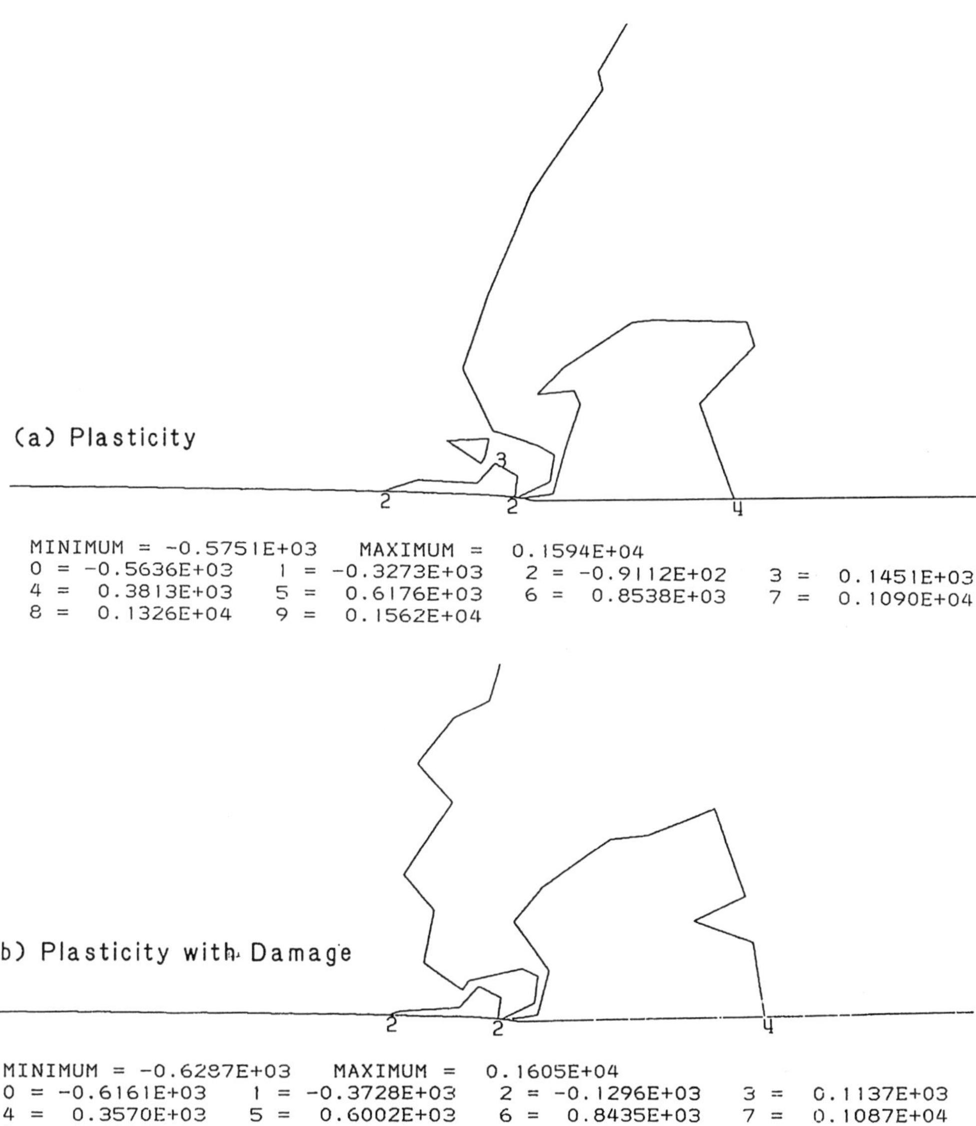

Figure 6.7   Distribution of σ_v around the Crack Tip.

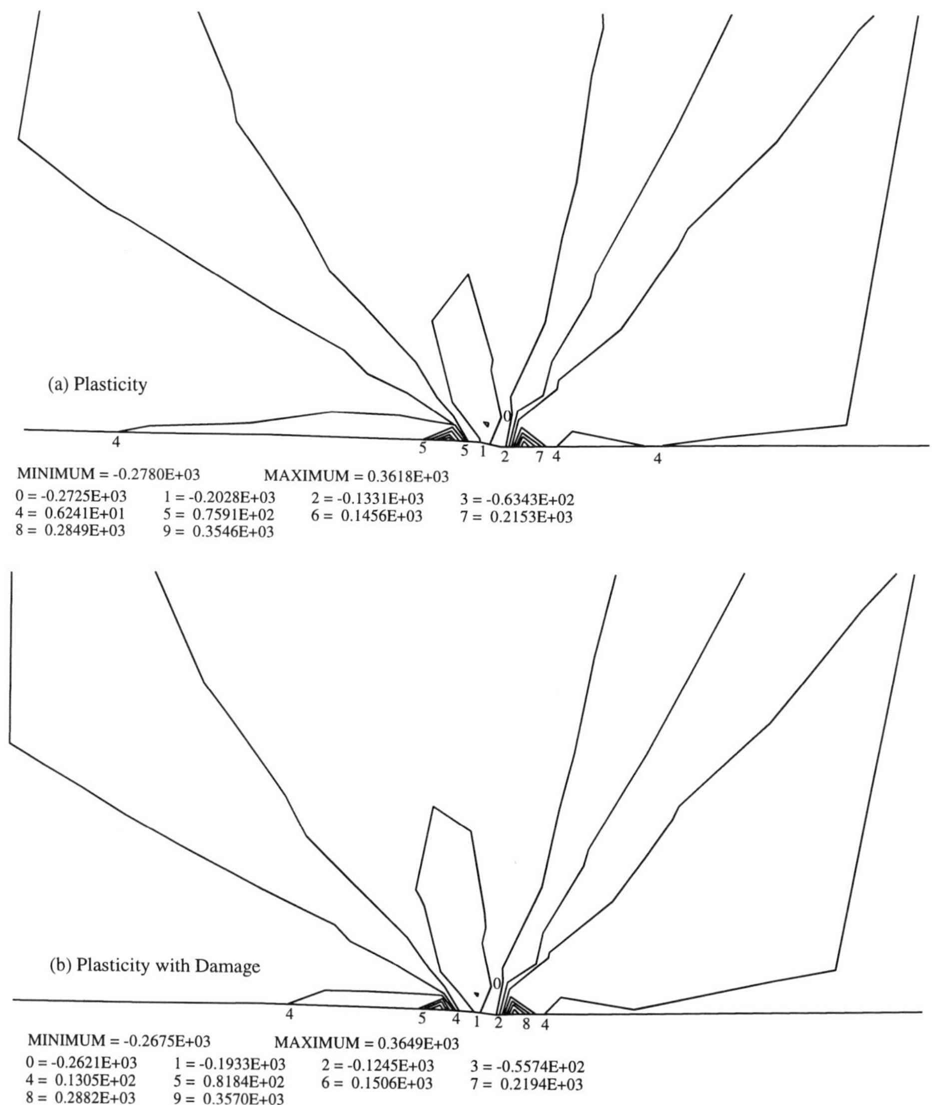

Figure 6.8   Distribution of σ_{xy} Around the Crack Tip

PLASTICITY

(a)

1   6   1

MINIMUM = 0.0000E+00   MAXIMUM = 0.1740E+01

0 = 0.0000E+00	1 = 0.1894E+00	2 = 0.3789E+00	3 = 0.5683E+00
4 = 0.7578E+00	5 = 0.9472E+00	6 = 0.1137E+01	7 = 0.1326E+01
8 = 0.1516E+01	9 = 0.1705E+01		

(b)

1   7   4   1

MINIMUM = -0.6646E+00   MAXIMUM = 0.1122E+02

0 = -0.6513E+00	1 = 0.6423E+00	2 = 0.1936E+01	3 = 0.3229E+01
4 = 0.4523E+01	5 = 0.5817E+01	6 = 0.7110E+01	7 = 0.8404E+01
8 = 0.9697E+01	9 = 0.1099E+02		

(c)

1   6   4   1

MINIMUM = -0.7827E+00   MAXIMUM = 0.3139E+02

0 = -0.7670E+00	1 = 0.2736E+01	2 = 0.6240E+01	3 = 0.9744E+01
4 = 0.1325E+02	5 = 0.1675E+02	6 = 0.2025E+02	7 = 0.2376E+02
8 = 0.2726E+02	9 = 0.3076E+02		

Figure 6.9   Distribution of Plastic Work κ Around the Crack Tip

PLASTICITY WITH DAMAGE

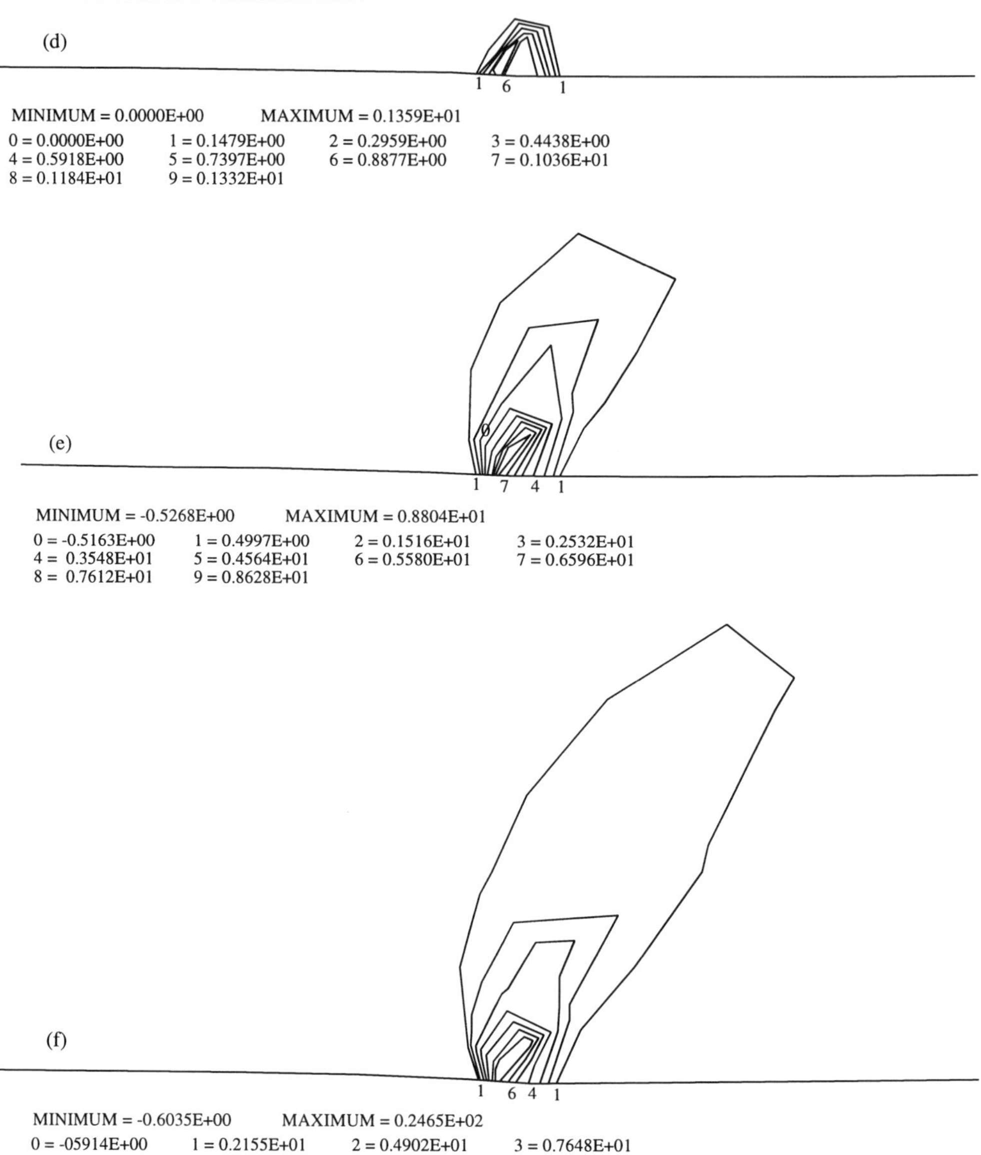

Figure 6.9 (continued)   Distribution of Plastic Work κ Around the Crack Tip

It is clear from the above results that the coupled model of damage and finite plasticity had resulted in smaller values for the strains, normal stresses and the plastic work done. Although slightly higher values for the shear stresses are obtained, the order of these values remains the same.

The critical value for the overall damage parameter ℓ is taken to be 0.115. This value is obtained experimentally from uniaxial tests performed by Chow and Wang [135]. The corresponding value for the load causing crack initiation is determined here to be 232 MPa. It should be noted that this value is dependent upon the appropriate choice of the constant μ. This is compared with the crack initiation load of 263.3 MPa obtained from experiments performed by Chow and Wang [135]. It should be mentioned that the material properties used in the experiments [135] are not specified. The discrepancy in the results can be explained by considering the material properties used in the finite element analysis. It is the authors' conviction that by using the appropriate material properties, one can obtain a better correspondence between the results.

In order to ascertain the accuracy of the finite element solution an alternate finite element mesh is used as shown in figure 6.10. This mesh comprises of 836 eight model isoparametric quadrilateral elements with a total of 2629 nodes. The results obtained for this mesh are shown in Figures 6.11 - 6.16. The results obtained from the two meshes are identical and confirm the accuracy of the solution.

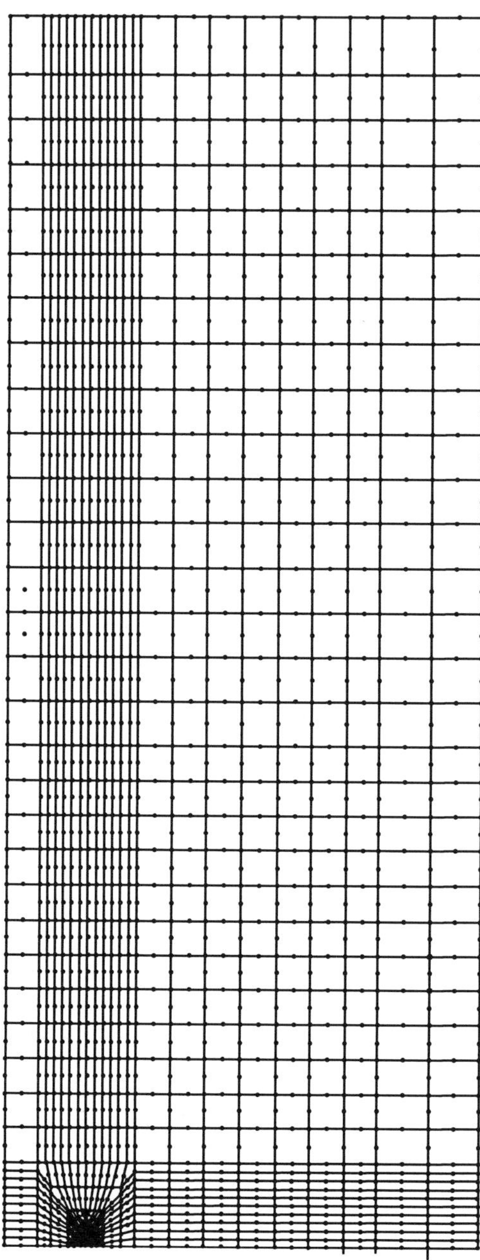

Figure 6.10   Alternate Finite Element Mesh

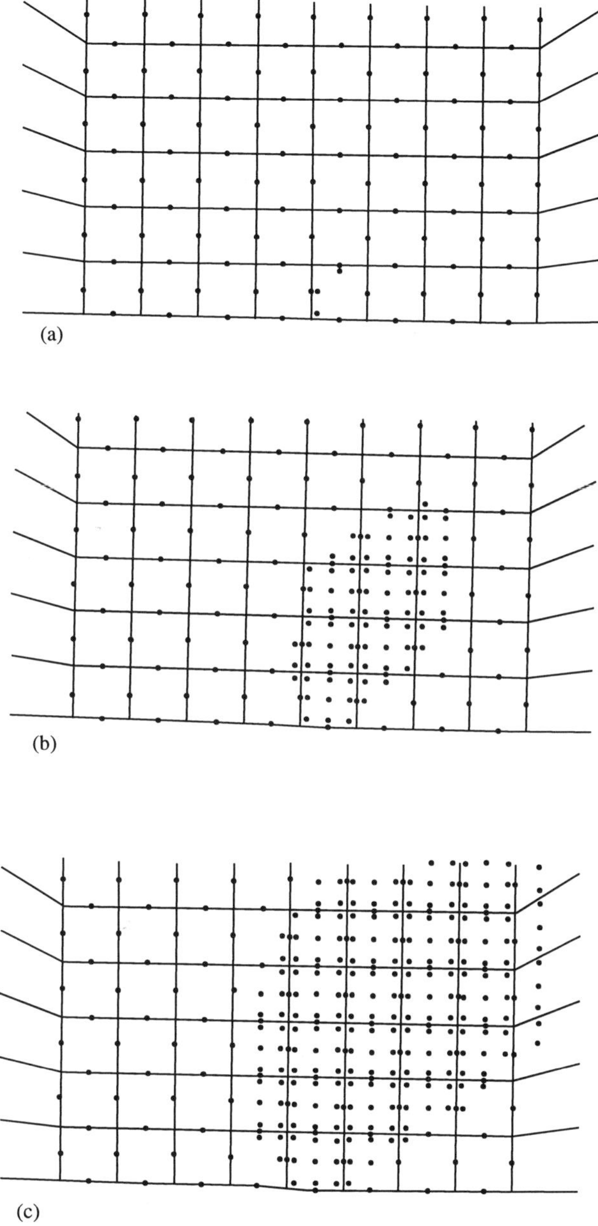

Figure 6.11   Development of Plastic Zone for the Damaged Model Alternate Mesh

152

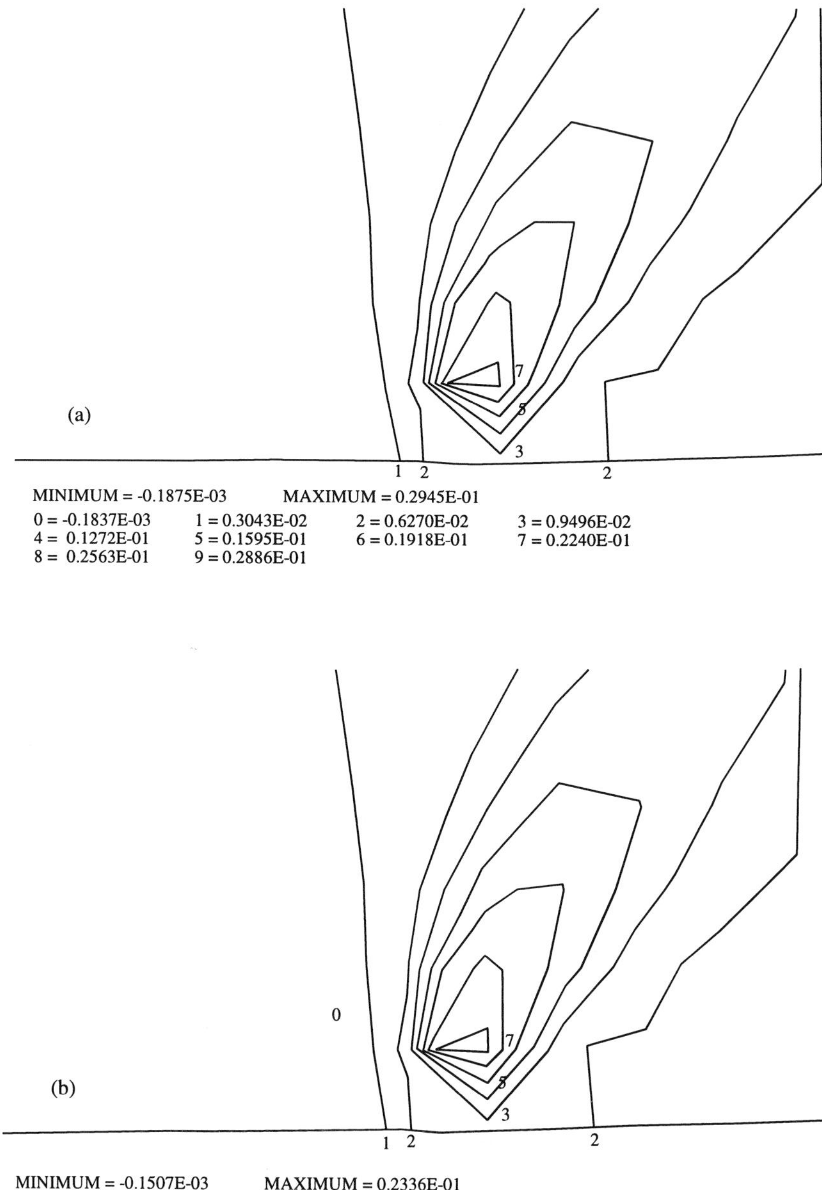

Figure 6.12   Distribution of ε_{yy} Around the Crack Tip
Alternate Mesh, (a) Plasticity, (b) Plasticity with Damage

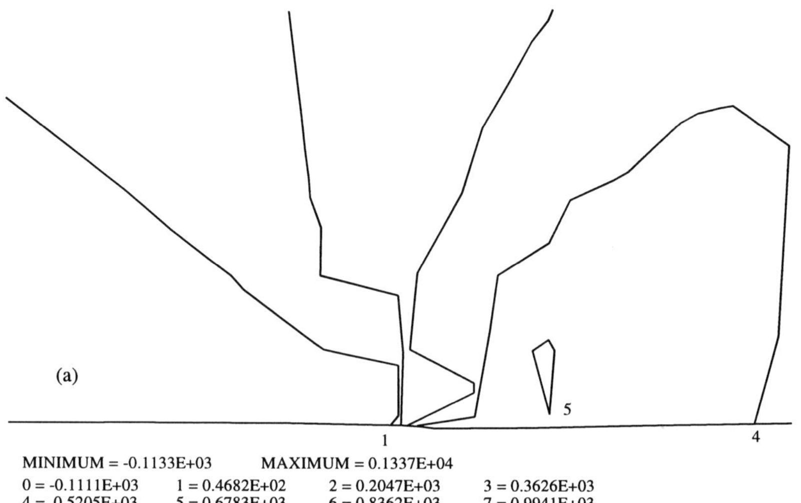

(a)

MINIMUM = -0.1133E+03   MAXIMUM = 0.1337E+04

0 = -0.1111E+03	1 = 0.4682E+02	2 = 0.2047E+03	3 = 0.3626E+03
4 = 0.5205E+03	5 = 0.6783E+03	6 = 0.8362E+03	7 = 0.9941E+03
8 = 0.1152E+04	9 = 0.1310E+04		

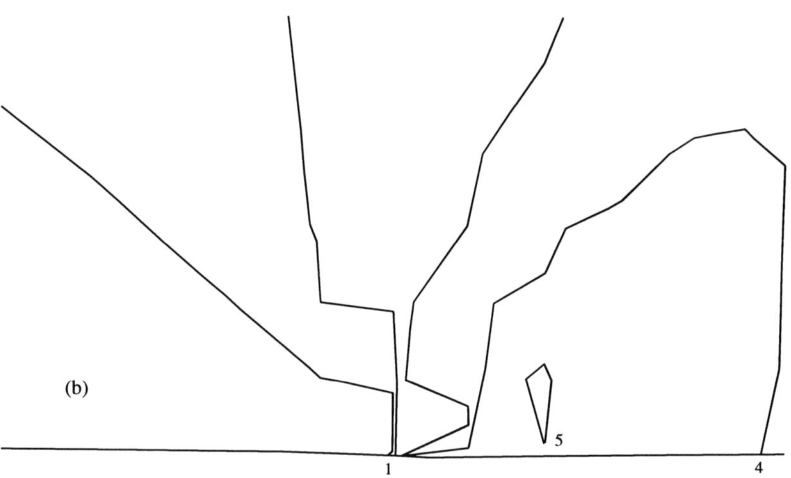

(b)

MINIMUM = -0.1078E+03   MAXIMUM = 0.1323E+04

0 = -0.1057E+03	1 = 0.5018E+02	2 = 0.2060E+03	3 = 0.3619E+03
4 = 0.5177E+03	5 = 0.6735E+03	6 = 0.8294E+03	7 = 0.9852E+03
8 = 0.1141E+04	9 = 0.1297E+04		

Figure 6.13   Distribution of σ_{yy} Around the Crack Tip
Alternate Mesh, (a) Plasticity, (b) Plasticity with Damage

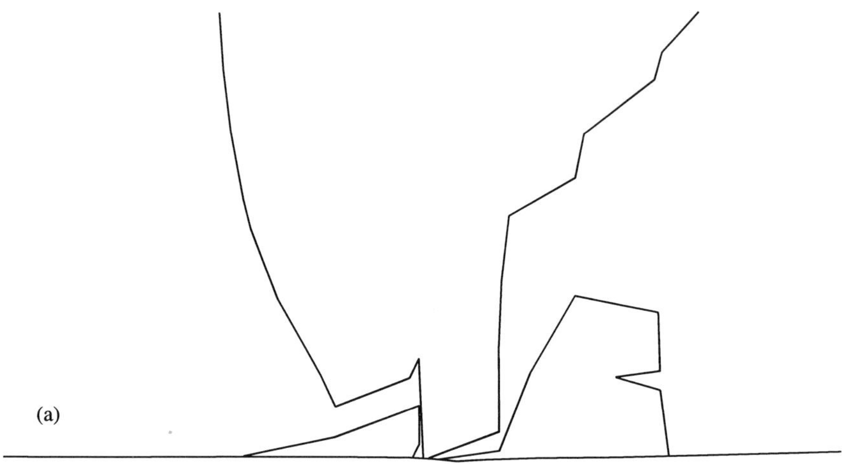

MAXIMUM = -0.2494E+03   MINIMUM = 0.1133E+04

0 = -0.2444E+03	1 = -0.9320E+02	2 = 0.5798E+02	3 = 0.2092E+03
4 = 0.3603E+03	5 = 0.5115E+03	6 = 0.6627E+03	7 = 0.8139E+03
8 = 0.9650E+03	9 = 0.1116E+04		

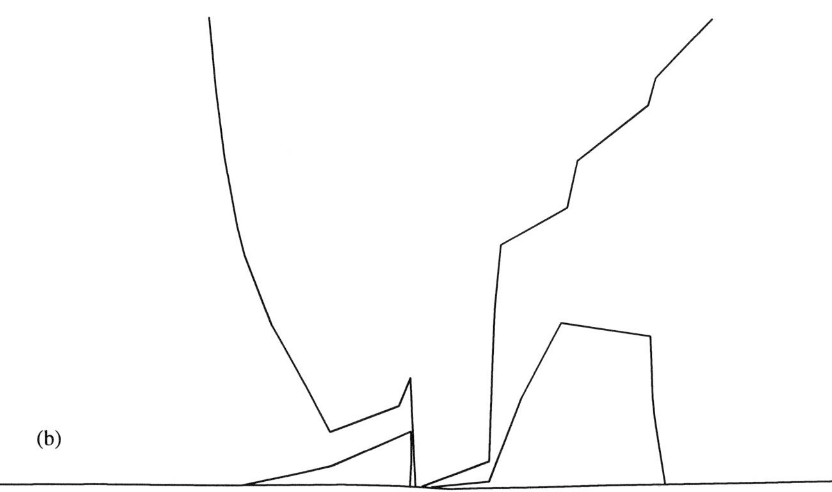

MAXIMUM = -0.2494E+03   MINIMUM = 0.1133E+04

0 = -0.2444E+03	1 = -0.9384E+02	2 = 0.5674E+02	3 = 0.2073E+03
4 = 0.3579E+03	5 = 0.5085E+03	6 = 0.6591E+03	7 = 0.8096E+03
8 = 0.9602E+03	9 = 0.1111E+04		

Figure 6.14   Distribution of σ_y Around the Crack Tip
Alternate Mesh, (a) Plasticity, (b) Plasticity with Damage

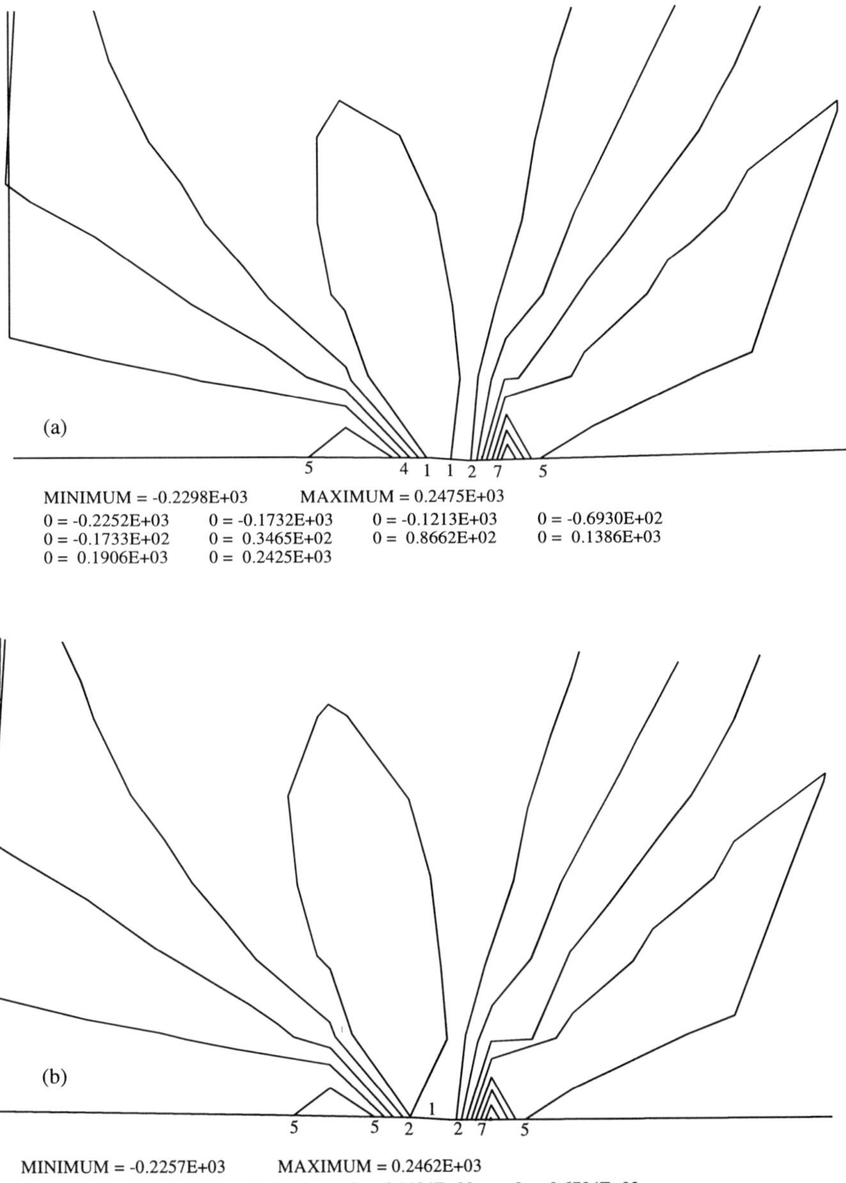

Figure 6.15   Distribution of σ_{xy} Around the Crack Tip
Alternate Mesh, (a) Plasticity, (b) Plasticity with Damage

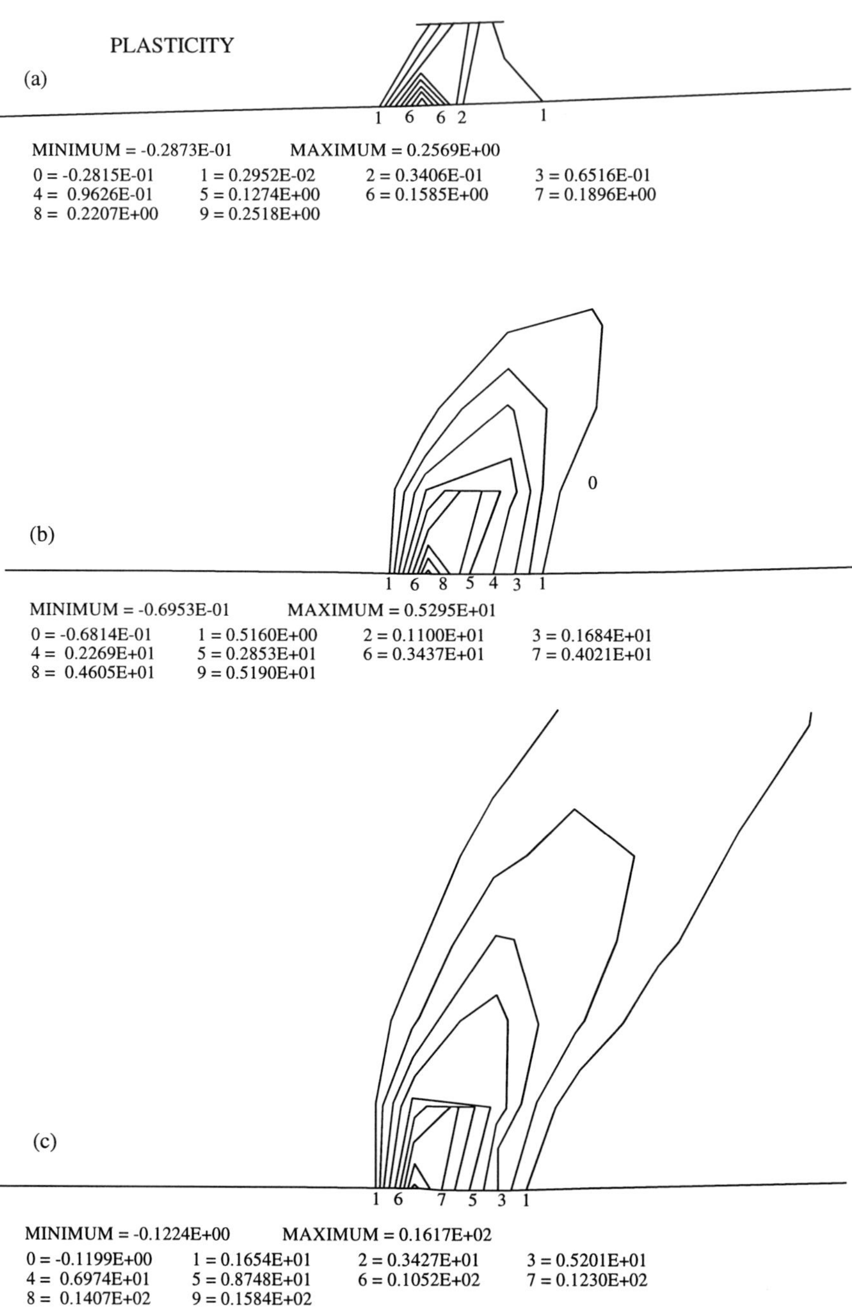

Figure 6.16 Distribution of Plastic Work κ Around the Crack Tip Alternate Mesh

PLASTICITY WITH DAMAGE

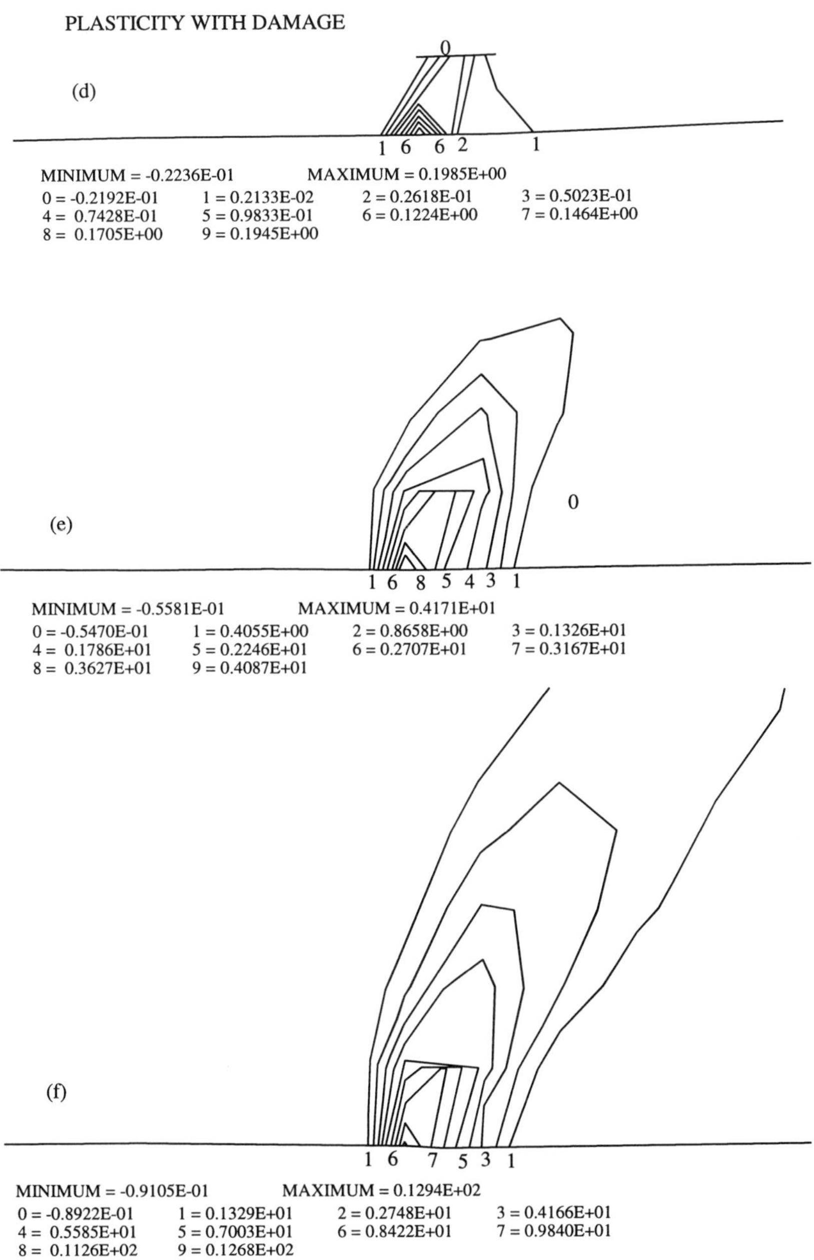

Figure 6.16 (continued)   Distribution of Plastic Work κ Around the Crack Tip Alternate Mesh

CHAPTER 7

METAL MATRIX COMPOSITES - OVERALL APPROACH

The concepts of damage mechanics are used with a micromechanical composite model to analyze damage and elasto-plastic deformation in fibrous composite materials with a ductile matrix. The main objective is to introduce appropriate expressions for the yield function and hardening rule in the damaged state in terms of an overall damage tensor M. The damage tensor M is assumed here to represent all types of damage that the composite system undergoes, such as nucleation and growth of voids in the matrix, fracture of fibers, debonding, and delamination. The end result of this approach is to introduce a new class of yield criteria, flow and hardening rules for the damaged composite system that can be used directly in the expression of the stiffness tensor to be used in numerical applications of the proposed model.

7.1 Preliminaries

A body of fiber-reinforced composite material is considered with elastic fibers and an elasto-plastic matrix. The fibers are assumed to be continuous and aligned. In the derivation, the strains and deformations are assumed to be small. In particular, the elastic strains are small compared with the plastic strain. Therefore, the existence of an elastic strain energy function can be assumed such that a linear relation can be used between the Cauchy stress tensor σ and the engineering elastic strain tensor ε'. The engineering elastic strain tensor ε' can be considered here instead of the elastic strain tensor due to the assumption of small deformations in general and small elastic strains in particular.

Let C_o be the initial undamaged and undeformed configuration of the composite body, and let C be the damaged and deformed configuration after the body is subjected to a set of external agencies. Consider a fictitious configuration of the body $\overline{C}$ that is obtained from C by removing all the damage that the composite body has undergone during the process of deformation. The fictitious configuration $\overline{C}$ is called the effective configuration because it is based on the effective stress concept that was originally proposed by Kachanov [1]. In fact, $\overline{C}$ is the configuration of the body after it had only deformed without damage.

Because the composite body considered here consists of matrix and fibers only, one denotes the matrix and fiber subconfigurations of C by C^M and C^F, respectively, where $C^M \subset C$, $C^F \subset C$, and

160

$C^M \cup C^F = C$. The corresponding matrix and fiber subconfigurations of $\overline{C}$ are denoted by $\overline{C}^M$ and $\overline{C}^F$, respectively, where $\overline{C}^M \subset \overline{C}$, $\overline{C}^F \subset \overline{C}$, and $\overline{C}^M \cup \overline{C}^F = \overline{C}$. Finally, all the subconfigurations C^M, C^F, $\overline{C}^M$ and $\overline{C}^F$ are assumed nonempty.

In the following formulation, the quantities are defined in the configuration C using spatial coordinates. Quantities that are based on the fictitious configuration $\overline{C}$ are denoted by a superposed bar. Matrix and fiber related quantities are denoted by a superscript M or F, appropriately. Finally, barred quantities with a superscript M or F refer to matrix or fiber related quantities that are defined in the fictitious configuration $\overline{C}$. For example, the Cauchy stress in C is denoted by σ, the effective Cauchy stress is $\overline{\sigma}$ in $\overline{C}$, the matrix and fiber stresses are σ^M and σ^F, respectively, in C^M and C^F and their effective counterparts are $\overline{\sigma}^M$ and $\overline{\sigma}^F$ in $\overline{C}^M$ and $\overline{C}^F$, respectively. Only Cartesian tensors are used in this work. The tensor components are denoted by lowercase Latin subscripts with the usual summation convention. Furthermore, no summation is assumed between a superscript and the corresponding identical subscript primarily because superscripts do not identify tensors but only the appropriate constituent of the composite.

The relations between the local (matrix and/or fibers) quantities and the overall (composite) quantities are initially cast in the configuration $\overline{C}$ because no damage effects are considered at this step. The initial analysis is based on the micromechanical composite model of Dvorak and Bahei-El-Din [68-70] and Bahei-El-Din and Dvorak [97]. In the configuration $\overline{C}$, the overall effective stress $\overline{\sigma}$ can be written in terms of the local effective stresses $\overline{\sigma}^M$ and $\overline{\sigma}^F$ as follows:

$$\overline{\sigma}_{ij} = \overline{c}^M \overline{\sigma}_{ij}^M + \overline{c}^F \overline{\sigma}_{ij}^F \tag{7.1}$$

where $\overline{c}^M$ and $\overline{c}^F$ are the matrix and fiber volume fractions given by $\overline{V}^M/\overline{V}$ and $\overline{V}^F/\overline{V}$, respectively. $\overline{V}^M$, $\overline{V}^F$, and $\overline{V}$ are the matrix, fiber, and overall volumes, where $\overline{V}^M + \overline{V}^F = \overline{V}$.

The local-overall relations for the stress tensor are assumed here for the matrix and fibers in the fictitious local and overall configurations as follows:

$$\overline{\sigma}_{ij}^M = \overline{B}_{ijk\ell}^M \, \overline{\sigma}_{k\ell} \tag{7.2}$$

$$\overline{\sigma}_{ij}^F = \overline{B}_{ijk\ell}^{FE} \, \overline{\sigma}_{k\ell} \tag{7.3}$$

and $\overline{B}_{ijk\ell}^M$ and $\overline{B}_{ijk\ell}^{FE}$ are the components of the fourth-rank plastic matrix and elastic fiber stress concentration factors, respectively. In the case of unloading or elastic loading, the tensor $\overline{B}_{ijk\ell}^{M}$ is replaced by the elastic matrix stress concentration factor $\overline{B}_{ijk\ell}^{ME}$. It follows from symmetry of the stress tensor and equation (7.2) that the concentration factor $\overline{B}^{M}$ is symmetric in the sense $\overline{B}_{ijk\ell}^{M} = \overline{B}_{jik\ell}^{M}$. The same type of symmetry can be shown to hold for $\overline{B}^{ME}$ and $\overline{B}^{FE}$. In general, the two tensors $\overline{B}^{M}$ and $\overline{B}^{ME}$ are different. Although $\overline{B}^{ME}(\overline{x})$ depends only on the undamaged coordinates $\overline{x}$, $\overline{B}^{M}(\overline{x}, \overline{\varepsilon})$ depends on both the undamaged coordinates $\overline{x}$ and the effective strain tensor $\overline{\varepsilon}$. The elastic fiber stress concentration factor $\overline{B}^{FE}(\overline{x})$ is similar to $\overline{B}^{ME}(\overline{x})$ in that it depends only on the undamaged coordinates $\overline{x}$. All the stress concentration tensors $\overline{B}^{ME}$, $\overline{B}^{FE}$ and $\overline{B}^{M}$ do not include the effects of damage as they are defined in the effective configurations $\overline{C}^{M}$, $\overline{C}^{F}$ and $\overline{C}$. There are many models available in the literature for the determination of the undamaged stress concentration factors just defined. The simplest method to use is based on the Voigt assumption where the matrix and fibers are assumed to deform equally. Another model in use is based on the Vanishing Fiber Diameter (VFD) model where the fibers are assumed to have vanishing diameters while occupying a finite volume fraction. Both these models are discussed in detail by Dvorak and Bahei-El-Din [68-70].

Substituting the relevant expressions from equations (7.2) and (7.3) into equation (7.1) and simplifying, one derives the following constraint relation for the elastic stress concentration factors for the matrix and fibers:

$$\overline{c}^{M}\,\overline{B}_{ijk\ell}^{ME} + \overline{c}^{F}\,\overline{B}_{ijk\ell}^{EF} = \delta_{ik}\,\delta_{j\ell} \tag{7.4}$$

where δ_{ij} are the components of the Kronecker delta. It is now clear that once the elastic matrix stress concentration factor $\overline{B}^{ME}$ is determined, one can use equation (7.4) to determine the corresponding fiber stress concentration factor $\overline{B}^{FE}$, and vice versa.

In a similar fashion, the local-overall relation for the strain rate tensor is assumed in the effective configurations as follows:

$$d\overline{\varepsilon}_{ij} = \overline{c}^{M}\,d\overline{\varepsilon}_{ij}^{M} + \overline{c}^{F}\,d\overline{\varepsilon}_{ij}^{F} \tag{7.5}$$

where $\overline{\varepsilon}_{ij}^{M}$ and $\overline{\varepsilon}_{ij}^{F}$ are the components of the effective matrix and fiber strain tensors, respectively. One also assumes an additive decomposition of the matrix and overall strain rates in $\overline{C}^{M}$ and $\overline{C}$, respectively, in the form:

$$d\overline{\varepsilon}_{ij} = d\overline{\varepsilon}_{ij}' + d\overline{\varepsilon}_{ij}'' \tag{7.6a}$$

$$d\bar{\varepsilon}_{ij}^{M} = d\bar{\varepsilon}_{ij}^{M'} + d\bar{\varepsilon}_{ij}^{M''} \tag{7.6b}$$

where $'$ indicates the elastic and $''$ indicates the plastic part of the tensor. In view of the assumption of small strains, equations (7.6) are justified. Because the fibers are assumed to deform elastically, only the overall fiber strain is totally the elastic fiber strain. Therefore, the fiber strain equation corresponding to equations (7.6) takes the following simple form:

$$\bar{\varepsilon}_{ij}^{F} = \bar{\varepsilon}_{ij}^{F'} \tag{7.7}$$

In the sequel, the elastic fiber strain will be denoted by $\bar{\varepsilon}^{F}$, where it is understood that it is comprised only of elastic strain according to equation (7.7).

In the local configurationS $\overline{C}^{M}$ and $\overline{C}^{F}$, and the overall configuration $\overline{C}$, one can write the local-overall relations for the effective strain tensors as follows:

$$\bar{\varepsilon}_{ij}^{M} = \overline{A}_{ijk\ell}^{M}\,\bar{\varepsilon}_{k\ell} \tag{7.8a}$$

$$\bar{\varepsilon}_{ij}^{F} = \overline{A}_{ijk\ell}^{F}\,\bar{\varepsilon}_{k\ell} \tag{7.8b}$$

where $\overline{A}_{ijkl}^{M}$ are the components of the plastic matrix strain concentration factor. Again, during unloading or elastic loading, the tensor $\overline{A}^{M}$ is replaced by the elastic matrix strain concentration factor $\overline{A}^{ME}$. In general, both $\overline{A}^{ME}(\bar{x})$ and $\overline{A}^{FE}(\bar{x})$ depend on the undamaged coordinates $\bar{x}$, while $\overline{A}^{M}(\bar{x},\,\bar{\varepsilon})$ depends on $\bar{x}$ as well as on the effective strain tensor $\bar{\varepsilon}$. Furthermore, the three tensors $\overline{A}^{ME}$, $\overline{A}^{FE}$, and $\overline{A}^{M}$ are symmetric in the sense $A_{ijkl}^{I} = A_{jikl}^{I}$, where I is replaced by ME, FE, or M, appropriately. The above strain concentration tensors can be determined through the Voigt or VFD models, as will be discussed later in this chapter. However, by substituting the relevant forms of equations (7.8) into equation (7.5), one can derive the following helpful constraint equation for the strain concentration factors:

$$\bar{c}^{M}\overline{A}_{ijk\ell}^{ME} + \bar{c}^{F}\overline{A}_{ijk\ell}^{FE} = \delta_{ik}\,\delta_{j\ell} \tag{7.9}$$

It is clear from equation (7.9) that once one of the elastic concentration factors is determined, the other one can be obtained directly.

In the analysis of plastic deformation and especially when considering a yield function, use is made of the deviatoric stresses instead of total stresses in the formulation as hydrostatic pressure has no effect on yielding in this work. Also, backstresses (or shift tensors) are used in the modeling of kinematic hardening. Therefore, it is necessary to derive local-overall relations for these quantities before one can proceed to formulate the constitutive model. The effective matrix deviatoric stress tensor $\bar{\tau}^M$ is then directly derived from equation (7.2) as follows:

$$\bar{\tau}_{ij}^M = \overline{P}_{ijk\ell}^M \, \bar{\sigma}_{k\ell} \tag{7.10}$$

where the fourth-rank tensor P^M is given by:

$$\overline{P}_{ijk\ell}^M = \overline{B}_{ijk\ell}^M - \frac{1}{3} \overline{B}_{ppk\ell}^M \delta_{ij} \tag{7.11}$$

During unloading or elastic loading, the tensor $\overline{P}^M$ is replaced by the tensor $\overline{P}^{ME}$, which is defined by a similar equation to (7.11) but with $\overline{B}^M$ replaced by $\overline{B}^{ME}$. Using equation (7.11), one can derive the following two useful identities that are used later in the analysis:

$$\overline{P}_{rrk\ell}^M = 0 \tag{7.12a}$$

$$\overline{P}_{ijk\ell}^M \, \overline{P}_{ijmn}^M = \overline{B}_{ijk\ell}^M \, \overline{P}_{ijmn}^M \tag{7.12b}$$

Similar relations to equations (7.12) can be shown to exist for the tensor $\overline{P}^{ME}$.

Because the constitutive model to be considered here involves kinematic hardening, one needs to derive local-overall relations for the backstress tensor β. One needs only to consider the matrix backstress because the fibers undergo elastic deformation only. It is assumed that the matrix stress concentration factor $\overline{B}^M$ holds for the matrix backstress tensor. Consequently, the sought relationship between the effective matrix backstress $\bar{\beta}^M$ and the effective overall backstress $\bar{\beta}$ takes the form:

$$\bar{\beta}_{ij}^M = \overline{B}_{ijk\ell}^M \, \bar{\beta}_{k\ell} \tag{7.13}$$

The corresponding local-overall relation for the effective deviatoric matrix backstress $\bar{\alpha}^M$ is then obtained from equation (7.13) and is given by:

$$\bar{\alpha}_{ij}^M = \overline{P}_{ijk\ell}^M \, \bar{\beta}_{k\ell} \tag{7.14}$$

164

The local-overall relations for the stresses, strains, and backstresses have now been presented in the respective effective configurations $\overline{C}^M$, $\overline{C}^F$ and $\overline{C}$. These equations can now be incorporated with a damage theory to formulate a suitable constitutive model.

7.2 Characterization of Damage

The principal objective of this chapter is to quantify damage in metal matrix composites that exhibit a ductile matrix behavior. The ultimate goal is to quantify damage in such a way that one can isolate and evaluate the different types of damage occurring in a composite system. These modes of damage can occur simultaneously or following each other. Some of these types are damage of the matrix, damage of the fiber, debonding, delamination, and so forth. In this chapter, the overall approach to modeling damage in composites is utilized. A procedure is outlined that eventually leads to a descriptive model for the quantification of damage in composite materials. As a first step, the quantification of an overall damage tensor is established in this chapter, and a procedure for isolating different types of damage is presented in the next chapter.

In this chapter, one considers damage in the overall composite system as a whole continuum. This is accomplished by first transforming the undamaged (effective) local quantities into undamaged (effective) overall quantities, then applying the equations of continuum damage mechanics to the overall configuration $\overline{C}$ to obtain the overall damaged quantities in the overall configuration C. (See Figure 7.1). In this approach, the resulting model reflects various types of damage mechanisms, such as void growth and coalescence in the matrix, fiber fracture, debonding and delimitation, and so forth. It should be noted that in this approach no distinction is made among all these types of damage as they will all be reflected through one overall damage variable.

The most important feature of this approach is that all the damage effects undergone by the composite system are lumped together and represented by one single damage variable. This variable is taken as the fourth-rank tensor M called here the overall damage effect tensor. It is defined as a linear transformation of the Cauchy stress space between the configurations $\overline{C}$ and C in the form:

$$\overline{\sigma}_{ij} = M_{ijk\ell}\, \sigma_{k\ell} \tag{7.15}$$

It then follows from the above equation that the overall effective deviatoric Cauchy stress rate $\overline{\tau}$ is given by (see chapter 6 for more details):

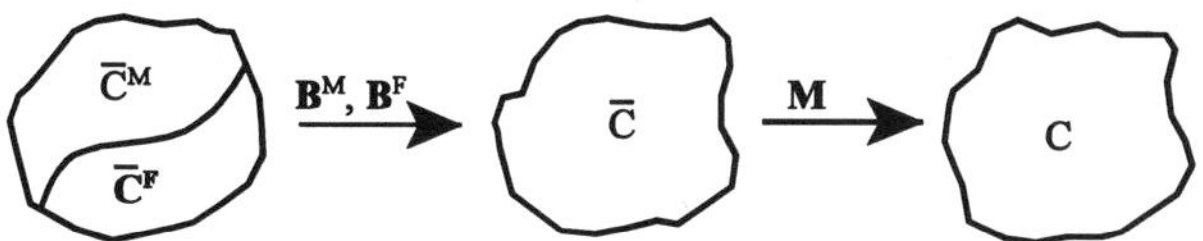

First transform the undamaged local quantities to undamaged
overall quantities, then apply the overall damage tensor to obtain the
damaged overall quantities, i.e.

$$\left.\begin{array}{c} \overline{\mathbf{D}}^{\mathbf{M}} \\ \overline{\mathbf{E}}^{\mathbf{F}} \end{array}\right\} \rightarrow \overline{\mathbf{D}} \rightarrow \mathbf{D}$$

Figure 7.1   Schematic Representation of the Overall Approach That Can Be
Followed to Derive the Constitutive Model

It then follows from the above equation that the overall effective deviatoric Cauchy stress rate $\overline{\tau}$ is given by (see chapter 6 for more details):

$$\overline{\tau}_{ij} = N_{ijk\ell}\, \sigma_{k\ell} \tag{7.16}$$

where the fourth-rank tensor N is given in terms of M as follows:

$$N_{ijk\ell} = M_{ijk\ell} - \frac{1}{3} M_{rrk\ell}\, \delta_{ij} \tag{7.17}$$

Certain useful identities follow directly from equations (7.16) and (7.17). The main two identities used here are listed below:

$$N_{rrk\ell} = 0 \tag{7.18a}$$

$$N_{ijk\ell}\, N_{ijmn} = M_{ijk\ell}\, N_{ijmn} \tag{7.18b}$$

The use of equation (7.15) and in particular the overall damage effect tensor M definitely accounts for all types of damage that the composite system undergoes. However, the main drawback of this approach is its nonlocality, that is, it does not take into consideration the local effects of damage in the matrix and fibers and their interface. Therefore, this approach is characterized by an overall description of damage. This is clearly illustrated in Figure 7.1 where it can be described by the series of transformations:

$$\overline{C}^{M},\ \overline{C}^{F} \longrightarrow \overline{C} \longrightarrow C$$

The fact that the damage effect tensor is introduced in the second step of the formulation accounts for the overall description in this approach. Therefore, this approach can be summarized in two steps:

$$\overline{B}^{M}.\ \overline{B}^{F} \qquad\qquad M$$

1.   Apply the concentration factors $\overline{B}^{M}$ and $\overline{B}^{F}$ to the effective local configurations $\overline{C}^{M}$ and $\overline{C}^{F}$.

2.   Apply the overall damage effect tensor M to the overall effective configuration $\overline{C}$.

The damage transformation equations for the elastic and plastic parts of the strain rate tensor according to this approach were derived previously in Chapter 6. They are presented here again as follows:

$$d\overline{\varepsilon}_{ij}^{\prime} = dM_{ijmn}^{-T}\,\varepsilon_{mn}^{\prime} + M_{ijk\ell}^{-T}\,d\varepsilon_{k\ell}^{\prime} \tag{7.19a}$$

$$d\overline{\varepsilon}_{ij}^{\prime\prime} = X_{ijk\ell}\,d\varepsilon_{k\ell}^{\prime\prime} + Z_{ij} \tag{7.19b}$$

where the tensors X and Z are given by:

$$X_{ijk\ell} = \frac{a_2}{a_1}\,M_{ijk\ell}^{-1} \tag{7.20a}$$

$$Z_{ij} = 3\,\frac{a_3}{a_1}\,M_{ijk\ell}^{-1}\,N_{pqk\ell}\,N_{pqmn}\,(\sigma_{mn} - \beta_{mn}) \tag{7.20b}$$

The relevant expressions for the scalars a_1, a_2, and a_3 along with the definition of the derivative dM are given in Appendix A-1.

In the next section, a constitutive model based on the proposed approach is formulated in detail. The constitutive model is based on analyzing elastic fibers embedded in an elastoplastic matrix. The classical equations of isotropic elasticity and elastoplasticity are used in the fictitious configuration $\overline{C}^F$ (for the fibers) and $\overline{C}^M$ (for the matrix), respectively. First, one considers the matrix constitutive equations and their respective transformations. The formulation is derived based on the overall approach.

7.3 Yield Criterion and Flow Rule

The elastoplastic constitutive model for the matrix is based on a von Mises type yield function $\bar{f}^M(\overline{\tau}^M, \overline{\alpha}^M)$ in the local configuration $\overline{C}^M$ that involves kinematic hardening through the evolution of the backstress tensor α^M. The yield function $\bar{f}^M(\overline{\tau}^M, \overline{\alpha}^M)$ is given in the configuration $\overline{C}^M$ by:

$$\bar{f}^M = \frac{3}{2}(\overline{\tau}_{k\ell}^M - \overline{\alpha}_{k\ell}^M)(\overline{\tau}_{k\ell}^M - \overline{\alpha}_{k\ell}^M) - \overline{\sigma}_o^{M^2} = 0 \tag{7.21}$$

where $\overline{\sigma}_o^M$ is a material constant denoting the uniaxial yield strength of the matrix material.

The plastic flow in the configuration $\overline{C}^M$ is described by the associated flow rule in the form:

$$d\overline{\varepsilon}_{ij}^{M''} = d\overline{\Lambda}^M \frac{\partial \bar{f}^M}{\partial \overline{\sigma}_{ij}^M} \tag{7.22}$$

where $d\overline{\Lambda}^M$ is a scalar function introduced as a Lagrange multiplier in the constraint thermodynamics equations for the matrix material, that is still to be determined. In the present formulation, one assumes that the associated flow rule of plasticity holds only in the local undamaged configuration $\overline{C}^M$. As will be seen later, a non-associated flow rule will be derived for the composite system.

In order to describe kinematic hardening for the matrix, the Prager-Ziegler evolution law [129] is used in the configuration $\overline{C}^M$, as follows:

$$d\overline{\alpha}_{ij}^M = d\overline{\mu}^M (\overline{\tau}_{ij}^M - \overline{\alpha}_{ij}^M) \tag{7.23}$$

where $d\overline{\mu}^M$ is a scalar function to be determined later. Before proceeding to derive the general inelastic constitutive equation for the composite system, it is necessary to derive some transformation equations that depend on the approach followed in this chapter.

First, the relation between the effective matrix stress tensor $\overline{\sigma}^{M}$ and the overall stress tensor σ is derived. This is done by substituting equation (7.15) into equation (7.2) where equation (7.2) is written in terms of total stresses and the elastic concentration factor. Therefore, one obtains:

$$\overline{\sigma}_{ij}^{M} = C_{ijk\ell}^{ME} \sigma_{k\ell} \tag{7.24}$$

where the fourth-rank tensor C^{ME} is given by:

$$C_{ijk\ell}^{ME} = \overline{B}_{ijmn}^{ME} M_{mnk\ell} \tag{7.25}$$

Using the symmetry of $\overline{B}^{ME}$ discussed earlier, it follows from equation (7.25) that the tensor C^{ME} is also symmetric in the sense $C_{ijk\ell}^{ME} = C_{jik\ell}^{ME}$.

Substituting equation (7.15) into equation (7.3) using total stresses, one obtains an expression for the effective fiber stress tensor $\overline{\sigma}^{F}$ in terms of the overall stress tensor σ:

$$\overline{\sigma}_{ij}^{F} = C_{ijk\ell}^{FE} \sigma_{k\ell} \tag{7.26}$$

where

$$C_{ijk\ell}^{FE} = \overline{B}_{ijmn}^{FE} M_{mnk\ell} \tag{7.27}$$

and C^{FE} satisfies the symmetry condition $C_{ijkl}^{FE} = C_{jikl}^{FE}$.

Substituting equation (7.15) into equation (7.10) using total stresses and elastic concentration factors, one obtains the following expressions for the effective matrix and fiber deviatoric stress tensors:

$$\tau_{ij}^{M} = R_{ijk\ell}^{ME} \sigma_{k\ell} \tag{7.28a}$$

$$\tau_{ij}^{F} = R_{ijk\ell}^{FE} \sigma_{k\ell} \tag{7.28b}$$

where the fourth-rank tensors $\boldsymbol{R}^{ME}$ and $\boldsymbol{R}^{FE}$ are given by:

$$R_{ijk\ell}^{ME} = \overline{P}_{ijmn}^{ME} M_{mnk\ell} \tag{7.29a}$$

$$R_{ijk\ell}^{FE} = \overline{P}_{ijmn}^{FE} M_{mnk\ell} \tag{7.29b}$$

It can be shown that the tensor $\overline{\boldsymbol{P}}^{ME}$ and $\boldsymbol{R}^{FE}$ are also symmetric in the sense $\overline{P}_{ijk\ell}^{ME} = \overline{P}_{jik\ell}^{ME}$ and $\overline{R}_{ijk\ell}^{ME} = \overline{R}_{jik\ell}^{ME}$. The same can be said about the tensors $\overline{\boldsymbol{P}}^{FE}$ and $\boldsymbol{R}^{FE}$. Furthermore, by substituting equation (7.11) into equation (7.29a), one can derive the following relation between the tensors $\boldsymbol{C}^{ME}$ and $\boldsymbol{R}^{ME}$:

$$R_{ijk\ell}^{ME} = C_{ijk\ell}^{ME} - \frac{1}{3} C_{rrk\ell}^{ME} \delta_{ij} \tag{7.30}$$

A similar relation can be shown to hold for the tensors $\boldsymbol{R}^{FE}$ and $\boldsymbol{C}^{FE}$. In addition, the tensor $\boldsymbol{R}^{ME}$ satisfies the two identities (7.12) and (7.18), namely, $R_{rrk\ell}^{ME} = 0$ and $R_{ijk\ell}^{ME} R_{ijmn}^{ME} = C_{ijk\ell}^{ME} R_{ijmn}^{ME}$. The same is also true for the tensor $\boldsymbol{R}^{FE}$.

The effective overall backstress tensor $\overline{\boldsymbol{\beta}}$ is assumed to be related to the overall backstress tensor $\boldsymbol{\beta}$ by the overall damage effect tensor $\boldsymbol{M}$. One may want to introduce a different damage effect tensor for the backstress but this will not be carried out at this stage to preserve simplicity and ease of implementation. Therefore, the following transformation equation for the backstress tensor is considered:

$$\overline{\beta}_{ij} = M_{ijk\ell} \beta_{k\ell} \tag{7.31}$$

Following the same procedure used for the stress tensor, the relevant equations for the different backstress tensors can now be derived. Therefore, the necessary backstress equations are listed below without proof:

$$\overline{\alpha}_{ij} = N_{ijk\ell} \beta_{k\ell} \tag{7.32a}$$

$$\overline{\beta}_{ij}^{ME} = C_{ijk\ell}^{ME} \beta_{k\ell} \tag{7.32b}$$

$$\overline{\alpha}_{ij}^{ME} = R_{ijk\ell}^{ME} \beta_{k\ell} \tag{7.32c}$$

Using the above transformation equations, one can transform the yield function, flow rule, and backstress evolution from the local configuration $\overline{C}^M$ to the overall configuration C.

The yield function of equation (7.21) can now be transformed to the overall composite configuration C. First, one uses equations (7.10) and (7.14) (with total stresses and elastic concentration factors) and substitutes them into equation (7.21). Simplifying the resulting equation and using the identity in equation (7.12b), one obtains the following expression for the yield function in the undamaged configuration $\overline{C}$:

$$\overline{f} = \frac{3}{2}\, \overline{B}^{ME}_{k\ell mn}\, \overline{P}^{ME}_{k\ell pq}\, (\overline{\sigma}_{mn} - \overline{\beta}_{mn})\, (\overline{\sigma}_{pq} - \overline{\beta}_{pq}) - \overline{\sigma}_o^{M^2} \tag{7.33}$$

The next step is to use the transformation equations (7.15) and (7.31) and substitute them into equation (7.33). Simplifying the resulting equation using the identities in equations (7.25) and (7.29a), one obtains the following expression for the yield function f in the overall composite configuration:

$$f = \frac{3}{2}\, H^{ME}_{ijk\ell}\, (\sigma_{ij} - \beta_{ij})\, (\sigma_{k\ell} - \beta_{k\ell}) - \overline{\sigma}_o^{M^2} \tag{7.34}$$

where the tensor H^{ME} is given by:

$$H^{ME}_{ijk\ell} = C^{ME}_{mnij}\, R^{ME}_{mnk\ell} \tag{7.35}$$

It is seen from equation (7.34) that the overall yield function in the damaged configuration resembles in form the anisotropic yield function of Hill.

Next, one transforms the associated flow rule to the configuration C. This is accomplished in two steps. First, one substitutes equations (7.2) and (7.8a) (using total stresses, strain rates and their corresponding concentration factors) into equation (7.22). Simplifying, one obtains the flow rule in the configuration $\overline{C}$:

$$d\overline{\varepsilon}''_{ij} = d\overline{\Lambda}_{ijk\ell}\, \frac{\partial \overline{f}}{\partial \overline{\sigma}_{k\ell}} \tag{7.36}$$

where the tensor $d\overline{\Lambda}$ is given by:

$$d\overline{\Lambda}_{ijk\ell} = d\overline{\Lambda}^M\, \overline{A}^{M^{-1}}_{mnij}\, \overline{B}^{ME^{-1}}_{mnk\ell} \tag{7.37}$$

It is noticed that the flow rule in equation (7.36) is associated. The next step is to substitute equations (7.15) and (7.19b) into equation (7.37) and simplify to obtain the flow rule in the configuration C:

$$d\bar{\varepsilon}_{ij}^{\prime\prime} = d\Lambda_{ijk\ell} \frac{\partial f}{\partial \sigma_{k\ell}} + \varepsilon_{ij}^{\prime\prime} \tag{7.38}$$

where the tensors $d\Lambda$ and $\varepsilon^{\prime\prime}$ are given by:

$$d\Lambda_{ijk\ell} = d\overline{\Lambda}_{mnpq} X_{mnij}^{-1} M_{pqk\ell}^{-1} \tag{7.39}$$

$$\varepsilon_{ij}^{\prime\prime} = - X_{ijk\ell}^{-1} Z_{k\ell} \tag{7.40}$$

It is noticed that the overall flow rule (7.38) of the composite system in the damaged configuration is nonassociated due to the presence of the additional term $\varepsilon^{\prime\prime}$. This additional term is primarily due to damage effects as shown in equation (7.40). This remark substantiates the fact that any damage theory, for metals or composites, that incorporates plastic deformation needs to consider a non-associated flow rule of the type given in equation (7.38).

This section is concluded by giving an explicit expression for the tensorial multiplier $d\Lambda$. Substituting equation (7.37) into equation (7.39), one obtains:

$$d\Lambda_{ijk\ell} = d\overline{\Lambda}^M X_{mnij}^{-1} \overline{A}_{rsmn}^{M^{-1}} \overline{B}_{rspq}^{ME^{-1}} M_{pqk\ell}^{-1} \tag{7.41}$$

In equation (7.41), the overall tensorial multiplier $d\Lambda$ is given in terms of the local scalar multiplier $d\overline{\Lambda}^M$, the concentration tensors $\overline{A}^M$, $\overline{B}^{ME}$ and the damage tensors X, M. Once an expression is determined for $d\overline{\Lambda}^M$, as will be seen later, one can use equation (7.41) to derive an appropriate expression for $d\Lambda$.

7.4 Kinematic Hardening in the Damaged Composite System

In this section, a kinematic hardening rule is developed for the composite system in the overall damaged configuration. Dvorak and Bahei-El-Din [69, 70] proposed a kinematic hardening rule for the composite system that did not include any damage effects. This rule is incorporated here in the fictitious undamaged configuration $\overline{C}$, which is then transformed into the damaged configuration C.

172

The overall kinematic hardening rule is now formulated consistently for the composite system. Subtracting equation (7.13) from equation (7.2), while using total stresses and elastic concentration factors, one obtains:

$$(\overline{\sigma}_{ij}^{M} - \overline{\beta}_{ij}^{M}) = \overline{B}_{ijk\ell}^{ME} (\overline{\sigma}_{k\ell} - \overline{\beta}_{k\ell}) \tag{7.42}$$

Next, one differentiates equation (7.42) with respect to time and substitutes for $\overline{\sigma}_{ij}^{M}$ from equation (7.2). Simplifying the resulting equation and solving for $\overline{\beta}_{ij}$, one obtains:

$$d\overline{\beta}_{ij} = (\delta_{ik}\delta_{j\ell} - \overline{B}_{mnij}^{ME^{-1}} \overline{B}_{mnk\ell}^{M}) d\overline{\sigma}_{k\ell} + \overline{B}_{k\ell ij}^{ME^{-1}} \overline{\beta}_{k\ell}^{M} \tag{7.43}$$

Equation (7.43) was originally derived by Dvorak and Bahei-El-Din [69, 70]. In equation (7.43), it is clear that the composite "undamaged" kinematic hardening rule consists of two parts: the matrix kinematic hardening rule given by the second term on the right-hand-side of equation (7.43) through $\overline{\beta}^{M}$, and the additional term in terms of $\overline{\sigma}$, which represents hardening due to the matrix-fiber interaction. It is also noticed that if kinematic hardening of the matrix is neglected, then overall kinematic hardening will still exist due to the interaction of the matrix and fibers, as can be clearly seen from equation (7.43).

At this step, one can substitute for $\overline{\beta}^{M}$ in equation (7.43) using the matrix Prager-Ziegler rule of equation (7.23). First, one rewrites equation (7.23) in terms of total overall stress and backstress tensors, instead of deviatoric matrix tensors. This is accomplished by substituting equations (7.10) and (7.14) (using total stresses and elastic concentration factors) into equation (7.23). Noting that $\overline{\beta}_{\ell\ell} = 0$, one obtains:

$$d\overline{\beta}_{ij}^{M} = d\overline{\mu}^{M} \overline{B}_{ijk\ell}^{ME} (\overline{\sigma}_{k\ell} - \overline{\beta}_{k\ell}) \tag{7.44}$$

Substituting equation (7.44) into equation (7.43), one obtains the overall kinematic hardening rule for the composite system in the undamaged configuration $\overline{C}$:

$$d\overline{\beta}_{ij} = (\delta_{ik}\delta_{j\ell} - \overline{B}_{mnij}^{ME^{-1}} \overline{B}_{mnk\ell}^{M}) d\overline{\sigma}_{k\ell} + d\overline{\mu}^{M} \overline{B}_{mnij}^{ME^{-1}} \overline{P}_{mnk\ell}^{ME} (\overline{\sigma}_{k\ell} - \overline{\beta}_{k\ell}) \tag{7.45}$$

The first step of the derivation is now complete. The next step is to transform equation (7.45) into the configuration C. This is done by substituting equations (7.15) and (7.31) into equation (7.45). Differentiating with respect to time, simplifying the resulting equations and solving for $d\,\beta$, one obtains:

$$d\beta_{ij} = (\Psi_{ijk\ell}\,\sigma_{k\ell} - \chi_{ijk\ell}\,\beta_{k\ell}) + \Pi_{ijk\ell}\,d\sigma_{k\ell} \tag{7.46}$$

where the fourth-rank tensors Ψ, χ and Π are given by:

$$\Psi_{ijk\ell} = M_{mnij}^{-1}(\delta_{mp}\delta_{nq} - \overline{B}_{rsmn}^{ME^{-1}}\overline{B}_{rspq}^{M})M_{pqk\ell} + d\overline{\mu}^{M}M_{mnij}^{-1}\overline{B}_{rsmn}^{ME^{-1}}\overline{P}_{rspq}^{ME}M_{pqk\ell} \tag{7.47a}$$

$$\chi_{ijk\ell} = M_{mnij}^{-1}(M_{mnk\ell} + d\overline{\mu}^{M}\overline{B}_{rsmn}^{ME^{-1}}\overline{P}_{rspq}^{ME}M_{pqk\ell}) \tag{7.47b}$$

$$\Pi_{ijk\ell} = M_{mnij}^{-1}(\delta_{mp}\delta_{nq} - \overline{B}_{rsmn}^{ME^{-1}}\overline{B}_{rspq}^{M})M_{pqk\ell} \tag{7.47c}$$

Equation (7.46) represents the overall kinematic hardening rule for the damaged composite system. It consists of a combination of a generalized Prager-Ziegler rule and a generalized Phillips-type rule for anisotropic materials. On the other hand, it also consists of the two types of kinematic hardening mentioned earlier, namely the matrix kinematic hardening (the terms containing $d\overline{\mu}^{M}$ in equations (7.47) and the kinematic hardening due to the interaction of the matrix and fibers (the terms that do not contain $d\overline{\mu}^{M}$ in equations (7.47)). Therefore, equation (7.46) can be rewritten in the following form:

$$d\beta_{ij} = d\beta_{ij}^{(M)} + d\beta_{ij}^{(I)} \tag{7.48}$$

where

$$d\beta_{ij}^{(M)} = d\overline{\mu}^{M}M_{mnij}^{-1}\overline{B}_{rsmn}^{ME^{-1}}\overline{P}_{rspq}^{ME}M_{pqk\ell}(\sigma_{k\ell} - \beta_{k\ell}) \tag{7.49a}$$

$$d\beta_{ij}^{(I)} = M_{mnij}^{-1}(\delta_{mp}\delta_{nq} - \overline{B}_{rsmn}^{ME^{-1}}\overline{B}_{rspq}^{M})(M_{pqk\ell}\sigma_{k\ell} + M_{pqk\ell}\,d\sigma_{k\ell}) \tag{7.49b}$$

Equations (7.48) and (7.49) show clearly the decomposition of kinematic hardening where $d\beta_{ij}^{(M)}$ indicates the contribution of the matrix, while $d\beta_{ij}^{(I)}$ indicates the contribution of the interaction between the matrix and fibers.

174

7.5 Constitutive Model

In this section, the inelastic damage model for the composite system is formulated. In order to obtain the overall elastoplastic stiffness tensor, two local linear elastic relations are assumed for the matrix and fibers in their respective undamaged local configurations $\overline{C}^M$ and $\overline{C}^F$ as follows:

$$d\overline{\sigma}_{ij}^M = \overline{E}_{ijk\ell}^M d\overline{\varepsilon}_{k\ell}^M \tag{7.50a}$$

$$d\overline{\sigma}_{ij}^F = \overline{E}_{ijk\ell}^F d\overline{\varepsilon}_{k\ell}^F \tag{7.50b}$$

where the tensors $\overline{E}^M$ and $\overline{E}^F$ are the constant fourth-rank elasticity tensors for the matrix and fibers, respectively. They are given by:

$$\overline{E}_{ijk\ell}^M = \lambda^M \delta_{ij} \delta_{k\ell} + G^M (\delta_{ik} \delta_{j\ell} + \delta_{i\ell} \delta_{jk}) \tag{7.51a}$$

$$\overline{E}_{ijk\ell}^F = \lambda^F \delta_{ij} \delta_{k\ell} + G^F (\delta_{ik} \delta_{j\ell} + \delta_{i\ell} \delta_{jk}) \tag{7.51b}$$

where λ^M, G^M, λ^F and G^F are Lamè's constants for the matrix and fibers. Using the transformation equation (7.8) (with elastic strains and elastic concentration factors), along with equation (7.50) and substituting them into equation (7.1), one obtains the overall elastic constitutive relation in the undamaged configuration $\overline{C}$:

$$d\overline{\sigma}_{ij} = \overline{E}_{ijk\ell} d\overline{\varepsilon}_{k\ell}' \tag{7.52a}$$

where

$$\overline{E}_{ijk\ell} = \overline{c}^M \overline{E}_{ijmn}^M \overline{A}_{mnk\ell}^{ME} + \overline{c}^F \overline{E}_{ijmn}^F \overline{A}_{mnk\ell}^{FE} \tag{7.52b}$$

In equation (7.52b), the "undamaged" overall elasticity tensor $\overline{E}$ is given in terms of the "undamaged" local elasticity tensors $\overline{E}^M$ and $\overline{E}^F$, and the strain concentration factors $\overline{A}^{ME}$ and $\overline{A}^{FE}$. In order to derive the overall inelastic constitutive equation in the configuration $\overline{C}$, one substitutes equations (7.6a) and (7.36) into equation (7.52a) to obtain:

$$d\overline{\sigma}_{ij} = \overline{E}_{ijk\ell} \left(d\overline{\varepsilon}_{k\ell} - d\overline{\Lambda}_{k\ell mn} \frac{\partial \overline{f}}{\partial \overline{\sigma}_{mn}} \right) \tag{7.53}$$

The tensorial multiplier $d\overline{\Lambda}$ is obtained from the consistency condition $d\overline{f} = 0$ applied in the configuration $\overline{C}$, such that

$$\frac{\partial \overline{f}}{\partial \overline{\sigma}_{ij}} d\overline{\sigma}_{ij} + \frac{\partial \overline{f}}{\partial \overline{\beta}_{ij}} d\overline{\beta}_{ij} = 0 \tag{7.54}$$

The partial derivatives $\partial \overline{f} / \partial \overline{\sigma}$ and $\partial \overline{f} / \partial \overline{\beta}$ are obtained from equation (7.35) as follows:

$$\frac{\partial \overline{f}}{\partial \overline{\sigma}_{ij}} = \overline{Q}_{ijmn} (\overline{\sigma}_{mn} - \overline{\beta}_{mn}) \tag{7.55a}$$

$$\frac{\partial \overline{f}}{\partial \overline{\beta}_{ij}} = - \overline{Q}_{ijmn} (\overline{\sigma}_{mn} - \overline{\beta}_{mn}) \tag{7.55b}$$

where

$$\overline{Q}_{ijmn} = \frac{3}{2} \left(\overline{B}^{ME}_{k\ell ij} \, \overline{P}^{ME}_{k\ell mn} + \overline{B}^{ME}_{k\ell mn} \, \overline{P}^{ME}_{k\ell ij} \right) \tag{7.55c}$$

Substituting equations (7.55), (7.52a), (7.45), (7.6a), and (7.36) into equation (7.54) and simplifying, one obtains:

$$\overline{Q}_{ijmn} (\overline{\sigma}_{mn} - \overline{\beta}_{mn}) \overline{B}^{ME^{-1}}_{rsij} \overline{B}^{M}_{rsk\ell} \overline{E}_{k\ell pq} \left(d\overline{\varepsilon}_{pq} - d\overline{\Lambda}_{pquv} \frac{\partial \overline{f}}{\partial \overline{\sigma}_{uv}} \right) -$$

$$d\overline{\mu}^{M} \overline{Q}_{ijmn} (\overline{\sigma}_{mn} - \overline{\beta}_{mn}) \overline{B}^{ME^{-1}}_{rsij} \overline{P}^{ME}_{rsk\ell} (\overline{\sigma}_{k\ell} - \overline{\beta}_{k\ell}) = 0 \tag{7.56}$$

The problem at this step is to find a relation between $d\overline{\mu}^{M}$ and $d\overline{\Lambda}$. This can be done indirectly by first deriving a relation between $d\overline{\mu}^{M}$ and $d\overline{\Lambda}^{M}$, and then using equation (7.37) to obtain the desired relation. In order to find an expression for $d\overline{\mu}^{M}$ in terms of $d\overline{\Lambda}^{M}$, one assumes that the projection of $d\overline{\alpha}^{M}$ on the gradient of the yield surface $\overline{f}^{M}$ in the "undamaged" matrix stress space is equal to $b d\overline{\varepsilon}^{M'}$ in the configuration $\overline{C}^{M}$, where b is a material parameter to be determined from the uniaxial tension test (see Voyiadjis [55] and Voyiadjis and Kattan [131]). This assumption is written as:

$$bd\overline{\varepsilon}_{k\ell}^{M''} = d\overline{\alpha}_{mn}^{M} \frac{\dfrac{\partial \overline{f}^{M}}{\partial \overline{\sigma}_{mn}^{M}}}{\dfrac{\partial \overline{f}^{M}}{\partial \overline{\sigma}_{pq}^{M}} \dfrac{\partial \overline{f}^{M}}{\partial \overline{\sigma}_{pq}^{M}}} \frac{\partial \overline{f}^{M}}{\partial \overline{\sigma}_{k\ell}^{M}} \tag{7.57}$$

Substituting for $d\overline{\alpha}^{M}$ and $d\overline{\varepsilon}^{M''}$ from equations (7.23) and (7.22), respectively, into equation (7.57) and post-multiplying the resulting equation by $\partial \overline{f}^{M} / \partial \overline{\alpha}_{kl}^{M}$, one obtains the required expression for $d\overline{\mu}^{M}$:

$$d\overline{\mu}^{M} = bd\overline{\Lambda}^{M} \frac{\dfrac{\partial \overline{f}^{M}}{\partial \overline{\sigma}_{k\ell}^{M}} \dfrac{\partial \overline{f}^{M}}{\partial \overline{\sigma}_{k\ell}^{M}}}{(\overline{\tau}_{mn}^{M} - \overline{\alpha}_{mn}^{M}) \dfrac{\partial \overline{f}^{M}}{\partial \overline{\sigma}_{mn}^{M}}} \tag{7.58}$$

Equation (7.58) is applicable for any matrix yield function $\overline{f}^{M}$. However, using the specific function $\overline{f}^{M}$ of equation (7.21), equation (7.58) reduces to the following simple form:

$$d\overline{\mu}^{M} = 3b \ d\overline{\Lambda}^{M} \tag{7.59}$$

Substituting equations (7.37), (7.55a), and (7.59) into equation (7.56), simplifying the resulting equation and solving for $d\overline{\Lambda}^{M}$, one obtains the following expression after some lengthy algebraic manipulations:

$$d\overline{\Lambda}^{M} = \overline{T}_{k\ell} \ \overline{E}_{k\ell mn} \ \overline{\varepsilon}_{mn} \tag{7.60}$$

where

$$\overline{T}_{k\ell} = \frac{\overline{Q}_{ijmn} (\overline{\sigma}_{mn} - \overline{\beta}_{mn}) \overline{B}_{rsij}^{ME^{-1}} \overline{B}_{rsk\ell}^{M}}{\overline{Q}_{ijmn} (\overline{\sigma}_{mn} - \overline{\beta}_{mn}) (\overline{\sigma}_{cd} - \overline{\beta}_{cd}) \overline{B}_{rsij}^{ME^{-1}} \left[\overline{B}_{rsef}^{M} \overline{E}_{efpq} \overline{A}_{abpq}^{M^{-1}} \overline{B}_{abuv}^{ME^{-1}} \overline{Q}_{uvcd} + 3b \overline{P}_{rsab}^{ME} \delta_{ac} \delta_{db} \right]} \tag{7.61}$$

Substituting for $d\overline{\Lambda}^{M}$ from equation (7.60) into equation (7.37), one obtains the following expression for $d\overline{\Lambda}$:

$$d\overline{\Lambda}_{k\ell mn} = \overline{T}_{ij} \ \overline{E}_{ijpq} \ \overline{A}_{rsk\ell}^{M^{-1}} \ \overline{B}_{rsmn}^{ME^{-1}} \ \overline{\varepsilon}_{pq} \tag{7.62}$$

Finally, one substitutes for $d\overline{\Lambda}$ from equation (7.62) into equation (7.53) to obtain the inelastic constitutive equation for the composite system in the configuration $\overline{C}$ as follows:

$$d\overline{\sigma}_{ijk\ell} = \overline{D}_{ijk\ell}\, d\overline{\varepsilon}_{k\ell} \tag{7.63}$$

where the overall "undamaged" elastoplastic stiffness tensor $\overline{D}$ is given by:

$$\overline{D}_{ijk\ell} = \overline{E}_{ijk\ell} - \overline{T}_{mn}\, \overline{E}_{ijpq}\, \overline{E}_{mnk\ell}\, \overline{A}_{abpq}^{M^{-1}}\, \overline{B}_{abrs}^{ME^{-1}}\, \overline{Q}_{rscd}\, (\overline{\sigma}_{cd} - \overline{\beta}_{cd}) \tag{7.64}$$

The tensor $\overline{D}_{ijkl}$ is the elastoplastic stiffness tensor for the overall system in the undamaged configuration $\overline{C}$.

Next, one is ready to undertake the second step in this approach, which consists of transforming the constitutive equation (7.63) from the configuration $\overline{C}$ to the configuration C. This step effectively introduces damage into the constitutive equations based on the overall approach considered in this chapter. The systematic approach in deriving this step has been given previously in Chapter 6. Following the same procedure detailed in section 6.4.3. One arrives at:

$$d\sigma_{ij} = D_{ijk\ell}\, d\varepsilon_{k\ell} + G_{ij} \tag{7.65}$$

where the "damage" overall elastoplastic stiffness tensor D and the additional tensor G are given by:

$$D_{ijk\ell} = O_{pqij}^{-1}\, \overline{D}_{pqmn}\, X_{mnk\ell} \tag{7.66}$$

$$G_{ij} = O_{pqij}^{-1}\, \overline{D}_{pqmn}\, Z_{mn} \tag{7.67}$$

and the fourth-rank tensor O is given by:

$$O_{ijpq} = M_{ijpq} + \cfrac{\dfrac{\partial M_{ijmn}}{\partial \varphi_{uv}}\, \dfrac{\partial g}{\partial \sigma_{pq}}\, \dfrac{\partial g}{\partial \sigma_{uv}}\, \sigma_{mn}}{\dfrac{\partial L}{\partial \ell} - \dfrac{\partial g}{\partial \varphi_{xy}}\, \dfrac{\partial g}{\partial \sigma_{xy}}} +$$

$$D_{ijk\ell}\left(M^{-T}_{k\ell mn}\, E^{-1}_{pqmn} \;-\; X_{k\ell mn}\, E^{-1}_{pqmn} \;-\; M^{-T}_{xyk\ell}\, \frac{\partial M^T_{xyuv}}{\partial \varphi_{mn}}\, \frac{\dfrac{\partial g}{\partial \sigma_{pq}}\,\dfrac{\partial g}{\partial \sigma_{mn}}}{\dfrac{\partial L}{\partial \ell} - \dfrac{\partial g}{\partial \varphi_{ab}}\,\dfrac{\partial g}{\partial \sigma_{ab}}}\, M^{-T}_{uvcd}\, \overline{E}_{rscd}\, \sigma_{rs} \right)$$

$$(7.68)$$

In equation (7.68), the function g represents the damage evolution criterion as given previously in equation (6.65). The constitutive equation (7.65) along with the elastoplastic stiffness tensor of equation (7.66) are derived in the configuration C of the composite system, thus including the effects of damage and plastic deformation. Finally, it must be emphasized that the focus of this chapter is to show appropriate expressions for yield functions, flow rules, and hardening rules in the damaged state described in terms of the overall damage tensor $M(\varphi)$. These results are summarized in Table 7.1. The end result of this chapter is to introduce a new class of yield criteria and hardening rules for the damaged composite system that can be used directly in the expression of the stiffness tensor to be used in numerical applications. The main features and results of the proposed model are summarized as follows:

1. An anisotropic yield function is derived for the composite system based on using a von Mises type yield criterion for the undamaged matrix material.

3. It is shown that a non-associated flow rule needs to be considered for the damaged composite although an associated flow rule is used for the matrix. In fact, an explicit form of the appropriate non-associated flow rule is derived in the context of the proposed model.

4. A generalized kinematic hardening rule is derived for the composite. This rule is shown to consist of a combination of a generalized Ziegler-Prager rule and a Phillips type rule. This rule results from the initial assumption of a Ziegler-Prager rule for the undamaged matrix.

5. An elastoplastic stiffness tensor for the whole composite system is derived in a closed form. This tensor incorporates the effects of damage in terms of the overall damage tensor M.

6. The formulation is based on an approach that utilizes the concept of an overall damage tensor. Another approach is presented in Chapter 8 that concentrates on formulating damage models for composite systems utilizing local (phase) damage tensors for the different phases of the system.

Table 7.1:   Explicit expressions for the yield function, flow rule, and kinematic hardening rule in the three configurations $\overline{C}^M$, $\overline{C}$, and C according to the overall approach.

Rule	Configuration		
	Local $\overline{C}^M$	Overall $\overline{C}$	Overall C
Yield Function	$\overline{f}^M = \dfrac{3}{2}(\overline{\tau}^M_{k\ell} - \overline{\alpha}^M_{k\ell})(\overline{\tau}^M_{k\ell} - \overline{\alpha}^M_{k\ell}) - \overline{\sigma}^{M^2}_o$	$\overline{f} = \dfrac{3}{2}\overline{B}^{ME}_{k\ell mn}\overline{P}^{ME}_{k\ell pq}(\overline{\sigma}_{mn} - \overline{\beta}_{mn})(\overline{\sigma}_{pq} - \overline{\beta}_{pq}) - \overline{\sigma}^{M^2}_o$	$f = \dfrac{3}{2}H_{ijk\ell}(\sigma_{ij} - \beta_{ij})(\sigma_{k\ell} - \beta_{k\ell}) - \overline{\sigma}^{M^2}_o$
Flow Rule	$d\overline{\varepsilon}^{M'}_{ij} = d\overline{\Lambda}^M \dfrac{\partial \overline{f}^M}{\partial \overline{\sigma}^M_{ij}}$	$d\overline{\varepsilon}''_{ij} = d\overline{\Lambda}_{ijk\ell} \dfrac{\partial \overline{f}}{\partial \overline{\sigma}_{k\ell}}$	$d\varepsilon''_{ij} = d\Lambda_{ijk\ell} \dfrac{\partial f}{\partial \sigma_{ij}} + \varepsilon''_{ij}$
Kinematic Hardening Rule	$d\overline{\alpha}^M_{ij} = d\overline{\mu}^M (\overline{\tau}^M_{ij} - \overline{\alpha}^M_{ij})$	$d\overline{\beta}_{ij} = (\delta_{ik}\delta_{j\ell} - \overline{B}^{ME^{-1}}_{mnij}\overline{B}^M_{mnk\ell})\,d\overline{\sigma}_{k\ell}$ $+ d\overline{\mu}^M \overline{B}^{ME^{-1}}_{mnij}\overline{P}^{ME}_{mnk\ell}(\overline{\sigma}_{k\ell} - \overline{\beta}_{k\ell})$	$d\beta_{ij} = (\Psi_{ijk\ell}\sigma_{k\ell} - \chi_{ijk\ell}\beta_{k\ell}) + \Pi_{ijk\ell}\,d\sigma_{k\ell}$

CHAPTER 8

METAL MATRIX COMPOSITES - LOCAL APPROACH

The approach followed in this chapter to characterize damage is termed local in the sense that damage is introduced at the constituent (or local) level. In the case of the composite material considered in this chapter, two damage tensors M^M and M^F are introduced for the two constituents (matrix and fibers) of the composite system. In general, for a composite system consisting of n constituents ($1 < n < \infty$), one needs to introduce n damage tensors $M^{(1)}$, $M^{(2)}$, ..., $M^{(n)}$ in order to locally characterize damage in a complete manner. These damage variables are then combined in a systematic way with an appropriate micromechemical constitutive model to develop the overall damage response of the composite system.

The matrix damage effect tensor M^M is assumed to reflect all types of damage that the matrix material undergoes, like nucleation and coalescence of voids and mircocracks. On the other hand, the fiber damage effect tensor M^F is considered to reflect all types of fiber damage, like fracture of fibers. These two tensors are then related through an overall damage effect tensor M for the whole composite system. In this respect, for example, the microvoid density may be used in the definition of M^M while the microcrack density may be used to define M. The problems of debonding and delamination may be conveniently represented through either M^M or M^F, depending on the desired complexity of the resulting constitutive model. For example, debonding may be linked to fiber damage and represented through M^F and likewise, delamination could be linked to M^M. Alternatively, one may wish to consider the interface material between the fibers and matrix as a separate constituent of the system and introduce an independent interfacial damage effect tensor. In this way, this tensor can be used to reflect debonding in a consistent way. However, this seems impractical at this time and is beyond the scope of this chapter.

8.1   Assumptions

The composite system considered consists of an elastoplastic ductile matrix reinforced by elastic aligned fibers. The strains undergone by the matrix and fibers are infinitesimal, thus restricting the formulation to small deformations. The initial configuration of the system is assumed to be undeformed and undamaged, and is denoted by C_o. The corresponding initial matrix and fiber subconfigurations are denoted by C_o^M and C_o^F, respectively.

When the composite system is acted upon by an external agency, it is assumed to undergo both damage and plastic deformation. The resulting overall configuration is denoted by C, while the matrix and fiber local subconfigurations are denoted by C^M and C^F, respectively. It is clear that $C^M \subset C$, $C^F \subset C$, and $C^M \cup C^F = C$ since the composite consists strictly of matrix and fibers. For the purpose of quantifying damage and using the concept originally proposed by Kachanov [1], one considers a fictitious configuration $\overline{C}$ of the composite system that is obtained from C by removing all the damage that the system (both matrix and fibers) has undergone during the process of deformation. The configuration $\overline{C}$ can be considered as the deformed configuration of the system without any damage effects. The corresponding fictitious (undamaged) matrix and fiber local configurations are noted by $\overline{C}^M$ and $\overline{C}^F$, respectively. It is also clear that $\overline{C}^M \subset \overline{C}$, $\overline{C}^F \subset \overline{C}$ and $\overline{C}^M \cup \overline{C}^F = \overline{C}$. In addition, all the local configurations C^M, C^F, $\overline{C}^M$ and $\overline{C}^F$ are assumed nonempty.

The intitial composite analysis is based on the micromechanical model of Dvorak and Bahei-El-Din [68-70] and Bahei-El-Din and Dvorak [97], where the overall effective Cauchy stress $\overline{\sigma}$ is related to the local effective Cauchy matrix and fiber stresses $\overline{\sigma}^M$ and $\overline{\sigma}^F$, respectively, by:

$$d\overline{\sigma} = \overline{c}^M d\overline{\sigma}^M + \overline{c}^F d\overline{\sigma}^F \tag{8.1}$$

where $\overline{c}^M$ and $\overline{c}^F$ are the matrix and fiber volume fractions, respectively, in the configurations $\overline{C}^M$ and $\overline{C}^F$.

$$\overline{c}^M = \frac{\overline{V}^M}{\overline{V}} \tag{8.2a}$$

$$\overline{c}^F = \frac{\overline{V}^F}{\overline{V}} \tag{8.2b}$$

and

$$\overline{V}^M + \overline{V}^F = \overline{V} \tag{8.2c}$$

In equations (8.2), $\overline{V}^M$ and $\overline{V}^F$ represent the matrix and fiber volumes, respectively, while $\overline{V}$ represents the overall volume. It is clear from equations (8.2) that $\overline{c}^M + \overline{c}^F = 1$.

Similarly, one assumes the following local-overall relation for the strain rate tensors in the effective configurations $\overline{C}^M$, $\overline{C}^F$ and $\overline{C}$:

$$d\bar{\varepsilon} = \bar{c}^M d\bar{\varepsilon}^M + \bar{c}^F d\bar{\varepsilon}^F \tag{8.3}$$

where $\bar{\varepsilon}^M$ and $\bar{\varepsilon}^F$ are the effective matrix and fiber strain tensors, respectively, and $\bar{\varepsilon}$ is the effective overall strain tensor. One also assumes an additive decomposition of the matrix and overall strain rates in $\bar{C}^M$ and $\bar{C}$, respectively as follows:

$$d\bar{\varepsilon} = d\bar{\varepsilon}' + d\bar{\varepsilon}'' \tag{8.4a}$$

$$d\bar{\varepsilon}^M = d\bar{\varepsilon}^{M'} + d\bar{\varepsilon}^{M''} \tag{8.4b}$$

$$d\bar{\varepsilon}^F = d\bar{\varepsilon}^{F'} \tag{8.4c}$$

where $'$ indicates the elastic part and $''$ indicates the plastic part of the tensor. Equations (8.4) are valid in this formulation because small (infinitesimal) strains are assumed throughout. The overall fiber strain consists only of the elastic part since the fibers can deform only elastically until they fracture. Therefore, the effective elastic (or total) fiber strain tensor is denoted by $\bar{\varepsilon}^F$.

8.2 Stress and Strain Concentration Factors

The matrix and fiber stress concentration factors $\overline{\boldsymbol{B}}^M$ and $\overline{\boldsymbol{B}}^{FE}$, respectively, are defined as fourth-rank tensors in the fictitious effective configurations as follows:

$$d\bar{\sigma}^M = \overline{\mathrm{B}}^M : d\bar{\sigma} \tag{8.5a}$$

$$d\bar{\sigma}^F = \overline{\mathrm{B}}^{FE} : d\bar{\sigma} \tag{8.5b}$$

The plastic matrix stress concentration factor $\overline{\boldsymbol{B}}^M$ is replaced by the elastic matrix stress concentration factor $\overline{\boldsymbol{B}}^{ME}$ when elastic loading or unloading takes place. It should be noted that the elastic tensors $\overline{\boldsymbol{B}}^{ME}$ and $\overline{\boldsymbol{B}}^{FE}$ depend only on the undamaged coordinates, while the plastic tensor $\overline{\boldsymbol{B}}^M$ depends on both the undamaged coordinates and the effective strain tensor. Since the above concentration factors are defined in the effective configurations, they are void of any damage effects and therefore can be determined using several available methods. The two simplest methods are the Voigt model and the VFD (Vanishing Fiber Diameter) model. The reader is referred to the recent work of Dvorak and Bahei-El-Din [70] for more details concerning these two assumptions. However, it should be noted that the proposed model is not restricted to these simple field theories. Other more sophisticated theories could be used in conjunction with the proposed model to determine the stress and strain concentration factors.

Substituting equations (8.5) (using elastic concentration factors) into equation (8.1) and simplifying, one obtains the following constraint relation:

$$\bar{c}^M \, \overline{\mathbf{B}}^{ME} + \bar{c}^F \, \overline{\mathbf{B}}^{FE} = \mathbf{I}_4 \tag{8.6}$$

It is clear that once one of the phase (matrix or fibers) stress concentration factors is determined, one can use equation (8.6) to find the other one.

Next, one uses the relations in equations (8.5) to derive similar relations in the configurations C^M, C^F and C. The corresponding relations take the following form:

$$d\bar{\boldsymbol{\sigma}}^M = \mathbf{B}^M : d\boldsymbol{\sigma} \tag{8.7a}$$

$$d\bar{\boldsymbol{\sigma}}^F = \mathbf{B}^{FE} : d\boldsymbol{\sigma} \tag{8.7b}$$

Equations (8.7) represent the current local-overall relations for the stresses in the damaged composite system. The concentration factors $\mathbf{B}^M$ and $\mathbf{B}^{FE}$ are fourth-rank tensors that include the local effects of damage. Equations (8.7) characterize the local nature of the approach adopted in this chapter in the sense that $\mathbf{B}^M$ contains the matrix damage effects, while $\mathbf{B}^{FE}$ contains fiber damage effects. Once suitable local damage tensors are defined (as will be seen in the next section), one can consistently derive the appropriate relations between the undamaged factors $\overline{\boldsymbol{B}}^M$, $\overline{\boldsymbol{B}}^{FE}$ and the damaged factors $\mathbf{B}^M$ and $\mathbf{B}^{FE}$. This process will be carried out in the next section where damage tensors are introduced. As before, the tensor $\mathbf{B}^M$ should be replaced by its elastic counterpart $\mathbf{B}^{ME}$ when unloading or elastic loading takes place.

Since equation (8.1) is valid in the configurations C^M, C^F and C when barred stresses are replaced by unbarred stresses, it can be shown that using equations (8.7), the damaged stress concentration factors satisfy a constraint equation similar to equation (8.6), namely:

$$c^M \mathbf{B}^{ME} + c^F \mathbf{B}^{FE} = \mathbf{I}_4 \tag{8.8}$$

where c^M and c^F are the phase volume fractions of the damaged composite system defined in a similar way as in equations (8.2), with $c^M + c^F = 1$.

This section is concluded with a similar but brief discussion of the matrix and fiber strain concentration factors $\overline{A}^M$ and $\overline{A}^{FE}$, respectively. These two factors are fourth-rank tensors defined in the fictitious (undamaged) configurations as follows:

$$d\overline{\varepsilon}^M = \overline{\mathbf{A}}^M : d\overline{\varepsilon} \tag{8.9a}$$

$$d\overline{\varepsilon}^F = \overline{\mathbf{A}}^{FE} : d\overline{\varepsilon} \tag{8.9b}$$

The remarks discussed earlier about the stress concentration factors apply also to the strain concentration factors, where one can also show that:

$$\overline{c}^M \overline{\mathbf{A}}^{ME} + \overline{c}^F \overline{\mathbf{A}}^{FE} = \mathbf{I}_4 \tag{8.10}$$

Similarly, one can define the damaged strain concentration factors $\mathbf{A}^M$ and $\mathbf{A}^{FE}$ as fourth-rank tensors in the damaged configurations as follows:

$$d\varepsilon^M = \mathbf{A}^M : d\varepsilon \tag{8.11a}$$

$$d\varepsilon^F = \mathbf{A}^{FE} : d\varepsilon \tag{8.11b}$$

The relation between the undamaged strain concentration factors $\overline{A}^M$, $\overline{A}^{FE}$ and their damaged counterparts A^M, A^{FE} will be derived in the next section using local damage tensors. Finally, one can easily show that:

$$c^M \mathbf{A}^{ME} + c^F \mathbf{A}^{FE} = \mathbf{I}_4 \tag{8.12}$$

Equation (8.12) is a constraint equation similar to equation (8.10), where one can determine one of the phase concentration factors once the other one is known. In the next section, explicit relations will be derived between the undamaged and damaged concentration factors for stresses and strains.

186

8.3 Matrix and Fiber Damage Analysis

In the local approach, one considers separately the local damage that the matrix and fibers undergo, such as nucleation and growth of voids and void coalescence for the matrix, and fracture of the fibers. Therefore, two fourth-rank matrix and fiber damage effect tensors $\mathbf{M}^M$ and $\mathbf{M}^F$ are introduced to reflect separately all types of damage that the matrix and fibers undergo. These two tensors are defined as linear transformations of the Cauchy stress space between the configurations $\overline{C}^M$, $\overline{C}^F$ and C^M, C^F as follows:

$$\overline{\boldsymbol{\sigma}}^M = \mathbf{M}^M : \boldsymbol{\sigma}^M \tag{8.13a}$$

$$\overline{\boldsymbol{\sigma}}^F = \mathbf{M}^F : \boldsymbol{\sigma}^F \tag{8.13b}$$

Using equations (8.13), one can directly derive two corresponding equations for the local effective deviatoric Cauchy stresses $\overline{\boldsymbol{\tau}}^M$ and $\overline{\boldsymbol{\tau}}^F$ in the form:

$$\overline{\boldsymbol{\tau}}^M = \mathbf{N}^M : \boldsymbol{\sigma}^M \tag{8.14a}$$

$$\overline{\boldsymbol{\tau}}^F = \mathbf{N}^F : \boldsymbol{\sigma}^F \tag{8.14b}$$

where the fourth-rank tensor $\mathbf{N}^M$ and $\mathbf{N}^F$ are given in terms of $\mathbf{M}^M$ and $\mathbf{M}^F$, respectively, as follows:

$$\mathbf{N}^M = \mathbf{M}^M - \frac{1}{3}\,\mathbf{I}_2 \otimes (\mathbf{I}_2 : \mathbf{M}^M) \tag{8.15a}$$

$$\mathbf{N}^F = \mathbf{M}^F - \frac{1}{3}\,\mathbf{I}_2 \otimes (\mathbf{I}_2 : \mathbf{M}^F) \tag{8.15b}$$

It should be noted that the deviatoric equations for the fibers (equations (8.14b) and (8.15b)) are not used in the formulation since the fibers are assumed to deform elastically until they fracture.

Some useful identities follow directly from equations (8.15). These identities are presented here for the sake of completeness. The main identities that will be used later are given as:

$$\mathbf{I}_2 : \mathbf{N}^M = 0 \quad , \quad \mathbf{I}_2 : \mathbf{N}^F = 0 \tag{8.16a}$$

$$\mathbf{N}^M : \mathbf{N}^M = \mathbf{M}^M : \mathbf{N}^M \quad , \quad \mathbf{N}^F : \mathbf{N}^F = \mathbf{M}^F : \mathbf{N}^F \qquad (8.16b)$$

The basic feature of the overall approach is that local effects of damage are considered whereby they are described separately by the matrix and fiber damage. Subsequent to this local damage description, the local-overall relations are used to transform the local damage effects to the whole composite system. Therefore, this approach represents a local damage description and is clearly shown in Figure 8.1. This can be presented by the two independent but parallel transformations:

$$\overline{C}^M \xrightarrow{\quad \mathbf{M}^M \quad} C^M$$

$$\overline{C}^F \xrightarrow{\quad \mathbf{M}^F \quad} C^F$$

which are followed by the transformation:

$$C^M, \; C^F \xrightarrow{\quad \mathbf{B}^M, \; \mathbf{B}^F \quad} C$$

where $\mathbf{B}^M$ and $\mathbf{B}^F$ are the damaged stress concentration factors.

Considering the above transformations, it becomes clear that the local nature of this approach stems from the fact that the damage effect tensors are introduced in the first step of the formulation.

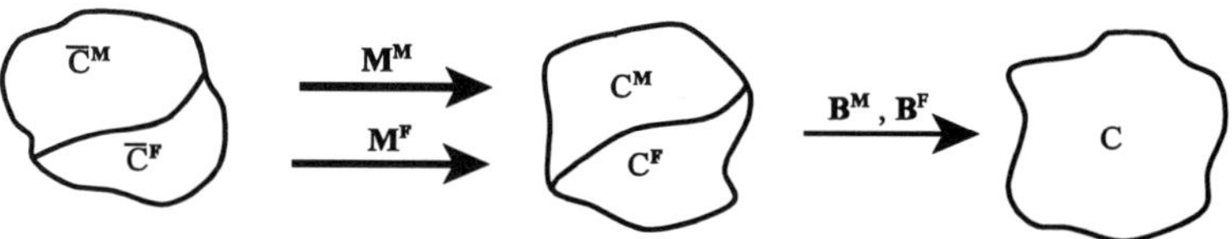

First apply the local damage effect tensors to transform the undamaged local quantities to damaged local quantities, then apply the damaged concentration factors to obtain the damaged overall quantities, i.e.

$$\left.\begin{array}{c} \overline{\mathbf{D}}^M \\ \overline{\mathbf{E}}^M \end{array}\right\} \longrightarrow \left.\begin{array}{c} \mathbf{D}^M \\ \mathbf{E}^M \end{array}\right\} \longrightarrow \mathbf{D}$$

Figure 8.1　Schematic Representation of the Local Approach that can be
Followed to Derive the Constitutive Model

188

Therefore, this approach can be summarized by the following two steps:

1. Apply the local damage effect tensors M^M and M^F to the local effective configurations $\overline{C}^M$ and $\overline{C}^F$.
2. Apply the damaged stress concentration factors B^M and B^F and the damaged strain concentration factors A^M and A^F to the local configurations C^M and C^F.

The tensor M^M encompasses all the pertinent damage related to the matrix while the tensor M^F reflects the damage pertinent to the fibers. However, no explicit account is made for such damage mechanisms as debonding and delamination. Nevertheless, these damage mechanisms can be conveniently incorporated in this theory where debonding can be introduced as part of the tensor M^F, while declamination can be represented by the tensor M^M. Alternatively, two new damage tensors can be defined to reflect these two mechanisms, but this will complicate the theory further and is beyond the scope of this chapter.

Considering the overall configurations C and $\overline{C}$, one can introduce an overall damage effect tensor M for the whole composite system. This tensor is defined in a way similar to the definitions of M^M and M^F, in the form:

$$\overline{\sigma} = \mathbf{M} : \sigma \tag{8.17}$$

The tensor M reflects all types of damage that the composite undergoes including the damage due to the interaction between the matrix and fibers. This tensor has been examined in detail in Chapter 6. A matrix representation was explicitly derived for this tensor where it was shown that M is symmetric. The symmetry property of the tensor M is used extensively in the derivations that follow. The same holds for the tensors M^M and M^F. Using equation (8.17), one can directly derive the following equations for the overall effective deviatorc Cauchy stress $\overline{\tau}$:

$$\overline{\tau} = \mathbf{N} : \sigma \tag{8.18}$$

where N is a fourth-rank tensor given by:

$$\mathbf{N} = \mathbf{M} - \frac{1}{3} \mathbf{I}_2 \otimes (\mathbf{I}_2 : \mathbf{M}) \tag{8.19}$$

Using the overall damage tensor M, one can develop an overall constitutive model to analyze damage in composite materials as was done in Chapter 7. Nevertheless, the tensor M, is introduced here because it is needed in the local approach as well. Furthermore, a very useful relation will be derived between the overall tensor M and the local tensors M^M and M^F.

One is now ready to derive explicit expressions for the damaged stress and strain concentration factors in terms of the undamaged factors and the damage tensors. Starting with equations (8.5) (using elastic loading, i.e. total stresses and elastic stress concentration factors), substituting equations (8.13) and (8.17) into equations (8.5), simplifying and comparing the result with equations (8.7), one obtains:

$$\mathbf{B}^{ME} = \mathbf{M}^{M^{-1}} : \overline{\mathbf{B}}^{ME} \cdot \mathbf{M} \tag{8.20a}$$

$$\mathbf{B}^{FE} = \mathbf{M}^{F^{-1}} : \overline{\mathbf{B}}^{FE} \cdot \mathbf{M} \tag{8.20b}$$

Equations (8.20) represent explicit formulae for the damaged stress concentration factors B^{ME} and B^{FE} in terms of the undamaged factors $\overline{B}^{ME}$ and $\overline{B}^{FE}$ and the damage tensors M^M, M^F and M. Therefore, once the undamaged stress concentration factors are determined (using the Voigt or VFD models), one can use equations (8.20) to find the damaged stress concentration factors.

In a similar fashion, one can repeat the above procedure to derive appropriate expressions for the damaged strain concentration factors. However, before attempting to do this, one needs to derive transformation equations for the local and overall strain tensors between the undamaged and damaged configurations. Utilizing the assumption of small elastic strains and the hypothesis of elastic energy equivalence, one can follow the procedure presented previously in Chapter 6 to derive the following equation for the overall effective elastic strain tensor $\overline{\varepsilon}'$:

$$\overline{\varepsilon}' = \mathbf{M}^{-T} : \varepsilon' \tag{8.21}$$

Applying the hypothesis of elastic energy equivalence to the matrix and fibers separately, one can derive the following transformation equations for the local elastic strain tensors:

$$\overline{\varepsilon}^{M'} = \mathbf{M}^{M^{-T}} : \varepsilon^{M'} \tag{8.22a}$$

$$\overline{\varepsilon}^{F} = \mathbf{M}^{F^{-T}} : \varepsilon^{F} \tag{8.22b}$$

190

Using total strains and elastic strain concentration factors, substituting equations (8.20) and (8.21) into equations (8.9), and simplifying and comparing the results with equations (8.11), one obtains the relation between the damaged and undamaged strain concentration factors:

$$\mathbf{A}^{ME} = \mathbf{M}^M : \overline{\mathbf{A}}^{ME} \cdot \mathbf{M}^{-1} \tag{8.23a}$$

$$\mathbf{A}^{FE} = \mathbf{M}^F : \overline{\mathbf{A}}^{FE} \cdot \mathbf{M}^{-1} \tag{8.23b}$$

Using equations (8.23), one can determine the damaged strain concentration factors $\boldsymbol{A}^{ME}$ and $\boldsymbol{A}^{FE}$ using the undamaged factors $\overline{\boldsymbol{A}}^{ME}$ and $\overline{\boldsymbol{A}}^{FE}$, and the damage tensors $\boldsymbol{M}^M$, $\boldsymbol{M}^F$ and $\boldsymbol{M}$.

In the next sections, the damage transformation equations developed previously will be used to derive the constitutive model including the yield criterion, flow rule, kinematic hardening rule and the damage-elastoplastic stiffness tensor.

8.4 Yield Criterion and Flow Rule

In this section, the yield function and flow rule are developed for the damaged composite system. The development is based on using a von Mises yield criterion and an associated flow rule for the undamaged matrix material. Considering the effective subconfiguration $\overline{C}^M$, one starts with a von Mises yield function $\overline{f}^M$ incorporating kinematic hardening in the form:

$$\overline{f}^M = \frac{3}{2}\,(\overline{\tau}^M - \overline{\alpha}^M) : (\overline{\tau}^M - \overline{\alpha}^M) - \overline{\sigma}_0^{M^2} = 0 \tag{8.24}$$

where $\overline{\sigma}_o^M$ is the uniaxial yield strength of the undamaged matrix material and $\overline{\boldsymbol{\alpha}}^M$ is the effective matrix deviatoric backstress tensor. The tensor $\overline{\boldsymbol{\alpha}}^M$ can be assumed related to the matrix backstress tensor $\boldsymbol{\beta}^M$ and the overall backstress tensor $\boldsymbol{\beta}$ using the same damage and stress concentration tensors that were used for the Cauchy stress. Therefore, in view of equations (8.14a) and (8.7a), one can write the following transformation equations for the backstress tensor:

$$\overline{\boldsymbol{\alpha}}^M = \mathbf{N}^M : \boldsymbol{\beta}^M \tag{8.25a}$$

$$d\boldsymbol{\beta}^M = \mathbf{B}^M : d\boldsymbol{\beta} \tag{8.25b}$$

where equation (8.25b) can be rewritten using the total backstresses $\boldsymbol{\beta}^M$ and $\boldsymbol{\beta}$ with the elastic stress concentration factor $\boldsymbol{B}^{ME}$.

In transforming the yield function to the damaged composite system, the first step is to transform $\bar{f}^M$ from the configuration $\overline{C}^M$ into a function f^M in the configuration C^M. This can be accomplished by substituting equations (8.14a) and (8.25a) into equation (8.24). Simplifying and using the first identity of equation (8.16b), one obtains the transformed yield function f^M in the damaged matrix configuration C^M:

$$f^M = \frac{3}{2}(\boldsymbol{\sigma}^M - \boldsymbol{\beta}^M) : (\mathbf{M}^M : \mathbf{N}^M) : (\boldsymbol{\sigma}^M - \boldsymbol{\beta}^M) - \overline{\sigma}_0^{M^2} = 0 \tag{8.26}$$

The second step involves transforming f^M from the configuration C^M into an overall yield function f in the configuration C. This is performed by substituting equations (8.7a) and (8.25b) (using total stresses and elastic concentration factors) into equation (8.26) and simplifying. One then obtains:

$$f = \frac{3}{2}(\boldsymbol{\sigma} - \boldsymbol{\beta}) : \mathbf{H}^M : (\boldsymbol{\sigma} - \boldsymbol{\beta}) - \overline{\sigma}_0^{M^2} = 0 \tag{8.27}$$

where the fourth-rank tensor $\boldsymbol{H}^M$ includes the effects of the matrix stress and damage, and is given by:

$$\mathbf{H}^M = \mathbf{M}^M : \mathbf{N}^M : \mathbf{B}^{ME} : \mathbf{B}^{ME} \tag{8.28}$$

The yield function of equation (8.27) is of the anisotropic type and represents the overall yield function for the damaged composite system.

Next, one studies the flow rule of plastic strain for the damaged composite system. One starts by postulating an associated flow rule for the matrix plastic strain in the undamaged configuration $\overline{C}^M$ as follows:

$$d\overline{\varepsilon}^{M''} = d\overline{\Lambda}^M \frac{\partial \bar{f}^M}{\partial \overline{\boldsymbol{\sigma}}^M} \tag{8.29}$$

where $d\overline{\Lambda}^M$ is a scalar multiplier to be determined from the consistency condition. The damage transformation equations for the plastic strain rate are shown in detail in Appendix A-2. Substituting for $d\overline{\varepsilon}^{M''}$ from equation (A17) into equation (8.29) and utilizing equation (8.13a) in the form:

192

$$\frac{\partial \bar{f}^M}{\partial \bar{\boldsymbol{\sigma}}^M} = \mathbf{M}^{M^{-1}} : \frac{\partial f^M}{\partial \boldsymbol{\sigma}^M} \tag{8.30}$$

one obtains:

$$d\,\varepsilon^{M''} = d\,\boldsymbol{\Lambda}^M : \frac{\partial f^M}{\partial \boldsymbol{\sigma}^M} + \overset{*}{\varepsilon}{}^{M''} \tag{8.31}$$

where the fourth-rank tensor $d\Lambda^M$ and the second-rank tensor $\overset{*}{\varepsilon}{}^{M''}$ are given by:

$$d\,\boldsymbol{\Lambda}^M = d\overline{\boldsymbol{\Lambda}}^M\,\mathbf{X}^{M^{-1}} : \mathbf{M}^{M^{-1}} \tag{8.32}$$

$$\overset{*}{\varepsilon}{}^{M''} = -\mathbf{X}^{M^{-1}} : \mathbf{Z}^M \tag{8.33}$$

Equation (8.31) represents the flow rule in the damaged matrix configuration C^M. It is clearly non-associated due to the presence of damage effects. Finally, substituting equation (8.11a) into equation (8.31) and utilizing equation (8.7a) in the form:

$$\frac{\partial f^M}{\partial \boldsymbol{\sigma}^M} = \mathbf{B}^{ME^{-1}} : \frac{\partial f}{\partial \boldsymbol{\sigma}} \tag{8.34}$$

one obtains:

$$d\,\varepsilon'' = d\boldsymbol{\Lambda} : \frac{\partial f}{\partial \boldsymbol{\sigma}} + \overset{*}{\varepsilon}{}'' \tag{8.35}$$

where the fourth-rank tensorial multiplier $d\Lambda$ and the second-rank strain rate tensor $\overset{*}{\varepsilon}{}''$ are give by:

$$d\,\boldsymbol{\Lambda} = \mathbf{A}^{M^{-1}} : d\,\boldsymbol{\Lambda}^M : \mathbf{B}^{ME^{-1}} \tag{8.36}$$

$$\overset{*}{\varepsilon}{}'' = \overset{*}{\varepsilon}{}^{M''} : \mathbf{A}^{M^{-1}} \tag{8.37}$$

The overall tensorial multiplier $d\Lambda$ can be written explicitly in terms of the matrix scalar multiplier $d\overline{\Lambda}^M$. This is done by substituting equation (8.32) into equation (8.36) and simplifying. Therefore, one obtains:

$$d\Lambda = d\overline{\Lambda}^M \, \mathbf{A}^{M^{-1}} : \mathbf{X}^{M^{-1}} : \mathbf{M}^{M^{-1}} : \mathbf{B}^{ME^{-1}} \tag{8.38}$$

Substituting equation (8.33) into equation (8.37), one obtains:

$$\overset{*}{\varepsilon}'' = -\mathbf{X}^{M^{-1}} : \mathbf{Z}^M : \mathbf{A}^{M^{-1}} \tag{8.39}$$

Equations (8.38) and (8.39) along with equation (8.35) represent the overall nonassociated flow rule for the damaged composite system. The non-associativity is primarily due to the inclusion of local damage effects through the tensors $\mathbf{M}^M$, $\mathbf{X}^M$ and $\mathbf{Z}^M$. In addition, the matrix stress and strain are also reflected in the nonassociated rule through the concentration factors $\mathbf{A}^M$ and $\mathbf{B}^{ME}$. The results obtained in this section are summarized in Table 8.1.

8.5 Kinematic Hardening

In this section, a kinematic hardening rule is derived for the damaged composite system. One first starts with a local kinematic hardening rule of the Prager-Ziegler type (Ziegler [129]) for the undamaged matrix in the form:

$$d\overline{\boldsymbol{\alpha}}^M = d\overline{\mu}^M \, (\overline{\boldsymbol{\tau}}^M - \overline{\boldsymbol{\alpha}}^M) \tag{8.40}$$

where $d\overline{\mu}^M$ is a local scalar multiplier to be determined shortly. The first step in the derivation is to transform the law in equation (8.40) into the damaged matrix configuration C^M. Substituting equations (8.14a) and (8.25a) (while taking the material time derivative of equation (8.25a)) into equation (8.40) and simplifying one obtains:

$$d\overline{\boldsymbol{\beta}}^M = d\overline{\mu}^M \boldsymbol{\sigma}^M - (d\overline{\mu}^M \mathbf{I}_2 \otimes \mathbf{I}_2 + \mathbf{N}^{M^{-1}} : d\mathbf{N}^M) : \boldsymbol{\beta}^M \tag{8.41}$$

where the material time derivative $d\mathbf{N}^M$ is defined in a manner similar to $d\mathbf{M}$, as given in equation (A16). The next step in the derivation involves transforming equation (8.41) into the damaged composite configuration C. One first starts by subtracting equation (8.25b) from equation (8.7a) (using total stresses and elastic concentration factors), and differentiating the resulting equation. Therefore, one obtains:

$$(d\boldsymbol{\sigma}^M - d\boldsymbol{\beta}^M) = \mathbf{B}^{ME} : (d\boldsymbol{\sigma} - d\boldsymbol{\beta}) + d\mathbf{B}^{ME} : (\boldsymbol{\sigma} - \boldsymbol{\beta}) \tag{8.42}$$

Substituting for $d\boldsymbol{\sigma}^M$ from equation (8.7a) into equation (8.42) and solving for $d\beta$, one obtains

$$d\,\mathbf{B} = (\mathbf{B}^{ME^{-1}} : \mathbf{B}^{ME} - \mathbf{B}^{ME^{-1}} : \mathbf{B}^M) : d\boldsymbol{\sigma} + \mathbf{B}^{ME^{-1}} : d\,\mathbf{B}^{ME} : \boldsymbol{\sigma}$$

$$- \mathbf{B}^{ME^{-1}} : d\,\mathbf{B}^{ME} : \mathbf{B} + d\,\mathbf{B}^M : \mathbf{B}^{ME^{-1}} \tag{8.43}$$

Finally, substituting equation (8.41) into equation (8.43) and simplifying, one obtains:

$$d\boldsymbol{\beta} = (\boldsymbol{\Psi} : \boldsymbol{\sigma} - \mathbf{X} : \boldsymbol{\beta}) + \boldsymbol{\Pi} : d\boldsymbol{\sigma} \tag{8.44}$$

where the fourth-rank tensors, $\boldsymbol{\Psi}$, χ and $\boldsymbol{\Pi}$ are given by:

$$\boldsymbol{\Psi} = \mathbf{B}^{ME^{-1}} : d\,\mathbf{B}^{ME} + d\,\bar{\mu}^M \mathbf{B}^{ME^{-1}} \tag{8.45a}$$

$$\chi = -\mathbf{B}^{ME^{-1}} : d\,\mathbf{B}^{ME} - \mathbf{B}^{ME^{-1}} : (d\,\bar{\mu}^M \mathbf{I}_2 \otimes \mathbf{I}_2 + \mathbf{N}^{M^{-1}} : d\,\mathbf{N}^M) \cdot \mathbf{B}^M \tag{8.45b}$$

$$\boldsymbol{\Pi} = \mathbf{I}_2 \otimes \mathbf{I}_2 - \mathbf{B}^{ME^{-1}} : \mathbf{B}^M \tag{8.45c}$$

Equation (8.44) represents the overall kinematic hardening rule in the damaged composite system. It consists of a combination of a generalized Prager-Ziegler rule (the term $\boldsymbol{\Psi}{:}\boldsymbol{\sigma} - \chi{:}\boldsymbol{\beta}$) and a Phillips-type rule (the term $\boldsymbol{\Pi}{:}d\boldsymbol{\sigma}$). This section is concluded with a brief discussion of how to determine the scalar multiplier $d\bar{\mu}^M$. One assumes that the projection of $d\bar{\boldsymbol{\alpha}}^M$ on the gradient of the yield surface $\bar{f}^M$ in the matrix configuration $\overline{C}^M$ is equal to $b\,d\bar{\varepsilon}^{M''}$, where b is a material parameter to be determined from the uniaxial test [55, 131]. This assumption is given by:

$$b\,d\bar{\varepsilon}^{M''} = \cfrac{d\bar{\boldsymbol{\alpha}}^M : \cfrac{\partial \bar{f}^M}{\partial \bar{\boldsymbol{\sigma}}^M}}{\cfrac{\partial \bar{f}^M}{\partial \bar{\boldsymbol{\sigma}}^M} : \cfrac{\partial \bar{f}^M}{\partial \bar{\boldsymbol{\sigma}}^M}} \; \cfrac{\partial \bar{f}^M}{\partial \bar{\boldsymbol{\sigma}}^M} \tag{8.46}$$

Substituting for $d\bar{\boldsymbol{\alpha}}^M$ and $d\bar{\varepsilon}^{M''}$ from equations (8.40) and (8.29), respectively, into equation (8.46) and post-multiplying the resulting equation by $\partial \bar{f}^M / \partial \bar{\boldsymbol{\alpha}}^M$, one obtains the following expression for $d\bar{\mu}^M$ in terms of $d\bar{\Lambda}^M$:

$$d\overline{\mu}^{M} = b\,d\overline{\Lambda}^{M}\ \frac{\dfrac{\partial \overline{f}^{M}}{\partial \overline{\sigma}^{M}} : \dfrac{\partial \overline{f}^{M}}{\partial \overline{\sigma}^{M}}}{(\overline{\tau}^{M} - \overline{\alpha}^{M}) : \dfrac{\partial \overline{f}^{M}}{\partial \overline{\sigma}^{M}}} \qquad (8.47)$$

The relation in equation (8.47) is applicable to any matrix yield function $\overline{f}^{M}$. If the yield function of equation (8.24) is used, then equation (8.47) reduces to the simple form:

$$d\overline{\mu}^{M} = 3b\,d\overline{\Lambda}^{M} \qquad (8.48)$$

The evolution law of equation (8.44) shows the complexity of kinematic hardening in the damaged composite system. Fortunately, it is possible to derive the damage constitutive model without using equation (8.44) directly. This is clearly shown in detail in the next section.

8.6 Constitutive Model

Derivation of the constitutive model for the damaged composite system is performed in two steps. The first step involves the derivation of separate constitutive equations for the matrix and fibers in their respective damaged configurations C^{M} and C^{F}. This is followed by the second step, which combines the two constitutive equations into one for the whole composite system. One first starts with the following two local linear elastic relations for the matrix and fibers in their respective undamaged configurations $\overline{C}^{M}$ and $\overline{C}^{F}$ as follows:

$$\overline{\sigma}^{M} = \overline{\mathbf{E}}^{M} : \overline{\varepsilon}^{M'} \qquad (8.49a)$$

$$\overline{\sigma}^{F} = \overline{\mathbf{E}}^{F} : \overline{\varepsilon}^{F'} \qquad (8.49b)$$

where $\overline{E}^{M}$ and $\overline{E}^{F}$ are the constant fourth-rank elasticity tensors for the matrix and fiber materials, respectively. During unloading or elastic loading, one can obtain the local response of each constituent in the following form [136]:

Table 8.1: Expressions for the yield function, flow rule, and kinematic hardening rule in the three configurations $\overline{C}^M$, C^M and C.

Rule	Configuration		
	$\overline{C}^M$	C^M	C
Yield Function	$\overline{f}^M = \dfrac{3}{2}\,(\overline{\tau}^M - \overline{\alpha}^M):$ $(\overline{\tau}^M - \overline{\alpha}^M) - \overline{\sigma}_0^{M2}$	$f^M = \dfrac{3}{2}(\sigma^M - \beta^M):(\mathbf{M}^M:\mathbf{N}^M):(\sigma^M - \beta^M)$ $-\,\overline{\sigma}_0^{M2}$	$f = \dfrac{3}{2}\,(\sigma - \beta):\mathbf{H}^M:(\sigma - \beta)$ $-\,\overline{\sigma}_0^{M2}$
Flow Rule	$d\,\overline{\varepsilon}^{M''} = d\overline{\Lambda}^M\,\dfrac{\partial \overline{f}^M}{\partial \overline{\sigma}^M}$	$d\,\varepsilon^{M''} = d\,\Lambda^M:\dfrac{\partial f^M}{\partial \sigma^M} + \overset{*}{\varepsilon}{}^{M''}$	$d\,\varepsilon'' = d\,\Lambda:\dfrac{\partial f}{\partial \sigma} + \overset{*}{\varepsilon}{}''$
Kinematic Hardening Rule	$d\,\overline{\alpha}^M = d\,\overline{\mu}^M\,(\overline{\tau}^M - \overline{\alpha}^M)$	$d\,\beta^M = d\,\overline{\mu}^M\,\sigma$ $-\,(d\,\overline{\mu}^M\mathbf{I}_2 \otimes \mathbf{I}_2 + \mathbf{N}^{M^{-1}}:d\,\mathbf{N}^M):\beta^M$	$\beta = (\psi:\sigma - \chi:\beta) + \Pi:d\sigma$

$$\sigma^M = \mathbf{E}^M : \varepsilon^{M'} \tag{8.50a}$$

$$\sigma^F = \mathbf{E}^F : \varepsilon^{F'} \tag{8.50b}$$

where the local "damaged" elasticity tensors E^M and E^F are given by:

$$\mathbf{E}^M = \frac{\bar{c}^M}{c^M} \mathbf{M}^{-1} : \overline{\mathbf{E}}^M \cdot \mathbf{M}^{M^{-1}} \tag{8.51a}$$

$$\mathbf{E}^F = \frac{\bar{c}^F}{c^F} \mathbf{M}^{-1} : \overline{\mathbf{E}}^F : \mathbf{M}^{F^{-1}} \tag{8.51b}$$

In this case, the overall response of the damaged composite system is given by [133]:

$$\sigma = \mathbf{E} : \varepsilon' \tag{8.52}$$

where the overall damaged elasticity tensor E is given by:

$$\mathbf{E} = c^M \mathbf{A}^M \cdot \mathbf{E}^M + c^F \mathbf{A}^{FE} \cdot \mathbf{E}^F \tag{8.53}$$

Before proceeding to formulate the system's response to plastic loading, one needs to adopt a certain evolution law for the damage tensor. The damage evolution criterion to be used here is that proposed by Lee, et al. [15] and is given by the function $g(y, L)$:

$$g(y, L) = \frac{1}{2} y : J : y - L(\ell) \tag{8.54}$$

where $L(\ell)$ is a scalar function of the overall scalar damage parameter ℓ and J is a constant fourth-rank tensor given previously in section 5.2. In equation (8.54), the second-rank tensor y is the generalized thermodynamic force associated with the second-rank damage tensor φ given by [8, 96]:

$$y = \frac{\partial U}{\partial \varphi} \tag{8.55}$$

where U is the free energy of the damaged material. This definition is made considering that $d\varphi{:}y$ is the power dissipated due to damage. Details of the derivations of the kinetic equation for the

198

damage tensor φ were given previously in Chapter 5 for metals. These derivations involve the use of the first and second laws of thermodynamics and can be applied here in the overall description.

Next, one is ready to undertake the second step in the derivation which involves the incorporation of damage in the matrix and fiber materials. One first starts by assuming that the undamaged matrix constitutive equation takes the form:

$$d\overline{\sigma}^M = \overline{D}^M : d\overline{\varepsilon}^M \tag{8.56}$$

where the damaged matrix elastoplastic stiffness tensor $\overline{D}^M$ is given by [96]:

$$\overline{D}^M = \overline{E}^M - \frac{1}{\overline{Q}}\left(\frac{\partial \overline{f}^M}{\partial \overline{\tau}^M} : \overline{E}^M\right) \otimes \left(\overline{E}^M : \frac{\partial \overline{f}^M}{\partial \overline{\sigma}^M}\right) \tag{8.57}$$

and the scalar quantity $\overline{Q}$ is given by:

$$\overline{Q} = \frac{\partial \overline{f}^M}{\partial \overline{\tau}^M} : \overline{E}^M : \frac{\partial \overline{f}^M}{\partial \overline{\tau}^M} - b\frac{\partial \overline{f}^M}{\partial \overline{\alpha}^M} : (\overline{\tau}^M - \overline{\alpha}^M)\frac{\dfrac{\partial \overline{f}^M}{\partial \overline{\sigma}^M} : \dfrac{\partial \overline{f}^M}{\partial \overline{\sigma}^M}}{(\overline{\tau}^M - \overline{\alpha}^M) : \dfrac{\partial \overline{f}^M}{\partial \overline{\sigma}^M}} \tag{8.58}$$

Considering the undamaged matrix constitutive law of equation (8.49a) and utilizing the damage theory for solids given in Chapter 6, one obtains the transformed damaged matrix constitutive equation in the form:

$$d\sigma^M = D^M : d\varepsilon^M + G^M \tag{8.59}$$

where the fourth-rank tensor D^M and the second-rank tensor G^M are given by:

$$D^M = O^{M^{-1}} : \overline{D}^M \cdot X^M \tag{8.60a}$$

$$G^M = O^{M^{-1}} : \overline{D}^M : Z^M \tag{8.60b}$$

In equations (8.60), the fourth-rank matrix tensor O^M is given by [99]:

$$O^M = \mathbf{M}^M + \frac{\dfrac{\partial \mathbf{M}^M}{\partial \varphi^M} : \dfrac{\partial g^M}{\partial \sigma^M} : \sigma^M \otimes \dfrac{\partial g^M}{\partial \sigma^M}}{\dfrac{\partial L^M}{\partial \ell^M} - \dfrac{\partial g^M}{\partial \varphi^M} : \dfrac{\partial g^M}{\partial \sigma^M}}$$

$$+ D^M \cdot \mathbf{M}^{M^{-T}} : \mathbf{E}^{M^{-1}} - \mathbf{X}^M : \mathbf{E}^{M^{-1}}$$

$$- \mathbf{M}^{M^{-T}} : \frac{\partial \mathbf{M}^{M^T}}{\partial \varphi^M} : \frac{\dfrac{\partial g^M}{\partial \sigma^M} \otimes \dfrac{\partial g^M}{\partial \sigma^M}}{\dfrac{\partial L^M}{\partial \ell^M} - \dfrac{\partial g^M}{\partial \varphi^M} : \dfrac{\partial g^M}{\partial \sigma^M}} \mathbf{M}^{M^{-T}} \otimes \sigma^M : \mathbf{E}^{M^{-1}} \qquad (8.61)$$

where g^M, φ^M, L^M and ℓ^M are the local matrix counterparts of g, φ, L, and ℓ, respectively. The nature of the sixth-rank tensor $\partial M^M / \partial \varphi^M$ is discussed in Appendix A-2. Equation (8.59) represents the constitutive law for the damaged matrix material, while equation (8.50b) represents the constitutive law for the damaged fiber material. The corresponding damage-elastic and damage-elastoplastic stiffness tensors for the matrix and fiber material are given, respectively, by equations (8.51b) and (8.60a). The final step in the formulation is to transform the local constitutive equation for the damaged matrix and fibers into one single overall constitutive law for the damaged composite system.

Taking the material time derivative of equation (8.50b), substituting the resulting equation along with equation (8.59) into equation (8.1) and simplifying, one obtains:

$$d\sigma = D : d\varepsilon + G \qquad (8.62)$$

where the overall damage-elastoplastic fourth-rank stiffness tensor D is given by:

$$D = c^M \mathbf{A}^M : D^M + c^F \mathbf{A}^{FE} : \mathbf{E}^F \qquad (8.63)$$

and the overall second-rank tensor G is given by:

$$G = c^M G^M + c^F \mathbf{A}^{FE} : d\mathbf{E}^F \tag{8.64}$$

Equation (8.62) represents the overall constitutive relation for the damaged composite system. The elastoplastic stiffness tensor D is given explicitly in terms of stiffness tensors of the two constituents and includes the effects of damage as clearly shown by the relevant expressions of D^M and E^F given in equations (8.60a) and (8.51b), respectively. It should be noted that the constitutive relation in equation (8.62) represents a nonlinear transformation which makes the numerical implementation of this model very difficult. This is mainly due to the additional term G which can be considered as some additional stress due to the damaging process. In fact, the term G is a function of the overall stress and strain tensors and is clearly illustrated in equations (8.60b), (8.64) and (A11). Therefore, the damage constitutive relation derived in the formulation is of the general type $d\sigma = f(\sigma, \varepsilon, d\varepsilon)$, whereas the corresponding constitutive relation for an undamaged material is of the general form $d\sigma = f(d\varepsilon)$. However, the constitutive law for a damaged homogeneous material (made of one phase only) is of the form $d\sigma = f(\sigma, d\varepsilon)$.

CHAPTER 9

EQUIVALENCE OF THE OVERALL AND LOCAL APPROACHES

The objective of this chapter is to demonstrate under similar assumptions that both the local and overall approaches give similar results when applied to fiber-reinforced metal matrix composites. Both elastic and inelastic composites are considered. The fibers are assumed to be continuous and perfectly aligned. In addition, a perfect bond is assumed to exist at the matrix-fiber interface. A consistent mathematical formulation is used to show the equivalence of the two approaches in this case. The elastic and plastic stiffness matrices are derived using both approaches and each is shown to be identical in both cases.

9.1   Elastic Behavior of Composites

The equivalence of the overall and local damage approaches will be shown first for elastic composites. For simplicity, the composite system is assumed to consist of a matrix reinforced with continuous fibers. Both the matrix and fibers are linearly elastic with different material constants. Let $\overline{C}$ denote the configuration of the undamaged composite system and let $\overline{C}^m$ and $\overline{C}^F$ denote the configurations of the undamaged matrix and fibers, respectively. Since the composite system assumes a perfect bond at the matrix-fiber interface, it is clear that $\overline{C}^m \cap \overline{C}^F = \varnothing$ and $\overline{C}^m \cup \overline{C}^F = \overline{C}$. In the overall approach, the problem reduces to transforming the undamaged configuration $\overline{C}$ into the damaged configuration C. In contrast, two intermediate configurations C^m and C^F are considered in the local approach for the matrix and fibers, respectively. In the latter approach, the problem is reduced to transforming each of the undamaged configurations $\overline{C}^m$ and $\overline{C}^F$ into the damaged configuration C^m and C^F, respectively.

In the case of elastic fiber-reinforced composites, the following linear relation is used for each constituent in its respective undamaged configuration $\overline{C}^k$:

$$\overline{\sigma}^k = \overline{E}^k : \overline{\varepsilon}^k \ , \quad k = M, F \tag{9.1}$$

where $\overline{\sigma}^k$ is the effective constituent stress tensor, $\overline{\varepsilon}^k$ is the effective constituent strain tensor, and $\overline{E}^k$ is the effective constituent elasticity tensor. For the case of an isotropic constituent, $\overline{E}^k$ is given by the formula:

202

$$\overline{\boldsymbol{E}}^k = \overline{\lambda}^k \, \boldsymbol{I}_2 \otimes \boldsymbol{I}_2 \, + \, 2\overline{\mu}^k \, \boldsymbol{I}_4 \qquad (9.2)$$

where $\overline{\lambda}^k$ and $\overline{\mu}^k$ are the effective constituent Lame's constants, $\boldsymbol{I}_2$ is the second-rank identity tensor and $\boldsymbol{I}_4$ is the fourth-rank identity tensor.

Within the framework of the micromechanical analysis of composite materials, the effective constituent stress tensor $\overline{\boldsymbol{\sigma}}^k$ is related to the effective composite stress tensor $\overline{\boldsymbol{\sigma}}$ by:

$$\overline{\boldsymbol{\sigma}}^k = \overline{\boldsymbol{B}}^k : \overline{\boldsymbol{\sigma}} \qquad (9.3)$$

The fourth-rank tensor $\overline{\boldsymbol{B}}^k$ is the constituent stress concentration tensor. It can be determined through several available models such as the Voigt and Mori-Tanaka models (Mori and Tanaka [106]; Lagoudas, et al., [108]; Weng [137]). The interested reader is referred to the work of Dvorak and Bahei-El-Din [70] and Voyiadjis and Kattan [104, 138] for a detailed examination. The effective constituent strain tensor $\overline{\boldsymbol{\varepsilon}}^k$ is determined in a similar way by the equation:

$$\overline{\boldsymbol{\varepsilon}}^K = \overline{\boldsymbol{A}}^K : \overline{\boldsymbol{\varepsilon}} \qquad (9.4)$$

where $\overline{\boldsymbol{\varepsilon}}$ is the effective composite strain tensor and $\overline{\boldsymbol{A}}^k$ is the fourth-rank strain concentration tensor.

Next, the overall and local approaches to damage in elastic composites are examined and compared.

9.1.1   Overall Approach:

In this approach, damage is incorporated in the composite system as a whole through one damage tensor called the overall damage tensor. The two steps needed in this approach are shown schematically in Figure 9.1 for a two-phase composite system consisting of a matrix and fibers. In the first step, the elastic equations are formulated in an undamaged composite system. This is performed here using the law of mixtures as follows:

$$\overline{\boldsymbol{\sigma}} = \overline{c}^M \, \overline{\boldsymbol{\sigma}}^M \, + \, \overline{c}^F \, \overline{\boldsymbol{\sigma}}^F \qquad (9.5)$$

where $\overline{c}^M$ and $\overline{c}^F$ are the effective matrix and fiber volume fractions, respectively.

In the effective composite configuration $\overline{C}$, the following linear elastic relation holds:

$$\overline{\sigma} = \overline{E} : \overline{\varepsilon}$$

(9.6)

where $\overline{E}$ is the fourth-rank constant elasticity tensor. Substituting equations (9.1), (9.4), and (9.6) into equation (9.5) and simplifying, one obtains the following expression for $\overline{E}$:

$$\overline{E} = \overline{c}^M \, \overline{E}^M : \overline{A}^M + \overline{c}^F \, \overline{E}^F : \overline{A}^F$$

(9.7)

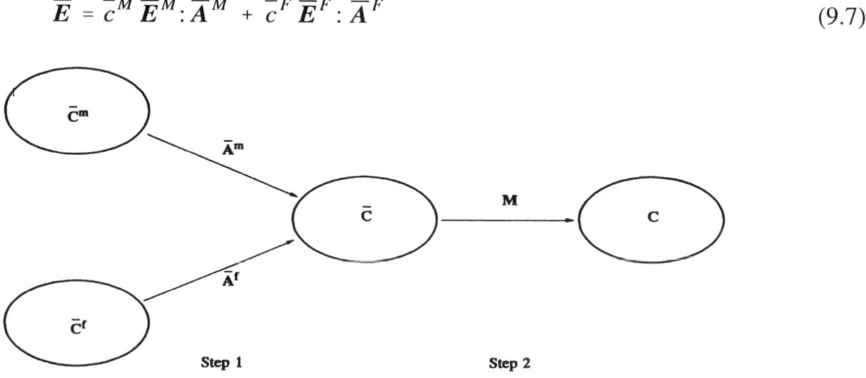

Figure 9.1 Overall Application for an Elastic Composite

In the second step of the formulation, damage is induced through the fourth-rank damage effect tensor **M** as follows:

$$\overline{\sigma} = M : \sigma$$

(9.8)

where σ is the composite stress tensor. Equation (9.8) represents the damage transformation equation for the stress tensor. In order to derive a similar relation for the strain tensor, one needs to use the hypothesis of elastic energy equivalence [16]. In this hypothesis, the elastic energy of the damaged system is equal to the elastic energy of the effective system. Applying this hypothesis to the composite system by equating the two elastic energies, one obtains:

$$\frac{1}{2}\,\varepsilon : \sigma = \frac{1}{2}\,\overline{\varepsilon} : \overline{\sigma}$$

(9.9)

where ε is the composite strain tensor. Substituting equation (9.8) into equation (9.9) and simplifying, one obtains the damage transformation equation for the strain tensor as follows:

$$\bar{\varepsilon} = M^{-T} : \varepsilon \qquad\qquad (9.10)$$

In order to derive the final elastic relation in the damaged composite system, one substitutes equations (9.8) and (9.10) into equation (9.6) to obtain:

$$\sigma = E : \varepsilon \qquad\qquad (9.11)$$

where the fourth-rank elasticity tensor E is given by:

$$E = M^{-1} : \bar{E} : M^{-T} \qquad\qquad (9.12a)$$

Substituting for $\bar{E}$ from equation (9.7) into equation (9.12a), one obtains:

$$E = M^{-1} : (\bar{c}^M \bar{E}^M : \bar{A}^M + \bar{c}^F \bar{E}^F : \bar{A}^F) : M^{-T} \qquad\qquad (9.12b)$$

The above equation represents the elasticity tensor in the damaged composite system according to the overall approach.

9.1.2 Local Approach:

In this approach, damage is introduced in the first step of the formulation using two independent damage tensors for the matrix and fibers. However, more damage tensors may be introduced to account for other types of damage such as debonding and delamination. The two steps involved in this approach are shown schematically in Figure 9.2. One first introduces the fourth-rank matrix and fiber damage effect tensor $\mathbf{M}^M$ and $\mathbf{M}^F$, respectively, as follows:

$$\bar{\sigma}^k = M^k : \sigma^k , \quad k = M, F \qquad\qquad (9.13)$$

The above equation can be interpreted in a similar way to equation (9.8), except that it applies at the constituent level. It also represents the damage transformation equation for each constituent stress tensor. In order to derive a similar transformation equation for the constituent strain tensor, one applies the hypothesis of elastic energy equivalence to each constituent separately as follows:

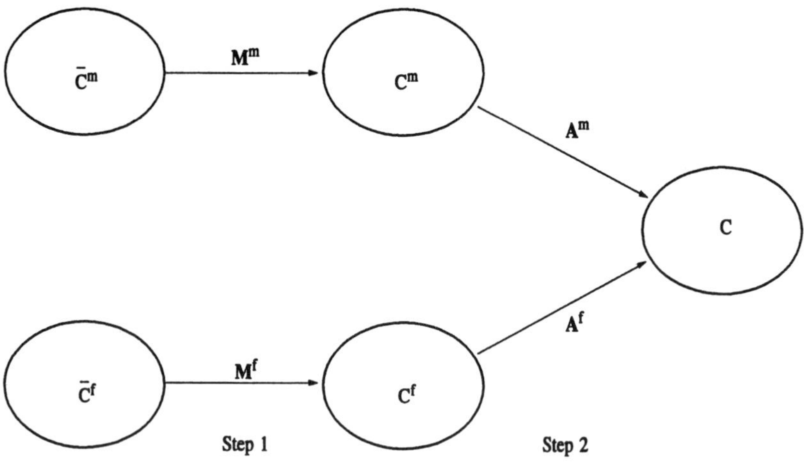

Figure 9.2 Local Approach for an Elastic Composite

$$\frac{1}{2}\,\varepsilon^k : \sigma^k = \frac{1}{2}\,\bar{\varepsilon}^k : \bar{\sigma}^k \tag{9.14}$$

In using equation (9.14), one assumes that there are no micromechanical or constituent elastic interactions between the matrix and fibers. This assumption is not valid in general. From micromechamical considerations, there should be interactions between the elastic energies in the matrix and fibers. However, such interactions are beyond the scope of this work as the resulting equations will be complicated and the sought equivalence relations will consequently be unattainable. It should be clear to the reader that equation (9.14) is the single most important assumption that is needed to prove the equivalence between the overall and local (constituent) approaches. In fact, the proposed proof hinges entirely on the assumption given by equation (9.14). Therefore, the sought equivalence between the overall and local (constituent) approaches is a very special case when equation (9.14) is valid.

Substituting equation (9.13) into equation (9.14) and simplifying, one obtains the required transformation equation for the constituent strain tensor:

206

$$\overline{\varepsilon}^k = M^{k^{-T}} : \varepsilon^k \tag{9.15}$$

The above equation implies a decoupling between the elastic energy in the matrix and fibers. Other methods may be used that include some form of coupling but they will lead to complicated transformation equations that are beyond the scope of this chapter.

Substituting equations (9.13) and (9.15) into equation (9.1) and simplifying, one obtains:

$$\sigma^k = E^k : \varepsilon^k \tag{9.16}$$

where the constituent elasticity tensor $\mathbf{E}^k$ is given by:

$$E^k = M^{k^{-1}} : \overline{E}^k : M^{k^{-T}} \tag{9.17}$$

Equation (9.16) represents the elasticity equation for the damaged constituents. The second step of the formulation involves transforming equation (9.17) into the whole composite system using the law of mixtures as follows:

$$\sigma = c^M \sigma^M + c^F \sigma^F \tag{9.18}$$

where c^M and c^F are the matrix and fiber volume fractions, respectively, in the damaged composite system. Before proceeding with equation (9.18), one needs to derive a strain constituent equation similar to equation (9.4). Substituting equations (9.10) and (9.15) into equation (9.4) and simplifying, one obtains:

$$\varepsilon^k = A^k : \varepsilon \tag{9.19}$$

where the constituent strain concentration tensor $\mathbf{A}^k$ in the damaged state is given by:

$$A^k = M^{k^T} : \overline{A}^k : M^{-T} \tag{9.20}$$

The above equation represents the damage transformation equation for the strain concentration tensor.

Finally, one substitutes equations (9.11), (9.16) and (9.19) into equation (9.18) and simplifies to obtain:

$$E = c^M E^M : A^M + c^F E^F : A^F$$

(9.21)

Equation (9.21) represents the elasticity tensor in the damaged composite system according to the local approach.

9.1.3 Equivalence of the Two Approaches:

In this section, it is shown that both the overall and local approaches are equivalent for elastic composites. This is performed by showing that both the elasticity tensors given in equations (9.12b) and (9.21) are the same. In fact, it is shown that equation (9.21) reduces to equation (9.12b) after making the appropriate substitutions.

First, one needs to find a damage transformation equation for the volume fractions. This is performed by substituting equations (9.8) and (9.13) into equation (9.5), simplifying and comparing the result with equation (9.18). One therefore obtains:

$$c^k I_4 = \overline{c}^k M^{-1} : M^k \quad , \quad k = M, F$$

(9.22)

Substituting equations (9.17) and (9.20) into equation (9.21) and simplifying one obtains:

$$E = (c^M M^{M^{-1}} : \overline{E}^M : \overline{A}^M + c^F M^{F^{-1}} : \overline{E}^F : \overline{A}^F) : M^{-T}$$

(9.23)

Finally, one substitutes equation (9.22) into equation (9.23) and simplifies to obtain:

$$E = M^{-1} : (\overline{c}^M \overline{E}^M : \overline{A}^M + \overline{c}^F \overline{E}^F : \overline{A}^F) : \overline{M}^{-T}$$

(9.24)

It is clear that the above equation is the same as equation (9.12b). Therefore, both the overall and local approaches yield the same elasticity tensor in the damaged composite system.

Equation (9.24) can be generalized to an elastic composite system with n constituents as follows:

$$E = M^{-1} : \left(\sum_{k=1}^{n} \overline{c}^k \overline{E}^k : \overline{A}^k \right) : M^{-T}$$

(9.25)

The two formulations of the overall and local approaches can be used to obtain the above equation for a composite system with n constituents. The derivation of equation (9.25) is similar to the derivation of equation (9.24); therefore, it is not presented here.

In the remaining part of this section, some additional relations are presented to relate the overall damage effect tensor with the constituent damage effect tensors. Substituting equation (9.3) into equation (9.5) and simplifying, one obtains the constraint equation for the stress concentration tensors. The constraint equation is generalized as follows:

$$\sum_{k=1}^{n} \overline{c}^k \, \overline{\boldsymbol{B}}^k = \boldsymbol{I}_4 \tag{9.26}$$

where $\boldsymbol{I}_4$ is the fourth-rank identity tensor. To find a relation between the stress concentration tensors in the effective and damaged states, one substitutes equation (9.8) and (9.13) into equation (9.3) and simplifies to obtain:

$$\sigma^k = \boldsymbol{B}^k : \sigma \ , \quad k = 1, 2, 3, \ldots.,n \tag{9.27}$$

where $\boldsymbol{B}^k$ is the fourth-rank stress concentration tensor in the damaged configuration and is given by:

$$\boldsymbol{B}^k = \boldsymbol{M}^{k^{-1}} : \overline{\boldsymbol{B}}^k : \boldsymbol{M} \ , \quad k = 1, 2, 3, \ldots.,n \tag{9.28}$$

Substituting equation (9.27) into equation (9.18) and simplifying, the resulting constraint is generalized as follows:

$$\sum_{k=1}^{n} c^k \, \boldsymbol{B}^k = \boldsymbol{I}_4 \tag{9.29}$$

Finally, substituting equation (9.28) into equation (9.29) and simplifying, one obtains:

$$\boldsymbol{M} = \left(\sum_{k=1}^{n} c^k \boldsymbol{M}^{k^{-1}} : \overline{\boldsymbol{B}}^k \right)^{-1} \tag{9.30}$$

Equation (9.30) represents the required relation between the overall and local (constituent) damage effect tensors.

9.2   Plastic Behavior of Composites

In this section, the overall and local approaches will be shown to be equivalent for the case of plastic composites. The composite system is assumed to consist of continuous fibers embedded in a matrix. Both the matrix and fibers are plastic obeying the von Mises yield criterion. For simplicity, it is assumed that both materials deform with no isotropic or kinematic hardening. These assumptions are made in order to obtain simple equivalence relations between the two approaches. However, more general cases could be used which will lead to more complex relations. In the derivations that follow, the formulation will be presented for a general case of a composite with n constituents. The case of the two-phase composite (matrix and fibers) can then be easily deduced.

Let C denote the configuration of the damaged composite system and let $\overline{C}^k$ denote the configuration of the undamaged constituent k, where k = 1, 2, 3, . . . ,n. Since a perfect bond is assumed to exist at the constituent interfaces, it is clear that $\overline{C}^k \cap \overline{C}^\ell = \varnothing$ for $k \neq \ell$, $\bigcap_{k=i}^{n} \overline{C}^{k} = \varnothing$, and $\bigcup_{k=i}^{n} \overline{C}^k = \overline{C}$, where $\overline{C}$ is the configuration of the undamaged composite. In the overall approach, the problem reduces to transforming the undamaged configuration $\overline{C}$ into the damaged configuration C. In contrast, there are n intermediate damaged configurations $\overline{C}^k$, k = 1, 2, 3, ..., n which must be considered in the local approach for the n constituents. In the latter approach, the problem is reduced to transforming each undamaged configuration $\overline{C}^k$ into the damaged configuration C^k, for k = 1, 2, 3, ..., n.

In the case of the plastic behavior of composites, the following incremental plastic constitutive relation is used for each constituent in its respective configuration $\overline{C}^k$:

$$d\overline{\sigma}^k = \overline{D}^k : d\overline{\varepsilon}^k \quad , \quad k = 1, 2, 3, \ldots, n \tag{9.31}$$

where $\overline{D}^k$ is the effective constituent elastoplastic stiffness tensor. In this chapter, the formulation is general and no specific form for the tensor $\overline{D}^k$ is used.

The effective constituent stress increment tensor $d\overline{\sigma}^k$ is related to the effective composite stress increment tensor $d\overline{\sigma}$ by a relation similar to equation (9.3) as follows:

$$d\overline{\sigma}^k = \overline{B}^k : d\overline{\sigma} \tag{9.32}$$

where the fourth-rank tensor $\overline{\boldsymbol{B}}^k$ in this case is the constituent plastic stress concentration factor. Similarly, the effective constituent strain increment tensor $d\overline{\varepsilon}^k$ is determined by an equation similar to equation (9.4) as follows:

$$d\overline{\varepsilon}^k = \overline{\boldsymbol{A}}^k : d\overline{\varepsilon} \qquad (9.33)$$

where $\overline{\boldsymbol{A}}^k$ in this case is the fourth-rank constituent plastic strain concentration tensor. The reader should keep in mind that the concentration tensors in equations (9.32) and (9.33) are not the same as those of equations (9.3) and (9.4). In the next two subsections, the overall and local approaches to damage in plastic composites are formulated, followed by the proof of their equivalence.

9.2.1   Overall Approach:

Figure 9.3 shows a schematic diagram of the overall approach for a plastic composite system that consists of n constituents. This approach consists of two steps. The first step involves formulating the plastic constitutive model in an undamaged composite system. This is performed using the law of mixtures for n constituents in the effective composite configuration as follows:

$$d\overline{\sigma} = \sum_{k=1}^{n} \overline{c}^k \, d\overline{\sigma}^k \qquad (9.34)$$

where $\overline{c}^k$ is the effective constituent volume fraction. In the effective composite configuration, the following elastoplastic constitutive equation holds:

$$d\overline{\sigma} = \overline{\boldsymbol{D}} : d\overline{\varepsilon} \qquad (9.35)$$

where $\overline{\boldsymbol{D}}$ is the fourth-rank effective elastoplastic stiffness tensor. Substituting equations (9.31), (9.33) and (9.35) into equation (9.34) and simplifying, one obtains the following expression for $\overline{\boldsymbol{D}}$:

$$\overline{\boldsymbol{D}} = \sum_{k=1}^{n} \overline{c}^k \overline{\boldsymbol{D}}^k : \overline{\boldsymbol{A}}^k \qquad (9.36)$$

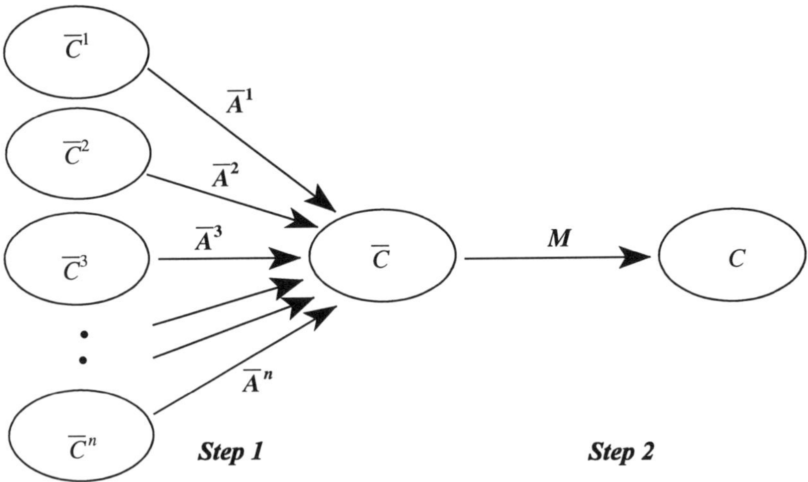

Figure 9.3   Overall Approach for a Plastic Composite

Damage is introduced in the second step of the formulation through the fourth-rank damage effect tensor **M** as follows:

$$\overline{\sigma} = M : \sigma \tag{9.37}$$

The above damage transformation equation for the stress tensor is exactly the same one used in the case of elastic composites (i.e. equation (9.8)). This is attributed mainly to the fact that the incorporation of damage in the constitutive model is basically independent of the material model used. This also makes the present formulation general in nature so that any viable material model can be used within the framework of this formulation. Therefore, in this chapter, no attempt will be made to describe the yield function or the flow rule as they have no bearing on the results obtained.

Since the plasticity formulation is incremental, one needs to derive the incremental form of equation (9.37). It should be noticed that the damage effect tensor **M** is not constant and one needs to evaluate its increment. The increment d**M** of the tensor **M** is determined from the model of damage evolution that is adopted. Again, no specific model for damage evolution will be discussed in this chapter in order to keep the formulation as general as possible. In this work, a general form of damage evolution is adopted by using the equation:

$$d\,\varphi = \boldsymbol{F} : d\,\boldsymbol{\sigma} \tag{9.38}$$

where φ is the second-rank damage tensor for the composite, and $\mathbf{F}$ is a fourth-rank tensor that depends on the rule of damage evolution used. As emphasized before, no specific form for the tensor $\mathbf{F}$ will be used, although the reader is referred to Chow and Wang [14] and Voyiadjis and Kattan [104] for some examples. See also the sections on damage evolution in the previous chapters.

The nature of the damage effect tensor has been studied before (section 6.3) where it is clear that $\mathbf{M}$ is a function of the damage tensor φ, i.e. $\mathbf{M} \equiv \mathbf{M}(\varphi)$. The increment of $\mathbf{M}$ is therefore given by:

$$d\,\boldsymbol{M} = \frac{\partial \boldsymbol{M}}{\partial \varphi} : d\,\varphi \tag{9.39}$$

where $\partial\mathbf{M}/\partial\varphi$ is a sixth-rank tensor. Substituting equation (9.38) into equation (9.39), post-multiplying the result by σ, and simplifying, one obtains:

$$d\,\boldsymbol{M} : \boldsymbol{\sigma} = C : d\,\boldsymbol{\sigma} \tag{9.40}$$

where the sixth-rank tensor $\mathbf{C}$ is given by:

$$C = \left(\frac{\partial \boldsymbol{M}}{\partial \varphi} : \boldsymbol{\sigma} \right) : \boldsymbol{F} \tag{9.41}$$

Taking the increment of equation (9.37), one obtains:

$$d\,\overline{\boldsymbol{\sigma}} = \boldsymbol{M} : d\,\boldsymbol{\sigma} + d\boldsymbol{M} : \boldsymbol{\sigma} \tag{9.42}$$

Substituting equation (9.40) into equation (9.42), and simplifying, one obtains:

$$d\,\overline{\boldsymbol{\sigma}} = (\boldsymbol{M} + \boldsymbol{C}) : d\,\boldsymbol{\sigma} \tag{9.43}$$

Equation (9.43) represents the incremental form of the damage transformation equation for the stress tensor. In order to derive a similar transformation equation for the strain tensor, one can show that the following relation is valid:

$$dM^T : \bar{\varepsilon} = C_o : d\bar{\varepsilon} \tag{9.44}$$

where the tensor C_o is given by:

$$C_o = \bar{\varepsilon} : \frac{\partial M^T}{\partial \varphi} : F : (M + C)^{-1} : \bar{D} \tag{9.45}$$

The similarities between equations (9.40) and (9.44) should be obvious. In deriving equation (9.44), we have used equations (9.39), (9.38), (9.43) and (9.35) exactly in this order.

Taking the increment of equation (9.10), one obtains:

$$d\varepsilon = M^T : d\bar{\varepsilon} + dM^T : \bar{\varepsilon} \tag{9.46}$$

Substituting equation (9.44) into equation (9.46) and simplifying, one obtains the desired transformation equation for dε:

$$d\bar{\varepsilon} = (M^T + C_o)^{-1} : d\varepsilon \tag{9.47}$$

The above equation represents the incremental form of the damage transformation equation for the strain tensor. Next, one proceeds with the second step of the formulation. In order to derive the final elastoplastic constitutive relation in the damaged composite system, one substitutes equations (9.43) and (9.47) into equation (9.35) to obtain:

$$d\sigma = D : d\varepsilon \tag{9.48}$$

where the fourth-rank elastoplastic tensor D is given by:

$$D = (M + C)^{-1} : \bar{D} : (M^T + C_o)^{-1} \tag{9.49a}$$

Substituting for $\bar{D}$ from equation (9.36) into equation (9.49a), one obtains:

$$D = (M + C)^{-1} : \left(\sum_{n=1}^{k} \bar{c}^k \, \bar{D}^k : \bar{A}^k \right) : (M^T + C_o)^{-1} \tag{9.49b}$$

214

The above equation represents the elastoplastic stiffness tensor in the damaged composite system. Equation (9.48) is the elastoplastic constitutive relation in the damaged composite system. Comparing equation (9.49b) with equation (9.25), one notices similarities between the two transformation equations.

9.2.2 Local Approach:

In Figure 9.4, a schematic diagram of the local approach for a plastic composite system that consists of n constituents is presented. There are two steps for the determination of the elastoplastic fourth-rank stiffness tensor **D**. The first step involves formulating the elastoplastic constituent stiffnesses $\mathbf{D}^k$ in the damaged state, for k = 1, 2, 3, ... , n. Let the constituent damage equation be given by expression (9.13). In incremental form, equation (9.13) is expressed as follows:

$$d\,\overline{\sigma}^k = dM^k : \sigma^k + M^k : d\sigma^k \quad , \quad k = 1, 2, 3, \ldots, n \tag{9.50}$$

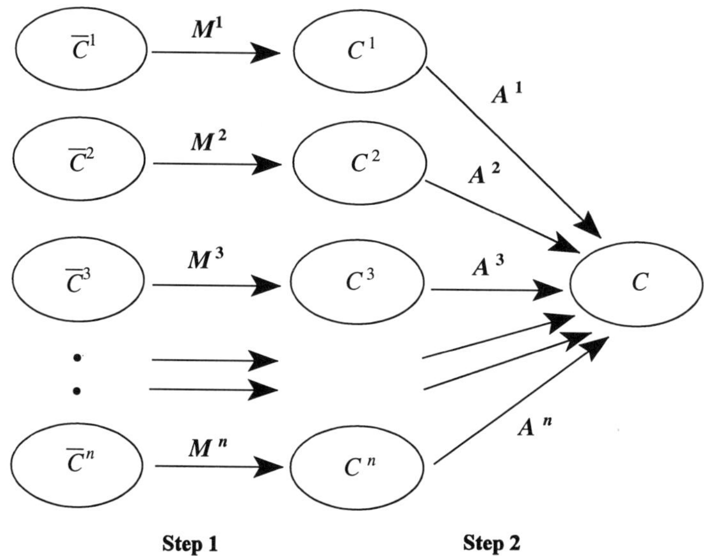

Figure 9.4   Local Approach for a Plastic Composite

Making use of equation (9.40), the first term on the right-hand-side of equation (9.50) may be expressed as follows:

$$dM^k : \sigma^k = C^k : d\sigma^k \quad , \quad k = 1, 2, 3, \ldots, n \tag{9.51}$$

where the sixth-rank tensor $\mathbf{C}^k$ is given as follows, similarly to that of equation (9.41):

$$C^k = \left(\frac{\partial M^k}{\partial \varphi^k} : \sigma^k \right) : F^k \quad , \quad k = 1, 2, 3, \ldots, n \tag{9.52}$$

and

$$d\varphi^k = F^k : d\sigma^k \quad , \quad k = 1, 2, 3, \ldots, n \tag{9.53}$$

In equation (9.53), φ^k is the constituent second-rank damage tensor and $\mathbf{F}^k$ is a fourth-rank tensor that depends on the rule of damage evolution used for the constituent k. Substituting equation (9.51) into expression (9.50), one obtains the relation:

$$d\overline{\sigma}^k = (M^k + C^k) : d\sigma^k \quad , \quad k = 1, 2, 3, \ldots, n \tag{9.54}$$

Following the derivation of equation (9.47), one can derive a similar transformation equation for the constituent incremental strain tensor $d\varepsilon^k$:

$$d\overline{\varepsilon}^k = \left(M^{k^T} + C_o^k \right)^{-1} : d\varepsilon^k \quad , \quad k = 1, 2, 3, , \ldots, n \tag{9.55}$$

where the tensor $\mathbf{C}^k_o$ is given by:

$$C_o^k = \overline{\varepsilon}^k : \frac{\partial M^{k^T}}{\partial \varphi^k} : F^k : (M^k + C^k)^{-1} : \overline{D}^k \quad , \quad k = 1, 2, 3, \ldots, n \tag{9.56}$$

and $\overline{D}^k$ is the fourth-rank elastoplastic constituent stiffness tensor in the undamaged configuration. In deriving equation (9.55), we have used:

$$dM^{k^T} : \overline{\varepsilon}^k = C_o^k : d\overline{\varepsilon}^k \quad , \quad k = 1, 2, 3, \ldots, n \tag{9.57}$$

The derivation of the above relation is similar to the derivation of equation (9.44) and is not shown here.

216

The constituent stiffness tensor appearing in equation (9.56) is defined by the constitutive equation:

$$d\bar{\boldsymbol{\sigma}}^k = \bar{\boldsymbol{D}} : d\bar{\boldsymbol{\varepsilon}}^k \quad , \quad k = 1, 2, 3, \ldots, n \tag{9.58a}$$

Substituting for $d\bar{\boldsymbol{\sigma}}^k$ and $d\bar{\boldsymbol{\varepsilon}}^k$ from relations (9.54) and (9.55), respectively, into equation (9.58a), one obtains:

$$d\boldsymbol{\sigma}^k = \boldsymbol{D}^k : d\boldsymbol{\varepsilon}^k \quad , \quad k = 1, 2, 3, \ldots, n \tag{9.58b}$$

where

$$\boldsymbol{D}^k = (\boldsymbol{M}^k + \boldsymbol{C}^k)^{-1} : \bar{\boldsymbol{D}}^k : (\boldsymbol{M}^{k^T} + \boldsymbol{C}_o^k)^{-1} \quad , \quad k = 1, 2, 3, \ldots, n \tag{9.59}$$

and $\mathbf{D}^k$ is the elastoplastic constituent stiffness tensor in the damaged configuration.

The second step in this formulation is to evaluate the elastoplastic stiffness tensor $\mathbf{D}$ of the composite in the damaged configuration. Let the incremental stress-strain relation be given as:

$$d\boldsymbol{\sigma} = \boldsymbol{D} : d\boldsymbol{\varepsilon} \tag{9.60}$$

Using the theory of mixtures, one obtains:

$$d\boldsymbol{\sigma} = \sum_{k=1}^{n} c^k d\boldsymbol{\sigma}^k \tag{9.61}$$

or

$$d\boldsymbol{\sigma} = \sum_{k=1}^{n} c^k \boldsymbol{D}^k : d\boldsymbol{\varepsilon}^k \tag{9.62}$$

Making use of equation (9.33) in the damaged configuration, in relation (9.62), one obtains:

$$d\boldsymbol{\sigma} = \sum_{k=1}^{n} c^k \boldsymbol{D}^k : \boldsymbol{A}^k : d\boldsymbol{\varepsilon} \tag{9.63}$$

Comparing equations (9.60) and (9.63), the stiffness tensor $\mathbf{D}$ is given by:

$$D = \sum_{k=1}^{n} c^{k} D^{k} : A^{k} \tag{9.64}$$

9.2.3 Equivalence of the Two Approaches:

The equivalence of the two approaches is obtained by showing that equation (9.64) is equivalent to equation (9.49b). Making use of equation (9.33) and substituting for $d\bar{\varepsilon}^{k}$ and $d\bar{\varepsilon}$ from equations (9.55) and (9.46), respectively, one obtains the following relation:

$$(M^{k^{T}} + C_{o}^{k})^{-1} : d\varepsilon^{k} = \bar{A}^{k} : (M^{T} + C_{o})^{-1} : d\varepsilon \quad , \quad k = 1, 2, 3, \ldots n \tag{9.65}$$

In the damaged configuration, one obtains:

$$d\varepsilon^{k} = A^{k} : d\varepsilon \quad , \quad k = 1, 2, 3, \ldots, n \tag{9.66}$$

Comparing equations (9.65) and (9.66), one obtains:

$$A^{k} = (M^{k^{T}} + C_{o}^{k}) : \bar{A}^{k} : (M^{T} + C_{o})^{-1} \quad , \quad k = 1, 2, 3, \ldots, n \tag{9.67}$$

The above equation represents an explicit relation between $\mathbf{A}^{k}$ and $\bar{\mathbf{A}}^{k}$.

One can also obtain a similar relation between c^{k} and $\bar{c}^{k}$. Making use of the theory of mixtures in the undamaged configuration, one obtains:

$$d\sigma = \sum_{k=1}^{n} \bar{c}^{k} d\bar{\sigma}^{k} \tag{9.68}$$

Substituting for $d\bar{\sigma}$ and $d\bar{\sigma}^{k}$ from equations (9.43) and (9.54) into equation (9.68), one obtains:

$$(M + C) : d\sigma = \sum_{k=1}^{n} \bar{c}^{k} (M^{k} + C^{k}) : d\sigma^{k} \tag{9.69}$$

or

218

$$d\boldsymbol{\sigma} = \sum_{k=1}^{n} \bar{c}^{k} (M + C)^{-1} : (M^{k} + C^{k}) : d\boldsymbol{\sigma}^{k} \qquad (9.70)$$

Comparing equations (9.70) and (9.61), one obtains a relation between c^{k} and $\bar{c}^{k}$ such that:

$$c^{k} I_{4} = \bar{c}^{k} (M + C)^{-1} : (M^{k} + C^{k}) \qquad (9.71)$$

Substituting for c^{k} and $\mathbf{A}^{k}$ from equations (9.71) and (9.67), respectively, into equation (9.64), one obtains:

$$\boldsymbol{D} = \sum_{k=1}^{n} \bar{c}^{k} (M + C)^{-1} : \overline{\boldsymbol{D}}^{k} : \overline{\boldsymbol{A}}^{k} : (M^{T} + C_{o})^{-1} \qquad (9.72)$$

which is identical to that of the overall approach given by equation (9.49b).

Both the overall and local approaches yield equivalent relations in terms of the elastic and elastoplastic stiffness matrices in the damaged configuration. Consistently derived overall-local relations are used to prove the equivalence of the two approaches. This equivalence allows one to use the less complex overall approach for the numerical analysis of boundary value problems but yet obtain the same level of accuracy as that of the local damage approach. However, at the same time, evolution laws for the local damage variables should be used and linked to the overall damage through the use of equation (9.30) for the case of the elastic behavior of composites. A similar equation could be obtained for the case of the plastic behavior of composites. Using this procedure allows one to use the simpler overall approach but nevertheless evaluate the damage associated with each constituent of the composite.

CHAPTER 10

METAL MATRIX COMPOSITES - LOCAL AND INTERFACIAL DAMAGE

A local approach is used in this chapter in the sense that damage is introduced at the constituent (local) level. Three fourth-rank damage tensors $\mathbf{M}^M$, $\mathbf{M}^F$ and $\mathbf{M}^D$ are used for the two constituents (matrix and fibers) of the composite system. The matrix damage effect tensor $\mathbf{M}^M$ is assumed to reflect all types of damage that the matrix material undergoes like nucleation and coalescence of voids and microcracks. The fiber damage effect tensor $\mathbf{M}^F$ is considered to reflect all types of fiber damage such as fracture of fibers. Finally, the interfacial damage effect tensor $\mathbf{M}^D$ represents the interfacial damage between the matrix and the fibers. The overall damage tensor $\mathbf{M}$ is introduced that accounts for all the separate damages $\mathbf{M}^M$, $\mathbf{M}^F$ and $\mathbf{M}^D$ that are present in a single lamina. The damage due to lamination is beyond the scope of this chapter and will be introduced in forthcoming research.

10.1 Assumptions:

The metal matrix composite used in this chapter consists of an elastoplastic ductile metal matrix reinforced with elastic aligned continuous fibers. The composite system is restricted to small deformations with infinitesimal strains. In the initial configuration C_0, the composite material is assumed to be undeformed and undamaged. The initial matrix and fiber subconfigurations are denoted by C_0^M and C_0^F, respectively. Due to applied loads, the composite material is assumed to undergo elastoplastic deformation and damage and the resulting overall configuration is denoted by C. The resulting matrix and fiber local subconfigurations are denoted by C^M and C^F, respectively. Damage is quantified using the concept proposed by Kachanov [1] whereby two kinds of fictitious configurations $\overline{C}$ and $\tilde{C}$ of the composite system are considered. The configuration $\overline{C}$ is obtained from C by removing all the damages, while the configuration $\tilde{C}$ is obtained from C by removing only the interfacial damage between the matrix and the fibers. In this chapter, $\overline{C}$ is termed the full effective configuration, while $\tilde{C}$ is termed the partial effective configuration.

A coupling formulation of plastic flow and damage propagation seems to be impossible due to the presence of the two different dissipative mechanisms that influence each other. For example, the position of slip planes affects the orientation of nucleated microcracks. One can, however, assume that the energy dissipated in the yielding and damaging processes be independent of each other and apply a phenomenological model of interaction. Use will be made of the concept of effective stress [5]. Assuming a fictitious undamaged system, the dissipation energy due to plastic flow in this undamaged system is assumed to be equal to the dissipation energy due to plastic flow in the real damaged system.

The basic feature of the approach presented in this chapter is that local effects of damages are considered whereby these affects are described separately by the matrix, fiber, and interfacial damage. This approach is schematically shown in Figure 10.1 where the undamaged matrix and fiber configurations $\overline{C}^M$ and $\overline{C}^F$, respectively, are transformed to their respective damaged configurations $\tilde{C}^M$ and $\tilde{C}^F$ through the fourth-rank damage effect tensors $\mathbf{M}^M$ and $\mathbf{M}^F$. The tensor $\mathbf{M}^M$ reflects damage in the matrix only and accordingly the tensor $\mathbf{M}^F$ reflects damage in the fibers only. Subsequent to this local damage description, the local-overall relations are used to transform the local damage effects to the whole composite system in the configuration $\tilde{C}$. This is accomplished through the stress concentration tensors $\tilde{\boldsymbol{B}}^M$ and $\tilde{\boldsymbol{B}}^F$ of the matrix and fibers, respectively. This approach clearly represents a local damage description which is depicted in Figure 10.1. The effect of interfacial damage between the fibers and the matrix is represented by a serial transformation $\mathbf{M}^D$ that transforms the configuration $\tilde{C}$ to the final damaged configuration C. Figure 10.1 represents two independent parallel transformations followed by one independent serial transformation. Considering the transformations outlined in Figure 10.1, the local nature of damage in this approach is clear whereby the different damages are separately isolated. This approach can be summarized in the following three steps. First, apply the local damage effect tensors $\mathbf{M}^M$ and $\mathbf{M}^F$ to the local effective configurations $\overline{C}^M$ and $\overline{C}^F$, respectively. This is followed by applying the damage stress concentration factors $\tilde{\boldsymbol{B}}^M$ and $\tilde{\boldsymbol{B}}^F$ to the local partial effective configuration $\tilde{C}^M$ and $\tilde{C}^F$ in order to obtain the overall partial effective configuration $\tilde{C}$. Finally, one applies the interfacial damage effect tensor $\mathbf{M}^D$ to the overall partial effective configuration $\tilde{C}$ to obtain the overall damaged configuration C.

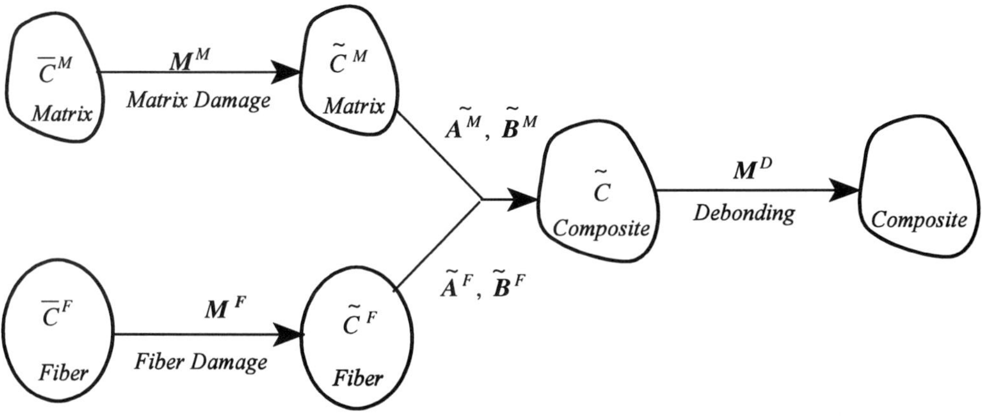

Figure 10.1

The tensor $\mathbf{M}^M$ encompasses all the pertinent damage related to the matrix while the tensor $\mathbf{M}^F$ reflects the damage pertinent to the fibers [138]. However, the interfacial damage $\mathbf{M}^D$ is related to the interfacial damage variable φ^D (second-rank tensor). An interfacial damage variable can be defined through the use of Figure 10.2 which shows an RVE (Representative Volume Element) such that:

$$\phi^D = \frac{S - \overline{S}}{S} \tag{10.1}$$

where S is the total interfacial length, between the fiber and the matrix, and $\overline{S}$ is the effective (net) resisting length corresponding to the total interfacial length in contact.

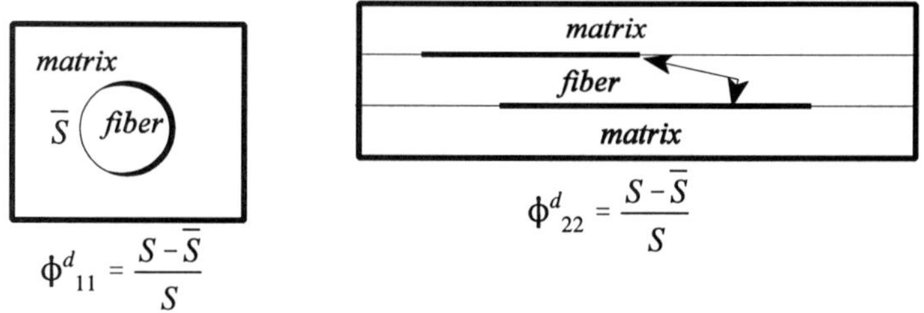

$\overline{S}$ = net resisting length   S = total interfacial length

Figure 10.2

10.2 Theoretical Formulation of the Damage Tensor M:

Considering the overall configurations C, $\tilde{C}$ and $\overline{C}$, one can introduce an overall damage effect tensor $\mathbf{M}$ for the whole composite system. This tensor is defined similarly to the definitions of $\mathbf{M}^M$, $\mathbf{M}^F$ and $\mathbf{M}^D$ such that:

$$\overline{\sigma} = \mathbf{M} : \sigma \tag{10.2}$$

The tensor $\mathbf{M}$ reflects all types of damages that the composite system undergoes including the damage due to the interaction between the matrix and fibers. This tensor has been studied previously in Chapter 6. A matrix representation was explicitly derived for this fourth-rank tensor by expressing the stresses in vector form. The tensor $\mathbf{M}$ was shown to be symmetric. The symmetry property of the tensor $\mathbf{M}$ is used extensively in the derivation that follows. The same holds true for the tensors $\mathbf{M}^M$, $\mathbf{M}^F$ and $\mathbf{M}^D$. Similar to $\mathbf{M}^D$, both tensors $\mathbf{M}^M$ and $\mathbf{M}^F$ could be represented in terms of second-rank tensors φ^M and φ^F, respectively [110, 138]. The overall effective Cauchy stress $\overline{\sigma}$ is related to the local effective Cauchy stresses $\overline{\sigma}^M$ and $\overline{\sigma}^F$ by making use of the micromechanical model proposed by Dvorak and Bahei-El-Din [68-70] and Bahei-El-Din and Dvorak [97] such that:

$$d\overline{\sigma} = \overline{c}^M \, d\overline{\sigma}^M + \overline{c}^F \, d\overline{\sigma}^F \tag{10.3}$$

The effective matrix Cauchy stress and the corresponding fiber Cauchy stress are defined as follows:

$$\overline{\sigma}^M = \mathbf{M}^M : \tilde{\sigma}^M \tag{10.4}$$

$$\overline{\sigma}^F = \mathbf{M}^F : \tilde{\sigma}^F \tag{10.5}$$

where $\tilde{\sigma}^M$ and $\tilde{\sigma}^F$ are the partial effective stresses in the configurations $\tilde{C}^M$ and $\tilde{C}^F$, respectively. These stresses are termed partial since the interfacial damage has not yet been incorporated into the formulation. The incremental relations for equations (10.4) and (10.5) are obtained through the material time differentiation of equations (10.4) and (10.5) such that:

$$d\overline{\sigma}^M = d\mathbf{M}^M : \tilde{\sigma}^M + \mathbf{M}^M : d\tilde{\sigma}^M \tag{10.6}$$

$$d\overline{\sigma}^F = d\mathbf{M}^F : \tilde{\sigma}^F + \mathbf{M}^F : d\tilde{\sigma}^F \tag{10.7}$$

Referring to Figure 10.1 and making use of the partial stress concentrations $\tilde{\boldsymbol{B}}^M$ and $\tilde{\boldsymbol{B}}^F$, the corresponding partial effective matrix Cauchy stress and partial effective fiber Cauchy stress are derived in the following form:

$$d\tilde{\sigma}^M = \tilde{\boldsymbol{B}}^M : d\tilde{\sigma} \tag{10.8}$$

$$d\tilde{\sigma}^F = \tilde{\boldsymbol{B}}^F : d\tilde{\sigma} \tag{10.9}$$

The partial effective overall composite Cauchy stress $\tilde{\sigma}$ is defined in terms of the Cauchy stress σ as follows:

$$\tilde{\sigma} = \boldsymbol{M}^D : \sigma \tag{10.10}$$

Taking the material time derivative of equation (10.10) results into the following expression:

$$d\tilde{\sigma} = d\boldsymbol{M}^D : \sigma + \boldsymbol{M}^D : d\sigma \tag{10.11}$$

Making use of equations (10.6) and (10.7) together with equation (10.3), one obtains the following expression:

$$d\overline{\sigma} = \overline{c}^M (d\boldsymbol{M}^M : \tilde{\sigma}^M + \boldsymbol{M}^M d\tilde{\sigma}^M) + \overline{c}^F (d\boldsymbol{M}^F : \tilde{\sigma}^F + \boldsymbol{M}^F : d\tilde{\sigma}^F) \tag{10.12}$$

Substituting into equation (10.12) for the partial effective matrix and fiber stress rates from equations (10.8) and (10.9), respectively, and making use of equation (10.11), the resulting equation is given as follows:

$$\begin{aligned}
d\overline{\sigma} = \overline{c}^M (d\boldsymbol{M}^M : \tilde{\sigma}^M + \boldsymbol{M}^M : \tilde{\boldsymbol{B}}^M : (d\boldsymbol{M}^D : \sigma + \boldsymbol{M}^D : d\sigma)) \\
+ \overline{c}^F (d\boldsymbol{M}^F : \tilde{\sigma}^F + \boldsymbol{M}^F : \tilde{\boldsymbol{B}}^F : (d\boldsymbol{M}^D : \sigma + \boldsymbol{M}^D : d\sigma))
\end{aligned} \tag{10.13}$$

Rearranging the terms in the above equation, one obtains:

$$\begin{aligned}
d\overline{\sigma} = \overline{c}^M (d\boldsymbol{M}^M : \tilde{\sigma}^M + \boldsymbol{M}^M : \tilde{\boldsymbol{B}}^M : d\boldsymbol{M}^D : \sigma) \\
+ \overline{c}^F (d\boldsymbol{M}^F : \tilde{\sigma}^F + \boldsymbol{M}^F : \tilde{\boldsymbol{B}}^F : d\boldsymbol{M}^D : \sigma) \\
+ (\overline{c}^M \boldsymbol{M}^M : \tilde{\boldsymbol{B}}^M : \boldsymbol{M}^D + \overline{c}^F \boldsymbol{M}^F : \tilde{\boldsymbol{B}}^F : \boldsymbol{M}^D) : d\sigma
\end{aligned} \tag{10.14}$$

Taking the time derivative of equation (10.2)

224

$$d\bar{\sigma} = d\mathbf{M} : \sigma + \mathbf{M} : d\sigma \tag{10.15}$$

and comparing terms with equation (10.14), one obtains the following relation:

$$\mathbf{M} = (\bar{c}^M \mathbf{M}^M : \tilde{\mathbf{B}}^M + \bar{c}^F \mathbf{M}^F : \tilde{\mathbf{B}}^F) : \mathbf{M}^D \tag{10.16}$$

The above expression defines the cumulative damage of the composite system as a function of the local matrix and fiber damages $\mathbf{M}^M$ and $\mathbf{M}^F$, respectively as well as the interfacial damage $\mathbf{M}^D$. However, it should be noted that by comparing relations (10.14) and (10.15), one also obtains the following expression:

$$d\mathbf{M} : \sigma = \bar{c}^M \, d\mathbf{M}^M : \tilde{\sigma}^M + \bar{c}^F \, d\mathbf{M}^F : \tilde{\sigma}^F + (\bar{c}^M \mathbf{M}^M : \tilde{\mathbf{B}}^M : d\mathbf{M}^D$$
$$+ \; \bar{c}^F \mathbf{M}^F : \tilde{\mathbf{B}}^F : d\mathbf{M}^D) : \sigma \tag{10.17}$$

10.3 Stress and Strain Concentration Factors:

The matrix and fiber stress concentration factors are defined as fourth-rank tensors. As composites undergo damage, the stress and strain concentration factors do not remain constant. The relations for the effective elastic stress concentration factors for the matrix and fibers in the configuration $\bar{C}$ are given by the following two relations, respectively:

$$\bar{\sigma}^M = \bar{\mathbf{B}}^{ME} : \bar{\sigma} \tag{10.18}$$

$$\bar{\sigma}^F = \bar{\mathbf{B}}^{FE} : \bar{\sigma} \tag{10.19}$$

The above stress concentration factors can be obtained using the Mori-Tanaka method [134] with the corresponding effective volume fractions $\bar{c}^M$ and $\bar{c}^F$ defined in section 10.5. Making use of equations (10.2) and (10.4) in expression (10.18), one obtains:

$$\sigma^M = (\mathbf{M}^{-M} : \bar{\mathbf{B}}^{ME} : \mathbf{M}) : \sigma \tag{10.20a}$$

or

$$\sigma^M = \mathbf{B}^{ME} : \sigma \tag{10.20b}$$

where

$$\boldsymbol{B}^{ME} = \boldsymbol{M}^{-M} : \overline{\boldsymbol{B}}^{ME} : \boldsymbol{M} \qquad (10.21)$$

In equation (10.21), the tensor $\mathbf{B}^{ME}$ is the elastic matrix stress concentration factor in the damaged configuration C. Similarly, the corresponding elastic fiber stress concentration factor in the damaged configuration C may be obtained such that:

$$\boldsymbol{B}^{FE} = \boldsymbol{M}^{-F} : \overline{\boldsymbol{B}}^{FE} : \boldsymbol{M} \qquad (10.22)$$

The variation of the stress concentration tensors $\mathbf{B}^{ME}$ and $\mathbf{B}^{FE}$ with damage is indirectly demonstrated in Figures 10.3 and 10.4. The material properties are shown in Table 10.1. This is demonstrated for a single lamina loaded axially along the fiber direction. Figure 10.3 shows the variation of the ratio of the axial stress in the fiber to the axial stress in the matrix with respect to the axial fiber damage ϕ_{11}^{F} while settling $\phi_{11}^{M} = 0$. In Figure 10.4, the variation of the ratio σ^{M}/σ^{F} or B^{ME}/B^{FE} is plotted with respect to ϕ_{11}^{M} while setting $\phi_{11}^{F} = 0$. In this simplistic loading case, the stress concentration reduces to a single scalar quantity B^{M} for the matrix and B^{F} for the fiber. A linear relation is observed in both Figure 10.3 and 10.4. Next, in Figure 10.5, the ratio between the local phase stress and the overall stress σ^{r}/σ is plotted with respect to the fiber damage ϕ_{11}^{F} (i.e. $\phi_{11}^{M} = 0$), and versus ϕ_{11}^{M} (i.e. $\phi_{11}^{F} = 0$) in Figure 10.6. A nonlinear relation is observed in both Figures 10.5 and 10.6.

226

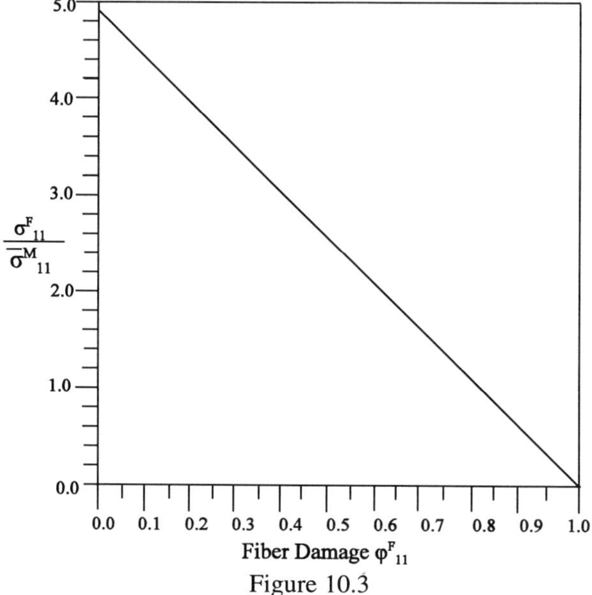

Fiber Damage φ^F_{11}

Figure 10.3

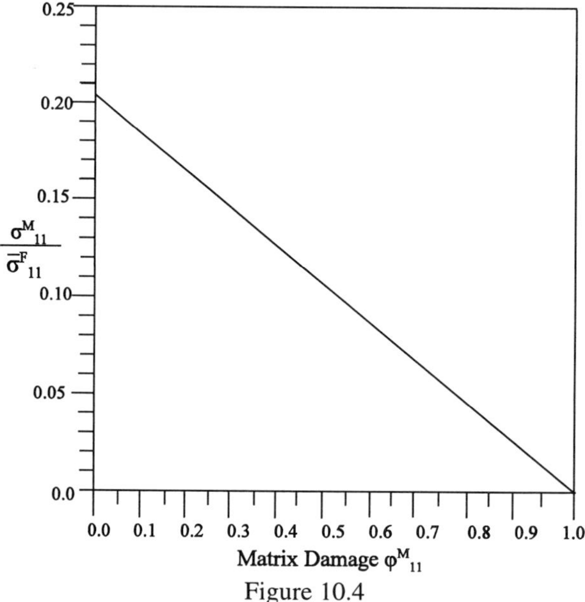

Matrix Damage φ^M_{11}

Figure 10.4

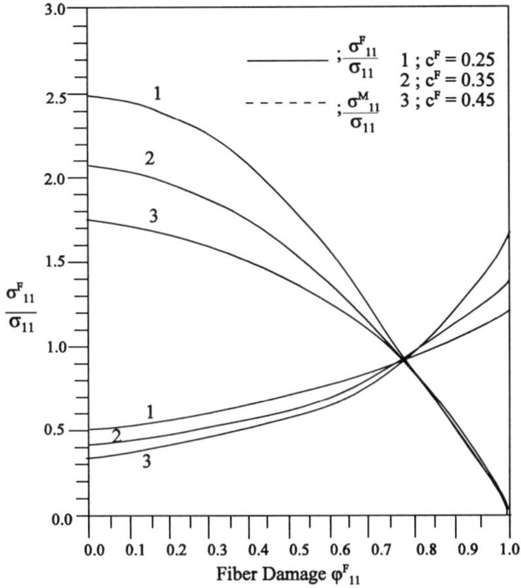

Figure 10.5   Strain Contours for (0/90)$_s$ layup (in %).

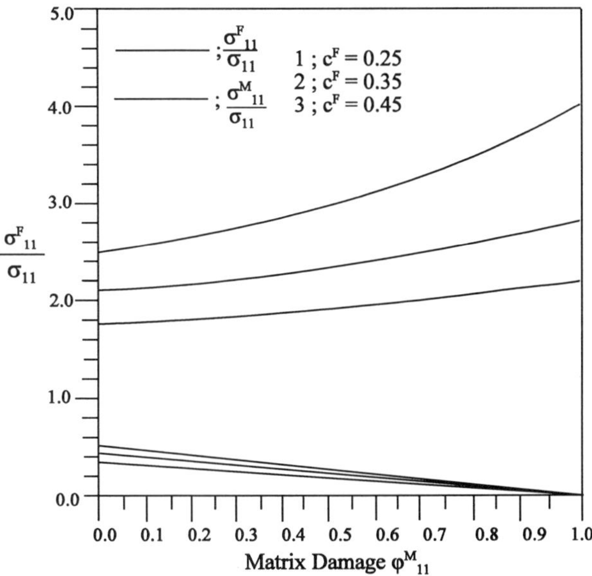

Figure 10.6   Stress contours for (±45)$_s$
layup (units are in MPa).

One assumes a similar relation for strains as that postulated for stresses given by equation (10.3) such that in the effective configurations $\overline{C}^M, \overline{C}^F$ and $\overline{C}$, one obtains:

$$d\overline{\varepsilon} = \overline{c}^M \, d\overline{\varepsilon}^M + \overline{c}^F \, d\overline{\varepsilon}^F \tag{10.23}$$

where $\overline{\varepsilon}^M$ and $\overline{\varepsilon}^F$ are the effective matrix and fiber strain tensors, respectively and $\overline{\varepsilon}$ is the effective overall strain tensor. One also assumes the additive decomposition of the matrix and overall strain rates in $\overline{C}^M$ and $\overline{C}$, respectively, as follows:

$$d\overline{\varepsilon} = d\overline{\varepsilon}' + d\overline{\varepsilon}'' \tag{10.24a}$$

$$d\overline{\varepsilon}^M = d\overline{\varepsilon}^{M'} + d\overline{\varepsilon}^{M''} \tag{10.24b}$$

and

$$d\overline{\varepsilon}^F = d\overline{\varepsilon}^{F'} \tag{10.24c}$$

where $'$ indicates the elastic part and $''$ indicates the plastic part of the tensor. Equations (10.24) are valid in this formulation because infinitesimal strains are assumed throughout here. The fiber strain consists only of the elastic part since the fibers can only deform elastically until they fracture. Consequently, the effective elastic (or total) fiber strain tensor is denoted by $\overline{\varepsilon}^F$.

Table 10.1: Material Properties

	Matrix (Ti-14Al-21 Nb)	Fibers (SiC)
Modulus	8×10^4 MPa	41×10^4 MPa
Poisson's Ratio	0.30	0.22
Initial Volume Fraction	0.65	0.35
Yielding Stress, $\overline{\sigma}_o^M$	550 MPa	-
Kinematic Hardening Parameter, b	90 MPa	-

In the case of the effective elastic strain concentration factors for the matrix and fibers in the configuration $\overline{C}$, one obtains the following expressions:

$$\overline{\varepsilon}^{M} = \overline{A}^{ME} : \overline{\varepsilon} \tag{10.25}$$

$$\overline{\varepsilon}^{F} = \overline{A}^{FE} : \overline{\varepsilon} \tag{10.26}$$

Making use of the following equations relating the effective elastic strains and the corresponding elastic strains [138]:

$$\overline{\varepsilon}' = M^{-1} : \acute{\varepsilon} \tag{10.27}$$

$$\overline{\varepsilon}^{M'} = M^{-M} : \varepsilon^{M'} \tag{10.28}$$

$$\overline{\varepsilon}^{F} = M^{-F} : \varepsilon^{F} \tag{10.29}$$

Using the above equations along with equations (10.25) and (10.26), one obtains the elastic strain concentration factors in the damaged configuration C. These are given by the following relations:

$$A^{ME} = M^{M} : \overline{A}^{ME} : M^{-1} \tag{10.30}$$

$$A^{FE} = M^{F} : \overline{A}^{FE} : M^{-1} \tag{10.31}$$

In order to obtain the corresponding plastic stress and strain concentration factors associated with plastic deformations, the material time differentiation is needed for the equations relating the effective stresses with their corresponding stresses:

$$d\overline{\sigma} = dM : \sigma + M : d\sigma \tag{10.32}$$

$$d\overline{\sigma}^{M} = dM^{M} : \sigma^{M} + M^{M} : d\sigma^{M} \tag{10.33}$$

$$d\overline{\sigma}^{F} = dM^{F} : \sigma^{F} + M^{F} : d\sigma^{F} \tag{10.34}$$

Similarly, the rate equations of (10.18) and (10.19) are required such that:

230

$$d\overline{\sigma}^{M} = \overline{\boldsymbol{B}}^{MP} : d\overline{\sigma} \tag{10.35}$$

$$d\overline{\sigma}^{F} = \overline{\boldsymbol{B}}^{FP} : d\overline{\sigma} \tag{10.36}$$

where $\overline{\boldsymbol{B}}^{MP}$ and $\overline{\boldsymbol{B}}^{FP}$ are the effective instantaneous plastic stress concentration factors. From equations (10.32), (10.33) and (10.35), one obtains:

$$d\boldsymbol{M}^{M} : \sigma^{M} + \boldsymbol{M}^{M} : d\sigma^{M} = \overline{\boldsymbol{B}}^{MP} : (d\boldsymbol{M} : \sigma + \boldsymbol{M} : d\sigma) \tag{10.37}$$

Assuming that

$$d\boldsymbol{M}^{M} : \sigma^{M} = \overline{\boldsymbol{B}}^{MP} : d\boldsymbol{M} : \sigma \tag{10.38}$$

and

$$\boldsymbol{M}^{M} : d\sigma^{M} = \overline{\boldsymbol{B}}^{MP} : \boldsymbol{M} : d\sigma \tag{10.39}$$

then it follows that

$$\sigma^{M} = (\boldsymbol{M}^{-M} : \overline{\boldsymbol{B}}^{MP} : \boldsymbol{M}) : \sigma \tag{10.40}$$

Therefore, we obtain the instantaneous stress concentration factor $\mathbf{B}^{MP}$ for the matrix

$$d\sigma^{M} = \boldsymbol{B}^{MP} : d\sigma \tag{10.41}$$

or

$$\boldsymbol{B}^{MP} = \boldsymbol{M}^{-M} : \overline{\boldsymbol{B}}^{MP} : \boldsymbol{M} \tag{10.42}$$

Similarly, one obtains the instantaneous stress concentration factor $\mathbf{B}^{FP}$ for the fibers such that:

$$\boldsymbol{B}^{FP} = \boldsymbol{M}^{-F} : \overline{\boldsymbol{B}}^{FP} : \boldsymbol{M} \tag{10.43}$$

Similarly, the plastic strain concentration factors can be determined which are given by the following relations:

$$A^{MP} = M^M : \overline{A}^{MP} : M^{-1} \qquad (10.44)$$

$$A^{FP} = M^F : \overline{A}^{FP} : M^{-1} \qquad (10.45)$$

10.4   The Damage Effect Tensor:

The concept of effective stress is used here as given by Kachanov [1] and generalized by Murakami [58]. A linear transformation is assumed between the Cauchy stress tensor σ and the effective Cauchy stress tensor $\overline{\sigma}$ as given previously in equation (10.2). The effective Cauchy stress need not be symmetric or frame-invariant under the given transformation. However, once the effective Cauchy stress is symmetrized, it can be shown that it satisfies the frame invariance principle [99]. Murakami [58] has shown that φ is symmetric and M can be represented by a 6 x 6 matrix as a function of $(I_2 - \varphi)$ in the form:

$$[M]^r = [M \, (I_2 - \phi^r) \,]^r \quad , \quad r = M, F \qquad (10.46)$$

where I_2 is the second-rank identity tensor. In conjunction with the matrix form of M given by equation (10.46), the stress tensor $\tilde{\sigma}^r$ is represented by a vector given by:

$$[\tilde{\sigma}]^r = [\tilde{\sigma}^r_{11} \ \tilde{\sigma}^r_{22} \ \tilde{\sigma}^r_{33} \ \tilde{\sigma}^r_{12} \ \tilde{\sigma}^r_{23} \ \tilde{\sigma}^r_{13}]^T \quad , \quad r = M, F \qquad (10.47)$$

The symmetrized $\overline{\sigma}^r$ used here is given by [127]:

$$\overline{\sigma}^r_{ij} = \frac{1}{2} [\tilde{\sigma}^r_{ik} \, (\delta_{kj} - \phi^r_{kj})^{-1} + (\delta_{il} - \phi^r_{il})^{-1} \tilde{\sigma}^r_{lj}] \quad , \quad r = M, F \qquad (10.48)$$

and

$$\tilde{\sigma}_{ij} = \frac{1}{2} [\sigma_{ik} (\delta_{kj} - \phi^D_{kj})^{-1} + (\delta_{il} - \phi^D_{il})^{-1} \sigma_{lj}] \quad , \quad r = M, F \qquad (10.49)$$

The stresses given by equations (10.48) and (10.49) are frame-independent. Using the symmetrization procedure outlined by equations (10.48) and (10.49), the corresponding 6 x 6 matrix form of the tensor M^r is given by Voyiadjis and Kattan [99] as follows: (see Chapter 11 for more details)

$$[M^r] = \frac{1}{2\nabla^r}
\begin{bmatrix}
2\Psi^r_{22}\Psi^r_{33} - 2\phi^{r2}_{23} & 0 & 0 \\
0 & 2\Psi^r_{11}\Psi^r_{33} - 2\phi^{r2}_{13} & 0 \\
0 & 0 & 2\Psi^r_{11}\Psi^r_{22} - 2\phi^{r2}_{12} \\
\phi^r_{13}\phi^r_{23} + \phi^r_{12}\Psi^r_{33} & \phi^r_{13}\phi^r_{23} + \phi^r_{12}\Psi^r_{33} & 0 \\
0 & \phi^r_{12}\phi^r_{13} + \phi^r_{23}\Psi^r_{11} & \phi^r_{12}\phi^r_{13} + \phi^r_{23}\Psi^r_{11} \\
\phi^r_{12}\phi^r_{23} + \phi^r_{13}\Psi^r_{22} & 0 & \phi^r_{12}\phi^r_{23} + \phi^r_{13}\Psi^r_{22}
\end{bmatrix}$$

$$
\begin{bmatrix}
2\phi^r_{13}\phi^r_{23} + 2\phi^r_{12}\Psi^r_{33} & 0 & 2\phi^r_{12}\phi^r_{23} + 2\phi^r_{13}\Psi^r_{22} \\
2\phi^r_{13}\phi^r_{23} + 2\phi^r_{12}\Psi^r_{33} & 2\phi^r_{12}\phi^r_{13} + 2\phi^r_{23}\Psi^r_{11} & 0 \\
0 & 2\phi^r_{12}\phi^r_{13} + 2\phi^r_{23}\Psi^r_{11} & 2\phi^r_{12}\phi^r_{23} + 2\phi^r_{13}\Psi^r_{22} \\
\Psi^r_{22}\Psi^r_{33} + \Psi^r_{11}\Psi^r_{33} - \phi^{r2}_{23} - \phi^{r2}_{13} & \phi^r_{12}\phi^r_{23} + \phi^r_{13}\Psi^r_{22} & \phi^r_{12}\phi^r_{13} + \phi^r_{23}\Psi^r_{11} \\
\phi^r_{12}\phi^r_{23} + \phi^r_{13}\Psi^r_{22} & \Psi^r_{11}\Psi^r_{33} + \Psi^r_{11}\Psi^r_{22} - \phi^{r2}_{13} - \phi^{r2}_{12} & \phi^r_{13}\phi^r_{23} + \phi^r_{12}\Psi^r_{33} \\
\phi^r_{12}\phi^r_{13} + \phi^r_{23}\Psi^r_{11} & \phi^r_{13}\phi^r_{23} + \phi^r_{12}\Psi^r_{33} & \Psi^r_{22}\Psi^r_{33} + \Psi^r_{11}\Psi^r_{22} - \phi^{r2}_{23} - \phi^{r2}_{12}
\end{bmatrix}$$

$$r = M, F \tag{10.50}$$

where v^r is given by

$$v^r = \Psi^r_{11}\,\Psi^r_{22}\,\Psi^r_{33} - \phi^{r2}_{23}\,\Psi^r_{11} - \phi^{r2}_{13}\,\Psi^r_{22} - \phi^{r2}_{12}\,\Psi^r_{33} - 2\phi^r_{12}\,\phi^r_{23}\,\phi^r_{13} \qquad (10.51)$$

and the notation Ψ^r_{ij} is used to denote $(\delta_{ij} - \phi^r_{ij})$. The variable ϕ^r_{ij} used in equations (10.50) and (10.51) represents ϕ^M_{ij} or ϕ^F_{ij} or ϕ^D_{ij} with respect to matrix damage, fiber damage, or interfacial damage accordingly. The physical characterization of the damage tensor φ is presented in section 10.8 and Chapter 12.

10.5 Effective Volume Fractions

Since the fictitious effective configuration is obtained by removing all the damages that the material has been subjected to, consequently it follows that the volume fractions in the effective configuration will differ from the initial volume fractions. However, the volume fractions of the configuration C are assumed to be equal to the initial volume fractions.

In order to obtain an evolution expression for the effective volume fractions, we first address the simple case of the one-dimensional damage model using the definition of the effective stress concept [1]. The effective local stresses for the matrix and fibers in the one-dimension case are defined by:

$$\bar{\sigma}^M = \frac{1}{1 - \phi^M}\,\sigma^M \qquad (10.52)$$

$$\bar{\sigma}^F = \frac{1}{1 - \phi^F}\,\sigma^F \qquad (10.53)$$

where

$$\phi^M = \frac{dA^M - d\bar{A}^M}{dA^M} \qquad (10.54)$$

$$\phi^F = \frac{dA^F - d\bar{A}^F}{dA^F} \qquad (10.55)$$

and dA^r is a differential area normal to the fiber direction, where r = M or F. The corresponding

234

volume fractions are defined as follows:

$$\overset{\circ}{c}{}^{r} = \frac{d\overset{\circ}{A}{}^{r}}{d\overset{\circ}{A}} \quad , \quad r = M, F \tag{10.56}$$

where

$$d\overset{\circ}{A} = d\overset{\circ}{A}{}^{M} + d\overset{\circ}{A}{}^{F} \tag{10.57}$$

Similarly, the effective volume fractions can be defined such as:

$$\overline{c}{}^{r} = \frac{d\overline{A}{}^{r}}{d\overline{A}} \quad , \quad r = M, F \tag{10.58}$$

where

$$d\overline{A} = d\overline{A}{}^{M} + d\overline{A}{}^{F} \tag{10.59}$$

From relations (10.54) and (10.55), one obtains respectively:

$$d\overline{A}{}^{M} = (1 - \phi^{M}) \, dA^{M} \tag{10.60}$$

$$d\overline{A}{}^{F} = (1 - \phi^{F}) \, dA^{F} \tag{10.61}$$

Substituting relations (10.60) and (10.61) into equation (10.58) and making use of the assumption $\overset{\circ}{c}{}^{r} = c^{r}$ $(r = M, F)$.

$$\frac{dA^{F}}{dA^{M}} = \frac{d\overset{\circ}{A}{}^{F}}{d\overset{\circ}{A}{}^{M}} = \frac{\overset{\circ}{c}{}^{F}}{\overset{\circ}{c}{}^{M}} \tag{10.62}$$

one obtains the following relations:

$$\overline{c}{}^{M} = \frac{(1 - \phi^{M})}{(1 - \phi^{M}) + (1 - \phi^{F}) \dfrac{\overset{\circ}{c}{}^{F}}{\overset{\circ}{c}{}^{M}}} \tag{10.63}$$

$$\overline{c}{}^{F} = \frac{(1 - \phi^{F})}{(1 - \phi^{F}) + (1 - \phi^{M}) \dfrac{\overset{\circ}{c}{}^{M}}{\overset{\circ}{c}{}^{F}}} \tag{10.64}$$

It is noticed that equations (10.62) and (10.63) satisfy the constraint

$$\bar{c}^{M} + \bar{c}^{F} = 1 \tag{10.65}$$

The variation of the effective volume fractions with matrix and fiber damage are shown in Figures 10.7 and 10.8, respectively, for the uniaxially loaded lamina. The initial fiber volume fraction is set equal to 0.45. The generalization of equations (10.63) and (10.64) to the three-dimensional damage model using the second-rank damage tensor φ, may be expressed as follows:

$$\bar{c}^{M} = \frac{(1 - \phi^{M}_{eq.})}{(1 - \phi^{M}_{eq.}) + (1 - \phi^{F}_{eq.})\,\dfrac{\overset{\circ}{c}\,^{F}}{\overset{\circ}{c}\,^{M}}} \tag{10.66}$$

$$\bar{c}^{F} = \frac{(1 - \phi^{F}_{eq.})}{(1 - \phi^{F}_{eq.}) + (1 - \phi^{M}_{eq.})\,\dfrac{\overset{\circ}{c}\,^{M}}{\overset{\circ}{c}\,^{F}}} \tag{10.67}$$

where

$$\phi^{M}_{eq} = \frac{\sqrt{\phi^{M} : \phi^{M}}}{\phi^{M}_{cr}} \tag{10.68a}$$

$$\phi^{F}_{eq} = \frac{\sqrt{\phi^{F} : \phi^{F}}}{\phi^{F}_{cr}} \tag{10.68b}$$

and ϕ^{M}_{cr} and ϕ^{F}_{cr} are the critical values of ϕ^{M}_{eq} and ϕ^{F}_{eq}, respectively, at failure.

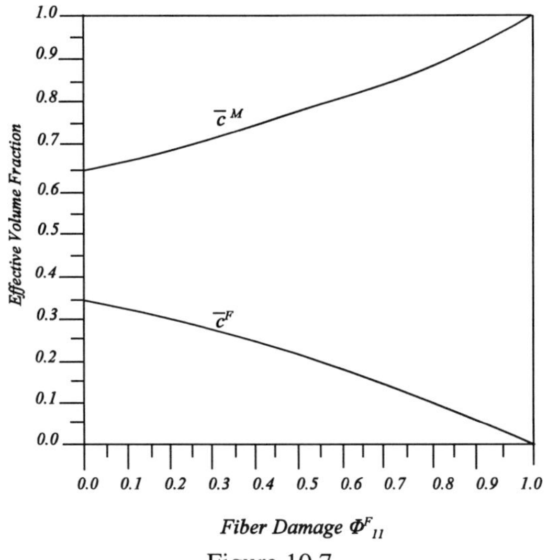

Figure 10.7

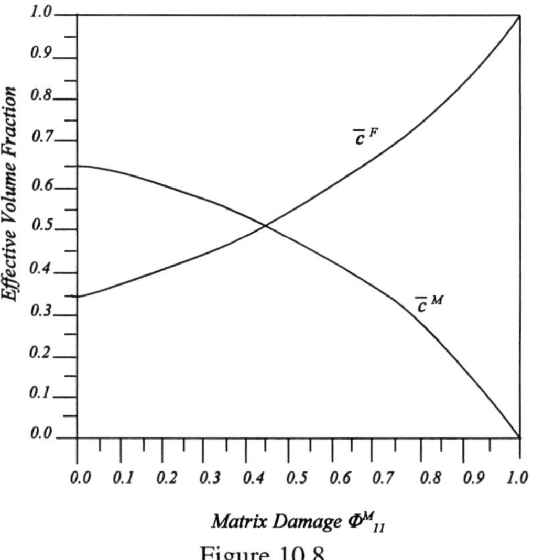

Figure 10.8

10.6 Damage Criterion and Damage Evolution

The damage mechanism for each of the constituents of the composite material is different from the other. The matrix undergoes ductile damage while the fiber undergoes brittle damage. The mechanism of interfacial damage is dependent on the fiber direction. It is clear that one single damage micro-mechanism cannot be considered for the three types of damage outlined above. We therefore consider damage evolution for each mechanism separately.

10.6.1 Damage Criterion

An anisotropic damage criterion is used in this work. In order to obtain a damage criterion for non-proportional loading, the anisotropy of damage increase (hardening) must be considered. This is accomplished by expressing the damage criterion in terms of a tensorial hardening parameter, **h**. The damage criterion used in this section is of the form suggested by Mroz [139] such that

$$g^{\,r} \equiv g^{\,r}(\mathbf{y}^r, \boldsymbol{h}^r) = 0 \quad , \quad r = M, F, D \tag{10.69}$$

where $\mathbf{y}^r$ is a generalized thermodynamic force conjugate to the damage tensor $\boldsymbol{\varphi}^{\,r}$ for each of the damage associated with the matrix, fibers, and debonding. Equation (10.69) represents an isotropic function of the tensors $\mathbf{y}^r$ and $\mathbf{h}^r$ such that

$$g^{\,r} \equiv p^{\,r}_{ijkl}\, y^{\,r}_{ij}\, y^{\,r}_{kl} - 1 = 0 \quad , \quad r = M, F, D \tag{10.70}$$

where

$$p^{\,r}_{ijkl} = h^{-r}_{ij}\, h^{-r}_{kl} \quad , \quad r = M, F, D \tag{10.71}$$

and $p^{\,r}_{ijkl}$ is equivalent to Hill's tensor for yield surfaces. The hardening tensor $\mathbf{h}^r$ is given by:

$$h^{\,r}_{ij} = (u^{\,r}_{ik})^{1/2}\, \phi^{\,r}_{kl}\, (u^{\,r}_{lj})^{1/2} + V^{\,r}_{ij} \quad , \quad r = M, F, D \tag{10.72}$$

Tensors **u** and **V** are defined here for orthotropic materials as follows:

$$[u^r] = \begin{bmatrix} \lambda_1^r \eta_1^r \left(\dfrac{k^r}{\lambda_1^r} \right)^{\xi_1^r} & 0 & 0 \\[2em] 0 & \lambda_2^r \eta_2^r \left(\dfrac{k^r}{\lambda_2^r} \right)^{\xi_2^r} & 0 \\[2em] 0 & 0 & \lambda_3^r \eta_3^r \left(\dfrac{k^r}{\lambda_3^r} \right)^{\xi_3^r} \end{bmatrix} , \quad r = M, F, D$$

$$(10.73)$$

$$[V^r] = \begin{bmatrix} \lambda_1^r {v_1^r}^2 & 0 & 0 \\[1em] 0 & \lambda_2^r {v_2^r}^2 & 0 \\[1em] 0 & 0 & \lambda_3^r {v_3^r}^2 \end{bmatrix} , \quad r = M, F, D \qquad (10.74)$$

The tensors $\mathbf{u}^r$ and $\mathbf{V}^r$ are generalizations to orthotropic materials of the scalar forms for isotropic materials originally proposed by Stumvoll and Swoboda [140]. In expressions (10.73) and (10.74), the scalar quantities λ_1^r, λ_2^r, λ_3^r, v_1^r, v_2^r, v_3^r, ξ_1^r, ξ_2^r, ξ_3^r, η_1^r, η_2^r and η_3^r are material parameters obtained by matching the theory with experimental results. The parameters λ_1^r, λ_2^r, λ_3^r, v_1^r, v_2^r and v_3^r are explicitly related to the physical properties of the material as shown below.

For elastic isotropic materials, the fourth-rank elasticity tensor $\overline{E}$ may be expressed by the following 6 x 6 matrix:

$$
[\overline{E}] = \begin{bmatrix}
\overline{E}_{1111} & \overline{E}_{1122} & \overline{E}_{1122} & 0 & 0 & 0 \\
\overline{E}_{1122} & \overline{E}_{1111} & \overline{E}_{1122} & 0 & 0 & 0 \\
\overline{E}_{1122} & \overline{E}_{1122} & \overline{E}_{1111} & 0 & 0 & 0 \\
0 & 0 & 0 & \frac{1}{2}(\overline{E}_{1111} - \overline{E}_{1122}) & 0 & 0 \\
0 & 0 & 0 & 0 & \frac{1}{2}(\overline{E}_{1111} - \overline{E}_{1122}) & 0 \\
0 & 0 & 0 & 0 & 0 & \frac{1}{2}(\overline{E}_{1111} - \overline{E}_{1122})
\end{bmatrix}
\tag{10.75}
$$

or

$$
[\overline{E}] = \begin{bmatrix}
\lambda + 2\mu & \lambda & \lambda & 0 & 0 & 0 \\
\lambda & \lambda + 2\mu & \lambda & 0 & 0 & 0 \\
\lambda & \lambda & \lambda + 2\mu & 0 & 0 & 0 \\
0 & 0 & 0 & \mu & 0 & 0 \\
0 & 0 & 0 & 0 & \mu & 0 \\
0 & 0 & 0 & 0 & 0 & \mu
\end{bmatrix}
\tag{10.76}
$$

It is clear from expressions (10.75) and (10.76) that:

$$\overline{E}_{1122} = \lambda \quad , \quad \overline{E}_{1212} = \mu \tag{10.77}$$

For an orthotropic material, a similar expression to equation (10.76) may be obtained such that:

$$[\overline{E}] = \begin{bmatrix} \overline{E}_{1111} & \overline{E}_{1122} & \overline{E}_{1133} & 0 & 0 & 0 \\ \overline{E}_{1122} & \overline{E}_{2222} & \overline{E}_{2233} & 0 & 0 & 0 \\ \overline{E}_{1133} & \overline{E}_{2233} & \overline{E}_{3333} & 0 & 0 & 0 \\ 0 & 0 & 0 & \overline{E}_{1212} & 0 & 0 \\ 0 & 0 & 0 & 0 & \overline{E}_{2323} & 0 \\ 0 & 0 & 0 & 0 & 0 & \overline{E}_{1313} \end{bmatrix} \tag{10.78}$$

or

$$[\overline{E}] = \begin{bmatrix} \lambda_1 + 2\mu_1 & \lambda_1 & \lambda_3 & 0 & 0 & 0 \\ \lambda_1 & \lambda_2 + 2\mu_2 & \lambda_2 & 0 & 0 & 0 \\ \lambda_3 & \lambda_2 & \lambda_3 + 2\mu_3 & 0 & 0 & 0 \\ 0 & 0 & 0 & \mu_1 & 0 & 0 \\ 0 & 0 & 0 & 0 & \mu_2 & 0 \\ 0 & 0 & 0 & 0 & 0 & \mu_3 \end{bmatrix} \tag{10.79}$$

By comparing expressions (10.78) and (10.79), similar relations to those obtained for isotropic materials given by expressions (10.77) are derived such that:

$$
\left.
\begin{array}{ll}
\overline{E}_{1122} = \lambda_1 \quad , & \overline{E}_{1212} = \mu_1 \\[2ex]
\overline{E}_{2233} = \lambda_2 \quad , & \overline{E}_{2323} = \mu_2 \\[2ex]
\overline{E}_{1133} = \lambda_3 \quad , & \overline{E}_{1313} = \mu_3
\end{array}
\right\}
\tag{10.80}
$$

We note from equations (10.80) that λ_1^r, λ_2^r and λ_3^r are generalized Lamé constants for an orthotropic material behavior.

In equation (10.74), v_1^r, v_2^r and v_3^r define the initial threshold against damage for the orthotropic material. It is obtained from the constraint that the onset of damage corresponds to the stress level at which the virgin material starts exhibiting nonlinearity.

Referring to equations (10.73), κ^r is a scalar hardening parameter given by:

$$
\kappa^r = \int_0^t - y_{ij}^r \, d\phi_{ij}^r \, dt \quad , \quad r = M, F\,D
\tag{10.81}
$$

As outlined by Stumroll and Swoboda [140], the damaging state is any state that satisfies $g = 0$. Four states are outlined here:

$$
\left\{
\begin{array}{lll}
g^r < 0 & \text{(elastic unloading)} & \tag{10.82a} \\[3ex]
g^r = 0 \,, \; \dfrac{\partial g^r}{\partial y_{ij}^r} \, dy_{ij}^r < 0 & \text{(elastic unloading)} & \tag{10.82b} \\[3ex]
g^r = 0 \,, \; \dfrac{\partial g^r}{\partial y_{ij}^r} \, dy_{ij}^r = 0 & \text{(neutral loading)} & \tag{10.82c} \\[3ex]
g^r = 0 \,, \; \dfrac{\partial g^r}{\partial y_{ij}^r} \, dy_{ij}^r > 0 & \text{(loading from a damaging state)} & \tag{10.82d}
\end{array}
\right.
$$

In this section, the anisotropic damage criterion g is defined by equation (10.70) as well as the loading conditions outlined by equations (10.82). The anisotropic damage criterion is defined through the second-rank tensors $\mathbf{u}^r$ and $\mathbf{V}^r$ and the damage tensor $\boldsymbol{\varphi}^r$ for each constituent of the composite material. In this work, we assume that the matrix and fibers are isotropic materials while the anisotropic damage criterion is used to describe the interfacial damage.

The authors have attempted to provide an anisotropic damage model in order to accurately predict the behavior of the material. The use of this versatile and general anisotropic model imposes six parameters for which the authors have not obtained direct physical correlation. However, some other parameters in the formulation have direct physical correlation such as λ_1^r, λ_2^r, λ_3^r, v_1^r, v_2^r and v_3^r. The authors are currently in the process of reducing the number of parameters thus providing physical significance but at the same time resulting in less generality in the application of the anisotropic damage criterion.

10.6.2 Damage Evolution of the Matrix:

The metal matrix exhibits two energy dissipation behaviors. Although the two dissipative mechanisms of plasticity and damage influence each other, it is assumed in this work that the energy dissipated due to plasticity and that due to damage are independent of each other. The power of dissipation for the matrix is given by:

$$\tilde{\Pi}^m = \tilde{\Pi}^{MD} + \tilde{\Pi}^{MP} \tag{10.83}$$

where $\tilde{\Pi}^{MP}$ is the plastic dissipation and $\tilde{\Pi}^{MD}$ is the corresponding damage dissipation. The power of plastic dissipation is given by:

$$\tilde{\Pi}^{MP} = \tilde{\sigma}^M : d\tilde{\varepsilon}^{M''} + \tilde{\alpha}^M : d\tilde{\beta}^M \tag{10.84}$$

where the term $\tilde{\alpha}^M : d\tilde{\beta}^M$ is associated with kinematic hardening. In this work, a small strain theory is assumed and the strain rate is assumed to be decomposed into an elastic component $\tilde{\varepsilon}^{M'}$ and a plastic component $\tilde{\varepsilon}^{M''}$, such that:

$$d\tilde{\varepsilon}^M = d\tilde{\varepsilon}^{M'} + d\tilde{\varepsilon}^{M''} \tag{10.85}$$

The associated power of damage dissipation is given by:

$$\tilde{\prod}^{MD} = y^M : d\phi^M + K^M d\kappa^M \qquad (10.86)$$

where $\mathbf{y}^M$ is a generalized thermodynamic force conjugate to the damage tensor $\boldsymbol{\varphi}^M$. The term $K^M d \kappa^M$ is associated with isotropic damage hardening. The fictitious undamaged material is characterized by the effective stress and the effective strain. Since in the full effective configuration, $\overline{C}^M$, the matrix has deformed with no additional damage, therefore, the dissipation energy in $\overline{C}^M$ consists only of the plastic dissipation.

$$\overline{\prod}^M = \overline{\prod}^{MP} \qquad (10.87)$$

and therefore

$$\overline{\prod}^M = \overline{\sigma}^M : d\overline{\varepsilon}^{M''} + \overline{\alpha}^M : d\overline{\beta}^{M''} \qquad (10.88)$$

This is because plastic yielding is assumed to be independent of the damage process. The plastic dissipation in the damaged matrix is equal to the corresponding plastic dissipation in the full effective configuration, $\overline{C}^M$. One concludes that:

$$\overline{\prod}^{MP} = \tilde{\prod}^{MP} \qquad (10.89)$$

which implies that:

$$\overline{line}\sigma^M : d\overline{\varepsilon}^{M''} + \overline{\alpha}^M : d\overline{\beta}^M = \tilde{\sigma}^M : d\tilde{\varepsilon}^{M''} + \tilde{\alpha}^M : d\tilde{\beta}^M \qquad (10.90)$$

From equation (10.90), it is assumed that:

$$\overline{\sigma}^M : d\overline{\varepsilon}^{M''} = \tilde{\sigma}^M : d\tilde{\varepsilon}^{M''} \qquad (10.91)$$

$$\overline{\alpha}^M : d\overline{\beta}^M = \tilde{\alpha}^M : d\tilde{\beta}^M \qquad (10.92)$$

This assumption is imposed in order to obtain equations (10.91) and (10.92) from equation (10.90) is an attempt to simplify the problem in order to obtain a closed form expression for the stiffness matrix. Without this assumption, the problem may not be solved. However, the good correlation between the experimental and numerical results provide a justification for this assumption.

Making use of equation (10.91) together with

$$\overline{\sigma}^M = M^M : \tilde{\sigma}^M \tag{10.93}$$

one obtains a transformation equation for the plastic strain rate such that

$$d\overline{\varepsilon}^{M''} = M^{-M} : d\tilde{\varepsilon}^{M''} \tag{10.94}$$

Making use of the calculus of functions of several variables, one introduces two Lagrange multipliers $\wedge_1^M$, and $\wedge_2^M$ in order to form the function Ω^M such that

$$\Omega^M = \tilde{\Pi}^M - \wedge_1^M \tilde{f}^M - \wedge_2^M g^M \tag{10.95}$$

In equation (10.95), $\tilde{f}^M (\tilde{\sigma}^M, \tilde{\alpha}^M)$ is the plastic yield function of the matrix and $\tilde{\alpha}^M$ is the backstress tensor. The function g^M is the damage potential which is a function of y^M. To extremize the function Ω^M, one uses the necessary conditions

$$\frac{\partial \Omega^M}{\partial \tilde{\sigma}^M} = 0 \tag{10.96}$$

$$\frac{\partial \Omega^M}{\partial y^M} = 0 \tag{10.97}$$

which give the corresponding plastic strain rate and damage rate evolution equations, respectively.

$$d\overline{\varepsilon}^{M''} = \wedge_1^M \frac{\partial \tilde{f}^M}{\partial \tilde{\sigma}^M} \tag{10.98}$$

$$d\phi^M = \wedge_2^M \frac{\partial g^M}{\partial y^M} \tag{10.99}$$

Equation (10.99) gives the increment of damage from the damage potential g^M. Using the consistency condition for the matrix damage g^M.

$$dg^M = 0 \tag{10.100}$$

One obtains the parameter $\wedge_2^M$. Equation (10.100) states that after an increment of damage, the volume element again must be in a damaging state. From equation (10.100), one obtains:

$$\Lambda_2^M = - \frac{\dfrac{\partial g^M}{\partial y^M} : y^M}{\dfrac{\partial g^M}{\partial \phi^M} : \dfrac{\partial g^M}{\partial y^M}} \tag{10.101}$$

Substituting equation (10.101) into equation (10.99), one obtains:

$$d\phi^M = \Psi^M : dy^M \tag{10.102}$$

where Ψ^m is a fourth-rank tensor defined as

$$\Psi^M = - \frac{\dfrac{\partial g^M}{\partial y^M} \otimes \dfrac{\partial g^M}{\partial y^M}}{\dfrac{\partial g^M}{\partial \phi^M} : \dfrac{\partial g^M}{\partial y^M}} \tag{10.103}$$

The generalized thermodynamic force $\mathbf{y}^M$ is assumed to be a function of the elastic component of the strain tensor $\tilde{\varepsilon}^{M'}$ and the damage tensor φ^M, or the stress tensor $\tilde{\sigma}^M$ and φ^M:

$$y^M \equiv y^M(\tilde{\varepsilon}^{M'}, \phi^M) \quad or \quad y^M = y^M(\tilde{\sigma}^M, \phi^M) \tag{10.104}$$

The evolution equation for $\mathbf{y}^M$ may be expressed as follows:

$$dy^M = \frac{\partial y^M}{\partial \tilde{\sigma}_{kl}^M} \, d\tilde{\sigma}_{kl}^M + \frac{\partial y^M}{\partial \phi_{ij}^M} \, d\phi_{ij}^M \tag{10.105}$$

Substituting for $\mathbf{y}^M$ from expression (10.105) into equation (10.102), one obtains the evolution equation for φ^M such that

$$d\phi_{kl}^M = \left(L_{ijkl}^{-M} \, \Psi_{ijrs}^M \, \frac{\partial y_{rs}^M}{\partial \tilde{\sigma}_{pq}^M} \right) d\tilde{\sigma}_{pq}^M \tag{10.106a}$$

or

$$d\phi^M = X^M : d\tilde{\sigma}^M \tag{10.106b}$$

where

$$L_{ijkl}^M = \frac{1}{2}(\delta_{ik}\delta_{jl} + \delta_{il}\delta_{jk}) - \Psi_{ijrs}^M \, \frac{\partial y_{rs}^M}{\partial \phi_{kl}^M} \tag{10.107}$$

246

The thermodynamic force associated with damage is obtained using the enthalpy of the damaged matrix where

$$V^M (\tilde{\sigma}^M, \phi^M) = \frac{1}{2} \tilde{\sigma}^M : \tilde{E}^{-M} (\phi^M) : \tilde{\sigma}^M - \Phi (\tilde{\alpha}^M) \tag{10.108}$$

where $\Phi (\tilde{\alpha}^M)$ is the specific energy due to kinematic hardening. In equation (10.108), $\tilde{E}^M$ is the damaged elastic stiffness of the matrix. The thermodynamic force of the matrix is defined by

$$y^M = - \frac{\partial V^M}{\partial \phi^M} \tag{10.109}$$

Using the energy equivalence principle [141], one obtains a relation between the damaged elastic compliance $\tilde{E}^{-M}$ for the matrix and its corresponding undamaged elastic compliance $\overline{E}^{-M}$ such that [99, 136]:

$$\tilde{E}_{ijkl}^{-M} (\phi^M) = M_{pqij}^M (\phi^M) \, \overline{E}_{pqrs}^{-M} \, M_{rskl}^M (\phi^M) \tag{10.110}$$

Making use of equations (10.108) and (10.109), the thermodynamic force for the matrix is obtained explicitly such that

$$y_{ij}^M = \frac{1}{2} (\tilde{\sigma}_{cd}^M \, \overline{E}_{abpq}^{-M} \, M_{pqkl}^M \, \tilde{\sigma}_{kl}^M + \tilde{\sigma}_{rs}^M \, M_{uvrs}^M \, \overline{E}_{uvab}^{-M} \, \tilde{\sigma}_{cd}^M) \, \frac{\partial M_{abcd}^M}{\partial \phi_{ij}^M} \tag{10.111}$$

10.6.3 Damage Evolution of the Fibers

The gradual degradation of the elastic stiffness of the fibers is caused only through damage and therefore no plastic dissipation occurs. One therefore has:

$$\tilde{\Pi}^F = \tilde{\Pi}^{FD} = y^F : d\phi^D + K^F \, d\kappa^F \tag{10.112}$$

$$\overline{\Pi}^F = 0 \tag{10.113}$$

Accordingly, the function Ω^F is given by:

$$\Omega^F = \tilde{\Pi}^F - \Lambda^F g^F \tag{10.114}$$

and

$$\phi^F = \Lambda^F \, \frac{\partial g^F}{\partial y^F} \tag{10.115}$$

Using the consistency condition for the damage criterion of the fibers

$$dg^F = 0 \tag{10.116}$$

one obtains the evolution equation for ϕ^F as follows:

$$d\phi^F = X^F : d\tilde{\sigma}^F \tag{10.117}$$

where X^F is a fourth-rank tensor similar in form to X^M expressed by equation (10.106). The thermodynamic force y^F is obtained in a similar approach to y^M and has a similar form, except that the superscript M is replaced with F.

10.6.4 Interfacial Damage Evolution

The interfacial damage can be defined as shown in Figure 10.2 in terms of a second-rank symmetric tensor ϕ^D that may be expressed as:

$$\phi^D \equiv \phi^D (S, \overline{S}) \tag{10.118}$$

More elaborate interfacial damage expressions could be derived based on the work of Levy [142]. The corresponding power of dissipation due to interfacial damage is given by:

$$\Pi^D = y^D : d\phi^D + K^D \, d\kappa^D \tag{10.119}$$

$$\tilde{\Pi}^D = 0 \tag{10.120}$$

The function Ω^D is expressed as:

$$\Omega^D = \Pi^D - \Lambda^D g^D \tag{10.121}$$

and

$$d\phi^D = \Lambda^D \frac{\partial g^D}{\partial y^D} \tag{10.122}$$

Using the consistency condition for the interfacial damage

$$dg^D = 0 \tag{10.123}$$

one obtains the evolution expression for φ^D such that:

$$d\phi^D = \mathbf{X}^D : d\sigma \tag{10.124}$$

Similar to the procedure outlined for the previous two types of damage, $\mathbf{y}^D$ can be easily obtained accordingly, such that

$$y_{ij}^D = \frac{1}{2}(\sigma_{cd}\, \tilde{E}_{abpq}^{-1}\, M_{pqkl}^D\, \sigma_{kl} + \sigma_{rs}\, M_{uvrs}^D\, \tilde{E}_{uvab}^{-1}\, \sigma_{cd})\, \frac{\partial M_{abcd}^D}{\partial \phi_{ij}^D} \tag{10.125}$$

10.7 Constitutive Model

Derivation of the elasto-plastic constitutive model for the damaged composite system is performed in three steps. The first step involves the derivation of separate constitutive equations for the matrix and fibers in their respective damaged configurations $\tilde{C}^M$ and $\tilde{C}^F$, respectively. This is followed by the second step which combines the two constitutive equations into one for the overall composite system in its partial effective configuration $\tilde{C}$. Finally, interfacial damage is incorporated in the last step to obtain the final constitutive equation that includes all the three types of damage in the damaged configuration C (see Figure 10.1).

One first starts with the elasto-plastic behavior of the matrix and the elastic behavior of the fibers in their respective effective configurations $\overline{C}^M$ and $\overline{C}^F$ as follows:

$$d\overline{\sigma}^M = \overline{D}^M : d\overline{\varepsilon}^M \tag{10.126}$$

$$d\overline{\sigma}^F = \overline{E}^F : d\overline{\varepsilon}^F \tag{10.127}$$

where $\overline{D}^M$ and $\overline{E}^F$ are the fourth-rank elastio - plastic stiffness tensor for the matrix and the elastic stiffness tensor for the fiber material, respectively.

The elasto-plastic effective stiffness for the matrix $\overline{D}^M$ is given by Voyiadjis and Kattan [136]:

$$\overline{D}^M = \overline{E}^M - \frac{1}{\overline{Q}^M} \left(\frac{\partial \overline{f}^M}{\partial \overline{\sigma}^M} : \overline{E}^M \right) \otimes \left(\overline{E}^M : \frac{\partial \overline{f}}{\partial \overline{\sigma}^M} \right) \tag{10.128}$$

where the scalar quantity $\overline{Q}^M$ is given by:

$$\overline{Q}^M = \frac{\partial \overline{f}^M}{\partial \overline{\sigma}^M} : \overline{E}^M : \frac{\partial \overline{f}^M}{\partial \overline{\sigma}^M} - b \frac{\partial \overline{f}^M}{\partial \overline{\alpha}^M} : (\overline{\sigma}^M - \overline{\alpha}^M) \frac{\dfrac{\partial \overline{f}^M}{\partial \overline{\sigma}^M} : \dfrac{\partial \overline{f}^M}{\partial \overline{\sigma}^M}}{(\overline{\sigma}^M - \overline{\alpha}^M) : \dfrac{\partial \overline{f}^M}{\partial \overline{\sigma}^M}} \tag{10.129}$$

where the matrix yield function $\overline{f}^m$ is of the form:

$$\overline{f}^M = \frac{3}{2} (\overline{\sigma}^M - \overline{\alpha}^M) : (\overline{\sigma}^M - \overline{\alpha}^M) - \overline{\sigma}_0^{M^2} \equiv 0 \tag{10.130}$$

A Prager-Ziegler kinematic hardening evolution law is used such that:

$$d\overline{\alpha}^M = d\overline{\mu}^M (\overline{\sigma}^M - \overline{\alpha}^M) \tag{10.131}$$

where $d\overline{\mu}^M$ is a local scalar multiplier.

The local damaged elastic stiffness tensors $\tilde{E}^M$ and $\tilde{E}^F$ in the configurations $\tilde{C}^M$ and $\tilde{C}^F$, respectively, are given by [135]:

$$\tilde{E}^M = M^{-M} : \overline{E}^M : M^{-M} \tag{10.132}$$

$$\tilde{E}^F = M^{-F} : \overline{E}^F : M^{-F} \tag{10.133}$$

The overall response of the composite system in the partial effective configuration $\tilde{C}$ is given by [138]:

250

$$d\tilde{\sigma} = \tilde{D} : d\tilde{\varepsilon} \tag{10.134}$$

The overall elasto-plastic stiffness tensor $\tilde{D}$ in the partial effective configuration $\tilde{C}$ is obtained by making use of the following relations:

$$d\tilde{\sigma} = \tilde{c}^M \, d\tilde{\sigma}^M + \tilde{c}^F \, d\tilde{\sigma}^F \tag{10.135}$$

$$d\tilde{\sigma}^M = \tilde{D}^M : d\tilde{\varepsilon}^M \tag{10.136}$$

$$d\tilde{\sigma}^F = \tilde{E}^F : d\tilde{\varepsilon}^F \tag{10.137}$$

$$d\tilde{\varepsilon}^M = \tilde{A}^{MP} : d\tilde{\varepsilon} \tag{10.138}$$

$$d\tilde{\varepsilon}^F = \tilde{A}^{FP} : d\tilde{\varepsilon} \tag{10.139}$$

The resulting equation for $\tilde{D}$ is given by:

$$\tilde{D} = \tilde{c}^M \, \tilde{D}^M : \tilde{A}^{MP} + \tilde{c}^F \, \tilde{E}^F : \tilde{A}^{FP} \tag{10.140}$$

where $\tilde{D}^M$ in the elasto-plastic stiffness for the damaged matrix constituent.

In order to obtain the damaged elasto-plastic stiffness of the matrix constituent, one needs to transform equation (10.126) from the undamaged matrix configuration $\overline{C}^M$ to the damaged matrix configuration $\tilde{C}^M$. This is performed through the use of equation (10.6) together with its strain rate counterpart obtained from equations (10.28), (10.29), and (10.94), such that:

$$d\bar{\varepsilon}^M = dM^{-M} : \tilde{\varepsilon}^{M'} + M^{-M} : d\tilde{\varepsilon}^M \tag{10.141}$$

The time rate of the matrix damage tensor used in equation (10.6) and its inverse used in equation (10.141) may be expressed as shown below by making use of equation (10.106b).

$$dM^M = \frac{\partial M^M}{\partial \phi^M} : X^M : d\tilde{\sigma}^M \tag{10.142}$$

$$dM^{-M} = \frac{\partial M^{-M}}{\partial \phi^M} : X^M : d\tilde{\sigma}^M \tag{10.143}$$

Making use of equations (10.6), (10.125), (10.139), (10.142), and (10.143), one obtains the resulting elasto-plastic stiffness relation for the damage matrix constituent:

$$\tilde{D}^{M} = N^{-M} : \overline{D}^{M} : M^{-M} \tag{10.144}$$

where the fourth-rank tensor $\mathbf{N}^{M}$ is given by:

$$N_{ijkl}^{M} = \frac{\partial M_{ijkl}^{M}}{\partial \phi_{pq}^{M}} X_{pqmn}^{M} \tilde{\sigma}_{mn}^{M} + M_{ijkl}^{M}$$

$$- \overline{D}_{ijkl}^{M} \frac{\partial M_{mnpq}^{M}}{\partial \phi_{rs}^{-M}} X_{rskl}^{M} \tilde{E}_{pqab}^{-M} \tilde{\sigma}_{ab}^{M} \tag{10.145}$$

The overall damage response of the composite system is obtained from equation (10.134) by applying the interfacial damage effect tensor $\mathbf{M}^{D}$. Using the following relations:

$$d\tilde{\sigma} = dM^{D} : \sigma + M^{D} : d\sigma \tag{10.146}$$

$$d\tilde{\varepsilon} = dM^{-D} : \varepsilon' + M^{-D} : d\varepsilon \tag{10.147}$$

one obtains the damage elasto-plastic constitutive relation including both the local damages, ϕ^{M} and ϕ^{F}, as well as the interfacial damage ϕ^{D}. Similarly, the rates of the debonding damage effect tensor M^{D} used in equation (10.146) and its inverse used in equation (10.147) are given as follows, by making use of equation (10.124):

$$dM^{D} = \frac{\partial M^{D}}{\partial \phi^{D}} : X^{D} : d\sigma \tag{10.148}$$

$$dM^{-D} = \frac{\partial M^{-D}}{\partial \phi^{D}} : X^{D} : d\sigma \tag{10.149}$$

Finally, one obtains the damage elasto-plastic constitutive relation including both the local damages, φ^{M} and φ^{F}, as well as the interfacial damage φ^{D}. Making use of equations (10.134) and (10.146) through (10.149), one obtains:

$$d\sigma = D : d\varepsilon \tag{10.150}$$

where the damage elasto-plastic stiffness of the material is given by:

$$D = N^{-D} : \tilde{D} : M^{-D} \tag{10.151}$$

and

$$N_{ijkl}^{D} = \frac{\partial M_{ijkl}^{D}}{\partial \phi_{pq}^{D}} X_{pqmn}^{D} \sigma_{mn} + M_{ijkl}^{D}$$

$$- \tilde{D}_{ijmn} \frac{\partial M_{mnpq}^{-D}}{\partial \phi_{rs}^{D}} X_{rskl}^{D} E_{pqab}^{-1} \sigma_{ab} \tag{10.152}$$

The elastic stiffness **E** for the damaged composite is given by:

$$E = M^{-D} : \tilde{E} : M^{-D} \tag{10.153}$$

where the elastic stiffness $\tilde{E}$ in the partial effective configuration $\tilde{C}$ is given by:

$$\tilde{E} = \tilde{c}^{M} \tilde{E}^{M} : \tilde{A}^{ME} + \tilde{c}^{F} \tilde{E}^{F} : \tilde{A}^{FE} \tag{10.154}$$

10.8 Physical Characterization of Damage

In this section, the physical interpretation of the damage tensor φ^{r} ($r = M, F$) is presented for the case of material damaged by micro-cracks. The tensor is evaluated experimentally for two different types of laminate layups. In each case, φ^{M} and φ^{F} are computed independently of each other.

Experimental investigations and procedures for the determination of damage are presented by Voyiadjis and Venson [143] for the macro- and micro-analysis of a SiC - titanium aluminide metal matrix composite. Furthermore, this is the subject of Chapter 12 in this book. The material properties are shown in Table 10.1. In this section, uniaxial tension tests are performed on laminate specimens of two different layups. Dogbone shaped flat specimens are fabricated from each of the layups. Specimens for the different layups are then loaded to various load levels ranging from the rupture load down to 70% of the rupture load at room temperature. Through this experimental procedure, damage evolution is experimentally evaluated through a quantitative micro-analysis technique. The micro-analysis is performed using scanning electron microscopy (SEM) on three mutually perpendicular representative cross-sections of all specimens for the qualitative and quantitative determination of damage. These representative cross-sections form a representative volume element (RVE) defined for the theoretical development of damage evolution.

A new damage tensor proposed by Voyiadjis and Venson [143] is defined here for a general state of loading based upon experimental observations of crack densities on three mutually perpendicular cross-sections of the specimens. The damage tensors φ^M, φ^F and φ^D are defined as second-rank tensors in the form:

$$[\phi^r] = \begin{bmatrix} \bar{\rho}_x^r \bar{\rho}_x^r & \bar{\rho}_x^r \bar{\rho}_y^r & \bar{\rho}_x^r \bar{\rho}_z^r \\ \bar{\rho}_y^r \bar{\rho}_x^r & \bar{\rho}_y^r \bar{\rho}_y^r & \bar{\rho}_y^r \bar{\rho}_z^r \\ \bar{\rho}_z^r \bar{\rho}_x^r & \bar{\rho}_z^r \bar{\rho}_y^r & \bar{\rho}_z^r \bar{\rho}_z^r \end{bmatrix} \tag{10.155}$$

where ρ_i^r ($i = x, y, z$; $r = M, F, D$) is the normalized crack density on a cross-section whose normal is along the i- axis. The crack density on the representative volume element (RVE) for the ith cross-section is calculated as follows:

$$\bar{\rho}_i^r = \frac{\rho_i^r}{m \rho_*^r} \tag{10.156}$$

where

$$\rho_i^r = \frac{\ell_i^r}{A_i^r} \tag{10.157}$$

where ℓ_i^r is the total length of the cracks on the i th cross-section for each constituent, A_i^r is the i th cross-sectional area for each constituent, m is a normalization factor chosen so that the values of the damage variable φ^r fall within the expected range $0 \le \phi_{ij}^r < 1$, and ρ_*^r is as defined below:

254

$$\rho_*^r = \sqrt{\rho_{x_{max}}^2 + \rho_{y_{max}}^2 + \rho_{z_{max}}^2} \tag{10.158}$$

where ρ_i^r is the value of ℓ_i^r / A_i^r at the maximum (rupture) load. The damage tensor obtained experimentally from equation (10.155) is then used in the constitutive equations to predict the mechanical behavior of the composite system. This procedure could be used independently to quantify each of the damages in the matrix and fibers.

The scanning electron microscope (SEM) is used in order to quantify the damage tensor φ^r expressed by equation (10.155). This is performed at various load levels ranging from the rupture load down to 70% of the rupture load at room temperature. The damage tensor $\boldsymbol{\varphi}^r$ is determined experimentally by Voyiadjis and Venson [143] (see Chapter 12) for two types of laminate layups (0/90)s and (±45)s, each consisting of four plies. These layers are examined in detail in Chapter 12 both numerically and experimentally.

The experimentally measured crack densities ($\rho_i^r = \ell_i^r / A_i^r$) are shown in Tables 10.2 and 10.3 for the (0/90)s and (±45)s layups, respectively. These values are used to calculate the normalized values $\overline{\rho}_i^r$ ($i = x, y, z$) for each layup using the method given above. These results are then used to calculate the values of the damage variable φ^r based on equation (10.155). In this way, damage - strain curves are generated for each layup orientation. These damage values can then be used in the constitutive model to accurately predict the mechanical behavior of metal matrix composites. The final results are presented in Chapter 12. In addition, Chapter 12 contains a more complete discussion on the physical characterization of the damage tensor φ.

Table 10.2: Local Crack Densities for (0/90)s Laminate

% Load	% Strain	$\rho_x^M \times 10^{-4}$ (mm/mm^2)	$\rho_x^F \times 10^{-4}$ (mm/mm^2)	$\rho_y^M \times 10^{-4}$ (mm/mm^2)	$\rho_y^F \times 10^{-4}$ (mm/mm^2)
70	0.3182	0.00	41.82	0.00	3.41
75	0.4487	0.00	70.32	0.00	36.40
80	0.4611	0.00	100.77	—	—
85	0.5202	0.00	106.24	0.00	56.43
90	0.5808	0.00	126.68	0.77	66.94

Table 10.3   Local Crack Densities for (±45)$_s$ Laminate

% Load	% Strain	ρ_x^M $x10^{-4}$ (mm/mm^2)	ρ_x^F $x10^{-4}$ (mm/mm^2)	ρ_y^M $x10^{-4}$ (mm/mm^2)	ρ_y^F $x10^{-4}$ (mm/mm^2)
70	0.2414	0.00	49.23	—	—
75	0.2779	0.00	49.32	0.00	42.44
80	0.4324	0.00	51.84	0.00	101.29
85	0.5268	0.00	52.99	0.00	117.01
90	0.5729	0.00	56.67	48.98	97.61

10.9 Numerical Solution of Uniaxially Loaded Symmetric Laminated Composites

The elasto-plastic damage stiffness tensor for a single lamina in its principal material coordinate system has been presented in equation (10.151). This stiffness tensor is transformed to the loading coordinate system and expressed as $[D]_k$ in matrix form (for the kth lamina). A symmetric stacking of plies is considered such that t is the thickness of the laminate consisting of n plies and t_k is the thickness of the kth lamina. The average stress is expressed as follows (in vector form):

$$\{d\sigma\}_{ave.} = \left[\frac{1}{t}\sum_{k=1}^{n}[D]_k\,t_k\right]\{d\varepsilon\} \tag{10.159}$$

Making use of equation (10.159), one can define the gross damage elasto-plastic stiffness for the laminated composite in matrix form as follows:

$$[D_g] = \left[\frac{1}{t}\sum_{k=1}^{n}[D]_k\,t_k\right] \tag{10.160}$$

Making use of the assumption of constant strain through the laminate thickness, the stresses in each lamina are calculated as follows:

$$\{d\sigma\}_k = [D]_k\{d\varepsilon\} \tag{10.161}$$

Two types of laminate layups are considered in this work, (0/90)s and (± 45)s, each consisting of four plies. The material used is a metal matrix composite (see Table 10.1). These layups are examined both numerically and experimentally in Chapter 12. The stress-strain curves for (0/90)s layup and (± 45)s layup are shown in Figures 10.9 and 10.10, respectively. The damage

256

parameters for the matrix, fibers, and debonding damage are selected such that the computed results present a best fit for the experimental data shown in Table 10.4. Very good correlation between the numerical and experimental results is shown in Figures 10.9 and 10.10. The implementation of the proposed theory by finite elements is discussed in the next section.

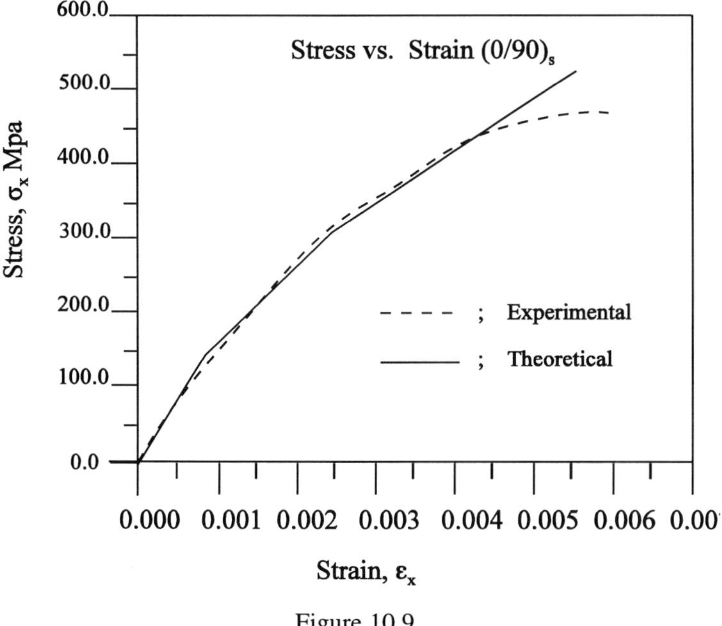

Figure 10.9

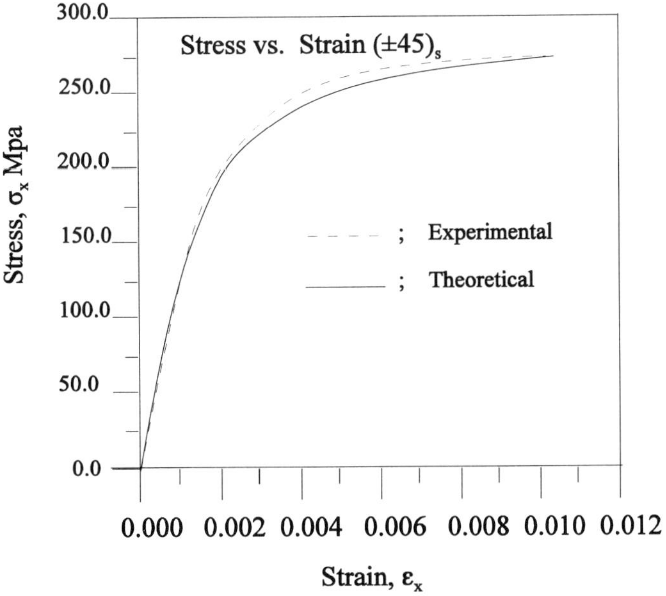

Figure 10.10

Table 10.4: Damage Parameters

	Matrix Damage	Fiber Damage	Interfacial Damage
η_1	0.08	0.06	0.075
η_2	0.08	0.06	0.073
η_3	0.08	0.06	0.073
ξ_1	0.55	0.52	0.55
ξ_2	0.55	0.52	0.53
ξ_3	0.55	0.52	0.53
v_1	0.0013	0.001	0.004
v_2	0.0013	0.001	0.003
v_3	0.0013	0.001	0.003

258

10.10 Finite Element Analysis

The general finite element implementation of the constitutive equation has been outlined in section 5.3. A full Newton-Raphson method is used to solve the system of nonlinear equations that arise from the equilibrium equations. A brief description of the method is given by Voyiadjis [144]. The steps involved in the nonlinear finite element procedure are outlined below:

1.   INCREMENT: Loop for each load increment.

    (1)   Calculate the load or applied displacement increment for the current incremental step or input the load/applied displacement increment.

    (2)   ITERATE: Loop for full Newton-Raphson iteration.

    1.   Compute the residual load vector for this iteration subtracting the equilibrium load from the load computed for the increment.

    2.   Rotate the appropriate loads and applied displacements such that the degrees of freedom at the skew boundary (a boundary condition that is not along the global coordinate system) are normal and tangential to the skew boundary.

    3.   Assemble the stiffness matrices and find the equivalent loads for the applied incremental displacements. Since explicit integration is difficult, Gaussian points are used to evaluate the integrals.

    4.   Solve for the incremental displacements using a linear solver.

    5.   Add the solved iterative incremental displacements to the applied incremental displacements to obtain the complete iterative incremental displacements.

    6.   Rotate back the complete iterative incremental displacements at the skew boundaries to the global coordinate system.

    7.   Cumulate the complete iterative incremental displacements to the total incremental displacements.

    8.   Find the stresses due to the iterative incremental displacements. From the iterative deformation gradient and the stresses updated, compute the updated constitutive matrix $[\mathbf{D}]$. From the total incremental displacements accumulated so far and the $[\mathbf{D}]$ matrix, calculate the equilibrium load vector.

    9.   Check if the convergence of the solution is met using a particular convergence criterion. If convergence has not occurred, go back to the step ITERATE.

    10.   If divergence occurs according to the convergence criterion, then reduce the load increment appropriately as specified by the user and start the iterative solution over again for that load increment.

    11.   If divergence occurs for a load increment that has been reduced "m" times (specified

by the user), then report "convergence not met" and leave the solution phase.

12.   If convergence has occurred, then perform the following operations before going for the next increment:

(1)   update the nodal positions by adding the currently obtained incremental displacements.

(2)   Transform the quantities pertaining to the material property to the present configuration.

(3)   Print out the appropriate quantities pertaining to the converged increment according to the user's specifications.

(4)   If the total load is not reached, go back to the step INCREMENT.

1 - Step 1. Retrieve σ_{ij}, σ_{ij}^r, ϕ_{ij}^r $(r = M, F)$. Retrieve also the information whether the previous loading was a damage loading or not (IDAMG) and plastic loading or not (IYILD).

(1)   If IDAMG = 0 when retrieved, then evaluate the incremental elastic-predictor stress σ_{ij}^P assuming that the loading is elastic. Use the undamaged elastic stiffness matrix for the calculations $(d\,\sigma_{ij}^P = \overline{E}_{ijkl}\,d\,\varepsilon_{kl})$.

(2)   If IDAMG $\neq$ 0 when retrieved, use $(d\sigma_{ij}^P = E_{ijkl}\,d\varepsilon_{kl})$.

(3)   Calculate the incremental elastic-predictor stress of matrix constituent $d\sigma_{ij}^{MP}$.

(4)   Check if the predicted stress state of matrix constituent is inside the yield surface or not.

(5)   If the stress state of matrix constituent is inside the yield surface, then:

1.   Assign elastic stiffness to the constitutive stiffness and the predictor stress increment to the actual computed stress increment.

2.   Set IYILD = 0 indicating the loading has taken place.

3.   Exit to Step 2. Otherwise go to the next step.

4.   Set IYILD = 1, then:

(1)   Calculate the elasto-plastic stiffness [**D**] (when IDAMG = 1) or $[\overline{D}]$ (when IDAMG = 0).

(2)   Update the quantities σ_{ij}, σ_{ij}^M, σ_{ij}^F, α_{ij}^M.

2 - Step 2.

(1)   Check the damage criterion using the updated quantity σ_{ij}^r (r = M, F).

(2)   If damage criterion $g^r < 0$, then IDAMG = 0. Exit from the routines.

(3)   If damage criterion $g^r > 0$. Then IDAMG = 1. Calculate the damage increment $d\varphi^r$ and update the damage quantity φ^r.

(4)   Store the updated quantities in a file.

The finite element method is used for solving a dog-bone shaped specimen and a center-cracked laminate plate that is subjected to inplane tension as shown in Figure 10.11. Due to symmetry in geometry and loading, one quarter of the plate needs to be analyzed. Two-dimensional plane stress analysis rather than three-dimensional analysis is used since the thickness of the plate is much smaller than the other dimensions. Applying the appropriate boundary conditions for the symmetry, both one quarter of the center-cracked laminate plate and the dog-bone shaped specimen are discretized using plane stress finite elements. The finite element meshes chosen for analyzing the problems are shown in Figure 10.12. The four-nodded quadrilateral element is used in both finite element analyses.

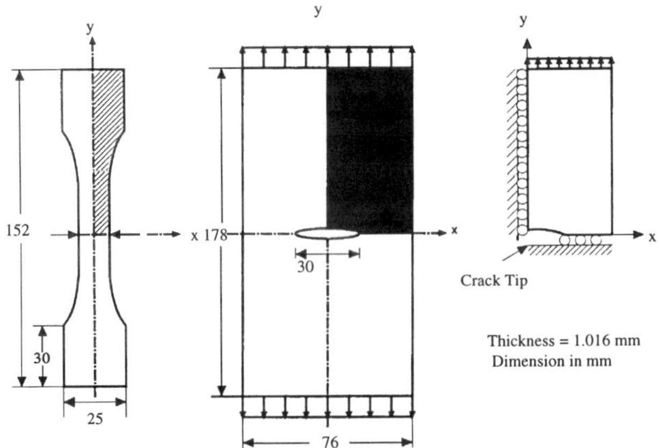

Figure 10.11 Dog-bone Shaped Specimen and Center-cracked Laminate Plate

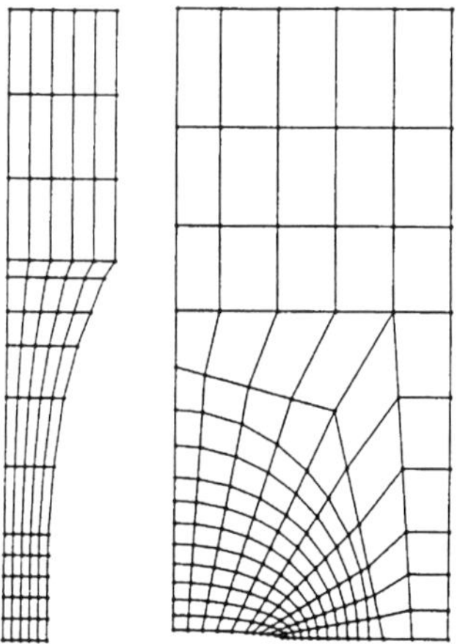

Figure 10.12 Finite Element Meshes

Two types of laminate layups (± 45)s and (0/90)s each consisting of four plies are used. The thickness of each ply is equal to 0.254 mm. Since both layups are symmetric, no curvature is assumed. Hence, the strain through the plate thickness is assumed to be constant. The material properties and damage parameters using the proposed constitutive model are listed in Tables 10.1 and 10.4, respectively.

The following convergence criterion is used in the analysis which is based on the incremental internal energy for each iteration in that incremental loading [113]. It represents the amount of work done by the out-of-balance loads on the displacement increments. Comparison is made with the initial internal energy increment to determine whether or not convergence has occurred. Convergence is assumed to occur if for an energy tolerance ε_E, the following condition is met:

$$\Delta U^{(i)} \left({}^{n+1}R - {}^{n+1}F^{(i-1)} \right) \leq \varepsilon_E \left(\Delta U^{(i)} \left({}^{n+1}R - {}^{n}F \right) \right) \tag{10.162}$$

where $\Delta U^{(i)}$ is the incremental displacement residual at the (i)th iteration, $\left({}^{n+1}R - {}^{n+1}F^{(i-1)} \right)$ is the out-of-balance force vector at the (i - 1) iteration, and $\left(\Delta U^{(i)} \left({}^{n+1}R - {}^{n}F \right) \right)$ is the internal energy term for the (i)th iteration in the (n + 1)th increment. Divergence is assumed to occur if the out-of-balance internal energy for the (i-1)th iteration is greater than the out-of-balance internal energy for the (i)th iteration.

The load is incremented with uniform load increments of 5 MPa until the principal maximum local damage value φ_p^r reaches 1.0 (*i.e.* $\varphi_p^r \geq 1.0$). The principal maximum local damage value φ_p^r is given by:

$$\varphi_p^r = \frac{\varphi_{11}^r + \varphi_{22}^r}{2} + \sqrt{\left(\frac{\varphi_{11}^r + \varphi_{22}^r}{2} \right)^2 + \varphi_{12}^{r2}} \quad , \quad r = M, F, D \tag{10.163}$$

Consequently, material failure at an integration point is assumed when $\varphi_p^r \geq 1$. The principal damage value of the integration point in all elements is monitored at each load increment since it is used to determine the onset of macro-crack initiation of the material.

The dog-bone shaped specimen failed when the final load of 270 MPa was reached for the (± 45)s layup and 480 MPa for the (0/90)s layup. These failure loads are close to the experimental failure loads of 276 MPa for the (± 45)s layup and 483 MPa for the (0/90)s layup [143]. The material failure for the center-cracked specimen occurs at the front of the crack tip when the final load of 80

MPa is reached for the (±45)s layup plate and 120 MPa for the (0/90)s layup plate.

The stress-strain curves from both the finite element analyses and experiments of the two types of layups of the dog-bone shaped specimens are shown in Figure 10.13. Good correlation is shown between the finite element analysis results and the experimental data obtained by Voyiadjis and Venson [143] (see Chapter 12 for details of the experimental procedure).

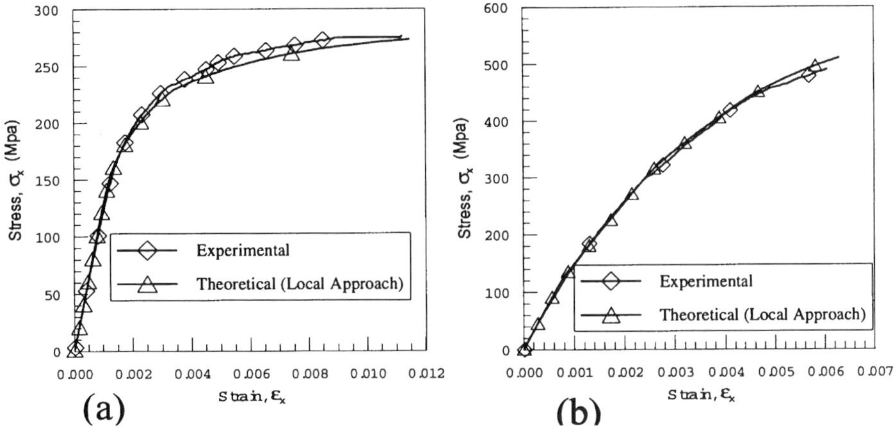

Figure 10.13 (a) Stress-strain Curves of [± 45]$_S$ Layup
(b) Stress-strain Curves of [0/90]$_S$ Layup

Strain contours for the (±45)s layup and (0/90)s layup of the center-cracked plates are shown in Figures 10.14 and 10.15, respectively. Since the two types of layups are symmetric, the strain in each laminate of the layup is constant. However, the stress and damage distributions are different for each lamina of the layup since each lamina has a different stiffness. Stress contours for each lamina are shown in Figure 10.16 for the (±45)s layup and Figure 10.17 for the (0/90)s layup. In Figures 10.18 and 10.19, a comparison is made between the damage analysis and the elastic analysis for the stress σ_{yy} contours around the crack tip. The damage analysis shows considerable stress reduction due to the damage around the crack tip. The stress σ_{yy} at the front of the crack tip as obtained from the elastic solution is higher than that of the material strength of the layup. However, in the damage elasto-plastic analysis, the stresses are reduced such that they are close to those of the material strength. The σ_{yy} stress reductions at the front of the crack tip are more than 50% for [± 45] plies, 40% for [0] ply, and 80% for [90] ply. Stress redistributions are clearly indicated in Figures 10.18 and 10.19. Primarily due to the stress reduction around the crack tip. The stress is transferred to the outer portion away from the crack tip. This is clearly indicated in Figure 10.19 where the stress reduction at the 90° ply is primarily due to considerable interfacial damage.

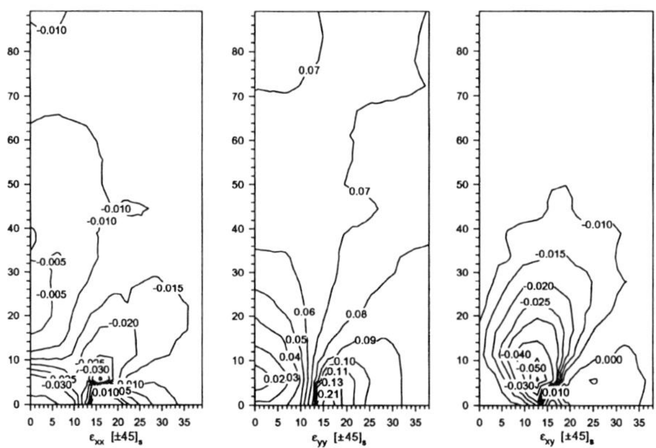

Figure 10.14 Strain Contours for [± 45]$_S$ Layup (in %)

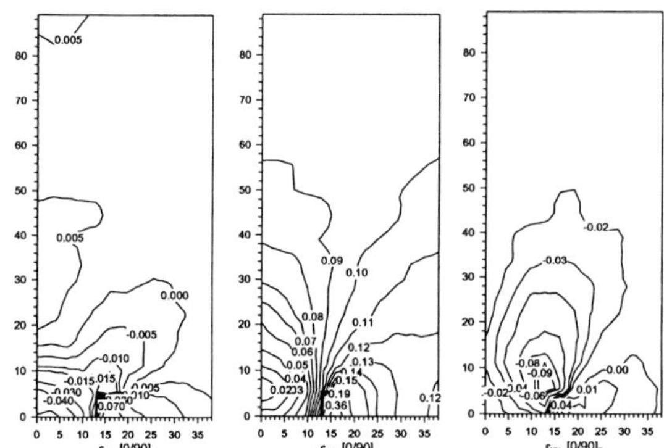

Figure 10.15 Strain Contours for [0/90]$_S$ Layup (in %)

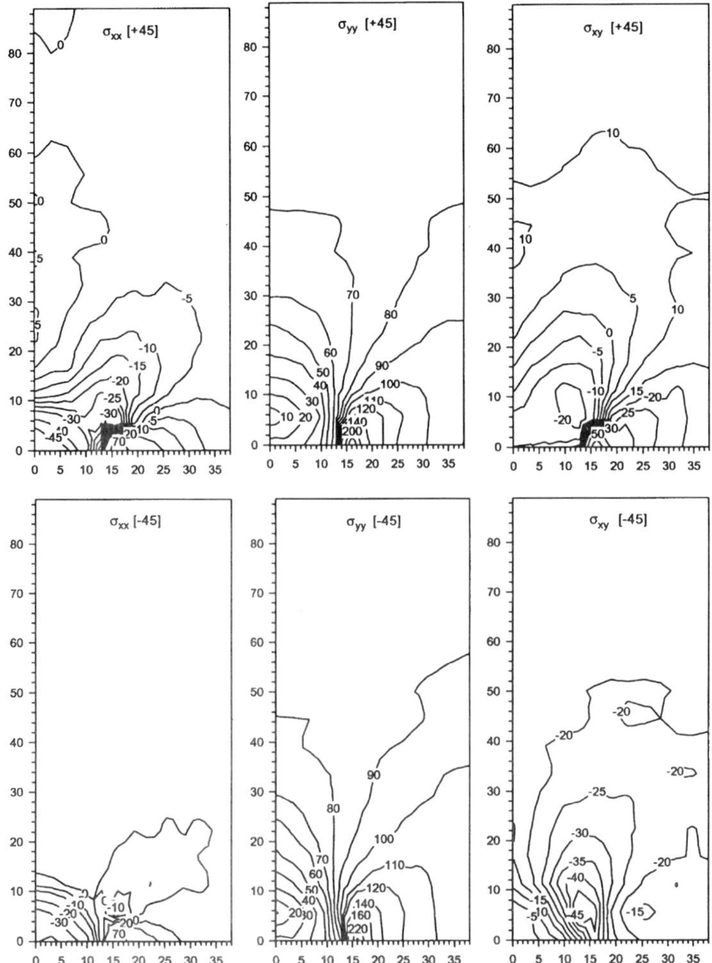

Figure 10.16 Stress Contours for [± 45]$_S$ Layup (units are in MPa)

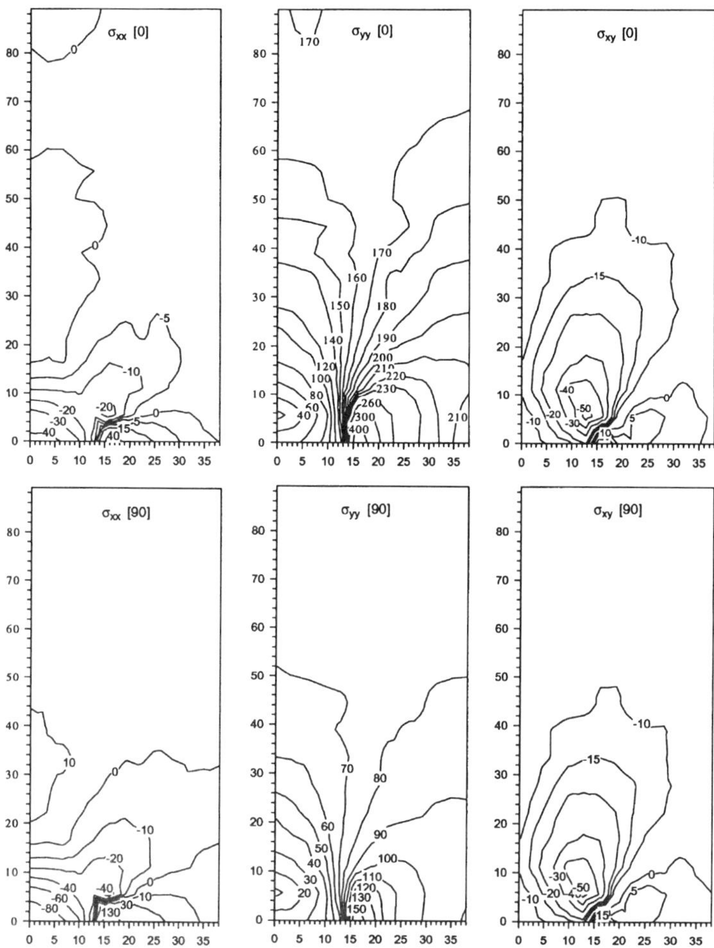

Figure 10.17 Stress Contours for [0/90]$_S$ Layup (units are in MPa)

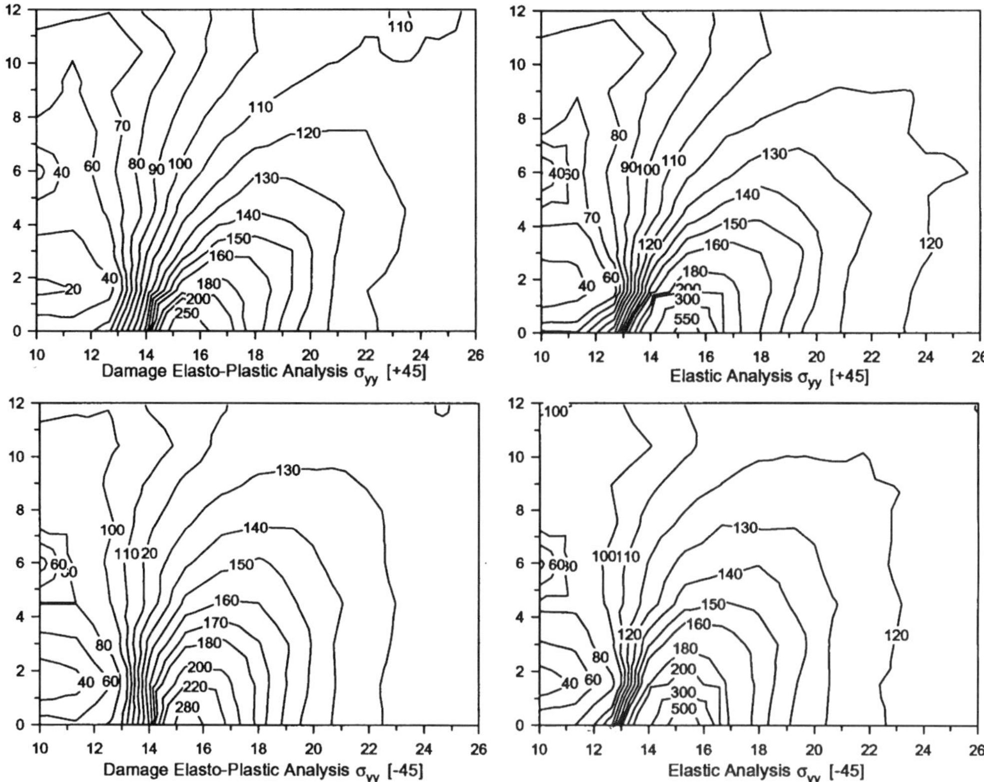

Figure 10.18   Comparison of damage elasto-plastic analysis with elastic analysis of stress σ_{yy} contours around the crack tip for $(\pm 45)_s$ layup (units are in MPa).

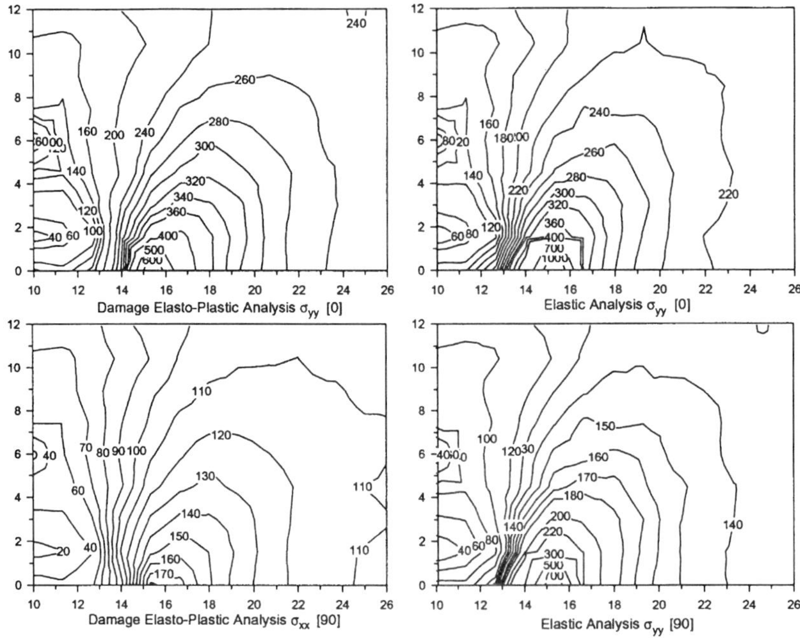

Figure 10.19 Comparison of Damage elasto-plastic analysis with elastic analysis of stress σ_{yy} contours around the crack tip for $(0/90)_S$ layup (units are in MPa)

The local damage contours around the crack tip are shown in Figures 10.20 - 10.23 for the failure loads in the case of [+45], [-45], [0], and [90] plies, respectively. For the [±45] layups, all types of damage such as matrix, fiber and interfacial damage, are developed. Fiber damage is considerably more spread in the [0] ply than interfacial damage. On the other hand, interfacial damage is more pronounced with the matrix damage for the [90] ply. However, fiber damage is much less developed in the case of the [90] ply. This is in line with the experimental results obtained by Voyiadjis and Venson [143] which are shown in Chapter 12.

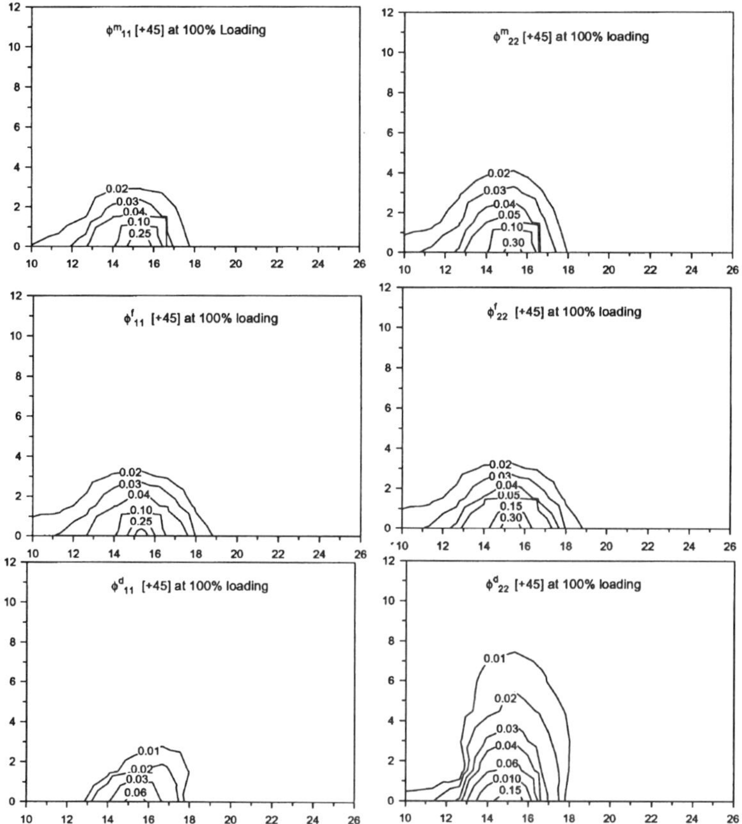

Figure 10.20 Damage contours around crack tip at the failure load for [+45] lamina

270

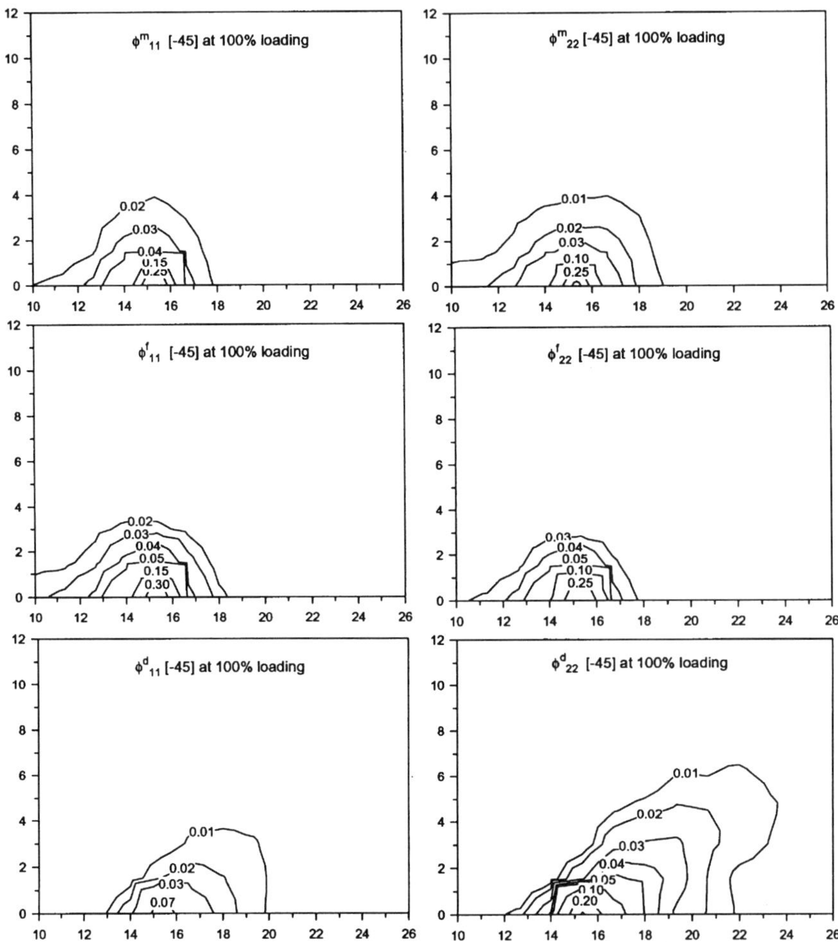

Figure 10.21 Damage contours around crack tip at the failure load for [- 45] lamina

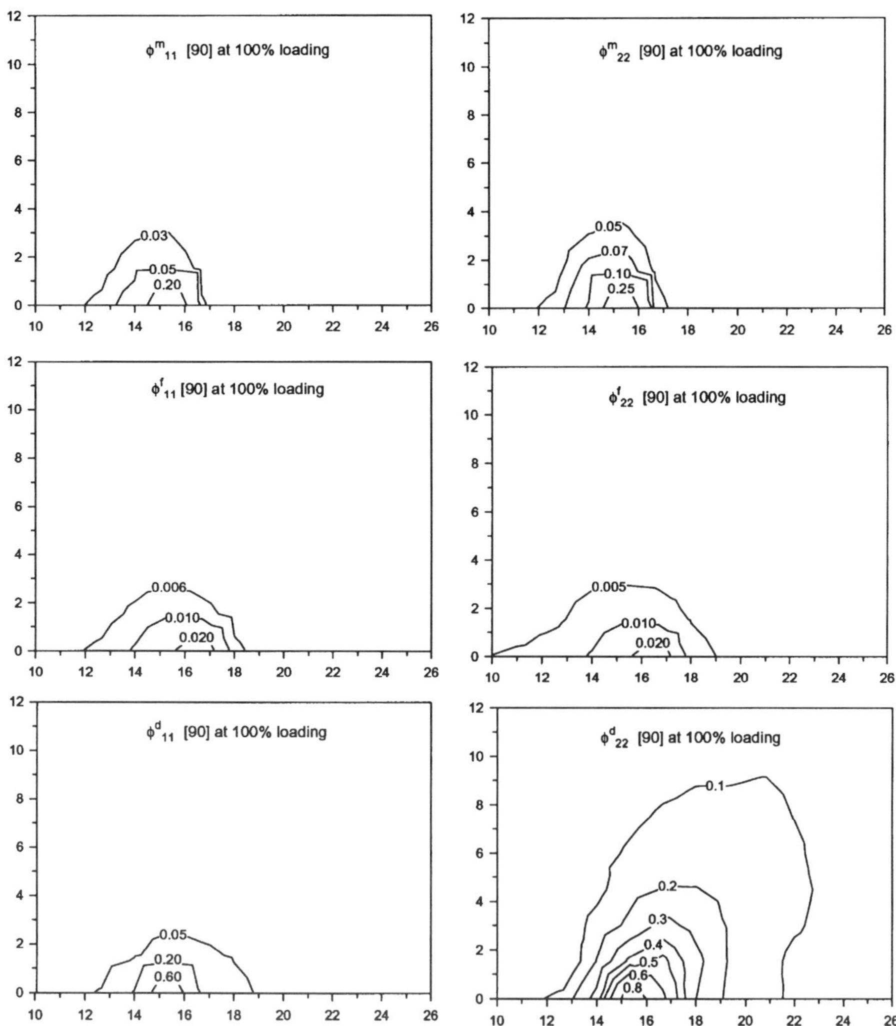

Figure 10.22 Damage contours around crack tip at the failure load for [0] lamina

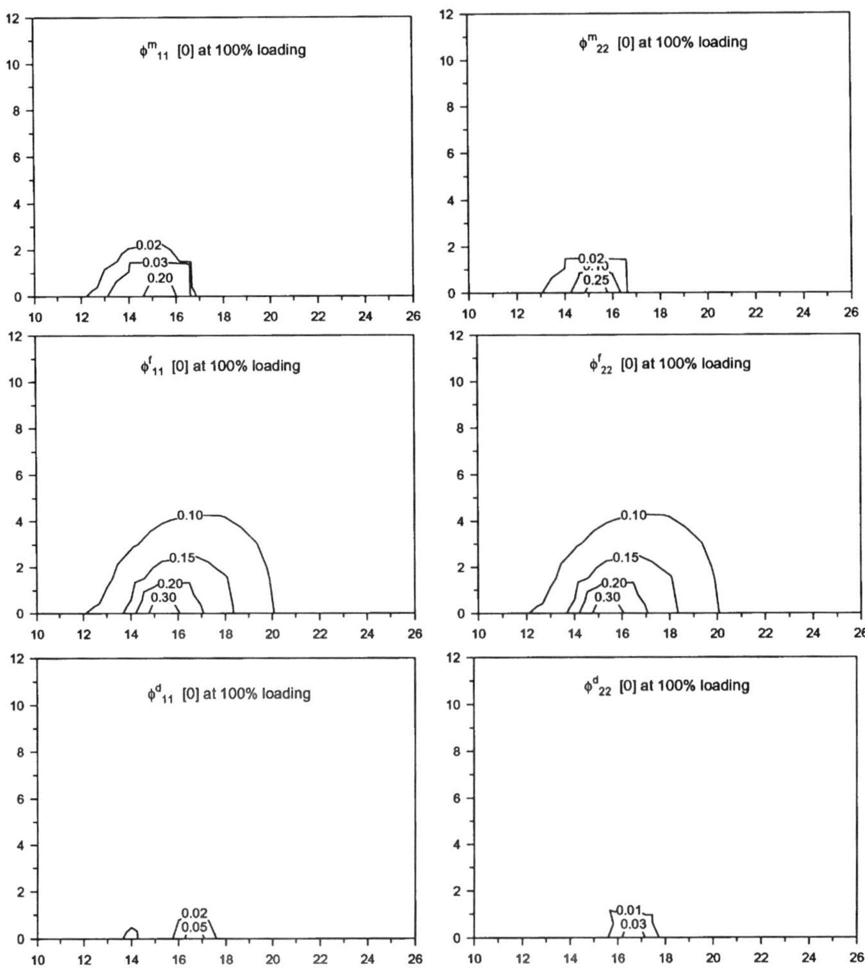

Figure 10.23 Damage contours around crack tip at the failure load for [90] lamina

CHAPTER 11

SYMMETRIZATION OF THE EFFECTIVE STRESS TENSOR

The effective stress tensor is examined within the framework of continuum damage mechanics. For a general state of deformation and damage, it is seen that the effective stress tensor is usually not symmetric. Therefore, its symmetrization is necessary for a continuum theory to be valid in the classical sense. Three such symmetrization methods are studied in detail. The three symmetrization methods are compared and certain recommendations are made regarding their suitability. Explicit matrix representations are obtained for two of the three techniques for a general case of deformation and damage. Explicit representation for the third technique is obtained only for the case of plane stress.

11.1 Preliminaries:

In a general state of deformation and damage, the effective stress tensor $\bar{\sigma}$ is related to the stress tensor σ by the following linear transformation:

$$\bar{\sigma} = M : \sigma \tag{11.1}$$

where M is a fourth-rank linear operator called the damage effect tensor. Depending on the form used for M, it is clear from equation (11.1) that the effective stress tensor $\bar{\sigma}$ is generally not symmetric.

Using a nonsymmetric stress tensor as given by equation (11.1) to formulate a constitutive model will result in the introduction of the equations of the Cosserat and micropolar continua. However, the use of such complicated mechanics can be easily avoided if the effective stress tensor is symmetrized. It turns out that the symmetrization of the stress tensor has been used in the literature [16,58] but lacks a systematic and consistent approach. It is the aim of this chapter to provide a solid mathematical basis for such symmetrization methods and justification for their use and validity.

In general, the use of an explicit form like $[M] \equiv [I - \varphi]^{-1}$ will still keep the effective stress tensor nonsymmetric. The matrices $[I]$ and $[\varphi]$ and 6 x 6 matrices denoting the identity and damage tensors, respectively. Using the following 3 x 3 matrix representations of the stress tensor σ and the

274

damage tensor $\boldsymbol{\varphi}$:

$$[\sigma] = \begin{bmatrix} \sigma_{11} & \sigma_{12} & \sigma_{13} \\ \sigma_{12} & \sigma_{22} & \sigma_{23} \\ \sigma_{13} & \sigma_{23} & \sigma_{33} \end{bmatrix} \tag{11.2}$$

$$[\varphi] = \begin{bmatrix} \varphi_{11} & \varphi_{12} & \varphi_{13} \\ \varphi_{12} & \varphi_{22} & \varphi_{23} \\ \varphi_{13} & \varphi_{23} & \varphi_{33} \end{bmatrix} \tag{11.3}$$

one can adopt several methods to symmetrize the effective stress tensor $\bar{\sigma}$. In this chapter, three such symmetrization methods are studied in detail. They are listed below using matrix notation:

$$[\bar{\sigma}] = \frac{1}{2}\left([\sigma][I - \varphi]^{-1} + [I - \varphi]^{-1}[\sigma]\right) \tag{11.4}$$

$$[\bar{\sigma}] = [I - \varphi]^{-\frac{1}{2}}[\sigma][I - \varphi]^{-\frac{1}{2}} \tag{11.5}$$

$$[\sigma] = \frac{1}{2}\left([\bar{\sigma}][I - \varphi] + [I - \varphi][\bar{\sigma}]\right) \tag{11.6}$$

It is clear that all the above methods produce symmetric effective stresses. It is also clear that the first two methods are explicit while the third one is implicit. However, as will be shown later in this chapter, it is possible to derive an explicit matrix representation of the damage effect tensor $\mathbf{M}$ for the first and third methods only. The second method does not lend itself easily to algebraic manipulations, therefore, the solution using the second method will not involve the derivation of a general explicit expression for $\mathbf{M}$. However, its representation for the case of plane stress is given. It should be mentioned that all three methods produce frame-indifferent symmetrized stress tensors [130]. Therefore, any one of the three formulations may be used in constitutive equations.

The key step in the derivation is the conversion of the representation of the stress tensors $\boldsymbol{\sigma}$ and $\bar{\boldsymbol{\sigma}}$ from 3 x 3 matrices to 6 x 1 vectors. In this work, these tensors are represented by vectors as follows:

$$\{\sigma\} = [\sigma_{11} \quad \sigma_{22} \quad \sigma_{33} \quad \sigma_{23} \quad \sigma_{13} \quad \sigma_{12}]^{T} \tag{11.7a}$$

$$\{\overline{\sigma}\} = [\overline{\sigma}_{11} \ \overline{\sigma}_{22} \ \overline{\sigma}_{33} \ \overline{\sigma}_{23} \ \overline{\sigma}_{13} \ \overline{\sigma}_{12}]^{T} \tag{11.7b}$$

Using the notation of equations (11.7), the damage transformation relation of equation (11.1) is represented in matrix notation by:

$$\{\overline{\sigma}\} = [M]\{\sigma\} \tag{11.8}$$

where [**M**] is the 6 x 6 matrix representation of the tensor **M**. The explicit form of the matrix [**M**] depends on the symmetrization method used. Its representation for the three symmetrization methods of equations (11.4), (11.5) and (11.6) is discussed next.

11.2 Explicit Symmetrization Method:

The first symmetrization method is the explicit scheme given in equation (11.4). It is possible to derive an explicit analytical expression for [**M**] using this method. One starts by substituting [φ] and [σ] from equations (11.3) and (11.2), respectively, into equation (11.4). In order to simplify the resulting equation, one needs to calculate the inverse of the matrix [**I** - φ]. The inverse matrix is obtained as follows*:

$$[\mathbf{I} - \varphi]^{-1} = \frac{1}{\Delta} \begin{bmatrix} \Psi_{22}\Psi_{33} - \varphi_{23}^2 & \varphi_{13}\varphi_{23} + \varphi_{12}\Psi_{33} & \varphi_{12}\varphi_{23} + \varphi_{13}\Psi_{22} \\ \varphi_{13}\varphi_{23} + \varphi_{12}\Psi_{33} & \Psi_{11}\Psi_{33} - \varphi_{13}^2 & \varphi_{12}\varphi_{13} + \varphi_{23}\Psi_{11} \\ \varphi_{12}\varphi_{23} + \varphi_{13}\Psi_{22} & \varphi_{12}\varphi_{13} + \varphi_{23}\Psi_{11} & \Psi_{11}\Psi_{22} - \varphi_{12}^2 \end{bmatrix} \tag{11.9}$$

where the determinant Δ is given by:

$$\Delta = \Psi_{11}\Psi_{22}\Psi_{33} - \varphi_{23}^2\Psi_{11} - \varphi_{13}^2\Psi_{22} - \varphi_{12}^2\Psi_{33} - 2\varphi_{12}\varphi_{23}\varphi_{13} \tag{11.10}$$

and the notation Ψ_{ij} is used to denote $\delta_{ij} - \varphi_{ij}$, where δ_{ij} is the Kronecker delta.

Substituting for $[\mathbf{I} - \varphi]^{-1}$ and [σ] from equation (11.9) and (11.2), respectively, into equation (11.4), simplifying and rewriting the result in the terminology of equation (11.8) (where the stresses are represented by 6 x 1 vectors), one obtains the following representation for **M**:*

---

   *   Most of the algebraic derivations that appear in this chapter where performed using the symbolic manipulation program REDUCE.

276

$$[M] = \frac{1}{2\Delta}
\begin{bmatrix}
2\Psi_{22}\Psi_{33} - 2\varphi_{23}^2 & 0 & 0 & 0 & 2\varphi_{12}\varphi_{23} + 2\varphi_{13}\Psi_{22} & 2\varphi_{13}\varphi_{23} + 2\varphi_{12}\Psi_{33} \\
0 & 2\Psi_{11}\Psi_{33} - 2\varphi_{13}^2 & 0 & 2\varphi_{12}\varphi_{13} + 2\varphi_{23}\Psi_{11} & 0 & 2\varphi_{13}\varphi_{23} + 2\varphi_{12}\Psi_{33} \\
0 & 0 & 2\Psi_{11}\Psi_{22} - 2\varphi_{12}^2 & 2\varphi_{12}\varphi_{13} + 2\varphi_{23}\Psi_{11} & 2\varphi_{12}\varphi_{23} + 2\varphi_{13}\Psi_{22} & 0 \\
0 & \varphi_{12}\varphi_{13} + \varphi_{23}\Psi_{11} & \varphi_{12}\varphi_{13} + \varphi_{23}\Psi_{11} & \Psi_{11}\Psi_{33} + \Psi_{11}\Psi_{22} - \varphi_{13}^2 - \varphi_{12}^2 & \varphi_{13}\varphi_{23} + \varphi_{12}\Psi_{33} & \varphi_{12}\varphi_{23} + \varphi_{13}\Psi_{22} \\
\varphi_{12}\varphi_{23} + \varphi_{13}\Psi_{22} & 0 & \varphi_{12}\varphi_{23} + \varphi_{13}\Psi_{22} & \varphi_{13}\varphi_{23} + \varphi_{12}\Psi_{33} & \Psi_{22}\Psi_{33} + \Psi_{11}\Psi_{22} - \varphi_{23}^2 - \varphi_{12}^2 & \varphi_{12}\varphi_{13} + \varphi_{23}\Psi_{11} \\
\varphi_{13}\varphi_{23} + \varphi_{12}\Psi_{33} & \varphi_{13}\varphi_{23} + \varphi_{12}\Psi_{33} & 0 & \varphi_{12}\varphi_{23} + \varphi_{12}\Psi_{33} & \varphi_{12}\varphi_{13} + \varphi_{23}\Psi_{11} & \Psi_{22}\Psi_{33} + \Psi_{11}\Psi_{33} - \varphi_{23}^2 - \varphi_{13}^2
\end{bmatrix}$$

$$(11.11)$$

The matrix representation of [**M**] given above must satisfy certain conditions. In particular, it must be valid for the special case of isotropic damage. In order to check this condition, one first needs to obtain the diagonalized form of [**M**] by letting $\varphi_{12} = \varphi_{13} = \varphi_{23} = 0$. After simplifying the resulting matrix, one obtains:

$$
[\boldsymbol{M}]_{diag.} =
\begin{bmatrix}
\dfrac{1}{1-\varphi_1} & 0 & 0 & 0 & 0 & 0 \\[2ex]
0 & \dfrac{1}{1-\varphi_2} & 0 & 0 & 0 & 0 \\[2ex]
0 & 0 & \dfrac{1}{1-\varphi_3} & 0 & 0 & 0 \\[2ex]
0 & 0 & 0 & \dfrac{(1-\varphi_3)+(1-\varphi_2)}{2(1-\varphi_3)(1-\varphi_2)} & 0 & 0 \\[3ex]
0 & 0 & 0 & 0 & \dfrac{(1-\varphi_3)+(1-\varphi_1)}{2(1-\varphi_3)(1-\varphi_1)} & 0 \\[3ex]
0 & 0 & 0 & 0 & 0 & \dfrac{(1-\varphi_2)+(1-\varphi_1)}{2(1-\varphi_2)(1-\varphi_1)}
\end{bmatrix}
$$

$$(11.12)$$

where φ_1, φ_2 and φ_3 are the principal values of the matrix $[\varphi]$ of equation (11.3). For the case of isotopic damage, one sets $\varphi_1 = \varphi_2 = \varphi_3 = \varphi$ to get the required isotropic matrix as follows:

$$
[M]_{isot.} = \frac{1}{1-\varphi} [I]
\tag{11.13}
$$

Next, the matrix representation of the damage effect tensor $\mathbf{M}$ is shown for the case of plane stress in the x_1 - x_2 plane. In this case, the stress components σ_{33}, σ_{13} and σ_{23} vanish. In addition, one also makes the assumption of plane damage, that is, one assumes that the damage variables φ_{33}, φ_{13} and φ_{23} also vanish (see reference [110]). The latter assumption is necessary in order to provide a consistent plane stress formulation.

In the case of plane stress, the vector representations of the stress tensor of equations (11.7) become:

$$
\{\sigma\} = [\sigma_{11}\ \sigma_{22}\ \sigma_{12}]^T
\tag{11.14a}
$$

$$
\{\bar{\sigma}\} = [\bar{\sigma}_{11}\ \bar{\sigma}_{22}\ \bar{\sigma}_{12}]^T
\tag{11.14b}
$$

In this case, the presentation of the damage effect tensor $\mathbf{M}$ becomes a 3 x 3 matrix. Using this method, the matrix in equation (11.11) reduces to:

$$[M] = \frac{1}{\Delta} \begin{bmatrix} \Psi_{22} & 0 & \varphi_{12} \\ 0 & \Psi_{11} & \varphi_{12} \\ \frac{1}{2}\varphi_{12} & \frac{1}{2}\varphi_{12} & \frac{\Psi_{11} + \Psi_{22}}{2} \end{bmatrix} \qquad (11.15a)$$

where Δ is given by:

$$\Delta = \Psi_{11}\,\Psi_{22} - \varphi_{12}^2 \qquad (11.15b)$$

The variations of M_{11}, M_{22} and M_{33} are shown in Figures 11.1, 11.2, and 11.3, respectively. Each of the damage effect components is plotted as a function of φ_{11} and φ_{22} based on the diagonal terms of the matrix in equation (11.15a). In Figure 11.1a, the variation of M_{11} is shown in terms of φ_{11} and φ_{22} for $\varphi_{12} = 0$. The value of φ_{12} is increased to 0.2, 0.4, and 0.5 in Figures 11.1b, 11.1c, and 11.1d, respectively. The distortional effect of increasing φ_{12} is very clear in Figure 11.1d when $\varphi_{12} = 0.5$. It should also be mentioned that when the authors tried to use a value of $\varphi_{12} = 0.6$, numerical instability occurred and the solution was discarded. Similar results are shown in Figures 11.2 and 11.3 for M_{22} and M_{33}.

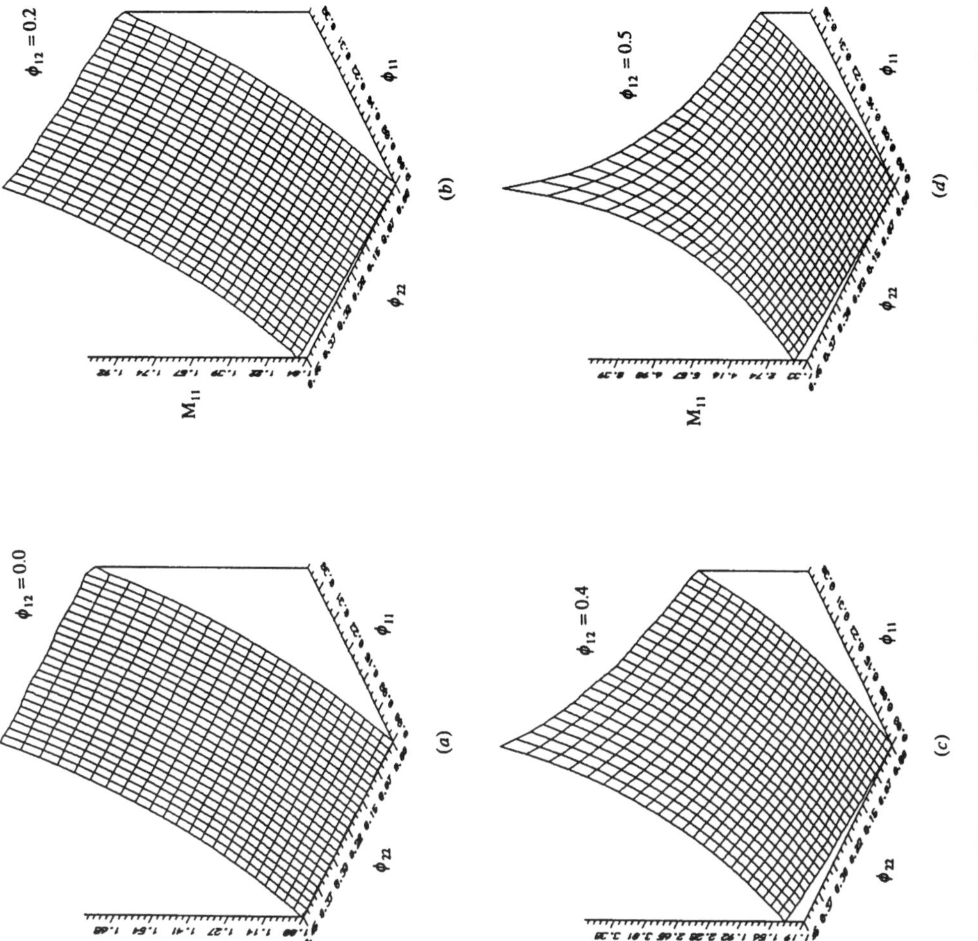

Figure 11.1: Variation of M_{11} vs. φ_{22} for the explicit symmetrization method.

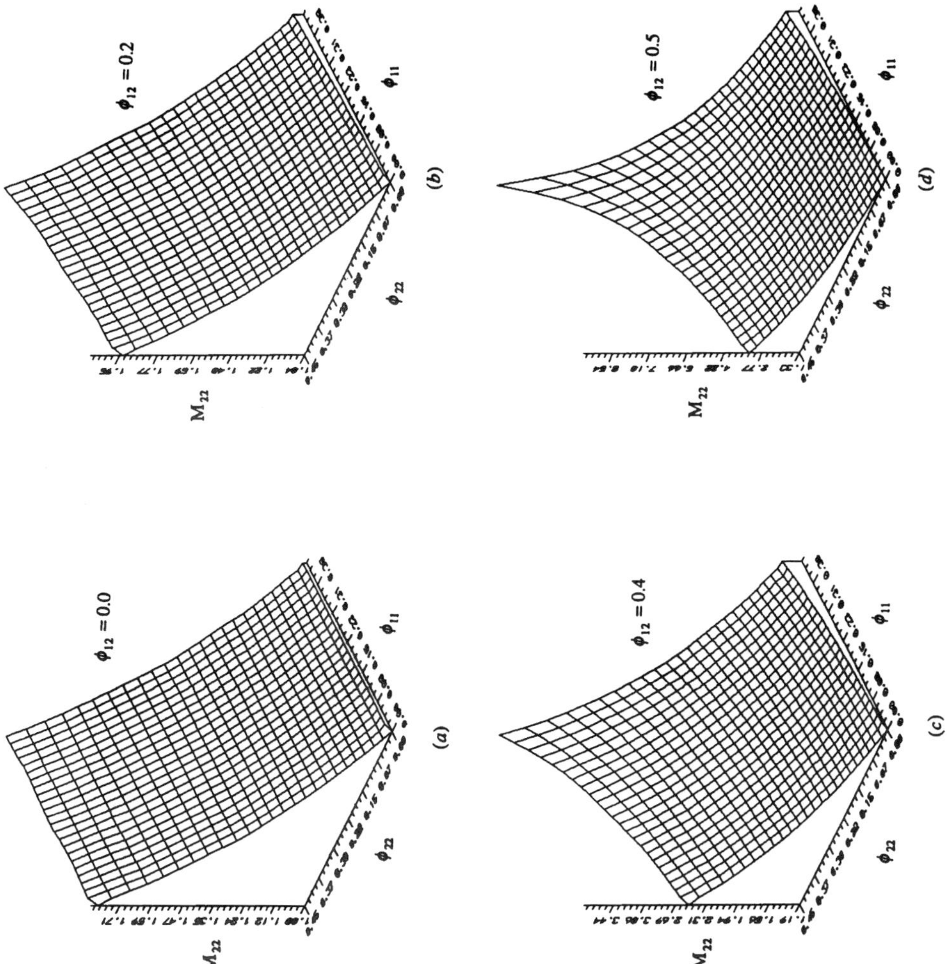

Figure 11.2: Variation of M_{22} vs. φ_{11} and φ_{22} for the explicit symmetrization method.

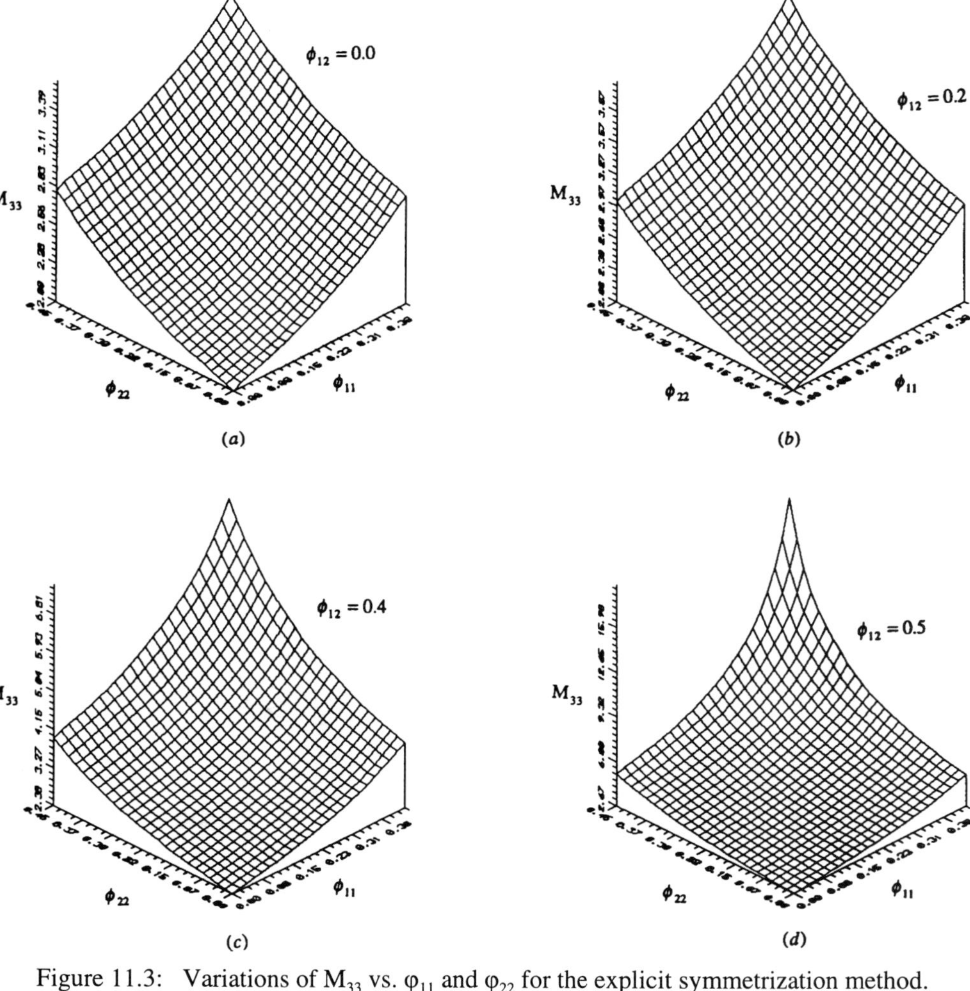

Figure 11.3: Variations of M_{33} vs. φ_{11} and φ_{22} for the explicit symmetrization method.

11.3 Square Root Symmetrization Method:

The symmetrization method of equation (11.5) is considered in this section. This is also an explicit method of symmetrization that is more sophisticated than the method in the previous section. Performing the algebraic operations in this case is lengthy if not impossible because of the square root of the matrix $[\mathbf{I} - \boldsymbol{\varphi}]$ that needs to be evaluated. Therefore, obtaining an explicit general matrix representation for $[\mathbf{M}]$ is not possible in this case. However, it turns out that a 3 x 3 explicit matrix representation for $[\mathbf{M}]$ can be obtained for the case of plane stress.

For the case of plane stress in the x_1 - x_2 plane, one assumes that a case of plane damage also exists. Examining equation (11.5), it is noticed that the square root of the following matrix is needed:

$$[I - \boldsymbol{\varphi}] = \begin{bmatrix} 1 - \varphi_{11} & -\varphi_{12} & 0 \\ -\varphi_{12} & 1 - \varphi_{22} & 0 \\ 0 & 0 & 1 - \varphi_{33} \end{bmatrix} \tag{11.16}$$

The procedure of finding the square root of a matrix involves diagonalization of the matrix and using the matrix of eigenvectors. However, the problem considered here may be simplified if one recognizes that the matrix in equation (11.16) is a block-diagonal matrix. Therefore, one needs only to evaluate the square root of the 2 x 2 submatrix:

$$[A] = \begin{bmatrix} 1 - \varphi_{11} & -\varphi_{12} \\ -\varphi_{12} & 1 - \varphi_{22} \end{bmatrix} \tag{11.17}$$

First, one calculates the eigenvalues of the above matrix. This is performed by solving the following equation for λ:

$$\begin{bmatrix} 1 - \varphi_{11} & -\varphi_{12} \\ -\varphi_{12} & 1 - \varphi_{22} \end{bmatrix} - \lambda \begin{bmatrix} 1 & 0 \\ 0 & 1 \end{bmatrix} = 0 \tag{11.18}$$

Equation (11.18) results in a quadratic equation in λ whose solution is obtained as follows:

$$\lambda_1 = 1 - \frac{1}{2}(\varphi_{11} + \varphi_{22}) + \frac{1}{2}\sqrt{(\varphi_{11} - \varphi_{22})^2 + 4\varphi_{12}^2} \tag{11.19a}$$

$$\lambda_2 = 1 - \frac{1}{2}(\varphi_{11} + \varphi_{22}) - \frac{1}{2}\sqrt{(\varphi_{11} - \varphi_{22})^2 + 4\varphi_{12}^2} \tag{11.19b}$$

The two expressions above are the eigenvalues of the matrix [**A**] given in equation (11.17). Therefore, the diagonalized form of [**A**] is given by:

$$[\Lambda] = \begin{bmatrix} \lambda_1 & 0 \\ 0 & \lambda_2 \end{bmatrix} \tag{11.20}$$

Next, one needs to obtain the matrix of eigenvectors [**B**] such that the following diagonalization formula holds:

$$[\Lambda] = [B]^{1} [A] [B] \tag{11.21}$$

Solving for the eigenvectors of [**A**], one finds out that $[1, (1 - \lambda_1 - \varphi_{11})/\varphi_{12}]^T$ and $[1, (1 - \lambda_2 - \varphi_{11})/\varphi_{12}]^T$ are the eigenvectors corresponding to the eigenvalues λ_1, and λ_2, respectively. Using these two vectors, one forms the matrix [**B**] as follows:

$$[B] = \begin{bmatrix} 1 & 1 \\ \dfrac{1 - \lambda_1 - \varphi_{11}}{\varphi_{12}} & \dfrac{1 - \lambda_2 - \varphi_{11}}{\varphi_{12}} \end{bmatrix} \tag{11.22}$$

It can be easily verified that the matrices [**A**] and [**B**] satisfy equation (11.21). It turns out that the matrix [**B**] of equation (11.22) is a submatrix of the required matrix to diagonalize [**I** - **φ**] of equation (11.16). This matrix is written as:

$$[P] = \begin{bmatrix} 1 & 1 & 0 \\ \dfrac{1 - \lambda_1 - \varphi_{11}}{\varphi_{12}} & \dfrac{1 - \lambda_2 - \varphi_{11}}{\varphi_{12}} & 0 \\ 0 & 0 & 1 \end{bmatrix} \tag{11.23}$$

Premultiplying [**I** - **φ**] by $[P]^{-1}$ and post multiplying it by [**P**], produces the diagonalized form of [**I** - **φ**] as follows:

284

$$[P]^{-1} [I - \varphi] [P] = \begin{bmatrix} \lambda_1 & 0 & 0 \\ 0 & \lambda_2 & 0 \\ 0 & 0 & 1 \end{bmatrix} = [D] \tag{11.24}$$

The square root of $[I - \varphi]$ is now obtained as follows:

$$[I - \varphi]^{\frac{1}{2}} = [P] \; [D]^{\frac{1}{2}} \; [P]^{-1} \tag{11.25}$$

The above equation can be easily verified by multiplying the right-hand-side by itself to obtain $[I - \varphi]$. The square root of the diagonal matrix $[D]$ is directly obtained by taking the square roots of the diagonal terms. Substituting equation (11.25) into the symmetrization formula (11.5), one finally obtains:

$$[\bar{\sigma}] = [P]^{-1} [D]^{-\frac{1}{2}} [P] [\sigma] [P]^{-1} [D]^{-\frac{1}{2}} [P] \tag{11.26}$$

where the matrix $[D]^{-1/2}$ is given by:

$$[D]^{-\frac{1}{2}} = \begin{bmatrix} \dfrac{1}{\sqrt{\lambda_1}} & 0 & 0 \\ 0 & \dfrac{1}{\sqrt{\lambda_2}} & 0 \\ 0 & 0 & 1 \end{bmatrix} \tag{11.27}$$

and the matrix $[\sigma]$ has only σ_{11}, σ_{22} and σ_{12} as the nonzero terms. The result is simplified and the coefficients of σ_{11}, σ_{22} and σ_{12} are extracted. These coefficients are the elements of the 3 x 3 matrix $[M]$ which is represented by:

$$[M] = \begin{bmatrix} M_{11} & M_{12} & M_{13} \\ M_{21} & M_{22} & M_{23} \\ M_{31} & M_{32} & M_{33} \end{bmatrix} \tag{11.28}$$

where

$$M_{11} = \frac{1}{\Delta}\left[\sqrt{\lambda_1}\,(\lambda_1 + \varphi_{11} - 1) - \sqrt{\lambda_2}\,(\lambda_2 + \varphi_{11} - 1)\right]^2 \qquad (11.29\text{a})$$

$$M_{12} = \frac{-1}{\Delta}\,(\lambda_1 + \varphi_{11} - 1)\,(\lambda_2 + \varphi_{11} - 1)\,(\sqrt{\lambda_1} - \sqrt{\lambda_2})^2 \qquad (11.29\text{b})$$

$$M_{13} = \frac{1}{\Delta}\,(\lambda_2 - \lambda_1)\left(\sqrt{\lambda_1} - \sqrt{\lambda_2}\right)\left[\sqrt{\lambda_1}\,(\lambda_1 + \varphi_{11} - 1) - \sqrt{\lambda_2}\,(\lambda_2 + \varphi_{11} - 1)\right]$$
$$(11.29\text{c})$$

$$M_{21} = M_{12} \qquad (11.29\text{d})$$

$$M_{22} = \frac{1}{\Delta}\left(\lambda_1 + \lambda_2 - 2\sqrt{\lambda_1 \lambda_2}\right)\left(1 - \varphi_{11} + \sqrt{\lambda_1 \lambda_2}\right)^2 \qquad (11.29\text{e})$$

$$M_{23} = \frac{1}{\Delta}\left(\lambda_2 - \lambda_1\right)\left(1 - \varphi_{11} + \sqrt{\lambda_1 \lambda_2}\right)\left(\sqrt{\lambda_1} - \sqrt{\lambda_2}\right)^2 \qquad (11.29\text{f})$$

$$M_{31} = \frac{1}{\Delta}\,(\lambda_1 + \varphi_{11} - 1)\left(\sqrt{\lambda_2} - \sqrt{\lambda_1}\right)\left[\sqrt{\lambda_1}\,(\lambda_1 + \varphi_{11} - 1) - \sqrt{\lambda_2}\,(\lambda_2 + \varphi_{11} - 1)\right]$$
$$(11.29\text{g})$$

$$M_{32} = \frac{1}{\Delta}\,(\lambda_1 + \varphi_{11} - 1)\left(\sqrt{\lambda_2} - \sqrt{\lambda_1}\right)\left[\sqrt{\lambda_2}\,(\lambda_1 + \varphi_{11} - 1) - \sqrt{\lambda_1}\,(\lambda_2 + \varphi_{11} - 1)\right]$$
$$(11.29\text{h})$$

$$M_{33} = \frac{1}{\Delta}\,(\lambda_2 - \lambda_1)\left(\sqrt{\lambda_2} - \sqrt{\lambda_1}\right)\left[\sqrt{\lambda_1 \lambda_2}\left(\sqrt{\lambda_2} + \sqrt{\lambda_1}\right) - (\lambda_1 + \varphi_{11} - 1)\left(\sqrt{\lambda_2} - \sqrt{\lambda_1}\right)\right]$$
$$(11.29\text{i})$$

and

$$\Delta = \lambda_1 \lambda_2 (\lambda_1 - \lambda_2)^2 \qquad (11.29\text{j})$$

In order to check the validity of equation (11.28) for the case of isotropic damage, one first sets $\varphi_{12} = 0$. In this case, one obtains the following form for [$\mathbf{M}$]:

$$[M] =
\begin{bmatrix}
\dfrac{1}{1-\varphi_{11}} & 0 \\[2em]
0 & \dfrac{1}{1-\varphi_{22}} \\[2em]
\dfrac{-1}{1-\varphi_{11}} + \dfrac{1}{\sqrt{(1-\varphi_{11})(1-\varphi_{22})}} & \dfrac{1}{1-\varphi_{22}} - \dfrac{1}{\sqrt{(1-\varphi_{11})(1-\varphi_{22})}}
\end{bmatrix}$$

$$\begin{bmatrix}
-\dfrac{1}{1-\varphi_{11}} + \dfrac{1}{\sqrt{(1-\varphi_{11})(1-\varphi_{22})}} \\[2em]
\dfrac{1}{1-\varphi_{22}} - \dfrac{1}{\sqrt{(1-\varphi_{11})(1-\varphi_{22})}} \\[2em]
\dfrac{1}{1-\varphi_{11}} + \dfrac{1}{1-\varphi_{22}} - \dfrac{1}{\sqrt{(1-\varphi_{11})(1-\varphi_{22})}}
\end{bmatrix}$$

$$(11.30)$$

It is noticed that the matrix representation of [M] given in the above equation is not diagonal. This is different from the result obtained in the previous section as well as that in the next section. It is easy to check isotropy by setting $\varphi_1 = \varphi_2 = \varphi$. Substituting this into equation (11.30), one obtains a diagonal matrix which is the analog of equation (11.13) for the case of plane stress, with [I] being the 3 x 3 identity matrix.

The three diagonal terms of the matrix in equation (11.28) are shown in Figures 11.4, 11.5, and 11.6. Their variation with φ_{11} and φ_{22} is shown while increasing the value of φ_{12}. In Figure 11.4, the variation of M_{11} is shown with φ_{12} taking the values 0.0, 0.2, 0.4, and 0.5 in Figures 11.4a, 11.4b, 11.4c, and 11.4d, respectively. Similar graphs are shown for M_{22} and M_{33} in Figures 11.5 and 11.6, respectively. Variations of these terms with values of φ_{12} greater than 0.5 are not shown since the damage effect becomes too large and numerical instability occurs.

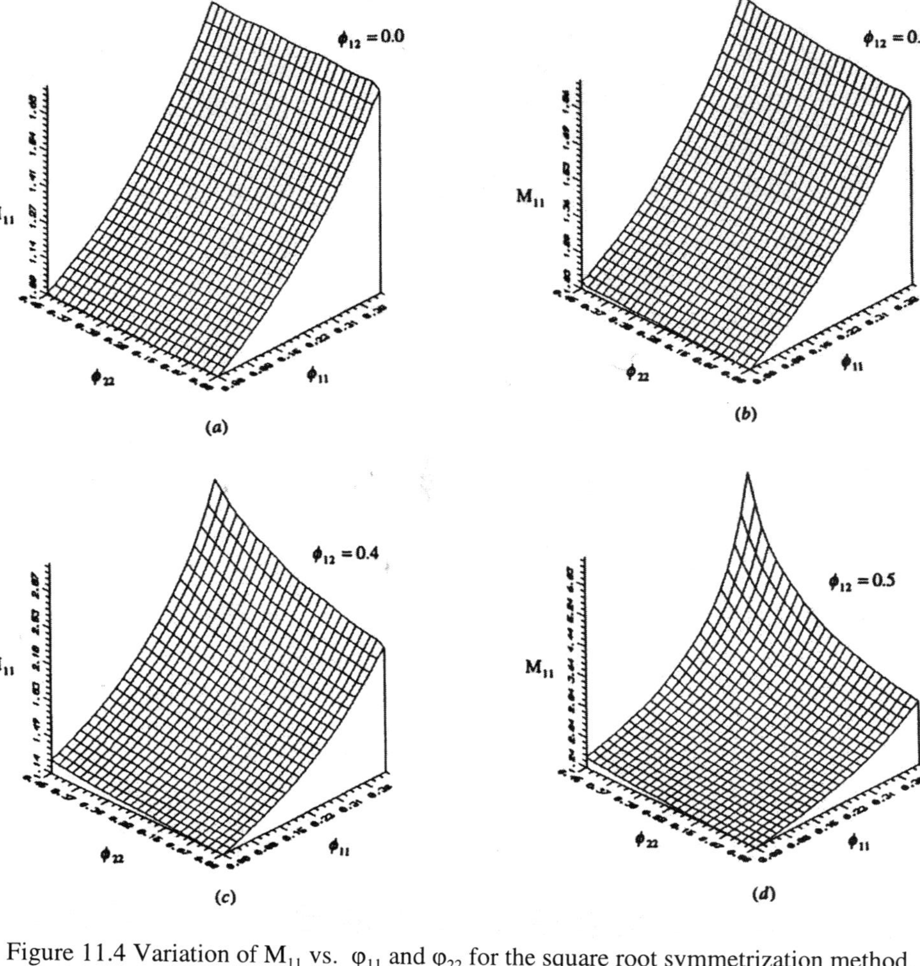

Figure 11.4 Variation of M_{11} vs. φ_{11} and φ_{22} for the square root symmetrization method.

288

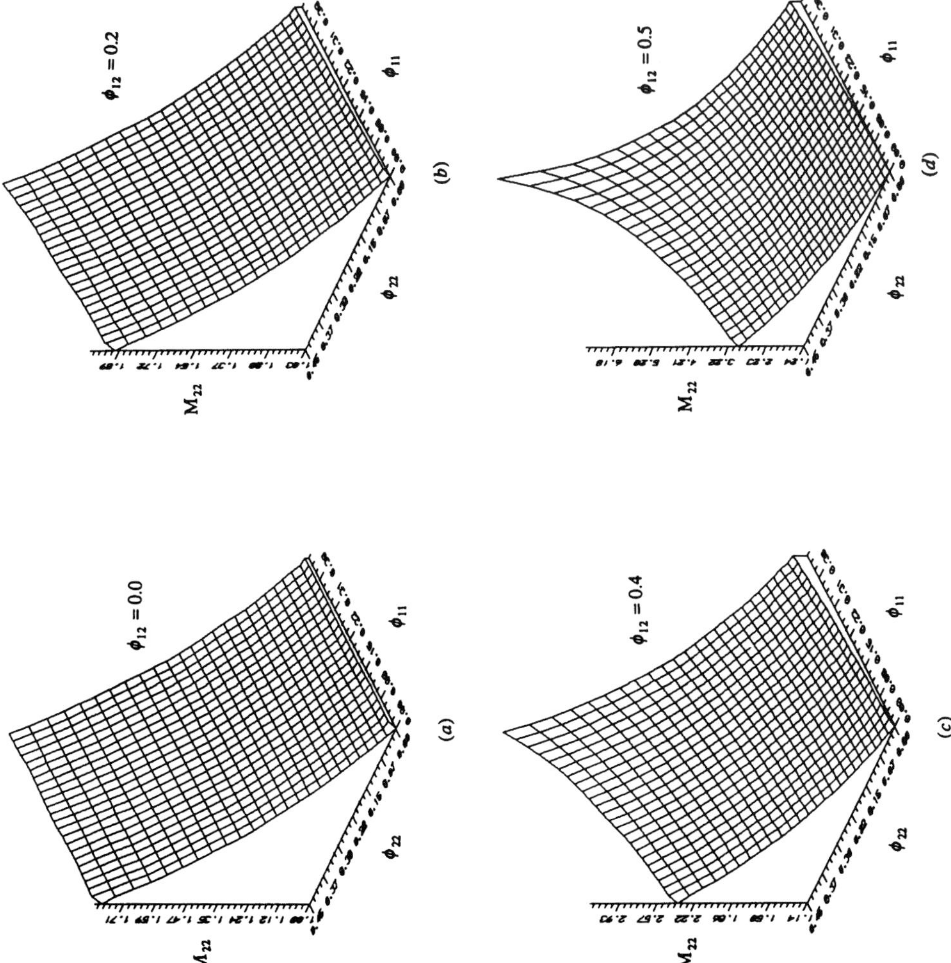

Figure 11.5 Variation of M_{22} vs. φ_{11} and φ_{22} for the square root symmetrization method.

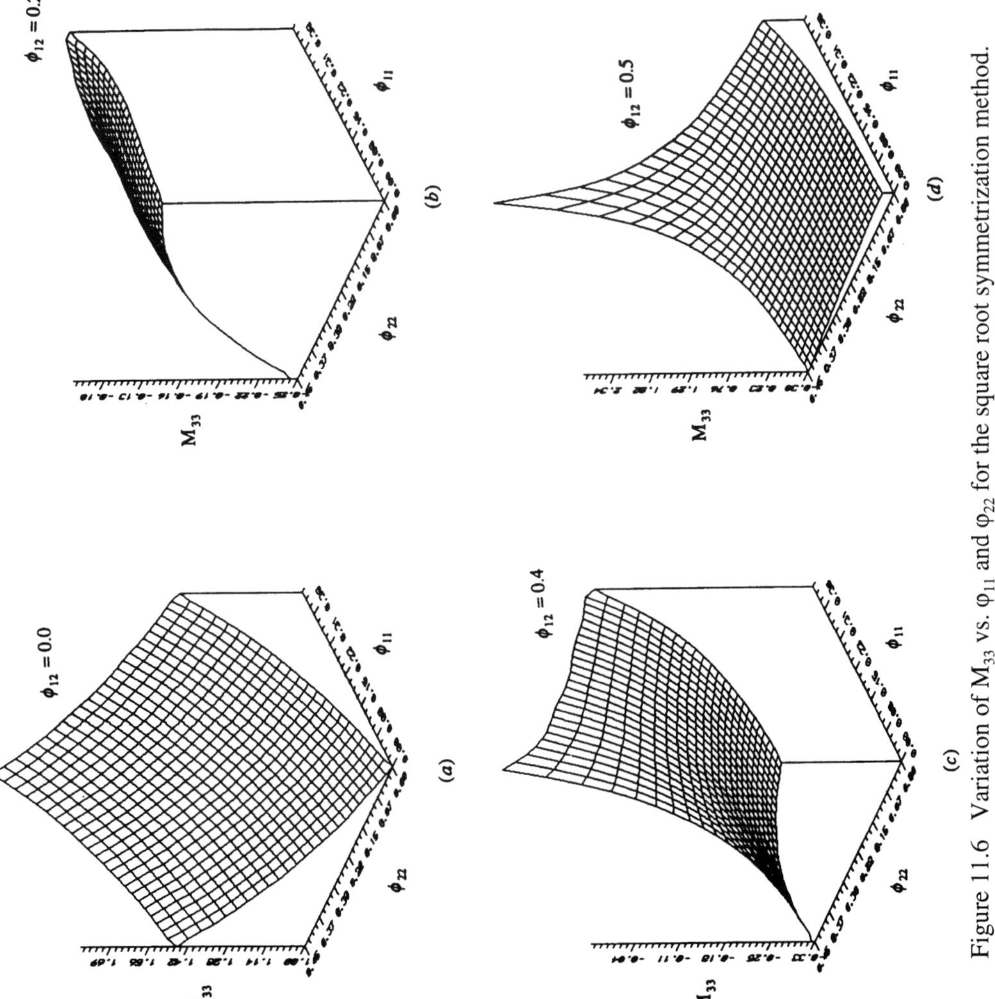

Figure 11.6 Variation of M_{33} vs. ϕ_{11} and ϕ_{22} for the square root symmetrization method.

11.4 Implicit Symmetrization Method:

The symmetrization scheme given in equation (11.6) is the third method used in this chapter. This is the only implicit symmetrization scheme discussed in this work. However, it produces an explicit representation for the matrix $[\mathbf{M}]^{-1}$ which can be easily inverted. In order to derive an expression for $[\mathbf{M}]$, one starts with the matrix $[\mathbf{I} - \boldsymbol{\varphi}]$ given by:

$$[\mathbf{I} - \boldsymbol{\varphi}] = \begin{bmatrix} 1 - \varphi_{11} & -\varphi_{12} & -\varphi_{13} \\ -\varphi_{12} & 1 - \varphi_{22} & -\varphi_{23} \\ -\varphi_{13} & -\varphi_{23} & 1 - \varphi_{33} \end{bmatrix} \tag{11.31}$$

Substituting equation (11.23) into equation (11.6), simplifying and rewriting the result in the terminology of equation (11.8) (where stresses are represented by 6 x 1 vectors), one obtains the following representation for $[\mathbf{M}]^{-1}$:

$$[\mathbf{M}]^{-1} = \begin{bmatrix} 1 - \varphi_{11} & 0 & 0 & 0 & -\varphi_{13} & -\varphi_{12} \\ 0 & 1 - \varphi_{22} & 0 & -\varphi_{23} & 0 & -\varphi_{12} \\ 0 & 0 & 1 - \varphi_{33} & -\varphi_{23} & -\varphi_{13} & 0 \\ 0 & -\frac{1}{2}\varphi_{23} & -\frac{1}{2}\varphi_{23} & 1 - \frac{1}{2}(\varphi_{22} + \varphi_{33}) & -\frac{1}{2}\varphi_{12} & -\frac{1}{2}\varphi_{13} \\ -\frac{1}{2}\varphi_{13} & 0 & -\frac{1}{2}\varphi_{13} & -\frac{1}{2}\varphi_{12} & 1 - \frac{1}{2}(\varphi_{11} + \varphi_{33}) & -\frac{1}{2}\varphi_{23} \\ -\frac{1}{2}\varphi_{12} & -\frac{1}{2}\varphi_{12} & 0 & -\frac{1}{2}\varphi_{13} & -\frac{1}{2}\varphi_{23} & 1 - \frac{1}{2}(\varphi_{11} + \varphi_{22}) \end{bmatrix}$$

$$\tag{11.32}$$

Inverting the above matrix, one obtains the 6 x 6 matrix representation for $\mathbf{M}$. The elements of the matrix $[\mathbf{M}]$ are give explicitly in Appendix A-3. The matrix representation given in Appendix A-3 is very large compared to that of equation (11.11). The authors would like to point out that the equations appearing in Appendix A-3 could not have been obtained without the use of the symbolic manipulation program REDUCE. However, it can be shown that the explicit expressions given in Appendix A-3 reduce to the isotropic damage effect matrix $[\mathbf{M}]_{\text{isot.}}$ of equation (11.13). First, set

$\varphi_{12} = \varphi_{13} = \varphi_{23} = 0$ and factor Δ as follows:

$$\Delta = \frac{1}{8}\Psi_{11}\,\Psi_{22}\,\Psi_{33}\,(\Psi_{11} + \Psi_{22})\,(\Psi_{11} + \Psi_{33})\,(\Psi_{22} + \Psi_{33}) \tag{11.33}$$

Using the expression of Δ given above in the equations and Appendix A-3, and simplifying, one obtains the following diagonalized form for the matrix $[\mathbf{M}]$:

$$[M]_{diag.} = \begin{bmatrix} \dfrac{1}{1-\varphi_1} & 0 & 0 & 0 & 0 & 0 \\[2ex] 0 & \dfrac{1}{1-\varphi_2} & 0 & 0 & 0 & 0 \\[2ex] 0 & 0 & \dfrac{1}{1-\varphi_3} & 0 & 0 & 0 \\[2ex] 0 & 0 & 0 & \dfrac{2}{(1-\varphi_3)+(1-\varphi_2)} & 0 & 0 \\[2ex] 0 & 0 & 0 & 0 & \dfrac{2}{(1-\varphi_3)+(1-\varphi_1)} & 0 \\[2ex] 0 & 0 & 0 & 0 & 0 & \dfrac{2}{(1-\varphi_2)+(1-\varphi_1)} \end{bmatrix}$$

$$\tag{11.34}$$

It is noticed that the first three diagonal terms in equation (11.34) are identical to those of equation (11.12). However, the last three diagonal terms are different because of the different symmetrization procedure. Using equation (11.34), it can be easily shown that by setting $\varphi_1 = \varphi_2 = \varphi_3 = \varphi$, one obtains the isotropic matrix representation given in equation (11.13).

For the case of plane stress, the matrix components given in Appendix A-3 reduce to:

292

$$[M] = \frac{1}{\Delta} \begin{bmatrix} \Psi_{22}^2 + \Psi_{11}\Psi_{22} - \varphi_{12}^2 & \varphi_{12}^2 & 2\varphi_{12}\Psi_{22} \\ \varphi_{12}^2 & \Psi_{11}^2 + \Psi_{11}\Psi_{22} - \varphi_{12}^2 & 2\varphi_{12}\Psi_{11} \\ \varphi_{12}\Psi_{22} & \varphi_{12}\Psi_{11} & 2\Psi_{11}\Psi_{22} \end{bmatrix} \qquad (11.35a)$$

where Δ is given by:

$$\Delta = (\Psi_{11} + \Psi_{22})(\Psi_{11}\Psi_{22} - \varphi_{12}^2) \qquad (11.35b)$$

Figure 11.7 shows the variation of M_{11} against φ_{11} and φ_{22} based on equation (11.35a). It is clear that these plots are similar to those in Figures 11.1 and 11.4 for the other two symmetrization procedures. The diagonal term M_{22} is plotted in Figure 11.8. It shows the same behavior noticed in Figures 11.2 and 11.5 for the previous two symmetrization methods. The last diagonal term M_{33} is plotted in Figure 11.9 and shows the same behavior noticed in Figure 11.3 for the explicit symmetrization method. However, when compared with Figure 11.6 for the square root symmetrization method some distinct differences appear. Although both explicit and implicit symmetrization methods show monotonic positive increases in the values of M_{33} with the increase in φ_{12}, this is not the case for the square root symmetrization method. The latter method shows an initial decrease in the value of M_{33} for the case of $\varphi_{12} = 0.2$ and $\varphi_{12} = 0.4$ and an increase in the magnitude of M_{33} for the case of $\varphi_{12} = 0.5$. This is not attributed to numerical instability but to the nature of the symmetrization procedure. This leads to the recommendation that only the explicit and implicit symmetrization procedures be used as they depict more accurately the physics of the material behavior.

It is also clear from Figures 11.7, 11.8 and 11.9 that again numerical instability occurs for large values of the damage variables especially for $\varphi_{12} > 0.5$. No such results are shown as the values of M_{11}, M_{22} and M_{33} become too large approaching infinity in many cases. This is mainly attributed to the fact that complete rupture occurs at large values of the damage variables. Looking at the isotropic matrix representation of damage in equation (11.13) provides a simple explanation since the term $(1 - \varphi)^{-1}$ approaches infinity as φ approaches 1. However, this is a very special case as the actual material behavior is more complicated. In fact, the values of M_{11}, M_{22} and M_{33} approach infinity as soon the value of φ_{12} increases beyond 0.5. Lemaitre [31] suggests the critical value of the damage variables to be in the range 0.2 to 0.8. Therefore, in a general state of deformation and damage, complete rupture occurs well before the extreme value of 1 is reached. These remarks apply to all three symmetrization procedures used.

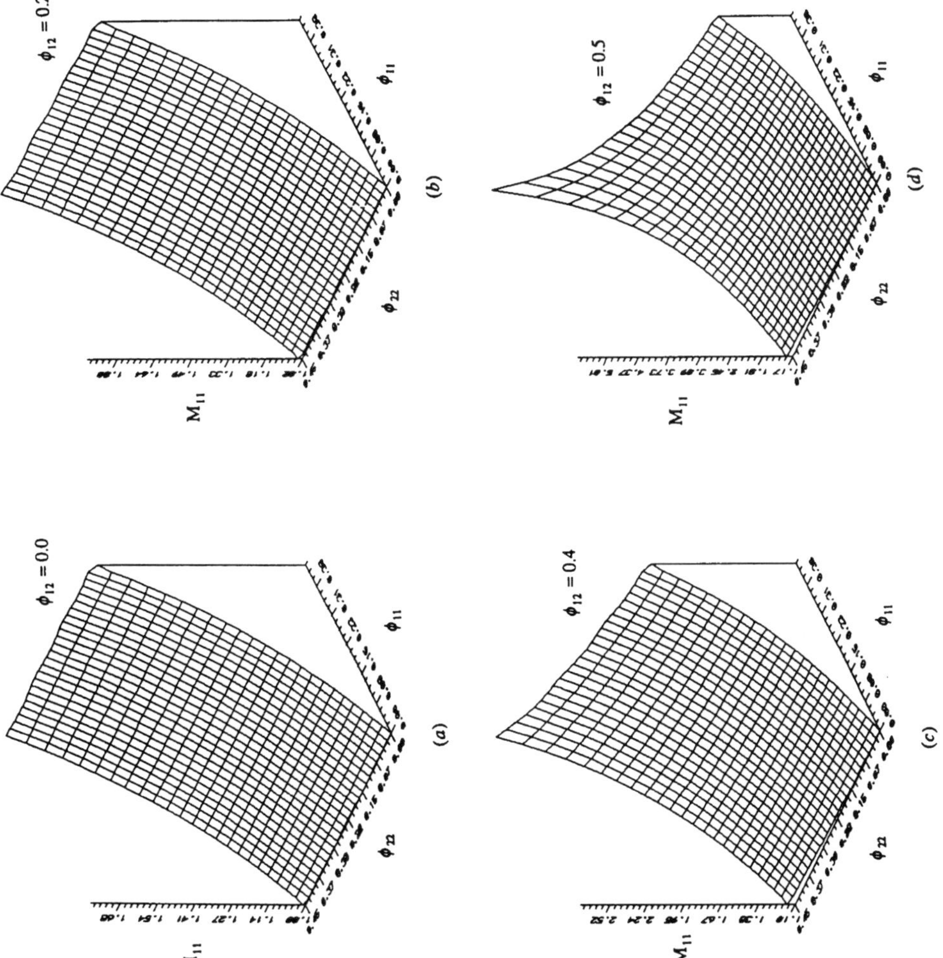

Figure 11.7 Variation of M_{11} vs. φ_{11} and φ_{22} for the implicit symmetrization method.

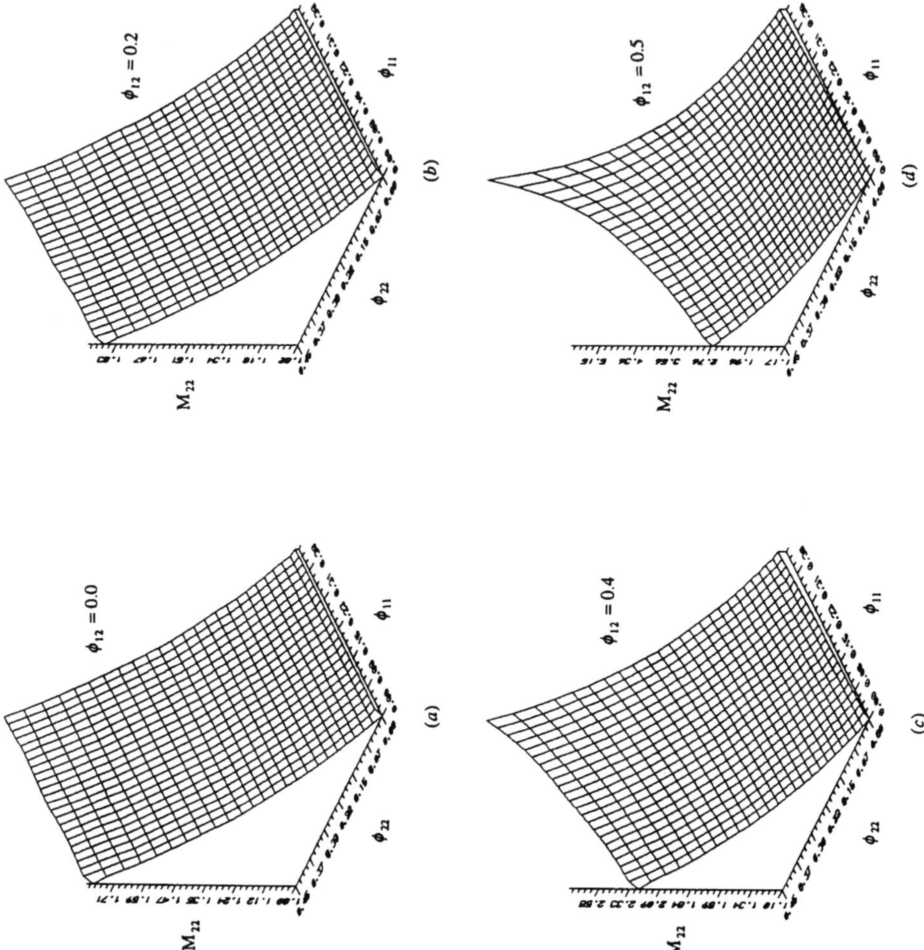

Figure 11.8 Variation of M_{22} vs. ϕ_{11} and ϕ_{22} for the implicit symmetrization method.

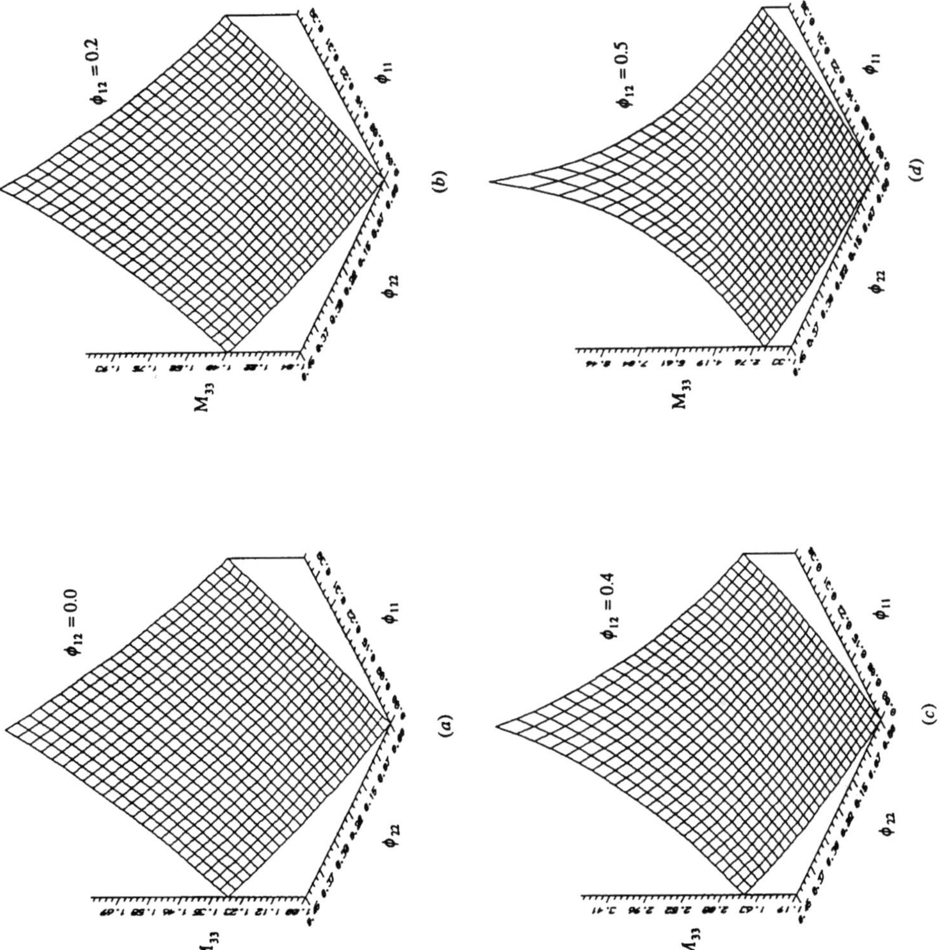

Figure 11.9 Variation of M_{33} vs. φ_{11} and $\varphi_{.22}$ for the implicit symmetrization method.

Comparing the values of M_{11} shown in Figures 11.1, 11.4 and 11.7 for the three symmetrization methods, it is clear that the explicit method produces the highest values followed by the square root method, then the implicit method. It is also clear that the explicit method produces the highest values of M_{22} as shown in Figures 11.2, 11.5 and 11.8. Similar observations are made about the values of M_{33} as can be seen from Figures 11.3, 11.6 and 11.9. It is concluded that the explicit method produces higher damage effect values thus resulting in higher effective stresses than the other two methods. On the other hand, the implicit method produces the lowest symmetrized stress values. Finally, it should be mentioned that all three symmetrization methods display qualitatively the same variation of the damage effect tensor. The only exception is the behavior of M_{33} for the square root method which is completely different from the other two methods as shown in Figure 11.6. This is mainly attributed to the effect of the square roots appearing in equations (11.29i) and (11.30). An oscillation is depicted in M_{33} values which initially decreases in magnitude when φ_{12} = 0.2 and 0.4 followed by an increase in M_{33} when φ_{12} = 0.5. This leads to the conclusion that only the explicit and implicit symmetrization methods depict more accurately the physics of the material damage behavior.

CHAPTER 12

EXPERIMENTAL DAMAGE INVESTIGATION

Fiber-reinforced composite materials, specifically those with a metal matrix, are having an increasing role in consideration for the design and manufacture of composite structures. This is a result of these types of materials having the ability to attain higher stiffness/density and strength/density ratios as compared to other materials. Along with this increased use comes the responsibility of designers to be able to understand and predict the behavior of metal matrix composite materials, especially that of damage initiation and evolution. Knowledge such as this can be obtained through experimental investigations. Although the literature contains an abundance of new developments in composite materials technology, it lacks a consistent analysis of damage mechanisms as well as damage evolution in composite materials.

A number of damage theories have been proposed with limited experimental investigation. These investigations are primarily confined to damage as a result of fatigue of fracture (Allix et al. [79], Poursatip et al. [80], Ladaveze et al. [81], and Wang [82]). Each of these investigations do not present damage evolution as a function of the measured physical damage over a load history. A more recent work by Majumdar et al. [83] provides a thorough examination and explanation on the microstructural evolution of damage. However, this work has not been extended to a constitutive theory for the quantification and evolution of physical damage. Recently, new experimental procedures have been introduced to quantify damage due to micro-cracks and micro-voids through X-ray diffraction tomography, etc. (Breunig et al. [84], Baumann et al. [85], and Benci et al. [86]). Nevertheless, these procedures need to be refined in order to differentiate between the different types of damages such as voids and cracks (radial, debonding, z-type). Additional experiments need to be performed in order to quantify the damage parameters as well as evaluate the proposed damage theory. Much of the work in this area has been done using a continuum approach with various schemes of measuring the damage. In each of the schemes, damage is a measure of ratio between an effective quantity and its respective damaged value. Lemaitre et al. [32] listed several methods of obtaining ratios for the damage parameter based on area of resistance, material density, and elasto-plastic modulus. Obtaining the damage parameter as a ratio of the elastic-plastic modulus is most widely used because of the ease in evaluating the damaged and undamaged elasto-plastic moduli. As previously mentioned, methods such as this cannot capture or predict the effect of local components on the overall damage evolution. Within this chapter, a method will be outlined to experimentally evaluate different types of damage in a metal matrix composite material that can be used in conjunction with a micromechanical damage theory. This is outlined through an overall damage quantification as well as a local damage quantification differentiating between damage in the matrix and in the fibers. Major topics covered are specimen design and preparation, mechanical testing (macro-analysis), Scanning Electron Microscope (SEM) analysis (micro-analysis), and evalaution of damage parameters based on the results of the micro-analysis.

12.1 Specimen Design and Preparation:

The material investigated in this chapter is a titanium aluminide composite reinforced with continuous SiC (SCS - 6) fibers. The SiC fibers are developed and produced by the manufacturer of the initial plate specimens. Typical properties of the SiC fibers, as provided by the manufacturer, are shown in Table 12.1. Additionally, the fibers have good wettability characteristics for metals, which should minimize the chances of voids being introduced during the manufacturing process. Also, these fibers are coated with a carbon rich coating that assists in protecting the inner SiC from damage during handling.

Table 12.1: Typical Properties of Silicon Carbide (SiC) Fibers

Diameter	0.14 mm (0.0056 in)
Density	3044 kg/mm^3 (0.11 lb/in^3)
Tensile Strength	3.44 GPa (500 ksi)
Young's Modulus	414 GPa (58 x 10^6 psi)
Poisson Ratio	0.22
CTE	2.3 x 10^{-6} ppm - °C at RT

(Provided by Textron Specialty Materials, Inc., Lowell, MA, USA)

The titanium aluminide foil is an α_2 phase material that has typical properties, provided by the manufacturer, as shown in Table 12.2. The manufacturer also provided properties of a composite lamina for 0^0 and 90° orientations obtained from experimental tests conducted on manufactured specimens. These values are as reported in Table 12.3.

Table 12.2: Typical Properties of Ti-14Al-21Nb (α_2) matrix

Composition	Ti	63.4%
	Al	14.4%
	Nb	22.1%
Tensile Strength	—	448 MPa (65 ksi)
Young's Modulus	—	84.1 GPa (12 x 10^6 psi)
Poisson Ratio	—	0.30

(Provided by Textron Specialty Materials, Inc., Lowell, MA, USA)

Table 12.3: Typical Properties of SiC-Ti-Al Lamina

$0°$ Tensile Strength	1.38 - 1.52 GPa
$90°$ Tensile Strength	103 - 206 MPa
Longitudinal Modulus	199 GPa
Transverse Modulus	136 GPa
Shear Modulus, G_{12}	52 GPa
Poisson Ratios	$\nu_{12} = 0.27$ $\nu_{21} = 0.185$ $\nu_{31} = 0.31$

(Provided by Textron Specialty Materials, Inc., Lowell, MA, USA)

Hand layup techniques are used to fabricate two different specimen layups [i.e. $(0/90)_s$ and $(\pm 45)_s$] from SCS-6 SiC fiber mats and Ti-14Al-21Nb (α_2) foils from rolled ingot material. Each of the layups contained four plies. Fibers, in the fiber mat, were held together with molybdenum wire. Consolidation is accomplished by hot-isostatic-pressing (HIP) in a steel vacuum bag at 1010 $°C \pm 25°$ under 103 MPa pressure for 2 hours. C-scans are performed on each specimen plate to evaluate the consolidation and fiber alignment of the finish product. Results indicate very good consolidation for the crossply specimen $(0/90)_s$ with some fiber misalignment along the plate edges. However, the $(\pm 45)_s$ plate has generally good consolidation with significant occurrences of fiber misalignment or fiber bundling on the interior of the plate as well as the edges.

As a result of fiber misalignment and differences in coefficients of thermal expansion for the fiber and matrix, noticeable warpage is found on each of the plate specimens. Much of the warpage was confined to the edges of each plate, with a maximum relative elevation difference of

2.24 cm for the (0/90)$_s$ plate and 1.30 cm for the (± 45)$_s$ plate. Of particular concern is whether or not this warpage will induce any detectable damage during the preparation of the actual test specimens.

Nevertheless, each of the laminates is machined to produce six test specimens with shape and dimensions as indicated in Figure 12.1. Specimen locations are selected in order to minimize the effects of the laminate warpage on the test specimens. The locations that were selected had the minimum amount of warpage, so that the level of prestress would be negligible during testing. They also exhibit no detectable evidence of damage to the fibers or matrix. This is verified through C-scans of the individual test specimens after machining. Sample C-scans for a typical specimen for each layup are shown in Figures 12.2 and 12.3 to illustrate this fact. These are gray scale images which are interpreted as the darker the image the better the consolidation and fiber alignment. The 3rd backwall echo represents the amplitude of the third return wave of the initial excitation frequency. Also, these scans correspond to the previous scans done on the initial plate specimen, which implies that machining of the test specimens did not induce any detectable damage.

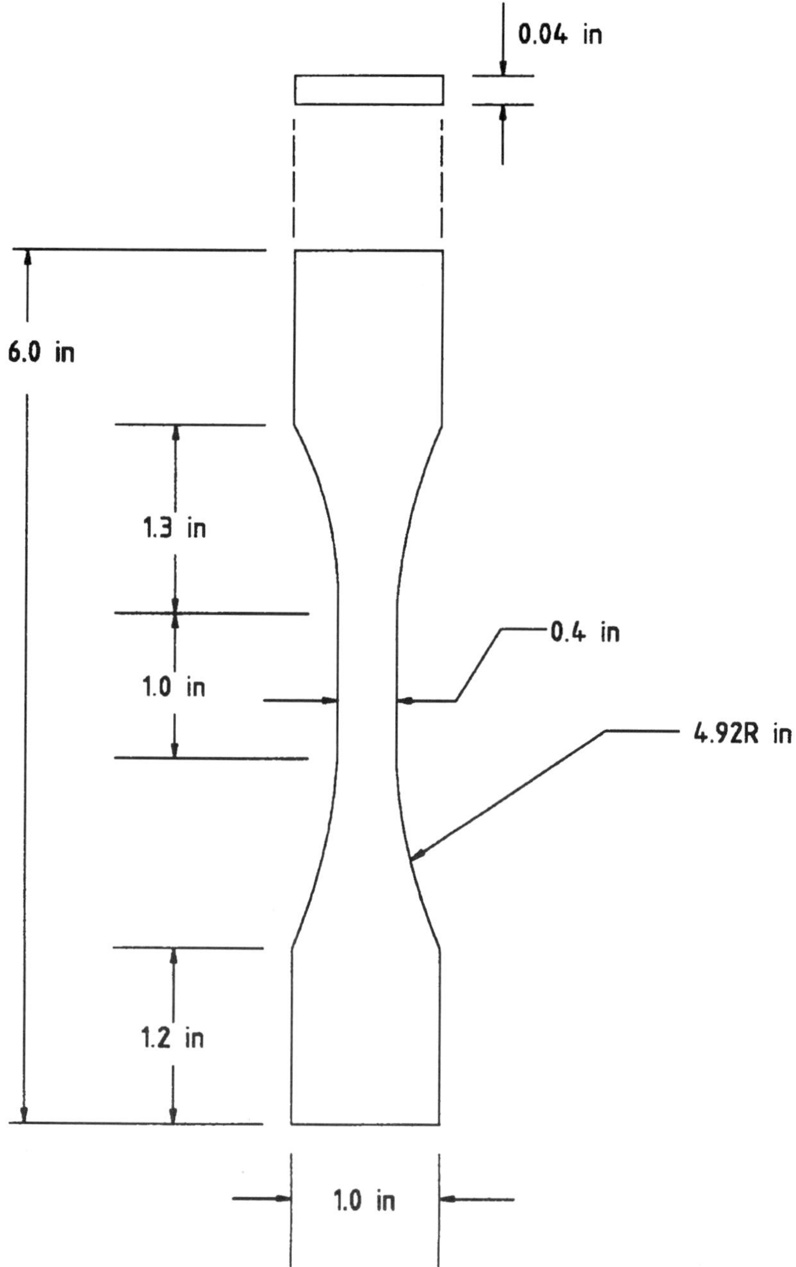

Figure 12.1 Dogbone Shaped Tensile Specimen

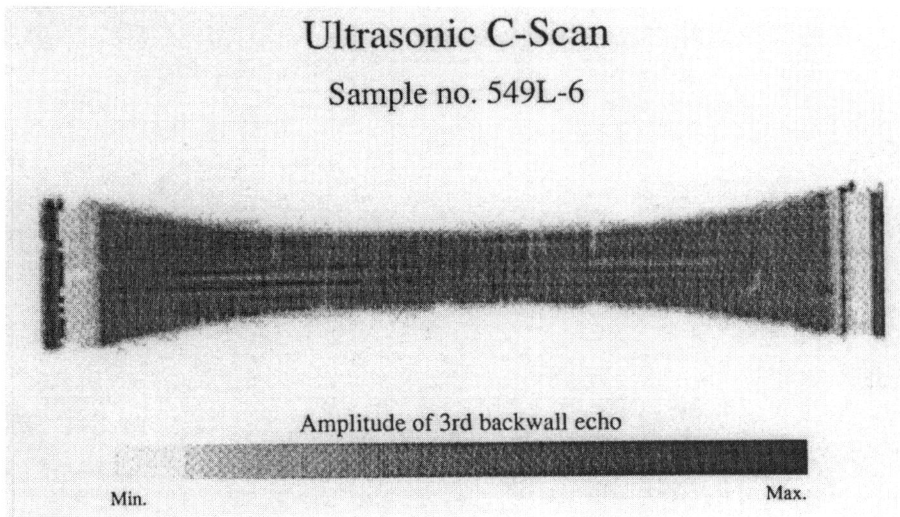

Figure 12.2   C-Scan of selected $(0/90)_s$ specimen.

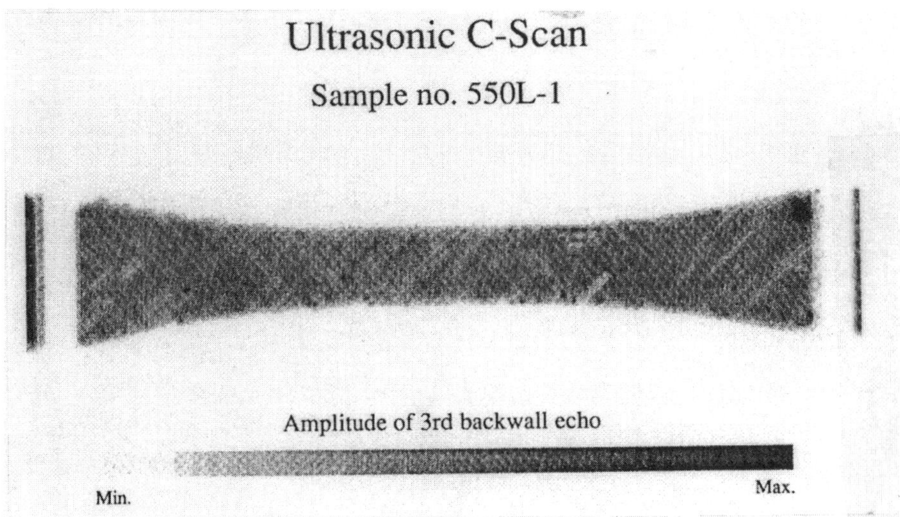

Figure 12.3   C-Scan of selected $(\pm45)_s$ specimen.

The dogbone type specimen has been used successfully by previous researches [91, 145] to ensure specimen failure within the gage section and not the grips. These specimens had aluminum tabs arc-welded onto the ends in order to prevent the mechanical grips from damaging the specimens. Welds are made on the extreme ends, producing local damage only in the vicinity of the weld.

12.2 Mechanical Testing of Specimens:

Before beginning the actual mechanical testing, much attention is given to specimen preparation, using the recommendations of Carlsson [146] and Tuttle [147], and the experimental data items sought as a guide. Quantitative information (stress and strain) is sought for use in the damage evolution model. Therefore, foil-resistance strain gages are used in obtaining the necessary strain data. Each of the dogbone type specimens has strain gages mounted on both faces, directly opposite one another. This is done to determine if eccentric loading occurs during the test, or, if the specimen contains any prestress as a result of geometric distortion, so that adjustments can be made to the raw data for these effects. Transverse and longitudinal gages are mounted on each face to monitor transverse and longitudinal strains.

All mechanical testing is done utilizing a computer-controlled testing machine with hydraulic grips. Specimens are loaded at a crosshead rate of 4.23 mm/hr to allow enough time to collect sufficient data during the test. Data is sampled continuously with all aspects of the test being controlled by a personal computer and data acquisition system once started. Calibration factors are obtained from all specimen strain gages before testing and used later during data reduction. An extensiometer was also attached to the specimen during testing with results being plotted on an oscilloscope for immediate feedback. Results from the extensiometer matched within $\pm 3\%$ the longitudinal results of the strain gages.

As a means of checking the prestress level resulting from manufacturing distortions, strain readings are taken during the process of gripping each end of the specimen in the testing machine. Strains obtained during this process from all specimens are considered negligible, with strain on the order of 120 $\mu\varepsilon$ for the dogbone type specimens. Thus, as mentioned previously, the effects of the warpage induced prestress are small and will be neglected.

Only one test specimen of each orientation of the dogbone type is loaded to rupture. The remaining five specimens are loaded at 90, 85, 80, 75 and 70% of rupture load. These five load

304

levels are used to measure the evolution of damage in the specimens through the progression of loading. Quantification of damage for each load level is obtained by sectioning each specimen and measuring damage features on a representative cross-section of the specimen. The actual process is explained fully in a subsequent section in this chapter. Stress-strain curves for selected specimens of orientations (0/90)$_s$ and (± 45)$_s$ are shown in Figures 12.4 and 12.5, respectively. The Nb in the matrix is added to improve ductility (Brindley [148], Mackey et al. [149]); however, it appears that ductility is also a function of fiber orientation for constant material properties. For example, the (0/90)$_s$ specimens have a maximum total longitudinal strain less than the (± 45)$_s$ specimens. A possible explanation for this observation is that there is an increased amount of mechanical interaction between the semi-ductile matrix and brittle fibers as the fiber orientation increased with respect to the loading direction. It is expected that there will be more physical damage in the matrix for the (± 45)$_s$ specimens than in the (0/90)$_s$ specimens. For each of the specimen layups shown, there is a slight variability in the response curve for different specimens with the same layup. It is proposed that this variability is due to the variable nature of composite materials and not a result of the damage evolution. Although, damage initiation may be different, the net effect for all specimens will be the same.

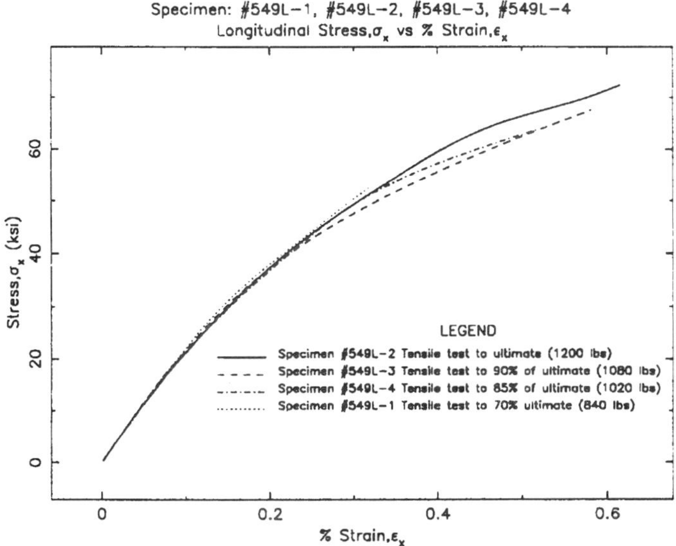

Figure 12.4   Stress-strain curves for selected (0/90)$_s$ Specimens.

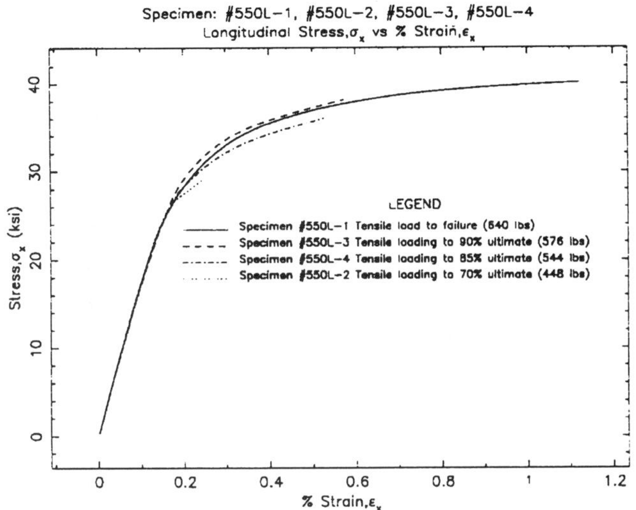

Figure 12.5   Stress-strain curves for selected $(\pm 45)_s$ specimens.

12.3 SEM and Image Analysis:

An SEM (Scanning Electron Microscope) analysis is done on a representative cross-section of all specimens in order to obtain a qualitative evaluation of damage in the specimens, as well as providing a means for measuring visible damage features later. Longitudinal and transverse sections are taken from all samples in the vicinity of the strain gages. The transverse cross-section investigated is at the midpoint of the specimen gage length, and the longitudinal sections are taken normal to this section. Information within two fiber diameters of the specimen edge on transverse sections is disregarded to eliminate any possible free edge effects in the analysis. This is not done for the longitudinal cross-sections since they are carefully taken from the middle of the specimen. All section surfaces are prepared by making the section cut with a low speed diamond saw, followed by grinding and polishing of the cut surface. The low speed diamond saw eliminates the possibility of introducing damage on the cross-section during sectioning. In addition, the grinding and polishing further eliminate any surface defects that can be introduced by the cutting operation. In short, this procedure ensures to a high degree that defects observed during the SEM analysis reflect damage as a result of the loading. Although the cross-section could contain radial cracks as a result of the fabrication cool down process, it is assumed that a well controlled manufacturing process is used such that the number of these cracks is low and can be neglected. Therefore all cracks measured are attributed to loading.

The scanning electron microscope is used to scan the entire cross-sectional area of the longitudinal and transverse sections at low magnification (< 1000X). Photographs are taken on an area of the cross-section that is 1% of the total area and contains an average representation of damage features for the complete cross-section. This area is defined as the representative cross-section with three mutually perpendicular areas of this type defining, the RVE (Representative Volume Element) that is later used to quantify damage evolution. Images are also investigated on the fracture surface of specimens loaded to rupture only as a means of qualitatively investigating the final deformation and failure mode. Results of this investigation showed fiber pull-out, with debonding occurring between the matrix and reaction zone surrounding the fiber. This implies that there is good fiber-matrix bonding. The fibers in the fiber mat are held in place with molybdenum wires to improve fiber alignment during the manufacturing process. Observations of the fracture surface showed a clean break where these wires crossed the surface. Thus, these wires tend to induce the matrix defect for loads normal to the fiber axis; otherwise, they tend to assist the matrix in transferring the load from fiber to fiber. These results also agree with the corresponding specimen stress-strain response in that information observed on the $(0/90)_s$ specimen shows very little deformation in the matrix and brittle failure of the longitudinal fibers, whereas the $(\pm 45)_s$ specimen shows a considerable amount of matrix deformation and a ragged fracture failure of the fibers. Deformation information on these surfaces is not quantified as damage, since it is due to processes other than damage evolution and is outside the valid range of damage mechanics.

Most of the SEM photos predominantly show damage in the fibers in the form of cracks. However, there is some local damage in the matrix in the form of cracks. Representative sections of the fracture surface are studied for the $(0/90)_s$ and $(\pm45)_s$ layups, respectively. The photos demonstrate the predominant brittle behavior of the fibers, in that the surfaces do not display any necking as would occur in ductile materials. This fact implies that the predominant damage feature at other sections will be in the form of fiber splitting/cracking and fiber-matrix interface debonding. However, on the $(\pm45)s$ specimen, the fracture surface is more jagged as a result of the increased fiber-matrix interaction. It is important to note the smooth surfaces left after fiber pullout on each of the layups indicating poor fiber-matrix bonding. This also demonstrates a weak interface bond with the matrix material. Wires normal to the direction of loading will serve as a defect in the matrix; otherwise, they tend to assist the matrix in transferring the load from fiber to fiber.

Other SEM photos were taken on representative cross-sections of the remaining specimens to investigate visible signs of damage. Some selected photos are shown in Figures 12-6 to 12-9 [143, 150]. Each of these photos were taken normal to the cross-section. Figures 12-6 illustrates matrix cracking on specimens with a $(0/90)_s$ layup for different load levels. The type and amount of damage shown in these figures are typical for specimens with this layup. However, specimens with a layup of $(\pm 45)_s$ displayed an increased amount of visible damage of different types, as indicated in Figures 12-8 and 12-9. Again, this is a result of the increased interaction between the fiber and matrix. Damage shown in these photos is typical for specimens with this layup.

The images shown in Figures 12-6 to 12-9 are indicative of the type and amount of damage features observed on all cross sections analyzed. The only measurable feature found for quantitative purposes was the crack length in the fiber and/or matrix. These crack lengths were obtained utilizing image analyzing equipment and software. Scanning of the SEM photos was done with an OmniMedic XRS-6c scanner at 600 dpi. A high resolution was selected to yield a TIFF image very close to the original photo. The scanned image was transferred to a UNIX-based Intergraph workstation (InterPro 360) for analysis with image analyzing software. Attempts were made to automate the process of measuring cracks on the image; however, available software was not successful in differentiating between defined damage features and noise features on the image. Therefore, it was decided to use a semi-manual technique to measure cracks. The Intergraph ISI-2 software allowed digitizing cracks on the image using a mouse. This software automatically computed the crack lengths with respect to the photo scale during digitization. Measured crack lengths were saved in a database for later processing with the damage characterization theory.

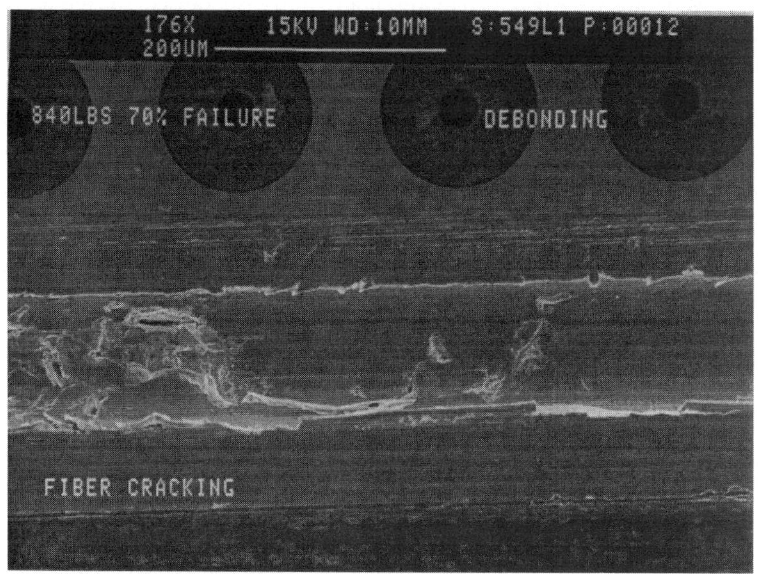

Figure 12.6 Micrograph Showing Micro-cracks in a MMC.

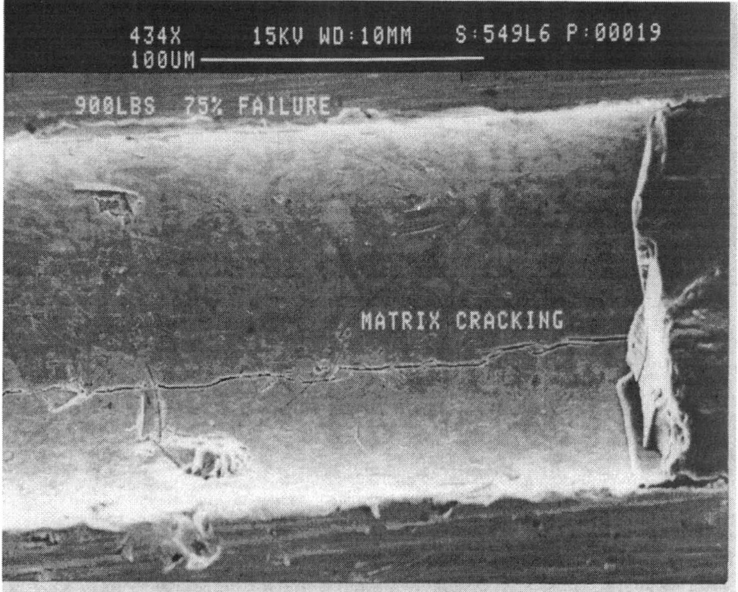

Figure 12.7   SEM photo of (0/90)$_s$ specimen at 75% of failure load showing matrix cracking.

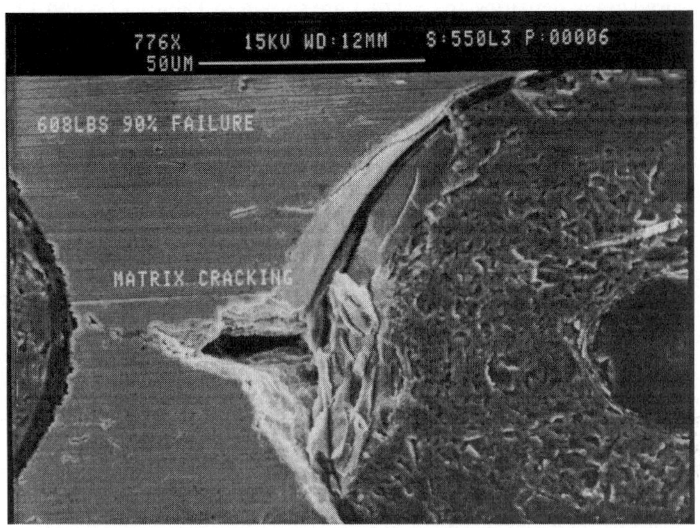

Figure 12.8 SEM photo of (±45)$_s$ specimen at 90% of failure load showing matrix cracking.

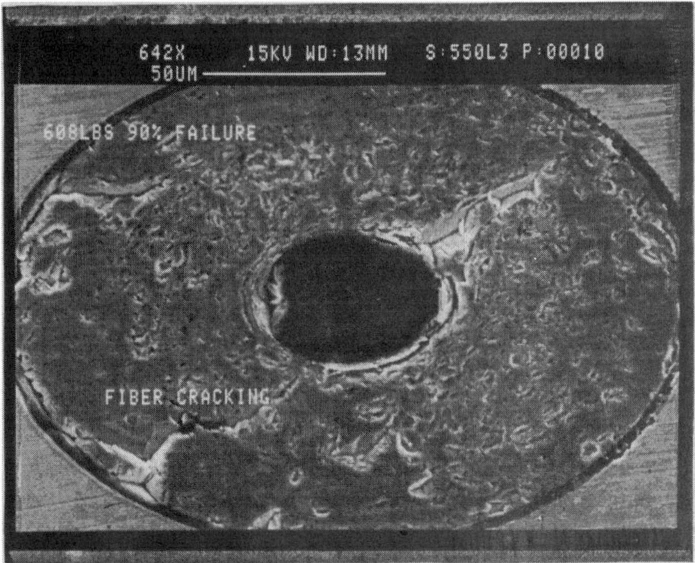

Figure 12.9 SEM photo of (±45)$_s$ specimen at 75% of failure load showing fiber cracking.

310

The measured crack densities ($\rho_i = l_i/A_i$) are shown in Tables 12.4 and 12.5 (Voyiadjis and Venson [143]) for the $(0/90)_s$ and $(\pm 45)_s$ layups, respectively. These values are used to calculate the normalized values $\overline{\rho}_i$ ($i = x, y, z$) for each layup using the four methods given in the next section. These results are then used to calculate the values of the damage variable ϕ. These damage values can then be used in the constitutive model to accurately predict the mechanical behavior of metal matrix composites.

Table 12.4: Measured Crack Densities for the $(0/90)_s$ layups

Specimen No.	Load (kN)	Percentage of Failure Load (%)	ρ_x	ρ_y
1	3.74	70	41.82	3.41
6	4.00	75	70.32	36.40
5	4.27	80	100.77	--
4	4.54	85	106.24	56.43
3	4.80	90	126.68	67.72
2	5.46	100	143.43	--

Table 12.5: Measured Crack Densities for the $(\pm 45)_s$ layups

Specimen No.	Load (kN)	Percentage of Failure Load (%)	ρ_x	ρ_y
2	2.14	70	49.23	--
6	2.28	75	49.32	42.44
5	2.42	80	51.84	101.29
4	2.56	85	52.99	117.01
3	2.70	90	56.67	146.59
1	2.86	100	78.82	--

12.4 Damage Characterization:

A new damage tensor is defined for the uniaxial state of loading based upon experimental observations of crack densities on three mutually perpendicular cross-sections of the specimens. The damage tensor is defined as a second-rank tensor in the form:

$$[\phi] = \begin{bmatrix} \bar{\rho}^2_x & 0 & 0 \\ 0 & \bar{\rho}^2_y & 0 \\ 0 & 0 & \bar{\rho}^2_z \end{bmatrix} \tag{12.1}$$

where $\bar{\rho}_i$ ($i = x, y, z$) is the crack density on a cross-section whose normal is along the i - axis. The crack density for the ith cross-section is calculated as follows:

$$\bar{\rho}_i = \frac{\rho_i}{m\rho^*} \tag{12.2}$$

$$\rho_i = \frac{l_i}{A_i} \tag{12.3}$$

where l_i is the total length of the cracks on the ith cross-section, A_i is the ith cross-sectional area, m is a normalization factor chosen so that the values of the damage variable ϕ fall within the expected range $0 \le \phi_{ij} < 1$, and ρ^* is as defined below. It is assumed that $\rho_z = \rho_y/2$ for computational purposes.

There are several techniques that can be used to choose an appropriate expression for ρ^*. The following are four methods that are used in this book:

$$(1) \quad \rho^* = \rho_{x_{max}} + \rho_{y_{max}} + \rho_{z_{max}}$$

$$(2) \quad \rho^* = \rho^2_{x_{max}} + \rho^2_{y_{max}} + \rho^2_{z_{max}}$$

$$(3) \quad \rho^* = \max(\rho_{x_{max}}, \rho_{y_{max}}, \rho_{z_{max}}) \tag{12.4}$$

$$(4) \quad \rho^* = \sqrt{\rho^2_{x_{max}} + \rho^2_{y_{max}} + \rho^2_{z_{max}}}$$

Where $\rho_{i_{max}}$ is the value of l/A_i at the maximum load. The damage tensor obtained experimentally from equation (12.1) is then used in the constitutive equations to predict the mechanical behavior of the composite system.

12.5 Application to Uniaxial Tension -Example:

In this section, explicit equations are developed to study damage in uniaxially loaded specimens of the two laminate layups discussed earlier. Consider a composite laminate subjected to uniaxial tension in the x-direction. Let ΔN_x be the incremental force resultant in the x-direction where $\Delta N_y = \Delta N_{xy} = 0$. Substituting this in the basic laminate constitutive relation (Jones [148]) and solving for the incremental laminate strain vector, one obtains:

$$\begin{Bmatrix} \Delta\varepsilon_x \\ \Delta\varepsilon_y \\ \Delta\varepsilon_{xy} \end{Bmatrix} = \begin{bmatrix} S_{11} & S_{12} & S_{13} \\ S_{12} & S_{22} & S_{23} \\ S_{13} & S_{23} & S_{33} \end{bmatrix} \begin{Bmatrix} \Delta N_x \\ 0 \\ 0 \end{Bmatrix} \tag{12.5}$$

where the matrix $[S]$ is the inverse of the matrix $[A]$, i.e. $[S] = [A]^{-1}$, and

$$A_{ij} = \frac{h}{n} \sum_{k=1}^{n} Q_{ij(k)} \tag{12.6}$$

where $Q_{ij(k)}$ is the stiffness of the kth lamina (Jones [151]). Simplifying equation (12.5), one can rewrite it in the following form:

$$\begin{Bmatrix} \Delta\varepsilon_x \\ \Delta\varepsilon_y \\ \Delta\varepsilon_{xy} \end{Bmatrix} = \begin{bmatrix} S_{11} \\ S_{12} \\ S_{13} \end{bmatrix} \Delta N_x \tag{12.7}$$

The remaining part of this section will be specific to each type of laminate layup. It is seen that the general laminate equations simplify for these two cases because of the layup symmetry.

12.5.1 Laminate Layup $(0/90)_s$:

The first type of laminate layup $(0/90)_s$ consists of four plies distributed symmetrically as shown in Figure 12-10(a). The angles $\theta_{(k)}$ for this layup are clearly given by:

$$\theta_{(1)} = \theta_{(4)} = 0^o \quad ; \quad \theta_{(2)} = \theta_{(3)} = 90^o \tag{12.8}$$

The values of $\theta_{(k)}$ given in equation (12.8) are used to calculate the transformation matrices for the laminas. After considerable algebraic manipulations, equation (12.7) reduces to:

$$\left\{ \begin{array}{c} \Delta \varepsilon_x \\ \Delta \varepsilon_y \\ \Delta \varepsilon_{xy} \end{array} \right\} = \frac{2h . \Delta N_x}{|A|} \left\{ \begin{array}{c} 2D_{33}(D_{11} + D_{22}) - (D_{13} - D_{23})^2 \\ -4D_{12}D_{33} - (D_{13} - D_{23})^2 \\ D_{23} - D_{13} \end{array} \right\} \tag{12.9}$$

where the determinant $|A|$ is given by:

$$|A| = 2h \, (D_{11} + D_{22} - 2D_{12}) \, [D_{33}(D_{11} + D_{22} + 2D_{12}) - (D_{13} - D_{23})^2] \tag{12.10}$$

The terms D_{ij} are the elements of the matrix representation of the fourth-rank tensor of equation (6.84a). In equation (12.9), one considers ΔN_x as the independent "time" variable Δt in order to solve the incremental system of equations. In the limit as $\Delta t \to 0$, the system of equations (12.9) can be reduced to a system of simultaneous differential equations in ε_x, ε_y and ε_{xy}. Therefore, the governing differential equations are given by:

$$\left\{ \begin{array}{c} \dfrac{d \varepsilon_x}{dt} \\[2ex] \dfrac{d \varepsilon_y}{dt} \\[2ex] \dfrac{d \varepsilon_{xy}}{dt} \end{array} \right\} = \frac{2h}{|A|} \left\{ \begin{array}{c} 2D_{33}(D_{11} + D_{22}) - (D_{13} - D_{23})^2 \\ -4D_{12}D_{33} - (D_{13} - D_{23})^2 \\ D_{23} - D_{13} \end{array} \right\} \tag{12.11}$$

The above system of ordinary differential equations is solved numerically using the IMSL routine DIVPRK. This solution subroutine uses the Runge-Kutta-Verner fifth-order and sixth-order methods for solving a system of simultaneous ordinary differential equations. It should be noted that the strain vector obtained in this way represents the laminate strain as well as the strain in each lamina.

The resulting stress-strain curve for the composite laminate is then plotted using the Mori-Tanaka method. Details about the numerical algorithm and comparisons with the experimental

314

measurements are discussed in Appendix A-4 and in section 12.6..

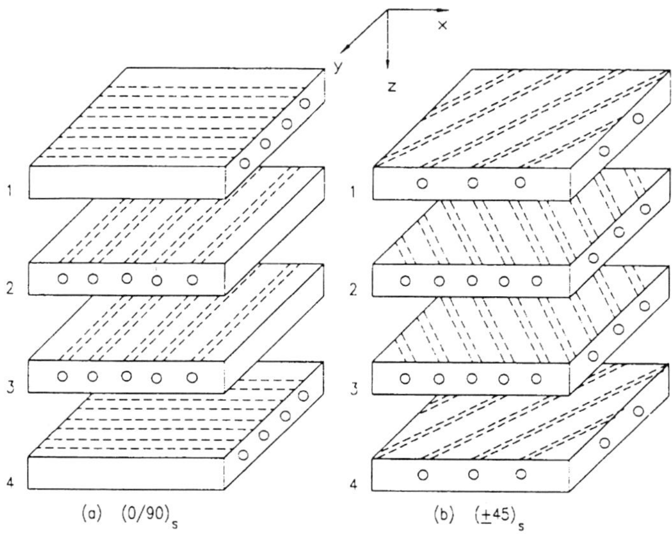

Figure 12.10   Laminate layups considered in this work.

12.5.2 Laminate Layup ($\pm$ 45)$_s$:

The second laminate layup considered in this chapter is ($\pm$45)$_s$ which consists of four cross-ply laminates distributed symmetrically as shown in Figure 12-10(b). The angles θ for this layup are clearly given by:

$$\theta_{(1)} = \theta_{(3)} = 45^o \; ; \; \theta_{(2)} = \theta_{(4)} = -45^o \tag{12.12}$$

After considerable algebraic manipulations, equation (12.7) reduces to:

$$\left\{ \begin{array}{c} \dfrac{d\varepsilon_x}{dt} \\[2ex] \dfrac{d\varepsilon_y}{dt} \\[2ex] \dfrac{d\varepsilon_{xy}}{dt} \end{array} \right\} = \dfrac{h}{|A|} \left\{ \begin{array}{c} (D_{11} + D_{22} - 2D_{12})\,(D_{11} + D_{22} + 2D_{12} + 4D_{33}) - 4\,(D_{13} - D_{23})^2 \\[1ex] -(D_{11} + D_{22} - 2D_{12})\,(D_{11} + D_{22} + 2D_{12} + 4D_{33}) + 4\,(D_{13} - D_{23})^2 \\[1ex] 4\,(D_{13} - D_{23})\,(D_{11} + D_{22} + 2D_{12}) \end{array} \right\}$$

$$(12.13)$$

where the determinant $|A|$ is given by:

$$|A| = 4h(D_{11} + D_{22} + 2D_{12})\,[D_{33}(D_{11} + D_{22} - 2D_{12}) - (D_{13} - D_{23})^2]$$

$$(12.14)$$

Equation (12.13) represents the governing differential system of ordinary differential equations for the strains ε_x, ε_y and ε_{xy}. The system is solved numerically using the IMSL routine DIVPRK. The resulting stress-strain curve for the composite laminate is then plotted using the Mori-Tanaka method. Details about the numerical algorithm and comparisons with the experimental measurements are discussed in Appendix A-4 and in section 12-6.

12.6 Theory vs. Experiment for Uniaxial Tension:

The theoretical model has been implemented numerically, in a stress-controlled algorithm. The flowchart in Figure 12-11 shows the sequence of steps used in the implementation. The load is incrementally increased from zero to the failure load for each type of ply orientation. At each load increment, the systems of differential equations (12.11) and (12.13) are solved for the strains in the laminate for the cases of $(0/90)_s$ and $(\pm 45)_s$ configurations, respectively. In the numerical algorithm, the kinematic hardening parameter b is taken as $b = A\,(B - \sigma/\sigma_f)$ where σ is the overall uniaxial average stress and σ_f is the overall uniaxial average stress at failure. The constants A and B are taken to be $A = 11,900$ ksi and $B = 1.0084034$. The expression used for b has the property that the value of b decreases as the stress increases.

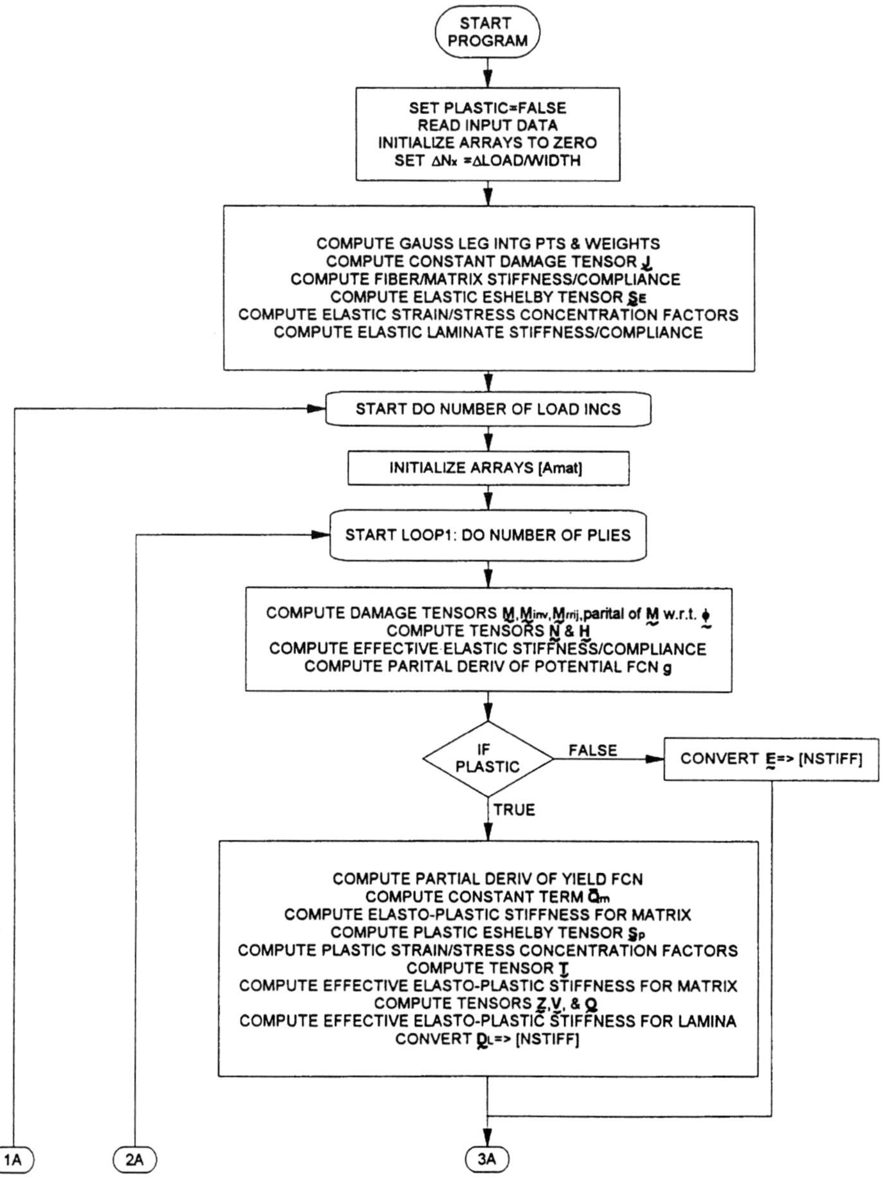

Figure 12.11   Flowchart of the numerical implementation of the theoretical model for the case of uniaxial tension.

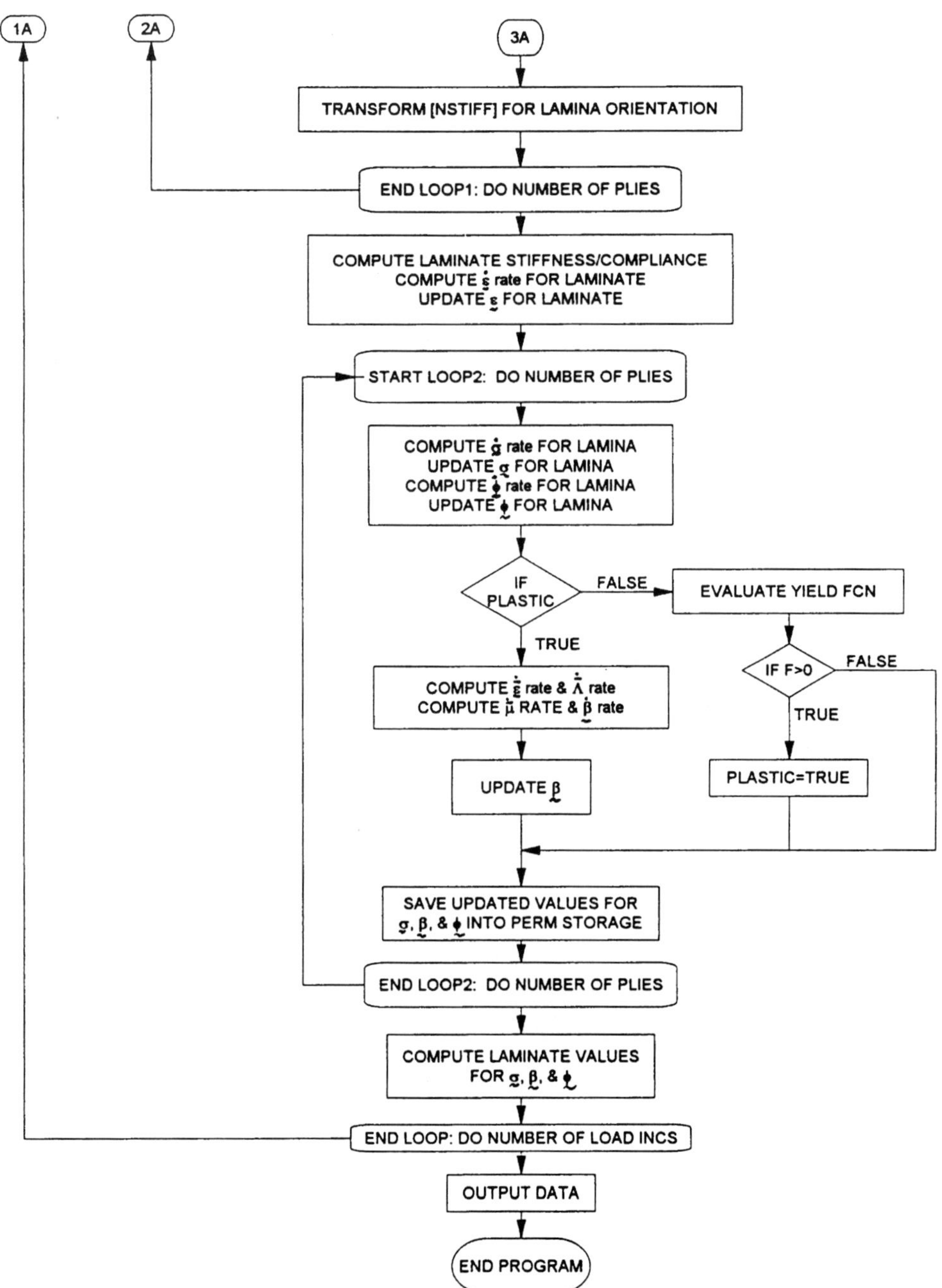

Figure 12.11 continued

The stress-strain curves of the numerical implementation are compared with the experimental results for both types of orientations as shown in Figures 12-12 and 12-13. It is clear from Figure 12-12 that the theoretical predictions closely match the experimental observations for the $(0/90)_s$ orientation. However, the results are not as good for the case of the $(\pm 45)_s$ orientation as shown in Figure 12-13. This figure shows good agreement in the elastic range only. The discrepancies in the outcome for this type of specimen may be attributed to several reasons. First, the elastic strains are not small compared with the plastic strains. Second, the deformation for the $(\pm 45)s$ layup appears to be matrix-dominated. In view of the assumptions of the proposed model concerning small elastic strains and fiber-dominated deformation, it is concluded that the theoretical model cannot be applied successfully to the $(\pm 45)_s$ orientation layup.

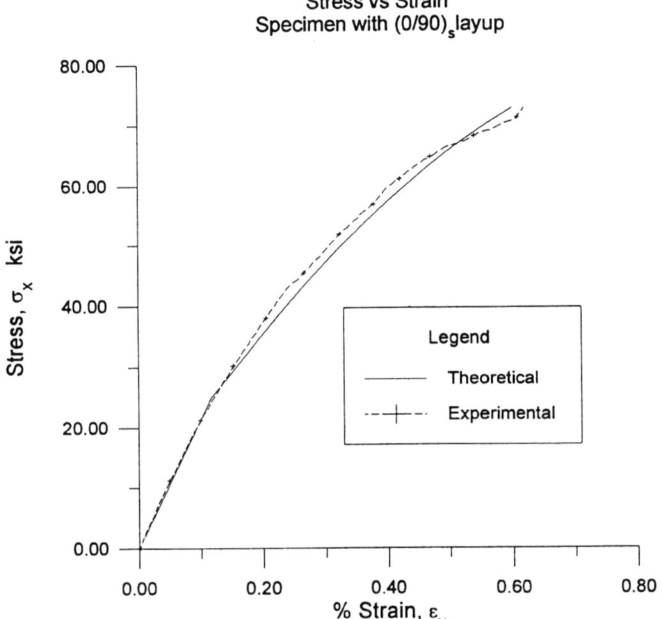

Figure 12.12 Stress-strain curves (σ_{11} vs ϵ_{11}) of the theoretical model and experimental measurements for the (0/90)$_s$ layup.

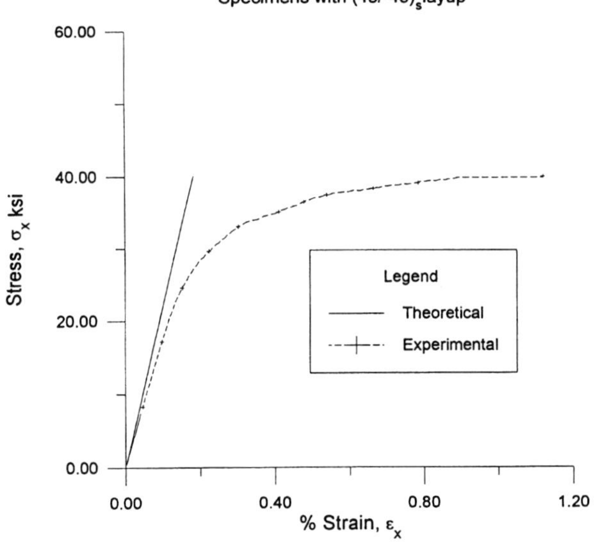

Figure 12.13 Stress-strain curves(σ_{11} vs. ϵ_{11}) of the theoretical model and experimental measurements for the (±45)$_s$ layup.

320

Figure 12-14 shows a comparison of the stress σ_{11} vs. the strain ε_1 for the first and second laminas in the (0/90)$_s$ layup. Also shown in this figure is the average stress in the laminate.

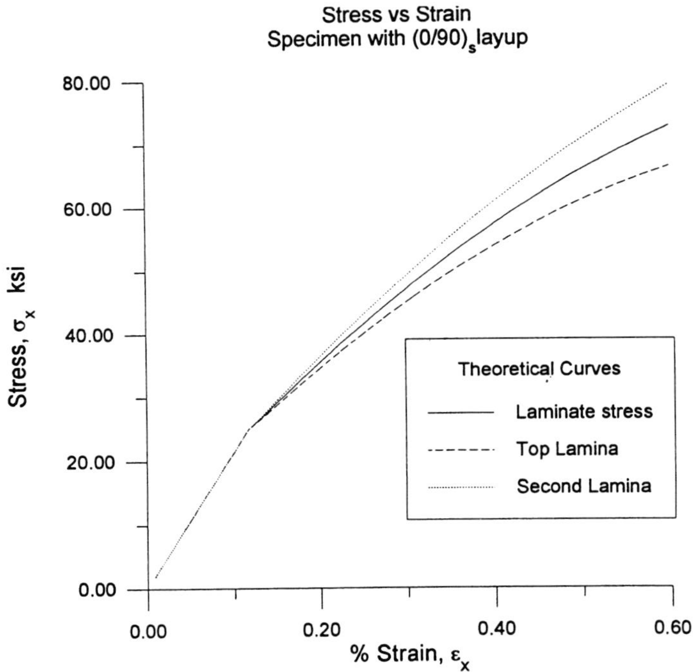

Figure 12.14   Stress-strain curves (σ_{11} vs. ϵ_{11}) of the theoretical model for the first and second laminas, and the laminate for the (0/90)$_s$ layup.

The theoretical and experimental results of the damage variable ϕ_{11} vs ε_{11} are shown in Figure 12-15 for the $(0/90)_s$ layup. In the experimental determination of the damage variable from the measured crack densities, the value $m = 30$ is used. Good agreement is obtained for this type of layup as shown in Figure 12-15. The theoretical predictions of the other components ϕ_{22} and ϕ_{12} of the damage tensor are shown in Figures 12-16 and 12-17 for both types of lamina layups.

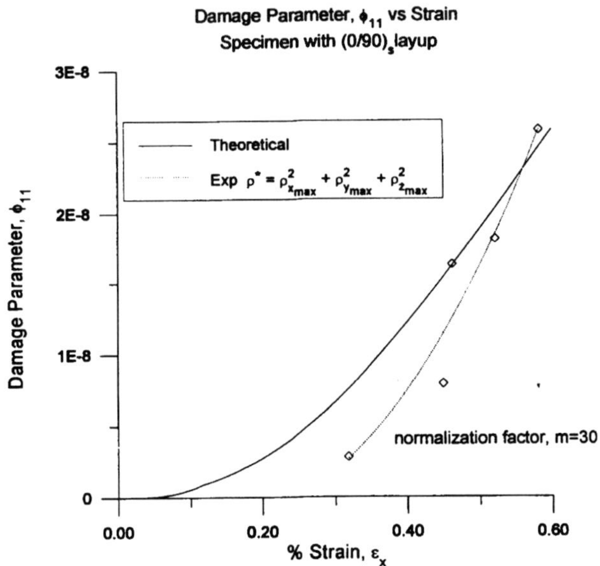

Figure 12.15   Damage variable ϕ_{11} vs. strain ϵ_{11} for the $(0/90)_s$ layup.

322

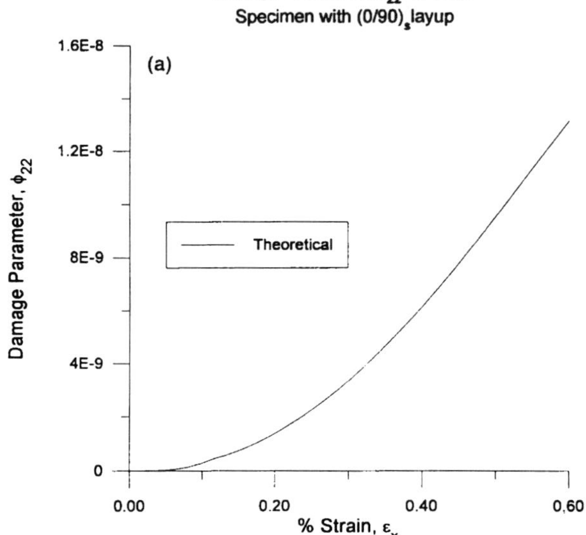

Figure 12.16a   Damage variable ϕ_{22} vs. strain ϵ_{11} for the (0/90)$_s$ layup.

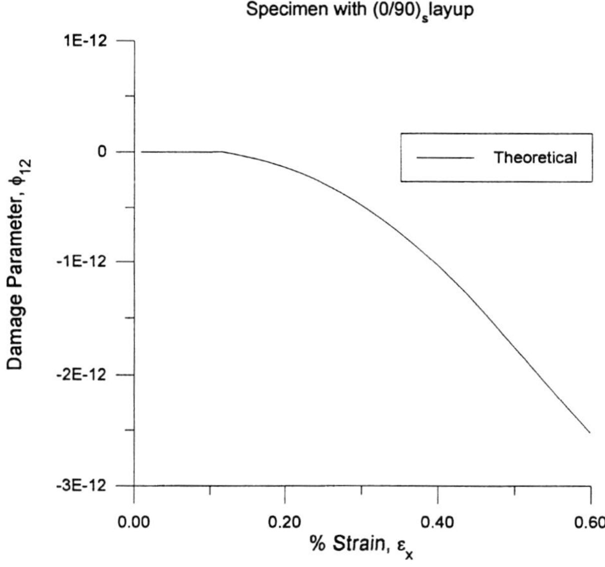

Figure 12.16b   Damage variable ϕ_{12} vs. strain ϵ_{11} for the (0/90)$_s$ layup.

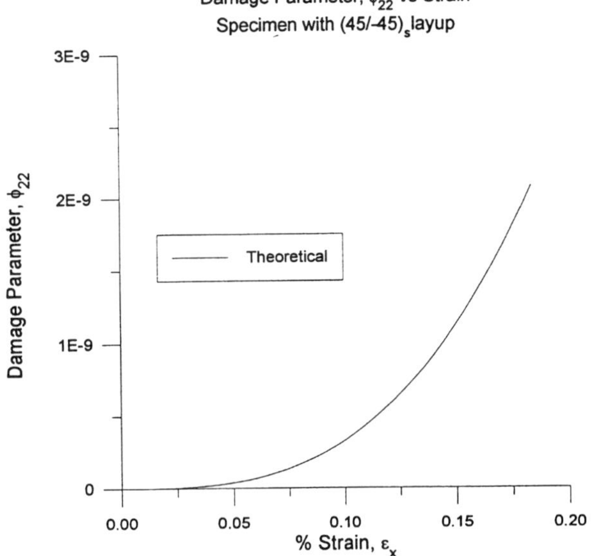

Figure 12.17a   Damage variable ϕ_{22} vs. strain ϵ_{11} for the $(\pm45)_s$ layup.

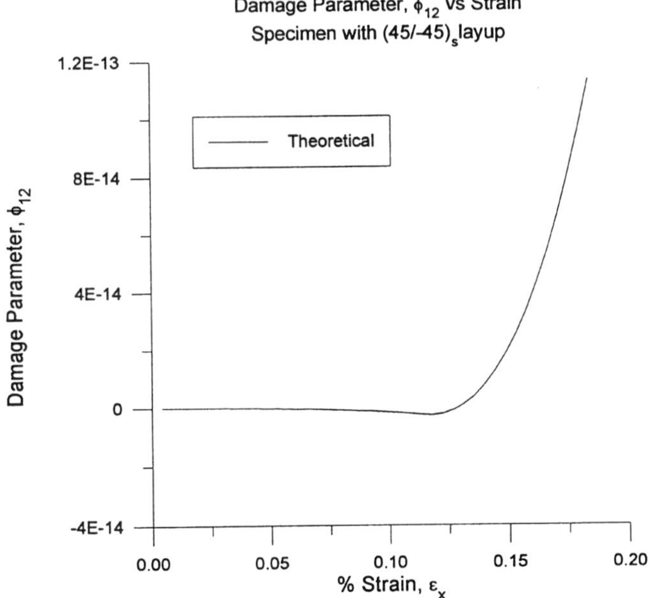

Figure 12.17b   Damage variable ϕ_{12} vs. strain ϵ_{11} for the $(\pm45)_s$ layup.

324

12.7 Evaluation of Damage Parameters:

The damage model developed in Chapter 6 defines a second-rank tensorial damage parameter ϕ whose eigenvalves are ϕ_{ii} (i = 1, 2, 3, no sum over i). Difficulty arises in being able to determine this damage tensor.

$$\overline{E}_{xx}^{t} = E_{xx}^{t} (1 - \phi_{xx})^2 \qquad (12.15)$$

$$\overline{E}_{yy}^{t} = E_{yy}^{t} (1 - \phi_{yy})^2 \qquad (12.16)$$

is used to investigate the dogbone shaped specimens. In the expressions $\overline{E}_{ii}^{t}$ (i = x, y, no sum over i) represents the current effective tangent modulus in the ith direction and E_{ii}^{t} (i = x, y, no sum over i) represents the initial tangent modulus or the elastic modulus in the ith direction. Using the experimentally obtained stress-strain curves, the tangent modulus is obtained by numerical differentiation based on cubic spline interpolation. Tangent moduli curves with $(0/90)_s$ and $(\pm 45)_s$ layups for selected specimens are shown in Figures 12-18 and 12-19, respectively. The damage parameter ϕ_{xx} is evaluated using equation (12.15) from the results of the tangent moduli curves and the results are shown in Figures 12-20 and 12-21. These curves behave as they should, in that the tangent moduli curves are an inverse mirror of the corresponding stress-stain curve, as a result of the inverse relationship between the two. The damage curves also mirror the stress-strain curves. Additionally, comparison of the magnitude of the damage parameter ϕ_{xx} in Figure 12-20 and 12-21 shows that the amount of damage in the $(\pm 45)_s$ specimens is greater than that in the $(0/90)_s$ specimens. This observation is verified both qualitatively and quantitatively from the SEM analysis of representative cross-sections of each of the specimen layups.

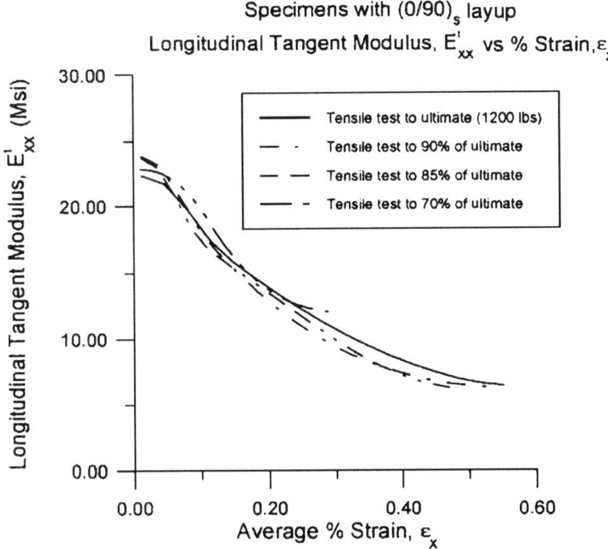

Figure 12.18   Tangent modulus, E^t_{xx} curves for (0/90)$_s$ specimens.

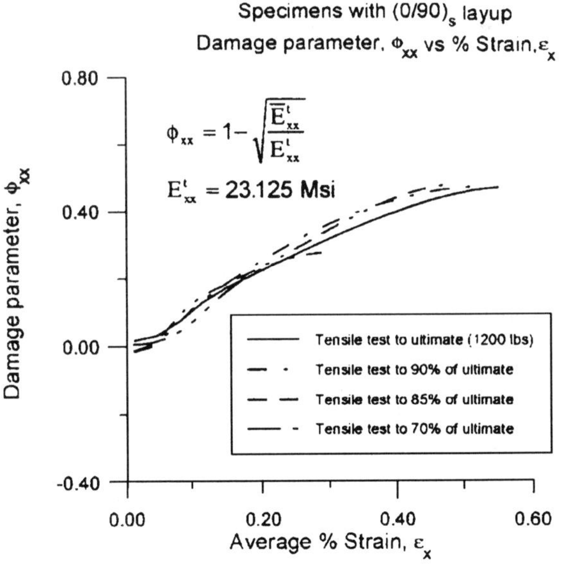

Figure 12.19   Damage parameter, ϕ_{xx} curves for (0/90)$_s$ specimens.

326

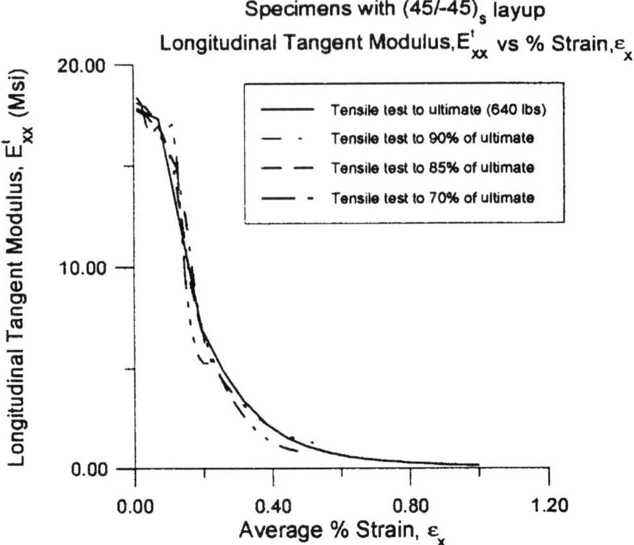

Figure 12.20   Tangent modulus, E^t_{xx} curves for $(\pm 45)_s$ specimens.

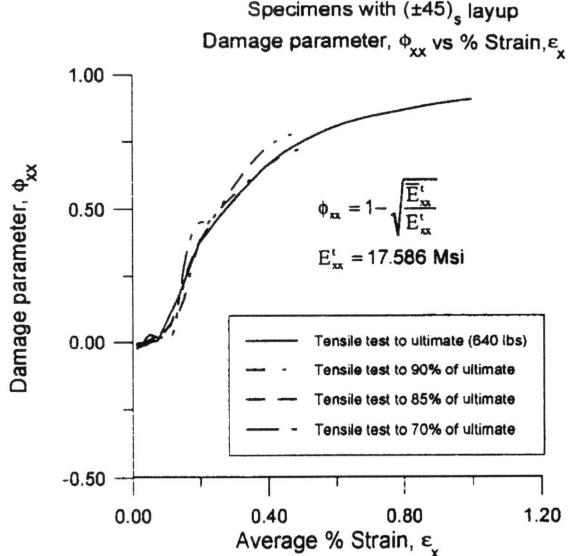

Figure 12.21   Damage parameter, ϕ_{xx} curves for $(\pm 45)_s$ specimens.

Although this approximation to obtain the damage parameter ϕ_{xx} is consistent and has magnitudes within the accepted range $0 \le \phi_{xx} < 1$, further use in the theoretical development is not warranted since this expression is only valid for a pure elastic response. It is proposed to handle an elasto-plastic response [152, 153] by defining the damage tensor ϕ as a function of the crack density as was done in section 12.4 on damage characterization. Damage is experimentally characterized by sectioning the uniaxial specimens perpendicular to the direction of loading and two additional plates mutually perpendicular to the loading direction. From the SEM photos, the crack densities are obtained on an RVE. In the first quantification of damage (overall type), no discrimination is made between the cracks in the matrix and the cracks in the fibers.

In the general case for off-axis laminates, the damage tensor takes the general form:

$$[\phi] = \begin{bmatrix} \overline{\rho}_x^2 & \overline{\rho}_x\overline{\rho}_y & \overline{\rho}_x\overline{\rho}_z \\ \overline{\rho}_y\overline{\rho}_x & \overline{\rho}_y^2 & \overline{\rho}_y\overline{\rho}_z \\ \overline{\rho}_z\overline{\rho}_x & \overline{\rho}_z\overline{\rho}_y & \overline{\rho}_z^2 \end{bmatrix} \tag{12.17}$$

The off-diagonal terms constitute damage introduced by the loads that are not parallel to the fiber direction. This implies that these terms represent damage due to the interaction of cracks on the three mutually perpendicular planes of the RVE. It also implies that shearing stresses impose this interactive damage.

12.7.1 Overall Quantification of Damage:

For the overall quantification of damage, the value $\overline{\rho}_i$ ($i = x, y, z$) represents the total crack density on the representative cross-sectional face of the RVE. In the current investigation, the values of $\overline{\rho}_i$ are evaluated for specimens loaded to the five load levels below the rupture load. Crack densities are not measured for specimens loaded to rupture since this load level produces damage features which are beyond the valid range of damage mechanics. Densities on the z-section are not measured and are assumed to be one half the magnitude of those on the respective y-section. Measured values of the crack densities for each of the layups are shown in Tables 12.6 and 12.7. These values are the crack density values obtained directly from the image analysis process without any normalization.

Table 12.6: Overall Crack Densities for $(0/90)_s$ laminate

% Load	% Strain	ρ_x x 10^{-4} (mm/mm^2)	ρ_y x 10^{-4} (mm/mm^2)
70	0.3182	41.82	3.41
75	0.4487	70.32	36.40
80	0.4611	100.77	--
85	0.5202	106.24	56.43
90	0.5808	126.68	67.72

Table 12.7: Overall Crack Densities for $(\pm 45)_s$ laminate

% Load	% Strain	ρ_x x 10^{-4} (mm/mm^2)	ρ_y x 10^{-4} (mm/mm^2)
70	0.2414	49.23	--
75	0.2779	49.32	42.44
80	0.4324	51.84	101.29
85	0.5268	52.99	117.01
90	0.5729	56.67	146.59

Using the relationship in equation (12.17), damage parameter curves for ϕ_{xx} with the $(0/90)_s$ and $(\pm 45)_s$ layups are developed and shown in Figures 12-22 and 12-23, respectively, and those for ϕ_{yy} are shown in Figures 12-24 and 12-25 for the $(0/90)_s$ and $(\pm 45)_s$ layups, respectively. All the curves shown are second-order polynomial fits of the measured data points with the normalization factor $m = 1$.

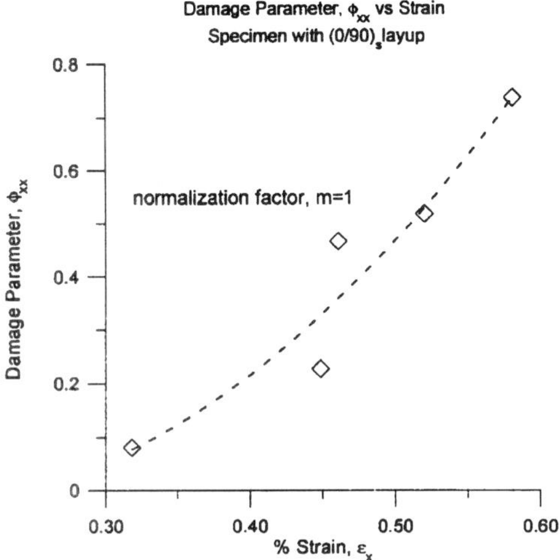

Figure 12.22 Experimentally measured damage parameter, ϕ_{xx} for $(0/90)_s$ laminate.

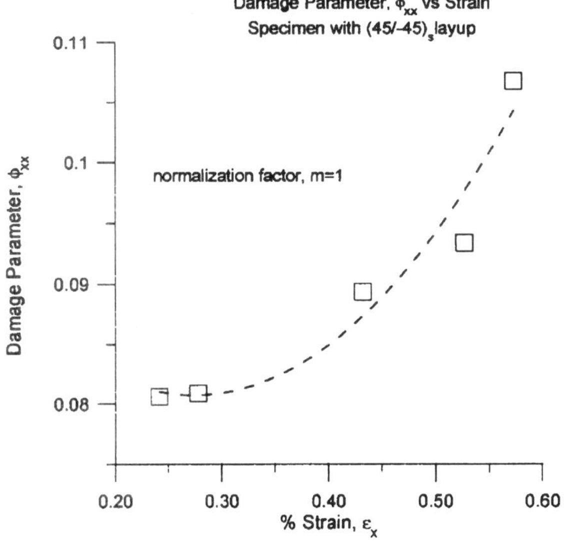

Figure 12.23 Experimentally measured damage parameter, ϕ_{xx} for $(\pm45)_s$ laminate.

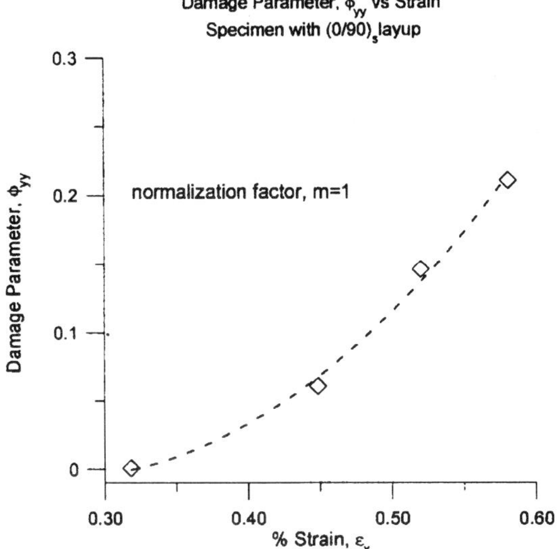

Figure 12.24   Experimentally measured damage parameter, ϕ_{yy} for $(0/90)_s$ laminate.

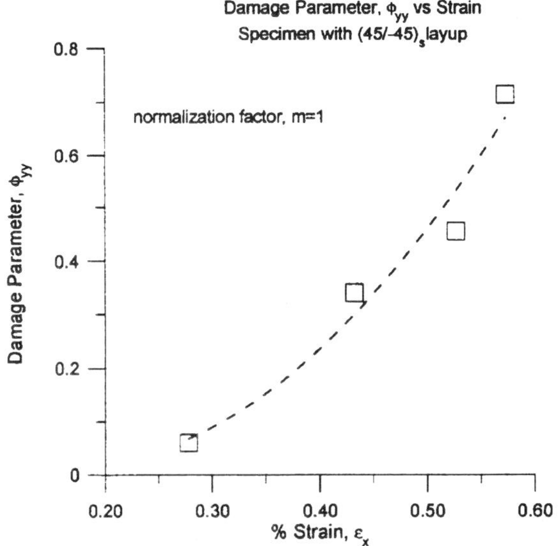

Figure 12.25   Experimentally measured damage parameter, ϕ_{yy} for $(\pm 45)_s$ laminate.

As shown in equation (12.17), the damage tensor ϕ is fully populated for anything other than unidirectional laminates loaded along the fiber direction as is the case for the specimens used during this investigation. Although damage parameter curves are not shown for the off-diagonal damage values, they were computed for each laminate layup at 90% of the rupture load and given as (using the fourth formula for ρ^* in equations (12.4)):

$$[\phi] = \begin{bmatrix} 0.73682 & 0.39388 & 0.19694 \\ & 0.21056 & 0.10528 \\ \text{symmetric} & & 0.052638 \end{bmatrix}^{90\%}_{(0/90)_s} \tag{12.18}$$

$$[\phi] = \begin{bmatrix} 0.10679 & 0.27622 & 0.13811 \\ & 0.71452 & 0.35725 \\ \text{symmetric.} & & 0.17862 \end{bmatrix}^{90\%}_{(\pm 45)_s} \tag{12.19}$$

Careful examination of the curve fits in Figures 12-22 to 12-25 show that the curves are consistent and well formed as compared to the theoretically generated curves of Voyiadjis and Kattan [104] for the case of a uniaxially loaded unidirectional lamina. With five data points the curve fits have an acceptable range of error; however, a much tighter fit could be obtained with more data points. The fit for the $(\pm 45)_s$ layup is considerably better than that of the $(0/90)_s$ layup as a result of the increased amount of damage information that is offered for specimens with this layup. Further examination of these curve fits for a particular ϕ_{ij} shows that they have very similar shapes and magnitudes. This supports the proposition made previously that damage evolution is independent of laminate layup and orientation. Additionally, as indicated by the damage tensors shown in equations (12.18) and (12.19) for loadings not in the direction of the fiber, the off-diagonal damage values are of a magnitude that cannot be neglected. Also, it is found that the selection of the average crack density ρ^* has an effect on the magnitude of the damage values. For example, using $\rho^* = (\sqrt{\rho_{x_{max}}} + \sqrt{\rho_{x_{max}}} + \sqrt{\rho_{z_{max}}})^2$, the damage values shown in equations (12.18) and (12.19) become

$$[\phi] = \begin{bmatrix} 3.9149 & 2.0928 & 1.0464 \\ & 1.1187 & 0.55936 \\ \text{symmetric} & & 0.27968 \end{bmatrix}^{90\%}_{(0/90)_s} \times 10^{-2} \tag{12.20}$$

332

$$[\phi] = \begin{bmatrix} 0.50804 & 1.3141 & 0.65708 \\ & 3.3992 & 1.6997 \\ \text{symmetric} & & 0.84984 \end{bmatrix}^{90\%}_{(\pm45)_s} \times 10^{-2} \qquad (12.21)$$

Comparing equations (12.18) with (12.20), and (12.19) with (12.21), a sizeable difference is noticed in the magnitude of the damage values. However, this difference can be nullified by redefining the normalization factor m to include this effect.

12.7.2 Local Quantification of Damage:

For the local quantification, damage is characterized the same way as for the overall quantification, with the exception that it is separated into matrix damage and fiber damage. Fiber damage is confined to cracks within the fiber. Damage as a result of matrix-fiber debonding is classified as a part of the matrix damage. Matrix damage is divided into three distinct types: radial cracks (cracks emanating from the reaction zone into the matrix), matrix cracks (within the matrix only), and matrix-fiber debonding. The total damage variable for the matrix is therefore defined as:

$$\phi_{ij}^m = w_1\,\phi_{ij}^{rm} + w_2\,\phi_{ij}^{mm} + w_3\,\phi_{ij}^{dm} \quad (i,j = 1, 2, 3) \qquad (12.22)$$

where ϕ_{ij}^{rm} is the damage variable resulting from radial cracks in the matrix, ϕ_{ij}^{mm} is the damage variable resulting from regular cracks in the matrix, and ϕ_{ij}^{dm} is the damage variable due to matrix-fiber debonding. The terms w_i $(i = 1, 2, 3)$ represent corresponding scalar weight functions that are determined through homogenization techniques.

Using the same technique as that used for the overall quantification of damage, crack densities were computed for the composite constituents. These results are shown in Tables 12.8 and 12.9 for the $(0/90)_s$ and $(\pm45)_s$ layups, respectively. As observed in Tables 12-8 and 12-___ the measurable damage in the matrix is practically nonexistent, which shows the dominant behavior of the fibers. Due to the lack of information, further precessing to obtain the local damage values is not performed. However, the information obtained agrees with the stress-strain response. The observed response for the $(0/90)_s$ layup is that of a specimen with high stiffness and low ductility, which implies that almost all of the load interaction with the laminate will be with the stiffer material, whereas the response for the $(\pm45)_s$ is that of a specimen with initial high stiffness and low ductility followed by a transition into a well defined plastic region.

Table 12.8: Local Crack Densities for $(0/90)_s$ Laminate

% Load	% Strain	$\rho_x^m \, x\,10^{-4}$ (mm/mm²)	$\rho_x^f \, x\,10^{-4}$ (mm/mm²)	$\rho_y^m \, x\,10^{-4}$ (mm/mm²)	$\rho_y^f \, x\,10^{-4}$ (mm/mm²)
70	0.3182	0.00	41.82	0.00	3.41
75	0.4487	0.00	70.32	0.00	36.40
80	0.4611	0.00	100.77	-	--
85	0.5202	0.00	106.24	0.00	56.43
90	0.5808	0.00	126.68	0.77	66.94

Table 12.9: Local Crack Densities for $(\pm 45)_s$ Laminate

% Load	% Strain	$\rho_x^m \, x\,10^{-4}$ (mm/mm²)	$\rho_x^f \, x\,10^{-4}$ (mm/mm²)	$\rho_y^m \, x\,10^{-4}$ (mm/mm²)	$\rho_y^f \, x\,10^{-4}$ (mm/mm²)
70	0.2412	0.00	49.23	-	--
75	0.2779	0.00	49.32	0.00	42.44
80	0.4324	0.00	51.84	0.00	101.29
85	0.5268	0.00	52.99	0.00	117.01
90	0.5729	0.00	56.67	48.98	97.61

CHAPTER 13

HIGH CYCLIC FATIGUE DAMAGE FOR UNI-DIRECTIONAL METAL MATRIX COMPOSITES

13.1   Cyclic/Fatigue Damage Models in the Literature

Reviewing the literature on the subject of fatigue in engineering materials reveals that the explanation of fatigue phenomena and the prediction of fatigue life have been the focus of immense research efforts for the last 50 years. The two major analytical approaches used are the phenomenological approach and the crack propagation approach. The former is concerned with lifetime prediction for complex loading histories using existing lifetime test data, mostly S-N data for constant amplitude cyclic loading. The second approach is concerned with predicting the growth of a dominant crack due to cyclic load which is not the case for metal matrix composites.

Almost all of the known fatigue damage models for composite materials are based on the models developed for their isotropic counterparts (Owen and Howe [154], Subramanyan [155], Srivatsavan and Subramanyan [156], Lemaitre and Plumtree [157], Fong [158], Hashin [159], Hwang and Han [160, 161], Whitworth [162]. Lack of theoretical knowledge and sufficient experimental tests on composite materials led to the application of known fatigue damage models to predict fatigue lifetime of such materials, despite the fact that the fatigue behavior of composite materials is quite different from that of isotropic materials, such as metals. With improvement in theoretical knowledge on composite materials and experimental equipment, a lot of studies have been conducted involving fatigue life and residual strength degradation, modulus degradation and residual life theories. However, it was soon recognized from the obtained models that the material structure of such composites has to be included in the development of fatigue damage models in order to arrive at more feasible and reliable models. Up to date there is no universal fatigue damage model based on the micro-structure of the composite materials capable of predicting fatigue life time for general fatigue loading with reasonable reliability.

Arnold and Kruch [163, 164] presented a phenomenological, isothermal transversely-isotropic differential continuum damage mechanics (CDM) model for fatigue of unidirectional composites. This model is based on the CDM fatigue models for isotropic materials developed at ONERA (Chaboche and Lesne [165], Chaboche [35, 36, 166], Lesne and Savalle [167], Lesne and

336

Cailletaud [168]. They considered the metal matrix composite as a pseudo homogeneous material with locally definable characteristics. Such local characteristics have been considered in the form of a directional tensor representing the fiber direction. Furthermore the concept of anisotropic failure surfaces has been introduced into the model based on deformation theories for high temperature metal matrix composites of Robinson et al. [169] and Robinson and Duff [170]. Despite the rigorous development the proposed model has two major drawbacks: (1) the expensive experimental setup and exhaustive experiments needed to obtain the material parameters used in the model equations, and (2) the employed scalar measure for the damage variable. Recently Wilt and Arnold [171] presented a fatigue damage algorithm which employs the fatigue damage model developed by Arnold and Kruch [163, 164]. They implemented their algorithm into the commercial finite element code MARC and used it to analyze a cladded MMC ring. Results were presented on a qualitative basis since no experimental results are available.

Nicholas [172] recently reviewed fatigue life time prediction models for TMC's which use fundamentally different approaches. His investigation showed that various models are based on a single parameter and have limited applicability. Two other models, a dominant damage model (Neu [173]) and a life fraction model, show applicability to various loading ranges, frequencies and temperature profiles. Neu [173] pointed out that despite the fact that there exist several damage mechanisms it is possible to consider the most dominant ones for modeling and include the influence of others in those since their behavior might be similar. His model was able to match experimental data for isothermal and thermo-mechanical fatigue for low-cycle fatigue experiments. The life fraction models, which are based on the fact that fatigue damage accumulates simultaneously due to independent mechanisms, are able to model only specific composite layups for which their parameters have been calibrated. Various other fatigue investigations have been performed but their focus is on specific ply-staking sequence of interest at the time of the investigations. In general it is found that even though micro-mechanical effects or mechanisms are considered and incorporated into the models there does not yet exist a true micro-mechanical fatigue damage model which considers the material behavior and damage evolution in the constituents individually. The following proposed micro-mechanical fatigue damage model is intended to exactly fill in this gap. It is considered as a first step along a consistent route to develop a universal micro-mechanical fatigue damage model capable of modeling various loading conditions including thermo-mechanical effects as well as environmental effects which occur during the service life of dynamically loaded composite structures.

13.2   Damage Mechanics Applied to Composite Materials

Kachanov [1] pioneered the idea of damage in the framework of continuum mechanics. For the case of isotropic damage and using the concept of effective stress, the damage variable is defined as a scalar in the following manner

$$\varphi = \frac{A - \overline{A}}{A} \tag{13.1}$$

where $\overline{A}$ is the effective (net) resisting area corresponding to the damaged area A. Using the hypothesis of elastic energy equivalence (Sidoroff [16]), the effective stress $\overline{\sigma}$ can be obtained from the above equation by equating the force acting on the hypothetical undamaged area with the force acting on the actual damaged area.

In a general state of deformation and damage, the scalar damage variable ϕ is replaced by a fourth-order damage effect tensor $\mathbf{M}$ which depends on a second-order damage tensor ϕ. In general, the effective stress tensor $\overline{\sigma}$ is obtained using the following relation

$$\overline{\sigma} = \mathbf{M} : \sigma \tag{13.2}$$

In general the analysis of composite materials falls into two categories. The first category consists of all approaches that employ the continuum concept (Talreja [174], Cristensen [62]), where the composite system is treated as one continuum and the equations of anisotropic elasticity are used in the analysis. The second category encompasses all approaches that use micro-mechanical models together with averaging procedures and homogenization techniques (Poursatip et al. [77]), Dvorak and Bahei-El-Din [69, 70], Dvorak and Laws [75], Dvorak et al. [74] to describe the material behavior. In these models, the composite is considered to be composed of a number of individual phases for which local equations are formulated. Employing a suitable homogenization procedure then allows one to analyze the material behavior of the entire composite system based on the local analysis.

Dvorak and Bahei-El-Din [69, 70] employed an averaging technique to analyze the elasto-plastic behavior of fiber-reinforced composites. They considered elastic fibers with an elasto-plastic matrix. However, no attempt was made to introduce damage in the constitutive equations. Voyiadjis and Kattan [104], Voyiadjis et al. [152], and Voyiadjis and Kattan [175] introduced a consistent and systematic damage theory for metal matrix composites utilizing the micro-mechanical composite model of Dvorak and Bahei-El-Din [70]. They introduced two approaches, referred to in the literature as the overall and the local approach, which allow for a consistent incorporation of the damage phenomenon in a composite material system.

The overall approach (Kattan and Voyiadjis [101]) to damage in composite materials employs one single damage tensor to reflect all types of damage mechanisms that the composite undergoes like initiation, growth and coalescence of micro-voids and micro-cracks. Voyiadjis and Park [176] improved the overall approach by including and adopting a general damage criterion for orthotropic materials by extending the formulation of Stumvoll and Swoboda [140] to MMC's. In this improved model all damage types are considered but the model lacks the consideration of local (constituent) as well as interfacial damage effects. In contrary to the overall approach the local approach (Voyiadjis and Kattan [138]) introduces two independent damage tensors, ϕ^M and ϕ^F, and hence two independent damage effect tensors, $\mathbf{M}^M$ and $\mathbf{M}^F$, to reflect appropriate damage mechanisms in the matrix and fibers, respectively. It is this latter approach which is employed in the proposed micro-mechanical fatigue damage model.

13.3 Stress and Strain Concentration Tensors

In the derivation of the model, the concept of effective stress (Rabotnov [177], Sidoroff [176]) is used. The effective stress is defined as the stress in a hypothetical state of deformation that is free of damage and is mechanically equivalent to the current state of deformation and damage. In a general state of deformation and damage, the effective Cauchy stress tensor $\bar{\sigma}$ is related to the current Cauchy stress tensor by the linear relation given in equation (13.2). In the case of composite materials, similar constituent (local) stress relations hold for the matrix and fiber stress tensors σ^M and σ^F, respectively.

$$\overline{\sigma^M} = M^M : \sigma^M \tag{13.3a}$$

$$\bar{\sigma}^F = M^F : \sigma^F \tag{13.3b}$$

where M^M and M^F are fourth-order local damage effect tensors for the matrix and fiber materials, respectively. The damage effect tensors $\mathbf{M}^M$ and $\mathbf{M}^F$ are dependent on second order damage variables ϕ^M and ϕ^F, respectively. These latter second order tensors quantify the crack density in the matrix and fibers, respectively (Voyiadjis and Venson [143]). The crack density tensors incorporate both, cracks in the fiber, matrix, as well as those due to fiber deponding. A complete discussion of these tensors is given in the work of Voyiadjis and Venson [143] and in Chapter 12.

In the proposed model the matrix is assumed to be elasto-plastic and the fibers are assumed to be elastic, continuous and aligned. Consequently, the undamaged (effective) incremental local (constituent) constitutive relations are given by:

$$d\,\overline{\sigma}^M = \overline{D}^M : d\overline{\varepsilon}^M \tag{13.4a}$$

$$d\overline{\sigma}^F = \overline{E}^F : d\overline{\varepsilon}^F \tag{13.4b}$$

The fourth-rank tensor $\overline{D}^M$ and $\overline{E}^F$ are the undamaged (effective) matrix elasto-plastic stiffness tensor and fiber elastic stiffness tensor, respectively. The incremental composite constitutive relation in the damaged state is expressed as follows

$$d\,\sigma = D : d\varepsilon \tag{13.5}$$

where $d\varepsilon$ is the incremental composite strain tensor.

In order to arrive at the local (constituent) relations, given by equations (13.4), a homogenization technique in the form of the Mori-Tanaka averaging scheme (Chen et al. [178]) is employed. Through the use of the so-called stress and strain concentration tensors, a relationship between the global applied effective composite stress, $\overline{\sigma}$, and the local effective stress in the constituents, $\overline{\sigma}^{(M,F)}$, is obtained as follows

$$\overline{\sigma}^M = \overline{B}^M : \overline{\sigma} \tag{13.6a}$$

$$\overline{\sigma}^F = \overline{B}^F : \overline{\sigma} \tag{13.6b}$$

where $\overline{B}^F$ and $\overline{B}^M$ represent the effective stress concentration tensors connecting the local effective stresses with the global effective stresses. In the damaged configuration the following relations are obtained:

$$\sigma^M = B^M : \sigma \tag{13.7a}$$

$$\sigma^F = B^F : \sigma \tag{13.7b}$$

Combining equations (13.2), (13.3), (13.6) and (13.7) one obtains the relation between the local stress concentration tensor and the local effective stress concentration tensor as follows

$$B^F = M^{-F} : \overline{B}^F : M \tag{13.8a}$$

$$B^M = M^{-M} : \overline{B}^M : M \tag{13.8b}$$

Similar relations may be obtained for the deformations in the effective (undamaged) configuration as follows

$$\overline{\varepsilon}^M = \overline{A}^M : \overline{\varepsilon} \tag{13.9a}$$

$$\overline{\varepsilon}^F = \overline{A}^F : \overline{\varepsilon} \tag{13.9b}$$

where $\overline{A}^F$ and $\overline{A}^M$ represent the effective strain concentration tensors connecting the local effective strains with the global effective strains. In the damaged configuration the relations are given by

$$\varepsilon^M = A^M : \varepsilon \tag{13.10a}$$

$$\varepsilon^F = A^F : \varepsilon \tag{13.10b}$$

and furthermore (Refer to Appendix A-4 for details.)

$$A^M = M^M : \overline{A}^M : M^{-1} \tag{13.11a}$$

$$A^F = M^F : \overline{A}^F : M^{-1} \tag{13.11b}$$

13.4 Effective Volume Fractions

During the process of damage evolution in the material another phenomenon has to be considered. As damage progresses within each constituent the effective load resisting area/volume changes while the gross area/volume remains the same. Since the distribution of forces/stresses to the constituents depends directly on the area/volume intact to resist an applied force/stress there is a change in the allocation of the external applied force/stress to the constituents. This redistribution of force/stress due to progressing damage can be accounted for by defining the so-called effective volume fractions which are based on the updated damage variable during each load/stress increment. Expressions for the effective volume fractions are given as

$$\bar{c}^{F} = \frac{1 - \phi_{eq}^{F}}{(1 - \phi_{eq}^{F}) + (1 - \phi_{eq}^{M}) \dfrac{c_{o}^{M}}{c_{o}^{F}}} \tag{13.13a}$$

and

$$\bar{c}^{M} = \frac{1 - \phi_{eq}^{M}}{(1 - \phi_{eq}^{M}) + (1 - \phi_{eq}^{F}) \dfrac{c_{o}^{F}}{c_{o}^{M}}} \tag{13.13b}$$

where c_0^{F} and c_0^{M} are defined as the volume fractions for the fiber and matrix and in the virgin material, respectively. The expressions for ϕ_{eq}^{M} and ϕ_{eq}^{M} are given us

$$\phi_{eq}^{F} = \frac{||\,\phi^{F}\,||_{2}}{||\,\phi_{crit}^{F}\,||_{2}} \tag{13.14a}$$

$$\phi_{eq}^{M} = \frac{||\,\phi^{M}\,||_{2}}{||\,\phi_{crit}^{M}\,||_{2}} \tag{13.14b}$$

with ϕ_{crit}^{F} and ϕ_{crit}^{M} defined as the critical damage tensors for the fibers and the matrix, respectively, and $||\cdot||_{2}$ defined as the L_{2} - *norm* of the quantity enclosed in the vertical bars.

13.5 Proposed Micro-Mechanical Fatigue Damage Model

The proposed fatigue damage criterion g is considered as a function of the applied stress σ, the damage parameter ϕ, the damage hardening parameter κ and a tensor quantity γ, which is explained below. The equation for g is defined by:

$$g = \mathfrak{S}^{n} - 1 \tag{13.15}$$

where $\mathfrak{S}$ is defined as

$$\mathfrak{S} = w_{ij}^{-1}\, w_{jk}^{-1}\, (Y_{kl} - \gamma_{kl})\, (Y_{li} - \gamma_{li}) \tag{13.16}$$

The term $(Y_{kl} - \gamma_{kl})$ represents the translation of the damage surface and therefore accounts for damage evolution during cyclic loading. The tensor $\mathbf{Y}$ represents the thermo-dynamical force conjugate to the damage variable ϕ and is defined as

$$Y_{ij} = \frac{1}{2} \left(\sigma_{cd} \, C_{abpq} \, M_{pqkl} \, \sigma_{kl} + \sigma_{pq} \, M_{uvpq} \, C_{uvab} \, \sigma_{cd} \right) \frac{\partial M_{abcd}}{\partial \phi_{ij}} \qquad (13.17)$$

with $C_{ijkl} = E_{ijkl}^{-1}$ while the quantity γ can be in principle be compared to the backstress in plasticity theory hence representing in this case the center of the damage surface in the thermo-dynamical conjugate force space $\mathbf{Y}$. Its evolution equation is given as follows

$$d\gamma_{ij} = c \, d\phi_{ij} \qquad (13.18)$$

similarly to the evolution equation for the backstress in plasticity. The tensor quantity w_{ij} accounts for the anisotropic expansion of the damage surface and is given as follows

$$w_{ij} = u_{ij} + V_{ij} \qquad (13.19)$$

where the tensor $\mathbf{u}$ is defined as

$$u_{ij} = \lambda_{(i)} \, \eta_{(i)} \left(\frac{\kappa}{\lambda_{(i)}} \right)^{\xi_{(i)}} \delta_{ij} \qquad \text{(no sum on i)} \qquad (13.20)$$

The tensor V_{ij} can be interpreted physically as the damage threshold tensor for the constituent material considered, while κ represents the effect of damage hardening and is defined as follows

$$\kappa = \int_{\phi 1}^{\phi 2} Y : d\phi = \int_{0}^{t} Y : \overset{\circ}{\phi} \; dt \qquad (13.21)$$

Damage hardening is based on the increase in the initial damage threshold due to micro-hardening occurring at a very local material level (Chow and Lu [179]). The parameter γ_{ij} in equation (13.16) adds to this hardening behavior due to the movement of the damage surface in the direction of the evolution of damage. The remaining variables n, λ_i, n_i, ξ_i and c are material parameters to be determined for each individual constituent. Especially the form of the variable ξ_i will be discussed below in the numerical implementation.

Based on the thermo-dynamical principles a potential function for each constituent is defined as

$$\Omega = \Pi^p + \Pi^d - d\Lambda_1 f - d\Lambda_2 g \tag{13.22}$$

where Π^p, Π^d, f and g represent the dissipation energy due to plasticity, the dissipation energy due to damage, the plasticity yield surface for the constituent material considered, and the damage surface, respectively. For loading in the elastic regime (high cycle fatigue) the terms involving plastic dissipation energy are neglected. The term Π^d representing the dissipation energy due to damage is given as

$$\Pi^d = Y_{ij}\, d\phi_{ij} + \kappa\, d\kappa \tag{13.23}$$

Applying the theory of calculus of several variables to solve for the coefficients $d\Lambda_1$ and $d\Lambda_2$ yields

$$\frac{\partial \Omega}{\partial Y_{ij}} = 0 \tag{13.24}$$

from which an expression for the damage increment is obtained as follows

$$d\phi_{ij} = d\Lambda_2 \frac{\partial g}{\partial Y_{ij}} \tag{13.25}$$

Hence $d\Lambda_2$ may be determined using the consistency condition

$$dg = \frac{\partial g}{\partial \sigma} : d\sigma + \frac{\partial g}{\partial \phi} : d\phi + \frac{\partial g}{\partial \kappa} d\kappa + \frac{\partial g}{\partial \gamma} : d\gamma = 0 \tag{13.26}$$

Substitution for the appropriate terms (equations (13.18) and (13.21) into equation (13.26)) yields

$$dg = \frac{\partial g}{\partial \sigma} : d\sigma + \frac{\partial g}{\partial \phi} : d\phi + \frac{\partial g}{\partial \kappa} Y : d\phi - c \frac{\partial g}{\partial Y} : d\phi = 0 \tag{13.27}$$

Replacing $d\phi$ with equation (13.25) an expression for $d\Lambda_2$ is obtained as follows

$$d\Lambda_2 = -\frac{\dfrac{\partial g}{\partial \sigma_{kl}} d\sigma_{kl}}{\left(\dfrac{\partial g}{\partial \phi_{ij}} + Y_{ij}\dfrac{\partial g}{\partial \kappa} - c\dfrac{\partial g}{\partial Y_{ij}}\right)\dfrac{\partial g}{\partial Y_{ij}}} \tag{13.28}$$

344

Backsubstitution of equation (13.28) into equation (13.25) yields an expression for the damage increment for the appropriate constituent in terms of a given stress increment as

$$d\phi_{mn} = - \frac{\dfrac{\partial g}{\partial \sigma_{kl}} \dfrac{\partial g}{\partial Y_{mn}} d\sigma_{kl}}{\left(\dfrac{\partial g}{\partial \phi_{ij}} + Y_{ij} \dfrac{\partial g}{\partial \kappa} - c \dfrac{\partial g}{\partial Y_{ij}} \right) \dfrac{\partial g}{\partial Y_{ij}}} \tag{13.29}$$

or

$$d\phi_{ij} = \Psi_{ijkl} \, d\sigma_{kl} \tag{13.30}$$

where

$$\Psi_{ijkl} = - \frac{\dfrac{\partial g}{\partial Y_{ij}} \dfrac{\partial g}{\partial \sigma_{kl}}}{\left(\dfrac{\partial g}{\partial \phi_{rs}} + Y_{rs} \dfrac{\partial g}{\partial \kappa} - c \dfrac{\partial g}{\partial Y_{rs}} \right) \dfrac{\partial g}{\partial Y_{rs}}} \tag{13.31}$$

and

$$Y_{rs} = \frac{1}{2} \left[\sigma_{cd} \overline{E}^{-1}_{abpq} M_{pqkl} \sigma_{kl} + \sigma_{pq} M_{uvpq} \overline{E}^{-1}_{uvab} \sigma_{cd} \right] \frac{\partial M_{abcd}}{\partial \phi_{rs}} \tag{13.32}$$

As stated elsewhere (Stumvoll and Swoboda [140]) a damaging state in a constituent is given if for any state the damage criterion is satisfied

$$g = 0 \tag{13.33}$$

$$[\Im = w_{ij}^{-1} w_{jk}^{-1} (Y_{kl} - \gamma_{kl}) (Y_{li} - \gamma_{li})]^n - 1 = 0 \tag{13.34}$$

for that specific constituent. In general four different loading states are possible

$$g < 0 \qquad\qquad\qquad\qquad\qquad \text{(non-damaging loading)} \tag{13.35}$$
$$g = 0 \qquad \frac{\partial g}{\partial Y_{ij}} dY_{ij} < 0 \qquad \text{(elastic unloading)} \tag{13.36}$$
$$g = 0 \qquad \frac{\partial g}{\partial Y_{ij}} dY_{ij} = 0 \qquad \text{(neutral unloading)} \tag{13.37}$$
$$g = 0 \qquad \frac{\partial g}{\partial Y_{ij}} dY_{ij} > 0 \qquad \text{(loading from damaging state)} \tag{13.38}$$

Using equation (13.29) the damage increment per fatigue cycle may be obtained by integration over one stress cycle as

$$\frac{d\phi_{ij}}{dN} = \int_{\sigma_{min}}^{\sigma_{max}} \Psi_{ijkl}\, d\sigma_{kl} + \int_{\sigma_{max}}^{\sigma_{min}} \Psi_{ijkl}\, d\sigma_{kl} \tag{13.39}$$

where Ψ_{ijkl} is given according to equation (13.31). The dependence of damage on the mean stress and the amplitude of the stress cycle is implicitly included through the integration of equation (13.39).

13.6   Return to the Damage Surface

In the numerical implementation of the model it appears that after calculating the damage increment $d\phi$ for the current stress increment $d\sigma$ and updating all the appropriate parameters depending on the damage variable ϕ, the damage surface is in general not satisfied. Therefore it is necessary to return the new image point to the damage surface by employing an appropriate return criterion.

At the beginning of the $(n + 1)^{st}$ increment we assume that the damage surface g is satisfied

$$g^{(n)}\left(\sigma^{(n)},\, \phi^{(n)},\, \kappa^{(n)}\, \gamma^{(n)}\right) = 0 \tag{13.40}$$

Applying the stress increment $d\sigma$ (assuming a damage loading) will result in a damage increment $d\phi$ which will be used to update the values for κ and γ. Checking the damage surface (equation 13.15) with the updated values for σ, ϕ, κ and γ will in general yield

$$g^{(n+1)}\left(\sigma^{(n+1)},\, \phi^{(n+1)},\, \kappa^{(n+1)}\, \gamma^{(n+1)}\right) > 0 \tag{13.41}$$

where

$$\sigma^{(n+1)} = \sigma^{(n)} + d\sigma^{(n+1)} \tag{13.42}$$

$$\phi^{(n+1)} = \phi^{(n)} + d\phi^{(n+1)} \tag{13.43}$$

$$\kappa^{(n+1)} = \kappa^{(n)} + d\kappa^{(n+1)} \tag{13.44}$$

$$\gamma^{(n+1)} = \gamma^{(n)} + d\gamma^{(n+1)} \tag{13.45}$$

Using a *Taylor series* expansion of order one expands the left hand side of equation (13.41) to yield

$$g^{(n+1)}\left(\sigma^{(n)} + d\sigma^{(n+1)},\ \phi^{(n)} + d\phi^{(n+1)},\ \kappa^{(n)} + d\kappa^{(n+1)},\ \gamma^{(n)} + d\gamma^{(n+1)}\right)$$

$$= g^{(n)}\left(\sigma^{(n)},\ \phi^{(n)},\ \kappa^{(n)},\ \gamma^{(n)}\right) + \frac{\partial g}{\partial \sigma}\Big|^{(n)} d\sigma^{(n+1)} + \frac{\partial g}{\partial \phi}\Big|^{(n)} d\phi^{(n+1)}$$

$$+ \frac{\partial g}{\partial \kappa}\Big|^{(n)} d\kappa^{(n)} + \frac{\partial g}{\partial \gamma}\Big|^{(n)} d\gamma^{(n+1)} > 0$$

$$(13.46)$$

Recalling the relationships in equation (13.18) and (13.21), relation (13.46) is given by

$$g^{(n+1)}\left(\sigma^{(n)} + d\sigma^{(n+1)},\ \phi^{(n)} + d\phi^{(n+1)}, \kappa^{(n)} + Y^{(n)} : d\phi^{(n+1)},\ \gamma^{(n)} + c\,d\phi^{(n+1)}\right) > 0$$

$$(13.47)$$

The return to the damage surface, hence $g^{(n+1)} = 0$, is now achieved by adjusting the damage increment $\mathbf{d\phi}$ using a linear coefficient α such that

$$g^{(n+1)}\left(\sigma^{(n)} + d\sigma^{(n+1)},\ \phi^{(n)} + \alpha d\phi^{(n+1)}, \kappa^{(n)} + \alpha Y^{(n)} : d\phi^{(n+1)},\ \gamma^{(n)} + \alpha c\,d\phi^{(n+1)}\right) = 0$$

$$(13.48)$$

Substitution of the appropriate expressions for the derivatives in equation (13.48) as well as equation (13.21) and (13.18) and setting the left hand side equal to zero allows one to solve for the unknown coefficient α such that

$$\alpha = - \frac{\left(g^{(n)} + \dfrac{\partial g}{\partial \sigma}\Big|^{(n)} d\sigma^{(n+1)}\right)}{\left(\dfrac{\partial g}{\partial \phi}\Big|^{(n)} + \dfrac{\partial g}{\partial \kappa}\Big|^{(n)} Y^{(n)} + c\,\dfrac{\partial g}{\partial \gamma}\Big|^{(n)}\right) d\phi^{(n+1)}}$$

$$(13.49)$$

13.7 Numerical Analysis - Application

The above model is implemented into a numerical algorithm and used to investigate the fatigue damage evolution in the individual constituents of a uni-directionally fiber reinforced metal matrix composite. No assumptions, except those implicitly included in the stress and strain concentration tensors based on the Mori-Tanaka averaging scheme (Chen et al. [178]) are made. The implementation is performed using a full 3-D modeling hence avoiding any assumptions to be made upon simplification of fourth order tensors to two-dimensional matrix representation. The Mori-Tanaka averaging scheme is implemented using the numerical algorithm according to Lagoudas et al [108]. Only an elastic analysis is performed at this time. Since no experimental data is yet available a parametric study is conducted in order to demonstrate the influence of various parameters on the damage evolution in the constituents. The constituents are assumed to consist of an isotropic material. The materials used in the analysis are given in Johnson et al. [180] and are shown in Table 13.1. The fatigue loading is applied in the form of a sinusoidal uni-axial loading given as

$$\sigma_{ij} = \sigma_{ij, mean} + \sigma_{ij, A} \sin\left(\frac{\theta}{2\pi}\right) \qquad (13.50)$$

where

$$\sigma_{11, mean} = 550\,MPa \quad and \quad \sigma_{ij, mean} = 0 \quad (for\ i, j \neq 1)$$

$$\sigma_{11, A} = 450\,MPa \quad and \quad \sigma_{ij, A} = 0 \quad (for\ i, j \neq 1)$$

Table 13.1   Material properties used in the analysis

	E (GPa)	ν	σ_u (MPa)	σ_y (MPa)	c (in %)
Matrix (Ti - 15 - 3)	92.4	0.35	933.6	689.5	67.5
Fiber (SCS - 6)	400.0	0.25	N/A	N/A	32.5

For the numerical integration scheme an adaptive alogrithm was implemented such that the stress increments where taken as

$$\Delta \sigma_{ij} = \frac{\sigma_{ij, mean}}{25} \quad if\ \sigma_{ij} < \sigma_{mean, ij} \qquad \text{(non-damage state)}$$

$$\Delta \sigma_{ij} = 1\,MPa \quad if\ \sigma_{ij} < \sigma_{mean, ij} \qquad \text{(damage state)}$$

348

during the loading phase to the mean stress and

$$\Delta \sigma_{ij} = \left[\sin \left(\frac{\theta + \Delta \theta}{2\pi} \right) - \sin \left(\frac{\theta}{2\pi} \right) \right] * \sigma_{ij, A} \qquad \text{(during cyclic loading)}$$

with

$$\Delta \theta = \frac{\pi}{50} \qquad \text{(during a non-damaging state)}$$

$$\Delta \theta = \frac{\pi}{900} \qquad \text{(during a damaging state)}$$

for the cyclic loading phase. Here θ represents simply the phase angle during the cyclic loading. The above limit values were adopted based on a numerical investigation which yielded satisfacotry behavior of the model using the above values.

The damage criterion is evaluated within each increment and a return criterion as described in equations (13.48) and (13.49) is applied if $|g^{(n+1)}| \geq 10^{-3}$. Except at the very first incident of damage this criterion shows satisfactory performance during the application of the return criterion (equations (13.41) - (13.49)). The numerical noise at the initiation of damage has been investigated and it is found that a reduction in the step size for the stress increment reduces the numerical error appropriately to fall within the specified bounds. This phenomenon is not observed at any other time during the analysis (Figures 13.1 - 13.3). It is attributed to the point of discontinuity in the damage criterion at the wake of damage. The flexibility of the model is demonstrated through a parametric study based on variations in the parameters λ and ξ. For the parametric study the values of all the parameters except for one are kept constant in order to study the effect of a single parameter on the model as shown in Table 13.2. The parameter ξ^F and ξ^M account for the variation in the damage evolution with respect to the number of cycles, especially the increase in the damage rate during the fatigue life of a material. The specific form of the parameters ξ^F and ξ^M is obtained from experimental curves, such as those shown in Figures 13.4 and 13.5, where the fatigue damage in the material is plotted versus the number of applied cycles. Since fatigue damage evolution for a specific stress ratio R is dependent on the applied mean stress as well as the stress amplitude, such experimental curves have to be obtained for different applied mean stresses and stress amplitudes.

The damage ϕ in the material during the fatigue life may be obtained by using the stiffness degradation or an equivalent method, such as sectioning and subsequent SEM evaluation of the specimens for damage quantification. Upon inspection of the obtained experimental curves it is observed that basically three different regions can be distinguished during the fatigue life of the material (Figures 13.4 and 13.5). These different regions pertain to the damage initiation phase (Phase I), the damage propagation phase (Phase II), and the failure phase (Phase III). A distinction for these regions may be made by specifying bounds in the form of the number of cycles such as N_1 and N_2, as indicated in Figure 13.4. This is done in general by visual inspection using engineering judgment and physical intuition. Using these curves an evolution equation for ξ with respect to the number of cycles N, the applied mean stress σ_{mean} and the stress ratio R maybe established. For the current analysis, since no such experimental data are available, the following forms for the parameters ξ^F and ξ^M in terms of N_1 and N_2 have been used and are given as

$$\xi_N^M = \frac{N_1^M - N}{N_1^M - 1} a^M + \left(1 - \frac{N_1^M - N}{N_1^M - 1} \right) b^M \qquad (1 \le N \le N_1^M) \tag{13.51}$$

$$\xi_N^M = \xi_0^M + \Delta\xi_1^M + \left(\frac{N - N_1^M}{N_2^M - N_1^M} \right) \Delta\xi_2^M \qquad (N_1^M < N \le N_2^M) \tag{13.52}$$

$$\xi_N^M = \xi_0^M + \Delta\xi_1^M + \left(\frac{N - N_1^M}{N_2^M - N_1^M} \right)^2 \Delta\xi_2^M \qquad (N > N_2^M) \tag{13.53}$$

$$\xi_N^F = \xi_0^F + \left(\frac{N - 1}{N_2^F - 1} \right) \Delta\xi_2^F \qquad (1 < N \le N_2^F) \tag{13.54}$$

$$\xi_N^F = \xi_0^F + \left(\frac{N - 1}{N_2^F - 1} \right)^2 \Delta\xi_2^F \qquad (N > N_2^F) \tag{13.55}$$

where

$$a^M = \xi_0^M + \sqrt[4]{\frac{N_1^M}{N}} \log\left(\frac{N}{N_1^M} \right) \Delta\xi_1^M \tag{13.56}$$

$$b^M = \xi_0^M + \Delta\xi_1^M + \left(\frac{N - N_1^M}{N_2^M - N_1^M} \right) \Delta\xi_2^M \tag{13.57}$$

Table 13.2 Model parameters used in the analysis

	V (Mpa)	λ (Mpa)	η	ξ	τ (Mpa)	n	Figure
Matrix (Ti - 15 - 3)	0.1	80000	1.0	refer to Eqs. (13.51) - (13.57)	1.0	1.0	13.6
Fiber (*SCS* - 6)	3	160000	1.0	refer to Eqs. (13.51) - (13.57)	1.0	1.0	13.6

	N_1	N_2	ξ_0	$\xi_{\cdot 1}$	ξ_2	Figure	
Matrix (Ti - 15 - 3)	10	110000	0.55	0.02	0.03	13.6	
Fiber (*SCS* - 6)	N/A	110000	0.56	N/A	0.03	13.6	

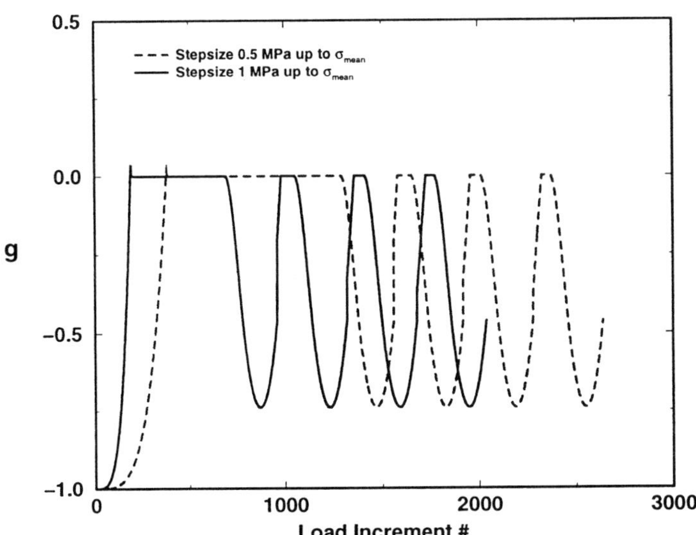

Figure 13.1 Validation of the employed return criterion.

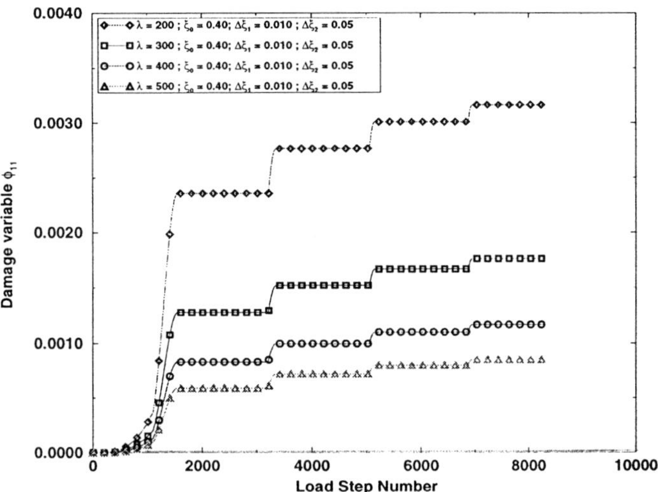

Figure 13.2   Variation in damage evolution for various values of λ.

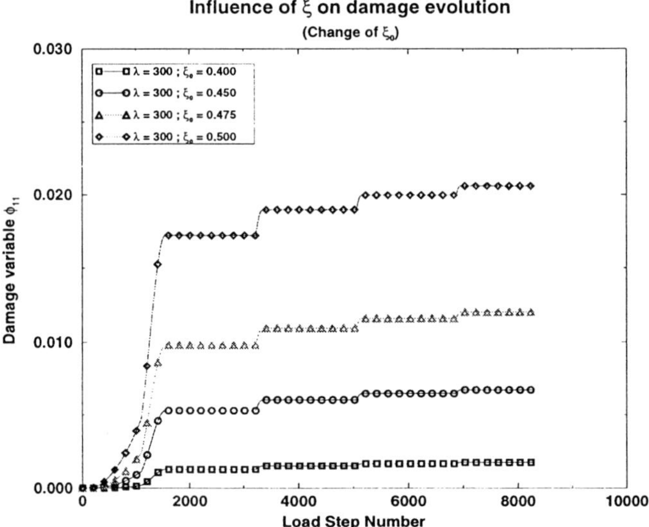

Figure 13.3   Variation in damage evolution for various values of ξ_0.

$R = const.$

$\sigma_{m_1} < \sigma_{m_2} < \sigma_{m_3} < \sigma_{m_4} < \sigma_{m_5}$

Figure 13.4 ϕ - N diagrams for determination of ξ for constant R.

$\sigma_m = const.$

$R_1 > R_2 > R_3 > R_4$

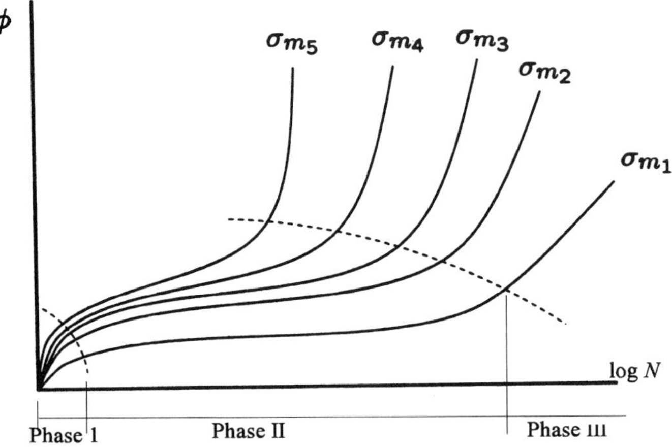

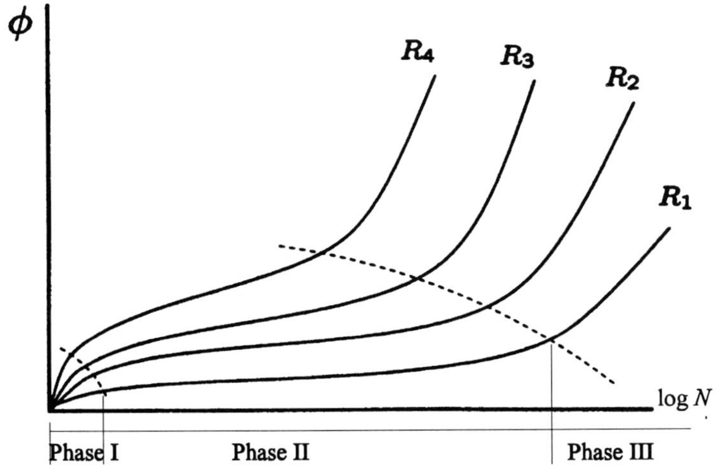

Figure 13.5 ϕ - N diagrams for determination of ξ for constant σ_{mean}.

The results for the parametric study in order to investigate the influence of the model parameter ξ_0 on the damage evolution in the matrix are shown in Figure 13.3 with all other parameters kept constant. Varying the value of the parameter λ and keeping ξ_0 constant will result in the curves shown in Figure 13.2. Only the damage variable ϕ_{11} is shown since the other components of ϕ are equal to zero or their value is smaller by a magnitude of 100. The reference frames of the damage tensor and the material system are identical, hence "1" representing the fiber direction while "2" and "3" indicate the transverse directions. For clarification it should be emphasized that the plateaus exhibited in Figures 13.2 and 13.3 represent the unloading phase in the cyclic loading where no further damage occurs.

Two sample analyses of complete fatigue simulations have been conducted to show the capabilities of the developed model. The result of such an analysis for the damage evolution in the matrix, in the fiber and the overall composite is shown in Figure 13.6. Failure of the entire composite occurs due to fiber failure at about 116000 cycles for the case of $\sigma_{11,\,max} = 1000$ MPa and a stress ratio $R = 0.1$. In a second complete fatigue simulation failure occurs at about 217000 fatigue cycles for $\sigma_{11,\,max} = 940$ MPa and a stress ratio $R = 0.1$. The obtained fatigue life in the two cases is compared with experimental results for a uni-directional composite (Johnson [181]) as shown in Figure 13.7. The results show satisfactory agreement which establishes the potential of the proposed model. The material appearing in this chapter has been outlined recently by Voyidajis and Echle [182].

354

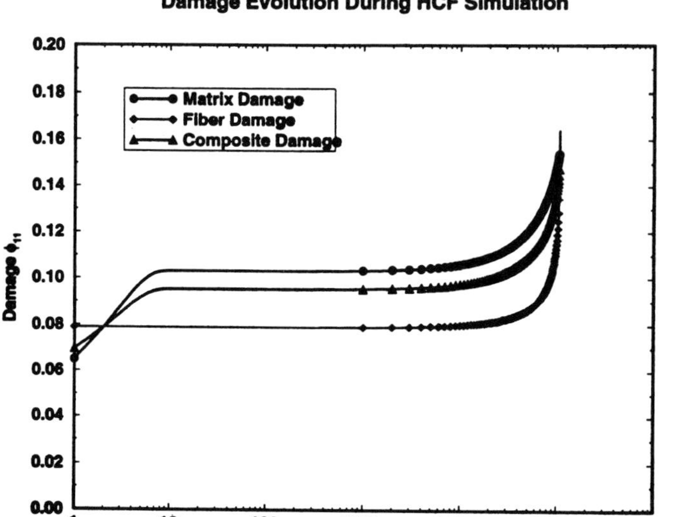

Figure 13.6 Fatigue damage evolution during a complete simulation.

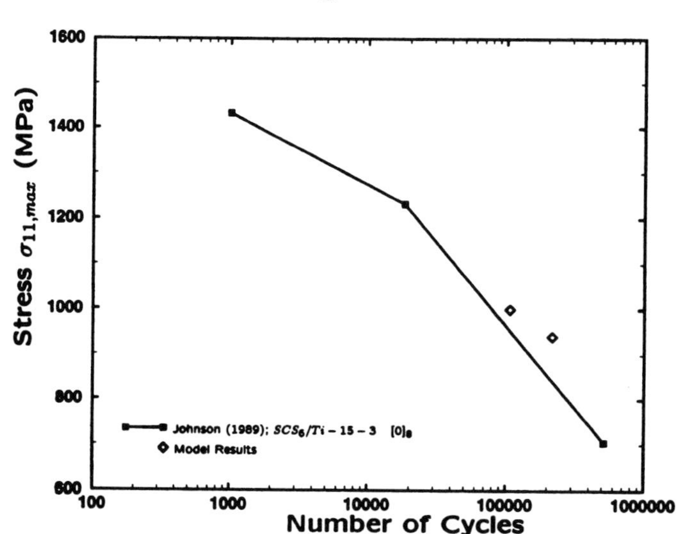

Figure 13.7   Comparison with experimental results (Johnson 1989).

CHAPTER 14

ANISOTROPIC CYCLIC DAMAGE-PLASTICITY MODELS FOR METAL MATRIX COMPOSITES

A new six parameter general anisotropic yield surface using the fourth order anisotropic tensor $\mathbf{M}$ is proposed here. This form is derived based on the physical behavior observed for the material under consideration - directionally reinforced metal matrix composites. Its validity is shown by proving its convexity and form under coordinate transformation. This form of the anisotropic yield function is general in nature which can be used for either pressure dependent or independent cases. By applying suitable conditions on the parameters, it can be reduced to the von-Mises and Tresca isotropic yield criteria. It can also be reduced to specific anisotropic models such as Hill's [183] pressure independent anisotropic yield function form and the Mulhern et al. [184] pressure independent yield criterion for transversely isotropic materials, which is used for the continuum description for yielding in metal matrix composites. The proposed surface compares well with the extensive experimental data of Dvorak et al. [185] and Nigam et al. [186] performed on boron-aluminum metal matrix composite. Based on the experimentally observed flow and hardening behavior, the elasto-plastic stiffness matrix is proposed.

14.1 Anisotropic Yield Surface Model for Directionally Reinforced Metal Matrix Composites

For a general orthotropic material it is assumed that the Cartesian Coordinate axes chosen coincide with the three principal axes of anisotropy. Before arriving at the form of the yield function it is necessary to first outline the criteria that the yield function must meet for the material under consideration here, namely the metal matrix composites.

Unlike isotropic materials where the shear strength can be expressed as a constant times the axial strength, anisotropic materials tend to have a shear strength independent of the axial strength of the material. Hence in addition to the three principal axial strengths along the three axes of anisotropy, one has three additional shear strengths. It was observed that the three axial strength parameters do not adequately represent the shear yielding. Hence three more shear strength parameters have been introduced. Also it is assumed at this stage, that the axial strength in compression must be the same as that in tension.

Another requirement is that when the principal material axes of anisotropy (also referred as local coordinate axes henceforth) coincide with the reference coordinate axes, the direction of shear stress is irrelevant and hence its strength is the same for either case. This is accounted for in the yield function by having only squared terms in the shear stress, in the local coordinate axes.

It is also seen that yielding in metal matrix composites is a pressure dependent phenomenon, (Zhao and Weng [187]). Most popular forms of the yield surface, like Hill's [183] anisotropic yield criterion and Mulhern et al. [184] criterion in-corporate a pressure independent form. It is proposed thus to use here a pressure dependent form.

A composite laminate is made up of a number of lamina, in which each lamina can have a local coordinate axes different from the reference coordinate axes. The yield surface equation in general must be valid for any orientation of the fiber. To reflect all the above criteria in the yield surface equation, one defines it as (Voyiadjis and Thiagarajan [188]),

$$F(\sigma_{ij}, d_{ij}, a_{ij} b_{ij}) = 0 \qquad (14.1)$$

where σ is the overall state of stress, d are the coefficients of the orthogonal transformation from axes of reference to principal material axes, a is a function of k_1, k_2, k_3 which are the magnitudes of the strength parameters in the three axial directions and b is a function of k_4, k_5, k_6 which are the magnitudes of the three shear strength parameters.

In the general case where the principal axes of anisotropy are not coincident with the reference coordinate axes one first writes the yield function in the local coordinate system as,

$$\overline{H}_{ijkl} \, \overline{\sigma}_{ij} \, \overline{\sigma}_{kl} - 1 = 0 \qquad (14.2)$$

where $\overline{H}$ and σ are expressed in terms of the local coordinate axes. From the stress transformation law, the stress in the local coordinate reference frame can be written in terms of the stress in the axes of reference frame as,

$$\overline{\sigma} = Q^T \sigma Q \qquad (14.3)$$

$$\overline{\sigma}_{ij} = d_{ip} \sigma_{pq} d_{qj} \qquad (14.4)$$

and assuming the fibers are aligned in the unidirectional composite single lamina system along the x-axis (1-direction) one can write,

$$d_{1j} = (\eta_1, \eta_2, \eta_3) \qquad (14.5)$$

where η, $i = 1,2,3$ are the direction cosines of the fiber in the reference coordinate system. A lamina of any arbitrary orientation is derived by rotating the principal axes of anisotropy about the z-axis

(3-direction) of the axes of reference. Hence in the transformation Q one obtains $d_{3j} = (0,0,1)$. The other three terms of d_{ij} namely d_{2j} are derived from the condition that applies to any orthogonal coordinate transformation namely,

$$d_{pi}\, d_{qi} = \delta_{pq} \tag{14.6}$$

Since η_i is known one can easily derive the components of the transformation matrix. One defines,

$$\overline{H} = \overline{H}\,(a,b) \tag{14.7}$$

Voyiadjis and Foroozeh [189], Eisenberg and Yen [190] and Voyiadjis et al. [191] had earlier proposed a form of the fourth order anisotropic tensor for an anisotropic distortional yield surface. However for metal matrix composites, since the subsequent yield surfaces do not appear to exhibit significant distortion and also due to the fact that the fibers and their direction impose certain constraints on the plastic behavior of the material, a new suitable form for $\overline{H}$ is derived here.

A suitable expression for $\overline{H}$, which will be termed the anisotropic yield tensor, has been derived to satisfy the properties that this yield surface must meet. $\overline{H}$ in local coordinate axes is expressed as (Voyiadjis and Thiagarajan [188]),

$$\overline{H}_{ijkl} = A\,a_{ij}\,a_{kl} + B\,a_{ik}\,a_{jl} + C\,a_{il}\,a_{jk} + D\,b_{ij}\,b_{kl} \tag{14.8}$$

where A, B, C and D are constants and,

$$a_{ij} = \begin{bmatrix} k_1 & 0 & 0 \\ 0 & k_2 & 0 \\ 0 & 0 & k_3 \end{bmatrix} \tag{14.9}$$

$$b_{ij} = \begin{bmatrix} 0 & k_4 & k_5 \\ k_4 & 0 & k_6 \\ k_5 & k_6 & 0 \end{bmatrix} \tag{14.10}$$

a_{ij} and b_{ij} are defined and determined along the principal axes of anisotropy and k_i, ($i=1,.....,6$) are the magnitudes of the six strength parameters (three normal and three shears).

Substituting (14.8) in (14.2) and expanding the terms, the yield equation in the local coordinate axes in component form can be expressed as the following form:

$$(A + B + C)\ (k_1^2\ \bar{\sigma}_{11}^2 + k_2^2\ \bar{\sigma}_{22}^2 + k_3^2\ \bar{\sigma}_{33}^2) +$$
$$(2A)\ (k_1\ k_2\ \bar{\sigma}_{11}\ \bar{\sigma}_{22} + k_1\ k_3\ \bar{\sigma}_{11}\ \bar{\sigma}_{33} + k_2\ k_3\ \bar{\sigma}_{22}\ \bar{\sigma}_{33}) +$$
$$\{2\ (B + C)k_1\ k_2 + 4\ D\ k_4^2\}\bar{\sigma}_{12}^2 + \{2\ (B + C)k_1\ k_3 + 4Dk_5^2\}\bar{\sigma}_{13}^2 +$$
$$\{2\ (B + C)k_2\ k_3 + 4Dk_6^2\}\bar{\sigma}_{23}^2 - 1 = 0$$

$$\text{(14.11)}$$

where the stresses are in the local coordinate system. The constants A, B, C and D are not material parameters. This combination is predetermined and chosen - as explained later in this work. Due to symmetry of stresses the constant B is equal to C.

Substituting (14.4) into (14.2) one can write the yield equation in the axes of reference system as follows,

$$\sigma_{ij}\ H_{ijkl}\ \sigma_{kl} - 1 = 0 \qquad\qquad \text{(14.12)}$$

where,

$$H_{ijkl} = \bar{H}_{pqrs}\ d_{ip}\ d_{jq}\ d_{kr}\ d_{ls} \qquad\qquad \text{(14.13)}$$

However a more practical form, for computational implementation, of the yield function in the case of single lamina is:

$$\sigma_{pq}\ d_{ip}\ d_{jp}\ \bar{H}_{ijkl}\ d_{km}\ d_{ln}\ \sigma_{mn} - 1 = 0 \qquad\qquad \text{(14.14)}$$

In general the backstress term is a very important component while expressing the yield surface in the above form. This will be introduced in a later section and the discussion is kept to the initial yield surface at present.

The general form of the anisotropic yield function can be reduced to be applicable for isotropic material, where $k_1 = k_2 = k_3 = k$ and since the shear stresses are dependent on the axial yield stresses, the dependence of H on k_4, k_5, k_6 is then eliminated. Hence (14.11) reduces to,

$$(A + B + C)\, k^2\, (\overline{\sigma}_{11}^2 + \overline{\sigma}_{22}^2 + \overline{\sigma}_{33}^2) +$$
$$(2A)k^2\, (\overline{\sigma}_{11}\,\overline{\sigma}_{22} + \overline{\sigma}_{11}\,\overline{\sigma}_{33} + \overline{\sigma}_{22}\,\overline{\sigma}_{33}) + \qquad (14.15)$$
$$2\,(B + C)k^2\, (\overline{\sigma}_{12}^2 + \overline{\sigma}_{13}^2 + \overline{\sigma}_{23}^2) - 1 = 0$$

It can easily be shown that (14.15) reduces to the familiar von-Mises and Tresca yield criterion under the following combinations of the constants A, B, C and D.

1.   For von-Mises (Isotropic) criterion (in the general stress-space)

$$A = -\frac{1}{9}, \; B = C = \frac{1}{6}$$

2.   For Tresca (isotropic) criterion (under plane-stress conditions)

$$A = -\frac{1}{4}, \; B = C = \frac{1}{4}.$$

Under plane-stress conditions in the $(\sigma_{11} - \sigma_{22} - \sigma_{12})$ space it reduces to

$$\left[\frac{\sigma_{11} - \sigma_{22}}{2}\right]^2 + \sigma_{12}^2 = \kappa^2 \qquad (14.16)$$

which is the Tresca yield criterion under plane stress conditions.

The equivalence between κ used here and k_o used for example in the von-Mises yield criterion is thus $\kappa = \dfrac{1}{k_o}$.

For the general orthotropic material suitable choice can be made and adopted. By choosing the first combination and letting $D = \dfrac{1}{6}$ the yield criterion can be written finally in component form as follows:

$$F = \frac{2}{9}\, (k_1^2\,\overline{\sigma}_{11}^2 + k_2^2\,\overline{\sigma}_{22}^2 + k_3^2\,\overline{\sigma}_{33}^2) - \frac{2}{9}\, (k_1 k_2\,\overline{\sigma}_{11}\,\overline{\sigma}_{22} + k_2 k_3\,\overline{\sigma}_{22}\,\overline{\sigma}_{33} + k_1 k_3\,\overline{\sigma}_{11}\,\overline{\sigma}_{33})$$
$$+ \frac{2}{3}\, (k_1 k_2 + k_4^2)\,\overline{\sigma}_{12}^2 + \frac{2}{3}\, (k_1 k_3 + k_5^2)\,\overline{\sigma}_{13}^2 \qquad (14.17)$$
$$+ \frac{2}{3}\, (k_2 k_3 + k_6^2)\,\overline{\sigma}_{23}^2 - 1.0$$

A necessary requirement for the yield surface to be valid is that it must be convex. This was demonstrated and proved by Drucker [192, 193] using the stability postulate. Convexity of the surface ensures that the material does not return to an elastic stage after undergoing plastic deformation, while the loading continues. Mathematically, the convexity of the yield surface F is demonstrated if it can be shown that the Hessian Matrix Z_{ij} of this function is positive semi-definite, i.e. its eigen values are all positive or zero. The Hessian matrix for the given function (14.17) is defined as

$$Z_{ij} = \frac{\partial^2 F}{\partial \overline{\sigma}_i \partial \overline{\sigma}_j} \tag{14.18a}$$

where $\overline{\sigma}_i = \{\sigma_x, \sigma_y, \sigma_z, \sigma_{xy}, \sigma_{xz}, \sigma_{yz}\}$ is the vector from of the stress tensor $\overline{\sigma}_{ij}$.

The Hessian matrix components using equation (14.18) and (14.17) is derived to be,

$$Z_{ij} = \begin{bmatrix} \dfrac{4k_1^2}{9} & -\dfrac{2k_1k_2}{9} & -\dfrac{2k_1k_3}{9} & 0 & 0 & 0 \\[2ex] -\dfrac{2k_1k_2}{9} & \dfrac{4k_2^2}{9} & -\dfrac{2k_2k_3}{9} & 0 & 0 & 0 \\[2ex] -\dfrac{2k_1k_3}{9} & -\dfrac{2k_2k_3}{9} & \dfrac{4k_3^2}{9} & 0 & 0 & 0 \\[2ex] 0 & 0 & 0 & \dfrac{4k_1k_2}{3}+\dfrac{4k_4^2}{3} & 0 & 0 \\[2ex] 0 & 0 & 0 & 0 & \dfrac{4k_1k_3}{3}+\dfrac{4k_5^2}{3} & 0 \\[2ex] 0 & 0 & 0 & 0 & 0 & \dfrac{4k_2k_3}{3}+\dfrac{4k_6^2}{3} \end{bmatrix}$$

$$\tag{14.18b}$$

Further the 6 eigen values of this Hessian matrix in terms of $k_1, ..., k_6$ are given by,

$$EIGEN1 = 0 \tag{14.19}$$

$$EIGEN2 = \frac{4k_2 k_3}{3} + \frac{4k_6^2}{3} \qquad (14.20)$$

$$EIGEN3 = \frac{4k_1 k_2}{3} + \frac{4k_4^2}{3} \qquad (14.21)$$

$$EIGEN4 = \frac{2k_1^2}{9} + \frac{2k_2^2}{9} + \frac{2k_3^2}{9} + \qquad (14.22)$$

$$\frac{2\sqrt{k_1^4 - k_1^2 k_2^2 - k_1^2 k_3^2 + k_2^4 - k_2^2 k_3^2 + k_3^4}}{9}$$

$$EIGEN5 = \frac{2k_1^2}{9} + \frac{2k_2^2}{9} + \frac{2k_3^2}{9} - \qquad (14.23)$$

$$\frac{2\sqrt{k_1^4 - k_1^2 k_2^2 - k_1^2 k_3^2 + k_2^4 - k_2^2 k_3^2 + k_3^4}}{9}$$

$$EIGEN6 = \frac{4k_1 k_3}{3} + \frac{4k_5^2}{3} \qquad (14.24)$$

It is now proved, that for all the eigenvalues to be either zero or positive the only condition that is mathematically imposed on $k_1, \ldots, k_6$ is that they must be all grater than zero, since they are the magnitudes of the strengths. This condition is satisfied physically too in their computation. Without any loss of generality it can be assumed that $k_1 > k_2 > k_3 > 0$ and that $k_4, k_5,$ and $k_6 > 0$ also.

From the above equations for the eigen values it is directly observed that EIGEN 1, EIGEN 2 and EIGEN 6 are all ≥ 0. For EIGEN 4 to be greater than zero, then one should show that

$$\sqrt{k_1^4 - k_1^2 k_2^2 - k_1^2 k_3^2 + k_2^4 - k_2^2 k_3^2 + k_3^4} > 0 \qquad (14.25)$$

then EIGEN4 would be greater than zero. Rearranging the expression within the root one obtains

$$k_1^4 + k_2^4 + k_3^4 > k_1^2 k_2^2 + k_1^2 k_3^2 + k_2^2 k_3^2 \tag{14.26}$$

$$(k_1^4 - k_1^2 k_2^2) + (k_2^4 - k_2^2 k_3^2) > k_1^2 k_3^2 - k_3^4 \tag{14.27}$$

Adding and subtracting $k_2^2 k_3^2$ to the right hand side of the above equation one obtains

$$k_1^2 (k_1^2 - k_2^2) + k_2^2 (k_2^2 - k_3^2) > k_1^2 k_3^2 - k_2^2 k_3^2 + k_2^2 k_3^2 - k_3^4 \tag{14.28}$$

$$k_1^2 (k_1^2 - k_2^2) + k_2^2 (k_2^2 - k_3^2) > k_3^2 (k_1^2 - k_2^2) + k_3^2 (k_2^2 - k_3^2) \tag{14.29}$$

Since $k_1 > k_3$ the first term on LHS is greater that the first term of the RHS. Also since $k_2 > k_3$ the second term of the LHS is greater than the second term of the RHS. This proves that the left hand side is greater than the right hand side and hence the term under the root is positive. This shows that EIGEN4 is positive.

To show that EIGEN5 is positive one should show that (Voyiadjis and Thiagarajan [188])

$$\frac{2 k_1^2}{9} + \frac{2 k_2^2}{9} + \frac{2 k_3^2}{9} > \frac{2\sqrt{k_1^4 - k_1^2 k_2^2 - k_1^2 k_3^2 + k_2^4 - k_2^2 k_3^2 + k_3^4}}{9} \tag{14.30}$$

Or squaring both sides one obtains,

$$k_1^4 + k_2^4 + k_3^4 + 2 k_1^2 k_2^2 + 2 k_1^2 k_3^2 + 2 k_2^2 k_3^2 > k_1^4 + k_2^4 + k_3^4 - k_1^2 k_2^2 - k_2^2 k_3^2 - k_1^2 k_3^2 \tag{14.31}$$

This is automatically proved since the quantity on the left hand side is positive whereas that on the right hand side is negative after similar terms are cancelled. This proves that EIGEN5 is positive. This completes the proof that all the eigen values are greater than or equal to zero.

It is observed from equation (2) that $\boldsymbol{H}$ is used here as a fourth order tensor. The basic form of this tensor has been derived from the idea of the general fourth order isotropic tensor. For an anisotropic material this form of $\boldsymbol{H}$ essentially represents the magnitude of the strength parameters of the material. The property of repeated indices of tensors is used here in this representation. In general an anisotropic tensor is defined as a tensor which is invariant under a certain specified group of transformations. Here H is defined and determined in the principal material axes of anisotropy. This can be termed as the strength tensor of the fourth order, analogous to the first and second order strength tensors defined by the Tsai-Wu [194] failure criteria for composites.

The tensor H is directly related to the magnitude of the strengths along the principal axes of anisotropy and is defined in the local coordinates of the lamina with reference to the fiber direction. For a general case where the fibers are oriented at an angle with respect to the general coordinate axes, it would necessitate the transformation of stresses to the local coordinate axes.

One now considers the form of the yield function under the condition of stress transformation. This would represent the case where the principal material axes are different from the general axes of reference (Figure 14.1). A common example of transformation is when the fibers are all aligned at an angle θ as shown in Figure 14.1. Mathematically, for the yield function to be independent of the direction of shear stress in the 1-2 axis, it should contain only squared terms of shear stress. This is ensured by choosing a suitable form of $\overline{H}$. One can express the fourth order form of $\overline{H}$ in the yield function, as shown in equation (14.2) by the equivalent second order form of $\overline{H}$ in the principal material axes as follows,

$$ F = \overline{H}_{ij} \overline{\sigma}_i \overline{\sigma}_j - 1 \quad (i,j = 1,,,6) \tag{14.32} $$

where $\overline{\sigma}_1 = \overline{\sigma}_{11}$, $\overline{\sigma}_2 = \overline{\sigma}_{22}$, $\overline{\sigma}_3 = \overline{\sigma}_{33}$, $\overline{\sigma}_4 = \overline{\sigma}_{12}$, $\overline{\sigma}_5 = \overline{\sigma}_{13}$, $\overline{\sigma}_6 = \overline{\sigma}_{23}$ and $\overline{H}_{ij}$ terms are derived from $\overline{H}_{ijkl}$ as (Voyiadjis and Thiagarajan [188])

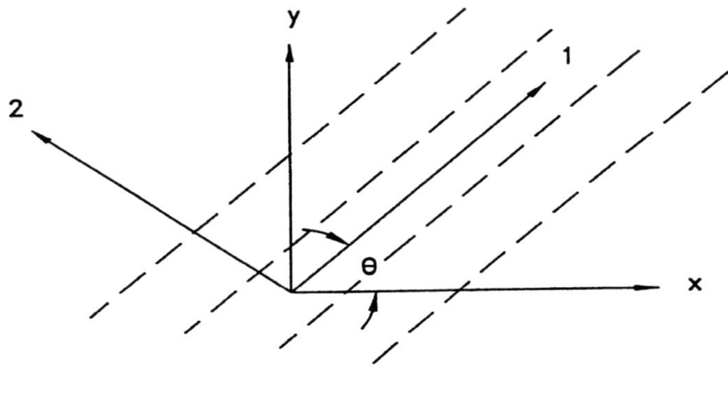

x–y: general axes of reference
1–2: principal material axes
— — direction of fiber

Fig. 14.1 Principal material axes and general reference axes

$$\begin{aligned}
\overline{H}_{11} &= \overline{H}_{1111} \\
\overline{H}_{22} &= \overline{H}_{2222} \\
\overline{H}_{33} &= \overline{H}_{3333} \\
\overline{H}_{12} &= \overline{H}_{1122} \\
\overline{H}_{21} &= \overline{H}_{2211} \\
\overline{H}_{13} &= \overline{H}_{1133} \\
\overline{H}_{31} &= \overline{H}_{3311} \\
\overline{H}_{23} &= \overline{H}_{2233} \\
\overline{H}_{32} &= \overline{H}_{3322} \\
\overline{H}_{44} &= \overline{H}_{1212} + \overline{H}_{1221} + \overline{H}_{2112} + \overline{H}_{2121} \\
\overline{H}_{55} &= \overline{H}_{1313} + \overline{H}_{1331} + \overline{H}_{3113} + \overline{H}_{3131} \\
\overline{H}_{66} &= \overline{H}_{2323} + \overline{H}_{2332} + \overline{H}_{3223} + \overline{H}_{3232}
\end{aligned} \tag{14.33}$$

The second order form of the yield surface in the general axes of reference can be expressed as follows

$$F = H_{ij}\,\sigma_i\,\sigma_j - 1 \quad (i, j = 1, 6) \tag{14.34}$$

where σ_i are vector form of stresses in this axes of reference. The stress-transformation law for stress expressed as a second order tensor is given as,

$$\overline{\sigma}_{ij} = d_{ik}\,d_{jl}\,\sigma_{kl} \tag{14.35}$$

where,

$$d_{ij} = \begin{bmatrix} m & n & o \\ -n & m & 0 \\ 0 & 0 & 1 \end{bmatrix} \tag{14.36}$$

and $m = cos\theta$ and $n = sin\theta$. Equation (14.30) can be expanded in component form as follows,

$$F = \overline{H}_{11}\,\overline{\sigma}_1^2 + \overline{H}_{22}\,\overline{\sigma}_2^2 + \overline{H}_{33}\,\overline{\sigma}_3^2 + (\overline{H}_{12} + \overline{H}_{21})\,\overline{\sigma}_1\,\overline{\sigma}_2 + (\overline{H}_{13} + \overline{H}_{31})\,\overline{\sigma}_1\,\overline{\sigma}_3$$
$$+ (\overline{H}_{23} + \overline{H}_{32})\,\overline{\sigma}_2\,\overline{\sigma}_3 + \overline{H}_{44}\,\overline{\sigma}_4^2 + \overline{H}_{55}\,\overline{\sigma}_5^2 + \overline{H}_{66}\,\overline{\sigma}_6^2 - 1 \tag{14.37}$$

Substituting vector form of (14.33) in (14.35) and rearranging the terms we can rewrite the yield equation
as,

$$F = H_{11}\,\sigma_1^2 + H_{22}\,\sigma_2^2 + H_{33}\,\sigma_3^2 + H_{12}\,\sigma_1\,\sigma_2 + H_{13}\,\sigma_1\,\sigma_3 + H_{23}\,\sigma_2\,\sigma_3$$
$$H_{14}\,\sigma_1\,\sigma_4 + H_{24}\,\sigma_2\,\sigma_4 + H_{34}\,\sigma_3\,\sigma_4 +$$
$$H_{44}\,\sigma_4^2 + H_{55}\,\sigma_5^2 + H_{66}\,\sigma_6^2 + H_{56}\,\sigma_5\,\sigma_6 - 1 \tag{14.38}$$

The terms in the square brackets represent the transformed **H** terms in the general axes of reference. These are functions of $\overline{H}$ and 'm' and 'n'. As there are terms in the above equation, which is in the general axes of reference, with single power of the shear stress, one concludes that the yield function is dependent on the direction of shear stress. It must be noted that there is an inherent exception which is that under pure shear conditions, the yield function is still independent of the shear stress direction.

The simplicity of this form and its ease of use are the attractive features of this representation.

14.2 Comparison with Other Anisotropic Yield Surfaces

It has been shown earlier that the proposed yield function reduces to the von-Mises and Tresca yield criteria (under plane stress conditions) for isotropic materials by making suitable assumptions. There are other well known yield criteria for anisotropic materials, with specific application to composites. Some of them are Hill's general anisotropic yield criterion [183] and Mulhern et al.'s [184] criterion for transversely isotropic composites. The proposed yield surface is compared and contrasted with these existing yield surface forms. By imposing suitable conditions on the parameters it is shown that it can be reduced to the corresponding forms. Hill proposed a pressure-independent yield criterion for general orthotropic materials. Hill described the yield function for the case where the axes of reference coincides with the principal axes of anisotropy as,

$$2f = P(\sigma_{yy} - \sigma_{zz})^2 + Q(\sigma_{zz} - \sigma_{xx})^2 + R(\sigma_{xx} - \sigma_{yy})^2 + \\ 2S\sigma_{yz}^2 + 2U\sigma_{xz}^2 + 2V\sigma_{xy}^2 - 1 \tag{14.39}$$

The proposed yield criterion here is a pressure dependent yield criterion. For the yield criterion expressed as in (14.2), to be pressure independent H the fourth order anisotropic tensor must satisfy the condition,

$$H_{iikk} = 0 \tag{14.40}$$

Applying this condition to the proposed H tensor one arrives at the condition,

$$k_1^2 + k_2^2 + k_3^2 - k_1 k_2 - k_1 k_3 - k_2 k_3 = 0 \tag{14.41}$$

The total stresses are now replaced by the deviatoric components of stresses in the proposed yield criterion to restrict the expression to be pressure independent. The resulting yield criterion is expanded and the condition given by equation (14.41) is imposed. Comparing it with Hill's criterion (14.39) one observes the following relationship between the parameters (Voyiadjis and Thiagarajan [188])

$$R + Q = \frac{2}{27} k_1^2$$

$$R + P = \frac{2}{27} k_2^2$$

$$P + Q = \frac{2}{27} k_3^2$$

$$2V = \frac{2}{3} (k_1 k_2 + k_4^2)$$

$$2U = \frac{2}{3} (k_1 k_3 + k_5^2)$$

$$2S = \frac{2}{3} (k_2 k_3 + k_6^2)$$

$$(14.42)$$

The first three terms in the above equation represent the correspondence of P, Q and R with k_1, k_2 and k_3, which are measured strength parameters along the three axes of anisotropy while the last three terms represent the corresponding shear strengths in both criteria. It is seen that the proposed criterion can be applied to both pressure dependent and pressure independent cases.

Mulhern, et al. [184] had proposed an elastic-plastic theory for materials reinforced by a single family of fibers. They assumed the material to be inextensible in the fiber direction and derived a pressure independent yield function wherein yielding was also independent of stress invariants as follows. In the case where the axes of reference coincide withthe local coordinate axes it is written as

$$f = \frac{1}{k_T^2} \left[\frac{1}{4} (\sigma_{22} - \sigma_{33})^2 + \sigma_{23}^2 \right] + \frac{1}{k_L^2} [\sigma_{12}^2 + \sigma_{13}^2] - 1 \qquad (14.43)$$

where k_T and k_L are shear yield stresses for shear on planes containing the fibers, in directions transverse and parallel to the fibers respectively.

To demonstrate the equivalence, one introduces the two assumptions made above, namely pressure independence and fiber direction stress independence (no yielding). These two constraints are introduced into the stress term by defining the 'extra-stress' term as, (Spencer [195]),

$$s_{ij} = \sigma_{ij} - T\eta_i \eta_j - p\delta_{ij} \qquad (14.44)$$

where T and p are constants determined by the constraints of inextensibility and incompressibility and η_i ($i = 1, 2, 3$) are the direction cosines of the fiber with respect to the axes of reference. Further it can be shown that the extra-stress term can be represented as

$$s_{ij} = \sigma_{ij} - \frac{1}{2}(\sigma_{kk} - \eta_r \eta_s \sigma_{rs})\delta_{ij} + \frac{1}{2}(\sigma_{kk} - 3\eta_r \eta_s \sigma_{rs})\eta_i \eta_j \tag{14.45}$$

For the case of $\eta = (1,0,0)$ the above equation reduces to,

$$s_{ij} = \sigma_{ij} - \frac{1}{2}(\sigma_{rr} - \sigma_{11})\delta_{ij} + \frac{1}{2}(\sigma_{rr} - 3\sigma_{11})\eta_i \eta_j \tag{14.46}$$

To incorporate the constraints in the yield function one introduces s_{ij} instead of σ_{ij} in (14.12) and write

$$F = s_{ij} H_{ijkl} s_{kl} - 1 \tag{14.47}$$

Since the Mulhern, Rogers and Spencer's criterion appears to be a shear yield stress form, one chooses the constants $A = -\frac{1}{4}, B = \frac{1}{4}, C = \frac{1}{4}$ and $D = 0$ and by reducing the proposed yield criterion for a transversely isotropic material, where k_1 is along the fiber direction and $k_2=k_3$ one can write (14.45) in component form as,

$$F = k_2^2 \left[\frac{1}{4}(s_{22} - s_{33})^2 + s_{23}^2 \right] + k_1 k_2 [s_{12}^2 + s_{13}^2] - 1 \tag{14.48}$$

The correspondence between the Mulhern et al.[184] parameters and the parameters used here are $k_2 = \frac{1}{k_T}$ and $k_1 k_2 = \frac{1}{k_L^2}$. The above equation shows the relationship with equation (14.41) with regard to the yield parameters in the two criteria and demonstrates that the generalized yield function can be reduced to Mulhern et al.[184] form by making the assumptions that they had made. In their form only two shear yield parameters need to be determined. Hence only two parameters from the generalized form are needed.

The yield surface presented here is based on a phenomenological approach for the metal matrix composite continuum. For a more accurate prediction of stresses in the metal matrix composite a microstructural characterization is required. In this section a microstructural representation of such a yield condition is presented. Usually the von Mises yield criterion is used for the matrix material (metal) when attempting to represent the yield criterion with a microstructural characterization. The matrix yield criterion F_m is given by

$$F^m \equiv \frac{3}{2} (\tau^m - \alpha^m)^T : I : (\tau^m - \alpha^m) - (\sigma_y^m)^2 = 0 \qquad (14.49)$$

where I is the matrix representation of the fourth order identity tensor and σ_y^m is the initial yield stress of the matrix material. The above equation is in terms of the deviatoric stresses and backstresses in the matrix phase of the material. Equation (14.47) may further be expanded in terms of the total stresses in the matrix as,

$$F^m \equiv \frac{3}{2} (\sigma^m - \beta^m)^T [c] [I] [c] (\sigma^m - \beta^m) - (\sigma_y^m)^2 = 0 \qquad (14.50)$$

where c is a 6 by 6 constant matrix defining,

$$\{\tau\} = [c]\{\sigma\} \qquad (14.51)$$

However it can be shown that

$$[c] [I] [c] = [c] \qquad (14.52)$$

Using the fourth order stress concentration tensor B^m one expresses the stresses in the composite and matrix material such that

$$\sigma^m = B^m : \sigma \qquad (14.53)$$

In vector form it is expressed as,

$$\{\sigma^m\} = [B^m]\{\sigma\} \qquad (14.54)$$

Substituting (14.52) in (14.48) one obtains,

$$F^m \equiv (\sigma - \beta)^T \left\{ \frac{3}{2(\sigma_y^m)^2} [B^m]^T [c] [B^m] \right\} (\sigma - \beta) - 1 = 0 \tag{14.55}$$

A relation similar to equation (14.53) was previously obtained by Dvorak and Bahel-El-Din [70]. The term in braces in the above equation will be defined as $[W]$ such that,

$$[W] = \frac{3}{2(\sigma_y^m)^2} [B^m]^T [c] [B^m] \tag{14.56}$$

or

$$[W] = [B^m]^T [h^m] [B^m] \tag{14.57}$$

where,

$$[h^m] = \frac{3}{2(\sigma_y^m)^2} [c] \tag{14.58}$$

The fourth order tensor H give by equation (14.13) is a more accurate in-situ representation of the tensor h^m for the matrix material than that given by equation (14.56) obtained through the utilization of the von-Mises yield criterion. It is propsoed here that using equation (14.56) is inappropriate to represent the matrix material since its confinement by the surrounding fibers is not accounted for. Therefore the tensor h^m should represent the in-situ characteristics of this matrix material and accordingly the expression given by equation (14.13) could be a possible appropriate representation of this tensor. This implies that the strength parameters k_i should more accurately predict the in-situ properties of the matrix material. Essentially the yield criterion of the matrix material can be only accurately represented if all the appropriate constraining conditions are accounted for.

14.3 Numerical Simulation of the Initial Anisotropic Yield Surface

In order to provide experimental validity, a numerical simulation is done to evaluate the values of the parameters and compare the experimental and model results. The experimental data from boron-aluminum composite tubular specimens having unidirectional lamina by Dvorak et al. [185] and Nigam. [186] are used here. The fibers in the tube are aligned parallel to the axis of the tube. The specimen is subjected to different load paths by applying axial force, torque and internal pressure in order to determine the yield surfaces in the σ_{11} - σ_{21} and σ_{22} - σ_{21} stress planes, where σ_{11} is the stress along the fiber direction, σ_{22} is the normal stress transverse to the fiber and σ_{21} is the longitudinal shear stress.

Since these were unidirectional specimens one has $\eta\,(1,0,0)$. Using equation (14.12) and reducing it to component form for transversely isotropic case where $k_1 = k_3$ one obtains

$$
\begin{aligned}
F &= \frac{2}{9}\,k_1^2\,\sigma_{11}^2 + \frac{2}{9}k_2^2\,(\sigma_{22}^2 + \sigma_{33}^2) - \frac{2}{9}k_1\,k_2\,\sigma_{11}\,(\sigma_{22} + \sigma_{33}) \\
&\quad -\frac{2}{9}\,k_2^2\,\sigma_{22}\,\sigma_{33} + \frac{2}{3}\,(k_1\,k_2 + k_4^2)\,(\sigma_{12}^2 + \sigma_{13}^2) + \frac{2}{3}\,(k_2^2 + k_6^2)\,\sigma_{23}^2 - 1
\end{aligned}
\tag{14.59}
$$

In the σ_{11}-σ_{21} stress space where except for these two stress components all the other stress components are zero the above equation can be expressed as follows,

$$
F = \frac{2}{9}\,k_1^2\,\sigma_{11}^2 + \frac{2}{3}\,(k_1\,k_2 + k_4^2)\,\sigma_{21}^2 - 1
\tag{14.60}
$$

A similar equation can be written in the σ_{22} - σ_{21} space. The parameter k_1 is determined from the yield stress along the σ_{11} axis and k_2 from that along the σ_{22} axis. From the third yield stress namely along the σ_{21} axis one can then determine k_4 from the above equation. From the experimental data, the following values of initial yield stress have been measured. They are

-    $\sigma_{11}^Y = 87.90\ MPa$

-    $\sigma_{22}^Y = 44.70\ MPa$

-    $\sigma_{21}^Y = 21.30\ MPa$

Using the above data the following values for k_1, k_2 and k_4 have been computed as,

-    $k_1 = \dfrac{1}{41.47}$

-    $k_2 = \dfrac{1}{21.09}$

-    $k_4 = \dfrac{1}{21.50}$

Figures 14.2 and 14.3 show the surfaces generated by using this model in comparison with the experimental data. The surface representing the Bimodal plasticity theory of Dvorak and Bahei-El-Din [70] are also shown in these figures.

Figures 14.2 and 14.3 show why it is necessary to have additional parameters to represent the additional effect of shear strength. The curves corresponding to a model having only the parameters k_1, k_2 and k_3 is shown along with the model corresponding to the six parameters. It is

372

observed that although the yielding along the axial directions are correctly simulated, the shear strength is overestimated. The introduction of the three shear strength parameters corrects this deficiency and allows for the correct representation of the observed phenomena.

Using the parameter values the yield surfaces have been generated. Figures 14.4 and 14.5 show the model generated yield surfaces along with the experimental points for the initial yield surfaces in this stress space.

Hardening is the parameter that determines the shape, size and location of the subsequent yield surfaces in the stress space. The experimental evidence used here indicates kinematic hardening to a great extent, along with some change in size. Distortion of the yield surface has not been observed for this case. For the determination of subsequent yield surfaces hardening is a very important aspect. The experimental evidence of Dvorak et al. [185] have indicated that the predominant mode of hardening is kinematic hardening. This is introduced into the yield function in the form of the backstress tensor α which represents translation of the yield surface as,

$$F\left(\sigma_{ij},\ \alpha_{ij},\ a_{ij},\ b_{ij},\ d_{ij}\right) = 0 \tag{14.61}$$

or

$$F = \left(\sigma_{ij} - \alpha_{ij}\right) H_{ijkl} \left(\sigma_{kl} - \alpha_{kl}\right) - 1 \tag{14.62}$$

The evolution equations for the backstress has also been observed to be of the Phillips form as

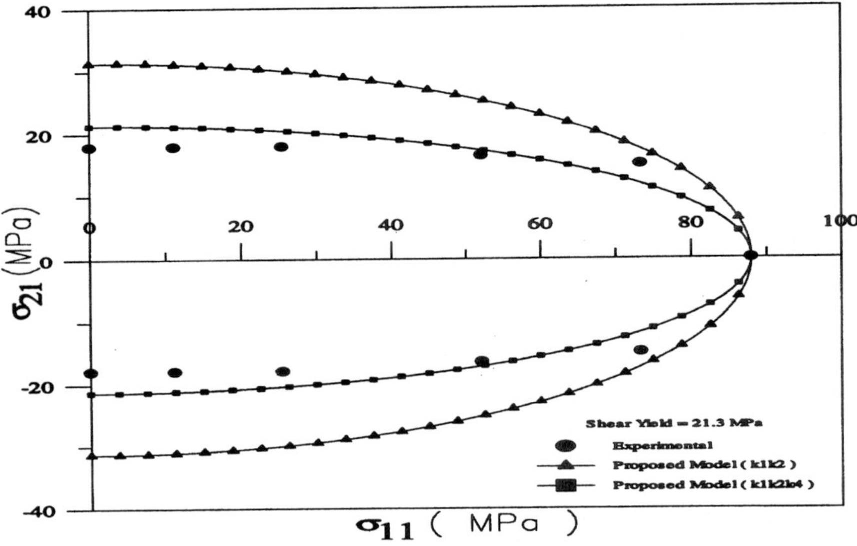

Fig. 14.2, Comparison of Initial Yield Surface in σ_{11} - σ_{12} space

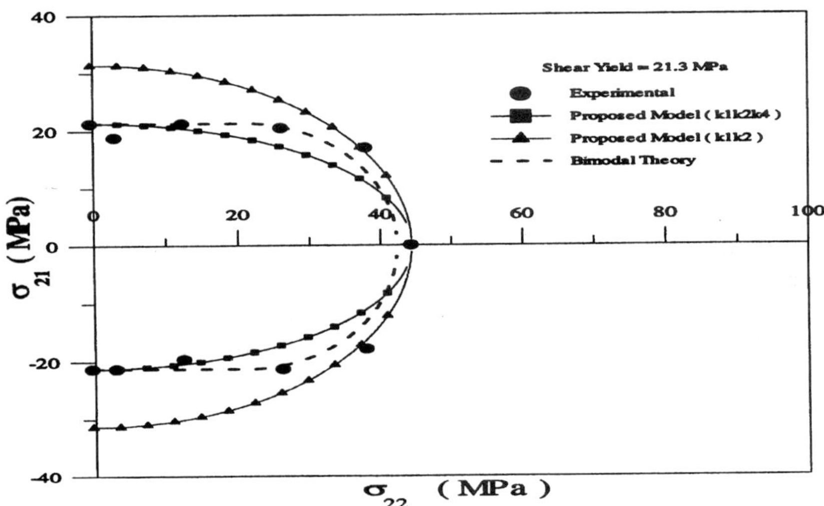

Fig, 14.3, Comparison of Initial Yield Surface in σ_{22} - σ_{12} space

$$d\alpha_{ij} = \dot{\mu} d\sigma_{ij} \tag{14.63}$$

that is the surface translates along the incremental stress vector. Nigam, et al. [186] have reported that the surface translates along the stress vector when loading takes place on the circular branches of the matrix dominated (MDM) yield surface of their bimodal plasticity surface. Along the flat branches of the MDM surface, the movement of the surface was found to be a translation along a longitudinal shear axes regardless of the loading direction.

The kinematic hardening rule that is adopted for this work is to be based on the above experimental observations. However at this juncture since the accent is on simulating the yield surfaces, the shape of the subsequent yield surfaces is shown in Figures 14.4, 14.5, 14.6 and 14.7. Figure 14.4 shows the comparison in the σ_{11} - σ_{12} space. Figures 14.5, 14.6 and 14.7 show the comparison for seven subsequent surfaces in the σ_{22} - σ_{12} space. These figures show the model generated surface in comparison to the experimental yield surface points reported by Nigam et al. [186] and the bimodal plasticity theory surfaces of Dvorak and Bahei-El-Din [70]. The centers of the respective surfaces have been taken from the table reported in the above experiments. The values of k_1 and k_2 have been determined from the axial yield strengths of the yield surfaces. Using the values of the parameter τ_o shown in the same paper, various values of the parameter k_4 have been determined. Since the translation here is only along the σ_{11} - σ_{21} plane, a constant value of k_2 is assumed, which is a reasonable assumption. Table 1 shows the numerical values corresponding to these parameters. Similarly Table 2 shows the numerical values of the parameters for the surfaces in the σ_{22} - σ_{12} space.

As Spencer [195] had pointed out, the current hardening (analogous to isotropic hardening) would require more than a single parameter for anisotropic materials and hence they termed it as 'proportional hardening'. This implies that in any given direction in the stress space the expansion is uniform, but differs with direction. In other words the length of the principal axes would not remain constant. Experimental evidence suggests that strain hardening does take place.

14.4. Cyclic Damage Models : Constitutive Modeling and Micromechanical Damage

A cyclic damage-plasticity model is used here for modeling the behavior of Metal-Matrix Composites (MMC's) under the behavior of cyclic multi-axial loading situations (Voyiadjis and Thiagarajan [197]). The damage theory proposed by Voyiadjis and Park [176] for monotonic loading is modified and extended by Voyiadjis and Thiagarajan [197] to incorporate damage combined together with the plasticity behavior under cyclic loading situations. Two different

Table 14.1, Parameter values for subsequent yield surfaces in σ_{11} - σ_{12} space

σ_{11}^{Y} (MPa)	k_1	σ_{11}^{Y} (MPa)	k_4
100.0	$\dfrac{1}{47.14}$	19.3	$\dfrac{1}{18.19}$
100.0	$\dfrac{1}{47.14}$	19.3	$\dfrac{1}{18.19}$
84.0	$\dfrac{1}{39.60}$	17.24	$\dfrac{1}{16.11}$
70.5	$\dfrac{1}{33.23}$	12.76	$\dfrac{1}{11.33}$
103.0	$\dfrac{1}{48.6}$	20.0	$\dfrac{1}{18.98}$
95.3	$\dfrac{1}{44.92}$	19.31	$\dfrac{1}{18.35}$
87.7	$\dfrac{1}{41.34}$	17.24	$\dfrac{1}{16.01}$

$$k_2 = \frac{1}{21.09}$$

Table 14.2, Parameter values for subsequent yield surfaces in σ_{22} - σ_{12} space

σ_{221}^{Y} (MPa)	k_2	σ_{12}^{Y} (MPa)	k_4
27.63	$\dfrac{1}{13.02}$	13.65	$\dfrac{1}{13.0}$
25.94	$\dfrac{1}{12.23}$	13.65	$\dfrac{1}{13.14}$
23.08	$\dfrac{1}{10.88}$	11.72	$\dfrac{1}{10.93}$
20.2	$\dfrac{1}{9.52}$	10.16	$\dfrac{1}{9.27}$
18.46	$\dfrac{1}{8.7}$	9.25	$\dfrac{1}{8.35}$
26.83	$\dfrac{1}{12.65}$	13.00	$\dfrac{1}{12.21}$
27.69	$\dfrac{1}{12.77}$	13.25	$\dfrac{1}{12.44}$

$$k_1 = \frac{1}{36.15}$$

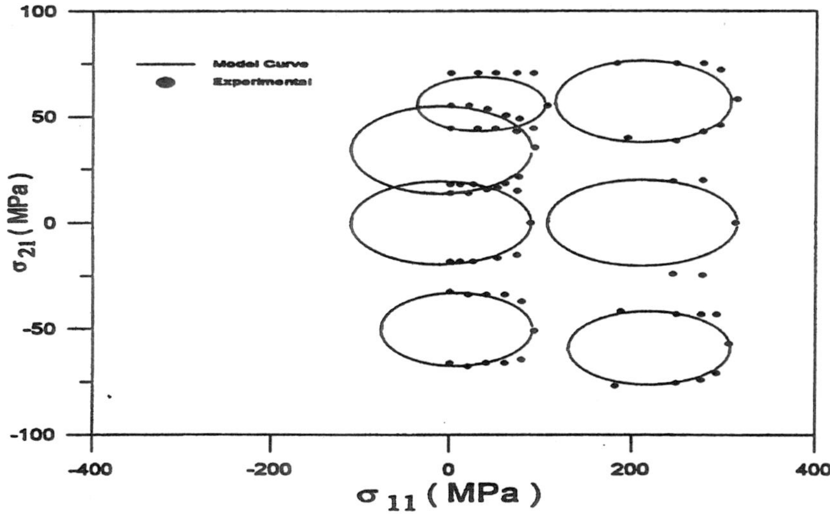

Fig. 14.4, Comparison of Subsequent Yield Surface in σ_{11} - σ_{12} space

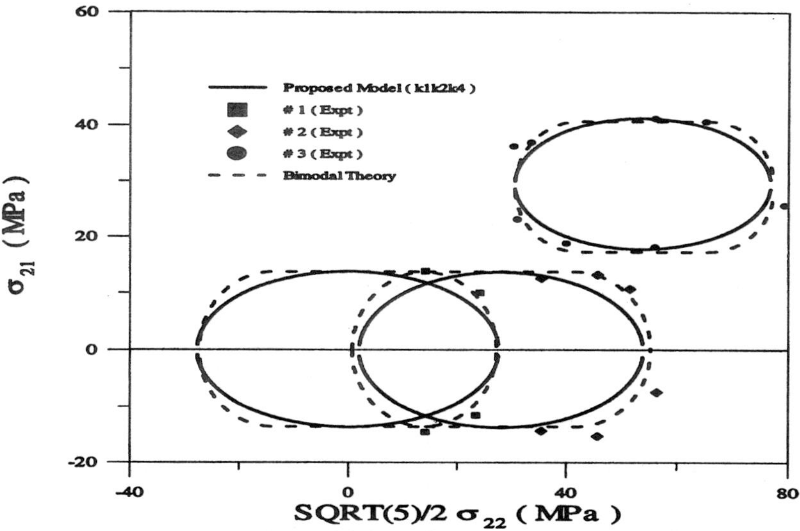

Fig. 14.5, Comparison of Subsequent Yield Surface in σ_{22} - σ_{12} space

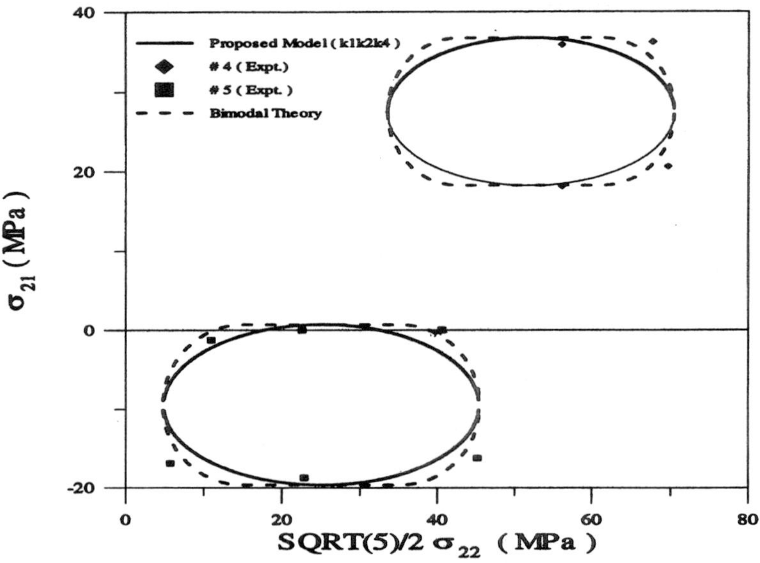

Fig. 14.6, Comparison of Subsequent Yield Surface in σ_{11} - σ_{12} space

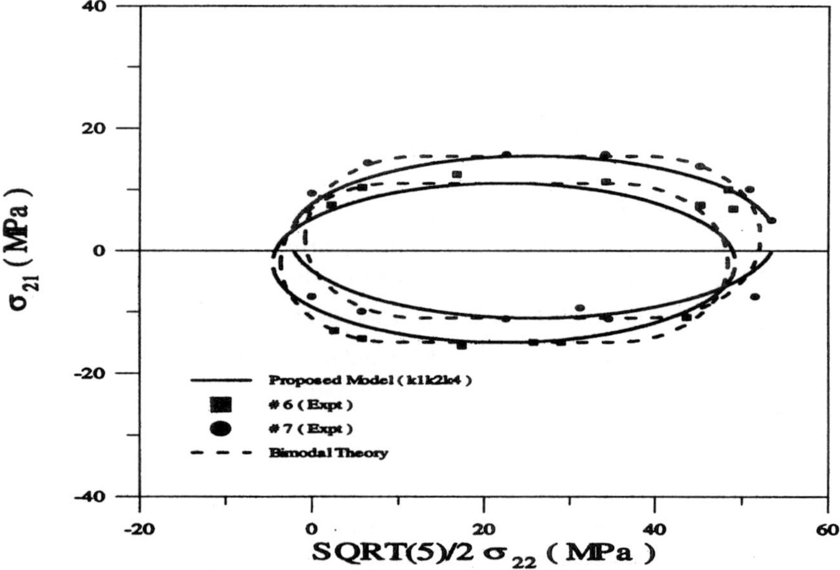

Fig. 14.7, Comparison of Subsequent Yield Surface in σ_{22} - σ_{12} space

approaches to model the damage behavior are presented here (Voyiadjis and Thiagarajan [197]). The first approach is a 'continuum-damage' model and the other is a 'micromechanical-damage' model. The 'continuum damage' model treats the composite as a separate combination of an 'in-situ' matrix and fibers. The matrix is termed 'in-situ' because it is assumed to behave differently in the presence of stiff fibers as compared to its behavior if it were present in a homogeneous medium. The effective stress concept of Kachanov[1] is used in a generalized form here to quantify the damage in the material. Overall damage is characterized through a fourth order tensor M (Voyiadjis and Venson [143], Voyiadjis and Kattan [138]).

The damage criterion of Voyiadjis and Park [176] is modified to account for damage under cyclic loading situations, by the incorporation of a term in the thermodynamic force Y space. The criterion itself is modified and simplified in its usage. A return criterion is also incorporated during the computation of the evolution of the damage variable ϕ to keep Y on the damage surface.

In general the metal matrix composite is assumed to consist of an elasto-plastic matrix with continuous aligned uni-directional elastic fibers. The composite system is restricted to small deformations with small strains. Two different approaches to model the cyclic damage behavior is presented here (Voyiadjis and Thiagarajan [196,197]). In both approaches the effective configuration is defined as a fictitious state with all damage removed, and the damaged configuration is the actual state of the material. The damaged configuration is termed as C whereas the fictitious undamaged configuration is termed as $\overline{C}$.

In the first approach the MMC is modeled using a 'Continuum damage' model, wherein the MMC is treated as a continuum. The elasto-plastic behavior of the continuum is modeled using the anisotropic cyclic plasticity model (Voyiadjis and Thiagarajan [197]), applied to the effective continuum material. The damage transformation of this fictitious undamaged continuum to the damaged configuration is then obtained using the proposed cyclic damage model. Figure 14.8 shows the schematic diagram of the states involved in this development.

In the second approach the MMC is treated as a micromechanical combination of an 'in-situ' plastic matrix and stiff elastic fibers. It is assumed that the in-situ behavior of the matrix material in the presence of the dense fibers is different from what it would be in the absence of fibers. Here only the in -situ plasticity behavior of the matrix is characterized by the continuum cyclic-plasticity-composite model proposed (Voyiadjis and Thiagarajan [196, 197]). The sub configurations $\overline{C}$ of the matrix and fibers are denoted by $\overline{C}^m$ and $\overline{C}^f$ respectively. All quantities based on the fictitious configuration $\overline{C}$ are denoted by a superposed bar and the fiber and matrix related quantities are denoted by a superscript m or f, respectively. Figure 14.9 shows the schematic development for this model.

In the effective undamaged configuration the effective Cauchy stress rate $\dot{\overline{\sigma}}$ is related to the local effective Cauchy stress rates $\dot{\overline{\sigma}}^m$ and $\dot{\overline{\sigma}}^f$ of the matrix and fiber respectively by making use of the micromechanical model proposed by Dvorak and Bahei-El-Din [70] such that (also indicated

in Chapter 8)

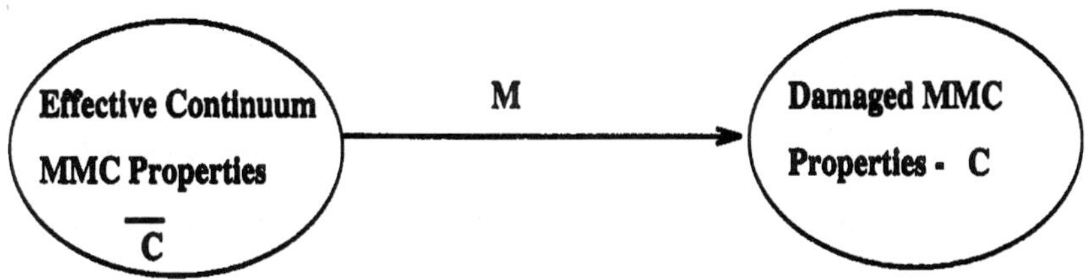

Fig. 14.8, Schematic Diagram of Continuum Model

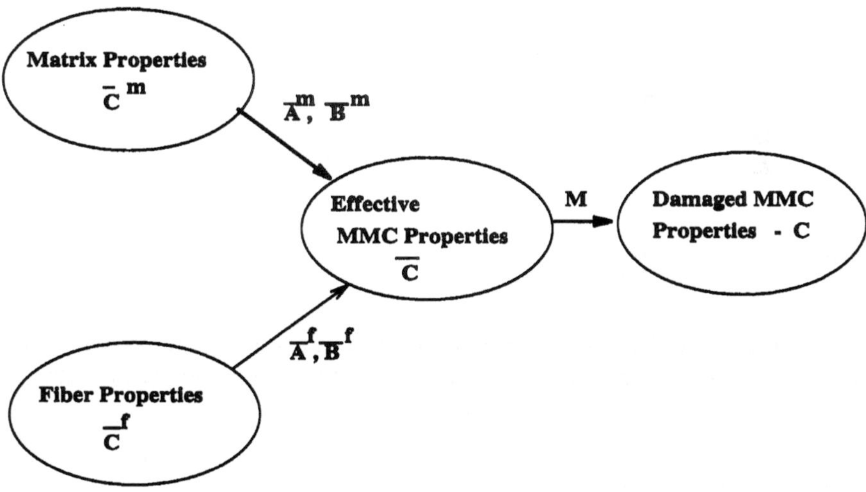

Fig. 14.9, Schematic Diagram of Micromechanical Damage Model

380

$$\dot{\bar{\sigma}}_{ij} = \bar{c}^m \dot{\bar{\sigma}}_{ij}^m + \bar{c}^f \dot{\bar{\sigma}}_{ij}^f \tag{14.64}$$

The superposed dot indicates material time differentiation and $\bar{c}^m$ and $\bar{c}^f$ are the matrix and fiber volume fractions, respectively. These volume fractions are assumed to remain same in both configurations. The equations shown below represent the local-overall relations that are used to transform the sub configurations $\bar{c}^m$ and $\bar{c}^f$ into the $\bar{c}$ effective configuration.

This is accomplished through the fourth order stress concentration tensors $\bar{B}^m$ and $\bar{B}^f$ of the matrix and fibers, respectively. The corresponding effective matrix Cauchy stress rate, $\dot{\bar{\sigma}}^m$ and fiber Cauchy stress rate, $\dot{\bar{\sigma}}^f$ are obtained from the following expressions

$$\dot{\bar{\sigma}}_{ij}^m = \bar{B}_{ijkl}^m \dot{\bar{\sigma}}_{kl} \tag{14.65}$$

$$\dot{\bar{\sigma}}_{ij}^f = \bar{B}_{ijkl}^f \dot{\bar{\sigma}}_{kl} \tag{14.66}$$

Substituting equations (14.64) and (14.65) into equation (14.63) one obtains the following constraint relation between the effective stress concentration tensors.

$$\bar{c}^m \bar{B}_{ijkl}^m + \bar{c}^f \bar{B}_{ijkl}^f = \frac{1}{2}(\delta_{ik}\delta_{jl} + \delta_{il}\delta_{jk}) \tag{14.67}$$

where δ_{ij} are the components of the Kronecker delta. A similar relation of strains is postulated such that,

$$\dot{\bar{\varepsilon}}_{ij} = \bar{c}^m \dot{\bar{\varepsilon}}_{ij}^m + \bar{c}^f \dot{\bar{\varepsilon}}_{ij}^f \tag{14.68}$$

where $\bar{\varepsilon}^m$ and $\bar{\varepsilon}^f$ are the effective matrix and fiber strain tensors respectively and $\bar{\varepsilon}$ is the effective overall strain tensor. The additive decomposition of the elasto-plastic matrix and overall strain rates is assumed in $\bar{c}^m$ and $\bar{c}$ respectively as,

$$\dot{\bar{\varepsilon}}_{ij} = \dot{\bar{\varepsilon}}_{ij}' + \dot{\bar{\varepsilon}}_{ij}'' \tag{14.68}$$

$$\dot{\bar{\varepsilon}}_{ij}^m = \dot{\bar{\varepsilon}}_{ij}^{m'} + \dot{\bar{\varepsilon}}_{ij}^{m''} \tag{14.70}$$

and since the fibers are only undergoing elastic behavior one obtains,

$$\dot{\bar{\varepsilon}}_{ij}^{f} = \dot{\bar{\varepsilon}}_{ij}^{f'} + \dot{\bar{\varepsilon}}_{ij}^{f''} \tag{14.71}$$

where $'$ indicates the elastic part and $''$ indicates the plastic part of the tensor. The local-overall relations for the effective strain rate tensors are given by the following relations

$$\dot{\bar{\varepsilon}}_{ij}^{m} = \bar{A}_{ijkl}^{m}\dot{\bar{\varepsilon}}_{kl} \tag{14.72}$$

$$\dot{\bar{\varepsilon}}_{ij}^{f} = \bar{A}_{ijkl}^{f}\dot{\bar{\varepsilon}}_{kl} \tag{14.73}$$

where $\bar{A}^{m}$ and $\bar{A}^{f}$ are fourth order effective strain concentration tensors for the matrix and fibers respectively. The elastic concentration factors are obtained using the Mori-Tanaka method (Weng, [198]), and plastic concentration factors are obtained by using the numerical method by Lagoudas et al. [108]. In the determination of the plastic strain concentration tensors the instantaneous elasto-plastic modulus of the in-situ matrix is used as the tangent modulus instead of the elastic modulus.

14.5 Overall Effective Elasto-Plastic Stiffness Tensor : Micromechanical Model

The constitutive relations for both the micromechanical and continuum damage models are presented here. The details of the derivation of the tensor in the effective (undamaged) configuration is outlined first in this section. In the micromechanical model the individual properties of the two different materials are considered separately for defining the constitutive equation, and the overall properties are then derived from these using a homogenization procedure (Voyiadjis and Thiagarajan [196]).

Effective Local Elastic Stress-Strain Relations for Fiber

The fibers are assumed to be elastic and isotropic with the elastic stiffness given by the following expression

$$\bar{E}_{ijkl}^{f} = \bar{\lambda}^{f}\delta_{ij}\delta_{kl} + \bar{\mu}^{f}(\delta_{ik}\delta_{jl} + \delta_{il}\delta_{jk}) \tag{14.74}$$

where $\bar{\lambda}^{f}$ and $\bar{\mu}^{f}$ are Lame's constants for the fiber. The incremental constitutive equation for the fiber is given by the following relation.

$$\dot{\bar{\sigma}}_{ij}^{f} = \bar{E}_{ijkl}^{f}\dot{\bar{\varepsilon}}_{kl}^{f} \tag{14.75}$$

Effective Local Elastic Stress-Strain Relations for In-Situ Matrix

Similarly for the case of the elastic behavior of the matrix, one obtains

$$\dot{\overline{\sigma}}_{ij}^{m} = \overline{E}_{ijkl}^{m}\,\dot{\overline{\varepsilon}}_{kl}^{m} \tag{14.76}$$

where $\overline{E}^{m}$ is the effective elastic stiffness of the matrix.

Then in-situ behavior of the matrix is obtained from the cyclic anisotropic plasticity model for metal matrix composites derived by Voyiadjis and Thiagarajan [197]. The matrix is a ductile material and due to the presence of fibers is constrained from yielding in the direction of the fibers and consequently a non-associated flow rule is introduced. Therefore the continuum elasto-plastic behavior of the composite is in fact the in-situ behavior of the matrix since the fibers are mainly elastic in their behavior.

The elastic behavior of the material, treated as a homogeneous continuum with transversely isotropic properties has been defined by Walpole [201]. The linear constitutive relation is given as follows

$$\overline{E}_{ijkl} = Kt_{ij}t_{kl} + El_{ij}l_{kl} + 2m' E_{ijkl}^{3} + 2pE_{ijkl}^{4} \tag{14.77}$$

where K is the plane-strain bulk modulus, m' is the transverse shear modulus, p is the axial shear modulus and E and v are Young's modulus and Poisson's ratio respectively, when the material is loaded in the fiber direction. For a transversely isotropic material the plane-strain bulk modulus can be defined in terms of the other four elastic constants. The tensors used in equation (14.77) are defined as follows.

$$t_{ij} = m_{ij} + 2vl_{ij} \tag{14.78}$$

$$l_{ij} = \eta_{i}\eta_{j} \tag{14.79}$$

$$m_{ij} = \delta_{ij} - \eta_{i}\eta_{j} \tag{14.80}$$

$$E_{ijkl}^{3} = \frac{1}{2}[m_{ik}m_{jl} + m_{jk}m_{il} - m_{ij}m_{kl}] \tag{14.81}$$

$$E_{ijkl}^{4} = \frac{1}{2}[m_{ik}l_{jl} + m_{il}l_{jk} + m_{jl}l_{ik} + m_{jk}l_{il}] \tag{14.82}$$

where η_{i} are the direction cosines of the fibers in the global coordinate system. Making use of the relations (14.65),(14.72),(14.75) and (14.76) the effective elastic stiffness of the matrix can be derived as follows,

$$\overline{E}^{m} = \frac{1}{\overline{c}^{m}}(\,\overline{E}{:}\overline{A}^{-m} - \overline{c}^{f}\,\overline{E}^{f}{:}\overline{A}^{f}{:}\overline{A}^{-m}\,) \tag{14.83}$$

In equation (14.83) $\overline{E}^f\overline{A}^f$ implied $\overline{E}_{ijkl}A^f_{klmn}$ and $\overline{A}^{-m}$ is the inverse of the tensor $\overline{A}^m$ such that

$$\overline{A}^m : \overline{A}^{-m} = I_4 \tag{14.84}$$

I_4 and is the fourth order identity tensor, given by the following expression

$$I_4 = \frac{1}{2}(\delta_{ik}\delta_{jl} + \delta_{il}\delta_{jk}) \tag{14.85}$$

Effective Local Elasto-Plastic Stiffness for In-Situ Matrix

The increment of stress in the matrix is computed from the total stress increment using the effective elasto-plastic stress concentration tensor. The effective undamaged local elasto-plastic constitutive relations for the matrix is given by the following expression

$$\dot{\overline{\sigma}}^m_{ij} = \overline{D}^m_{ijkl}\dot{\overline{\varepsilon}}^m_{kl} \tag{14.86}$$

where $\overline{D}^m$ is the in-situ effective elasto-plastic stiffness tensor of the matrix.

The in-situ effective elasto-plastic constitutive model for the matrix is based on anisotropic yield function such that (similar to expression (14.60)).

$$\overline{F}^m \equiv (\overline{\sigma}^{m'}_{ij} - \overline{\alpha}^{m'}_{ij})\overline{H}^m_{ijkl}(\overline{\sigma}^{m'}_{kl} - \overline{\alpha}^{m'}_{kl}) - 1 = 0 \tag{14.87}$$

where $\overline{H}^m$ is the fourth order anisotropy yield tensor for the matrix and is expressed by equation (14.8).

The yield equation (14.87) can be expressed in the global axes of reference through the following transformation equations

$$\overline{\sigma}^{m'}_{ij} = d_{ip}\overline{\sigma}^m_{pq}d_{qj} \tag{14.88}$$

and

$$\overline{\alpha}^{m'}_{ij} = d_{ip}\overline{\alpha}^m_{pq}d_{qj} \tag{14.89}$$

where d_{ij} are the coefficients of the orthogonal transformation matrix. Assuming that the fibers are aligned along the x-axis (1-direction) one can write,

$$d_{1j} = (\eta_1, \eta_2, \eta_3) \tag{14.90}$$

where η_i, (i = 1, 2, 3) are the direction consines of the fiber in the global coordinate system.

The evolution equation for the back stress $\overline{\alpha}_{ij}^m$ is based on the Phillips rule and can be expressed as follows

$$\dot{\overline{\alpha}}_{ij}^m = \dot{\overline{\mu}}^m \dot{\overline{\sigma}}_{ij}^m \tag{14.91}$$

or

$$\dot{\overline{\alpha}}_{ij}^m = \left\| \dot{\overline{\alpha}}_{ij}^m \right\| l_{ij} \tag{14.92}$$

where,

$$l_{ij} = \dot{\overline{\sigma}}_{ij}^m / \left\| \dot{\overline{\sigma}}^m \right\| \tag{14.93}$$

and $\dot{\overline{\mu}}^m$ is obtained from the consistency condition. The norm of $\dot{\alpha}$ is found from the consistency condition as follows

$$\frac{\partial \overline{F}^m}{\partial \overline{\sigma}_{ij}^m} \dot{\overline{\sigma}}_{ij}^m + \frac{\partial \overline{F}^m}{\partial \overline{\alpha}_{ij}^m} \dot{\overline{\alpha}}_{ij}^m = 0 \tag{14.94}$$

A non-associated flow rule is used for the matrix in the undamaged state, in this work such that (Voyiadjis and Thiagarajan [188])

$$\dot{\overline{\varepsilon}}_{ij}'' = \dot{\Lambda} \frac{\partial \overline{G}^m}{\partial \overline{\sigma}_{ij}^m} \tag{14.95}$$

where $\overline{G}^m$ is the plastic potential function for the definition of plastic strains in the matrix such that

$$\overline{G}^m = \omega \, \overline{F}^m + (1.0 - \omega) \overline{g}^m, \qquad 0 \leq \omega \leq 1.0 \tag{14.96}$$

In equation (14.95) the function $\overline{g}^m$ is defined using the fourth order anisotropic tensor $\overline{H}$ and a constrained stress term $\overline{r}_{ij}^m$ such that,

$$\overline{g}^m = \overline{r}_{ij}^m \overline{N}_{ijkl}^m \overline{r}_{kl}^m - 1 \tag{14.97}$$

The constraint that is introduced in the stress tensor is that the plastic strain increment is independent of the component of stress along a specified direction (defined by η_i). Using the procedure outlined by Spencer [200] the constraint is incorporated into the stress term such that

$$\bar{r}_{ij}^{m} = \bar{\sigma}_{ij}^{m} - T\eta_i \eta_j \tag{14.98}$$

where $T\,\eta \otimes \eta$ is the reaction to an inextensibility constraint along the direction of η_i of the fibers such that

$$T = \bar{\sigma}_{rs}^{m} \eta_r \eta_s \tag{14.99}$$

The non-associativity of the flow rule is built into the definition of the potential function through the factor ω. Based on the flow rule the second order tensor representing the direction of plastic strain is given by

$$\bar{n}_{ij}^{m} = \omega\, \bar{n}_{ij}^{mF} + (1.0 - \omega\,)\bar{n}_{ij}^{mg} \tag{14.100}$$

where,

$$\bar{n}_{ij}^{mF} = \frac{\partial \bar{F}^{m}}{\partial \bar{\sigma}_{ij}^{m}} \bigg/ \left\| \frac{\partial \bar{F}^{m}}{\partial \bar{\sigma}_{rs}^{m}} \right\| \tag{14.101}$$

$$\bar{n}_{ij}^{mg} = \frac{\partial \bar{g}^{m}}{\partial \bar{\sigma}_{ij}^{m}} \bigg/ \left\| \frac{\partial \bar{g}^{m}}{\partial \bar{\sigma}_{rs}^{m}} \right\| \tag{14.102}$$

The experimental work of Dvorak et al. [185] and Nigam et al.[186], indicate that the plastic strains are predominantly along the shear direction. This is incorporated into the flow rule by using a value of ω between 0 and 1. A value of $\omega = 1$ gives a purely associative rule whereas a value $\omega = 0$ results in the use of the function g^{m} in equation (14.96). The parameter ω here is treated as a constant, however, a more elaborate expression for this term will be presented by the authors in the future in terms of the direction of fibers and the loading directions.

The effective matrix elasto-plastic stiffness for this cyclic anisotropic-plasticity model is given by (Voyiadjis and Thiagarajan [197])

$$\bar{D}^{m} = \bar{E}^{m} - \frac{(\bar{E}^{m}:\bar{n}^{m})(\bar{n}^{m}:\bar{E}^{m})}{H + (\bar{n}^{m}:\bar{E}^{m}:\bar{n}^{m})} \tag{14.103}$$

where H is the plastic modulus based on the bounding surface model derived by Voyiadjis and Thiagarajan [196].   Substituting for $\dot{\bar{\sigma}}_{ij}^{m}$ and $\dot{\bar{\sigma}}_{ij}^{f}$ from equations (14.86) and (14.75) respectively into equation (14.64), one obtains the following relation

$$\dot{\bar{\sigma}}_{ij} = \bar{D}_{ijkl}\dot{\bar{\varepsilon}}_{kl} \tag{14.104}$$

where

$$\overline{D} = \overline{c}^{\,m}\,\overline{D}^{\,m} : \overline{A}^{\,m} + \overline{c}^{\,f}\,\overline{E}^{\,f} : \overline{A}^{\,f} \qquad\qquad (14.105)$$

is the elasto-plastic stiffness of the composite in the effective undamaged configuration.

14.6 Overall Effective Elasto-Plastic Stiffness Tensor : Continuum-Damage Model

For the continuum-damage model the effective undamaged elasto-plastic relationship is given by the stiffness generated by the anisotropic cyclic plasticity model of Voyiadjis and Thiagarajan [197]. $\overline{D}$ is the effective undamaged elasto-plasto stiffness. This is based on equation (14.102) above, with the only difference being that it is now applied to the overall continuum material instead of the in-situ matrix, as shown below

$$\overline{D} = \overline{E} - \frac{(\overline{E}:\overline{n})(\overline{n}:\overline{E})}{H + (\overline{n}:\overline{E}:\overline{n})} \qquad\qquad (14..106)$$

14.7 Damage

Damage is characterized in the overall composite system as a whole effective continuum. The derivation of the stiffness of the effective continuum in two approaches (continuum versus micromechanical) has been described earlier. The equations of continuum damage mechanics are then applied to the overall configruation $\overline{C}$ in order to obtain the damaged quantities in the overall configuration C. The resulting model reflects various types of damage mechanisms such as void growth and coalescence in the matrix, fiber fracture, debonding, delamination, etc. Here, however, all the damage will be reflected through only one damage variable.

The damage criterion is given in terms of the tensorial damage hardening parameter h and the generalized termodynamic force Y conjugate to the damage tensor ϕ and a term γ which is defined in the thermodynamic force space such that

$$g \equiv (Y_{ij} - \gamma_{ij})P_{ijkl}(Y_{kl} - \gamma_{kl}) - 1 = 0 \qquad\qquad (14.107)$$

The fourth order tensor P is expressed in terms of the second order tensors h such that

$$P_{ijkl} = h_{ij}^{-1} h_{kl}^{-1} \qquad\qquad (14.108)$$

A new and simplified form of the tensor h is given in terms of the second order tensors u, V and ϕ as follows

$$h_{ij} = u_{ij} + V_{ij} \qquad\qquad (14.109)$$

The tensors u and V were originally proposed by Stumvoll and Swoboda [140] as scalars. The tensors are given as follows

$$u = \begin{bmatrix} \lambda_1 q(\dfrac{\kappa}{\lambda_1})^r & 0 & 0 \\ 0 & \lambda_2 q(\dfrac{\kappa}{\lambda_2})^r & 0 \\ 0 & 0 & \lambda_3 q(\dfrac{\kappa}{\lambda_3})^r \end{bmatrix} \tag{14.110}$$

and

$$V = \begin{bmatrix} \lambda_1 v_1^2 & 0 & 0 \\ 0 & \lambda_2 v_2^2 & 0 \\ 0 & 0 & \lambda_3 v_3^2 \end{bmatrix} \tag{14.111}$$

The material parameters λ_1, λ_2, and λ_3, are Lame's constants for anisotropic materials and are related to the elasticity tensor E for an orthotropic material expressed by the 6x6 matrix shown by Voyiadjis and Park [176]. The material parameters v_1, v_2 and v_3 define the initial threshold against damage for the orthotropic material. These are obtained from the onset of damage at a stress level. The scalar damage hardening parameter κ is given by

$$\kappa = \int_0^t (-Y_{ij} \dot{\phi}_{ij}) dt \tag{14.112}$$

Finally the material parameters r and q are obtained by comparing the theory with experimental results.

A new term γ has been introduced here in the definition of the damage criterion g in equation (14.107). This term is analogous to the backstress term in the stress-space yield criterion. It represents the translation of the damage surface as loading progress akin to kinematic hardening.

The evolution of the term γ in the anisotropic damage criterion equation is needed in order to account for the motion of the damage surface in the Y space. This is dependent on the evolution of damage itself. Hence it can be expressed mathematically as follows,

$$\dot{\gamma}_{ij} = c\dot{\phi}_{ij} \tag{14.113}$$

Since Y is negative γ too has to be negative. It has been found that it is suitable to adopt a value of -1 for the value of c.

As outlined in Chapter (10) on can similarly obtain the evolution expression for damage, such that (Voyiadjis and Thiagarajan, [187])

388

$$\dot{\phi}_{ij} = \Psi'_{ijkl} \dot{Y}_{kl} \tag{14.114}$$

where

$$\Psi_{ijkl} = -\frac{\dfrac{\partial g}{\partial Y} \otimes \dfrac{\partial g}{\partial Y}}{\dfrac{\partial g}{\partial \phi} : \dfrac{\partial g}{\partial Y} - \dfrac{\partial g}{\partial \kappa} Y : \dfrac{\partial g}{\partial Y} - c \dfrac{\partial g}{\partial Y} : \dfrac{\partial g}{\partial Y}} \tag{14.115}$$

The generalized thermodynamic free energy Y is assumed to be a function of the elastic-component of the strain tensor ε' and the damage tensor ϕ, or the stress σ and ϕ

$$Y = Y(\overline{\sigma}, \overline{\phi}) \tag{14.116}$$

Making use of the evolution equation for Y

$$\dot{Y}_{ij} = \frac{\partial Y_{ij}}{\partial \sigma_{mn}} \dot{\sigma}_{mn} + \frac{\partial Y_{ij}}{\partial \phi_{kl}} \dot{\phi}_{kl} \tag{14.117}$$

One obtains the evolution expression for the damage ϕ such that (Voyiadjis and Park [176])

$$\dot{\phi}_{kl} = [L^{-l}_{ijkl} \Psi_{ijrs} \frac{\partial Y_{rs}}{\partial \sigma_{mn}}] \dot{\sigma}_{mn} \tag{14.118}$$

or

$$\dot{\phi}_{kl} = T_{klmn} \dot{\sigma}_{mn} \tag{14.119}$$

where

$$L_{ijkl} = \frac{1}{2} (\delta_{ik} \delta_{jl} + \delta_{il} \delta_{jk}) - \Psi_{ijrs} \frac{\partial Y_{mn}}{\partial \phi_{kl}} \tag{14.120}$$

The thermodynamic force associated with damage is obtained using the enthalpy of the damage material, V,

$$V(\overline{\sigma}, \overline{\phi}) = \sigma_{ij} \varepsilon'_{ij} - W \tag{14.121}$$

or

$$V = \frac{1}{2}\sigma_{mn}E^{-1}_{mnkl}(\phi)\sigma_{kl} \tag{14.122}$$

where W is the specific energy and E is the damaged elasticity tensor. The thermodynamic force Y is given by

$$Y_{ij} = -\frac{\partial V}{\partial \phi_{ij}} \tag{14.123}$$

or

$$Y_{ij} = -\frac{\partial V}{\partial M_{abcd}}\frac{\partial M_{abcd}}{\partial \phi_{ij}} \tag{14.124}$$

Making used of the energy equivalence principle, one obtains a relation between the damaged elasticity tensor E and the effective undamaged elasticity tensor $\overline{E}$ such that (Voyiadjis and Kattan [138])

$$E^{-1}_{mnkl}(\phi) = M_{uvmn}(\phi)\overline{E}^{-1}_{uvpq}M_{pqkl}(\phi) \tag{14.125}$$

Through the use of equation (14.122) and (14.124) the thermodynamic force is given explicitly as follows

$$Y_{ij} = -\frac{1}{2}(\sigma_{cd}\overline{E}^{-1}_{abpq}M_{pqkl}\sigma_{kl} + \sigma_{rs}M_{uvrs}\overline{E}^{-1}_{uvab}\sigma_{cd})\frac{\partial M_{abcd}}{\partial \phi_{ij}} \tag{14.126}$$

The stiffness tensor D for the damaged material is now derived for isothermal conditions and in the absence of rate dependent effects. Making use of the incremental form one obtains

$$\dot{\overline{\sigma}}_{ij} = \dot{\overline{M}}_{ijkl}\sigma_{kl} + M_{ijkl}\dot{\sigma}_{kl} \tag{14.127}$$

Through the additive decomposition of the effective strain rate one obtains

$$\dot{\overline{\varepsilon}}_{ij} = \dot{M}^{-1}_{ijkl}\varepsilon^{-1}_{kl} + M^{-1}_{ijkl}\dot{\varepsilon}_{kl} \tag{14.128}$$

The rates of the damage effect tensor and its inverse maybe expressed ad follows

$$\dot{M}_{ijkl} = \frac{\partial M_{ijkl}}{\partial \phi_{pq}}T_{pqmn}\dot{\sigma}_{mn} \tag{14.129}$$

$$= Q_{ijklmn}\dot{\sigma}_{mn} \tag{14.130}$$

390

and

$$\dot{M}_{ijkl}^{-1} = \frac{\partial M_{ijkl}^{-1}}{\partial \phi_{pq}} T_{pqmn} \dot{\sigma}_{mn} \tag{14.131}$$

$$= R_{ijklmn} \dot{\sigma}_{mn} \tag{14.132}$$

The resulting elastoplastic stiffness relation in the damaged configuration is obtained as follows:

$$\dot{\sigma}_{ij} = D_{ijkl} \dot{\varepsilon}_{kl} \tag{14.133}$$

where

$$D = O^{-1} : D : M^{-1} \tag{14.134}$$

and

$$O_{ijkl} = Q_{ijmnkl} \sigma_{mn} + M_{ijkl} - \overline{D}_{ijmn} R_{mnpqkl} E_{pqab}^{-1} \sigma_{ab} \tag{14.135}$$

In the case of no damage, both tensors Q and R reduce to zero and M becomes a fourth order identity tensor.

14.8 Numerical Simulation and Discussions

The above described models have been implemented in a computer program to generate the stress-strain curves. In the development of the damage elasto-plastic stiffness matrix the effective undamaged stiffness matrix is computed. The damage criterion is then checked for and the damage transformation is then applied to this matrix.

The computation of the inverse of the fourth order damage tensor M and its first derivative with respect to the second-order damage tensor ϕ, which results in a sixth order tensor was the biggest challenge in these computations. The symbolic programming language MAPLEV was used to compute symbolically the individual terms of the inverse tensor. It was further programmed to generate the terms of the sixth order derivative tensor in term by term fashion. MAPLEV was also used to generate automatically a FORTRAN code for these tensors.

Another important numerical consideration is the overshooting of the damage surface in the theormodynamic force Y space. Hence a return criterion has been adopted to prevent this. This is done using a first-order technique as described below, while maintaining the current stress level. Correction is applied to Y and the increment of the damage-effect tensor is recomputed based on the new value of the thermodynamic force. All other parameters that are affected are then recomputed.

From numerical considerations, the starting and subsequent values of Y after the application of the stress increment can be expressed as

$$g(Y_{ij}) < 0 = g_1 \tag{14.136}$$

$$g(Y_{ij} + \Delta Y_{ij}) > 0 = g_2 \tag{14.137}$$

From Taylor's series expansion and retaining only the first order terms it can be shown that

$$\frac{\partial g}{\partial Y_{ij}} = g_2 - g_1 \tag{14.138}$$

Using the correction factor ζ one obtains

$$g(Y_{ij} + \zeta \Delta Y_{ij}) = 0 \tag{14.139}$$

$$g(Y_{ij} + \Delta Y_{ij} + (\zeta - 1)\Delta Y_{ij}) = g_2 + (\zeta - 1)(g_2 - g_1) \tag{14.140}$$

From the above derivation it can be shown that

$$\zeta = - \frac{g_1}{g_2 - g_1} \tag{14.141}$$

It is assumed that the correction is equal in all directions for simplicity and it has been found from the implementation that this is a reasonable assumption.

The physical significance of the tensor V, which is used as a measure of damage initiation in th damage criterion equation, is shown below. To understand it in a simplified form, one considers an isotropic material for which the tensors reduce to the following form

$$V_{ij} = \lambda \, v^2 \delta_{ij} \tag{14.142}$$

$$u_{ij} = \lambda \, q \left(\frac{\kappa}{\lambda} \right)^2 \delta_{ij} \tag{14.143}$$

where λ is the first Lame's constant. At the outset when damage is about to occur, $\kappa = 0$ and thus

$$h_{11} = h_{22} = h_{33} = \lambda \, v^2 \tag{14.144}$$

$$P_{1111} = \frac{1}{h_{11}^2} = \frac{1}{\lambda^2 v^4} \tag{14.145}$$

The other nonzero terms of the fourth order tensor P are $P_{2222} = P_{3333} = P_{1122} = P_{2211} = P_{1133} = P_{3311} = P_{2233} = P_{3322} = P_{1111}$. Also it can be shown that at the initiation of damage

$$Y_{11} = -\frac{\sigma_{11d}^2}{E_{11}^2} \tag{14.146}$$

$Y_{22}=Y_{33}=0$. The other nonzero terms are as follows

$$Y_{21} = Y_{12} = -0.25(\frac{\sigma_{11d}^2}{E_{2111}}) \tag{14.147}$$

$$Y_{13} = Y_{31} = -0.25(\frac{\sigma_{11d}^2}{E_{3111}}) \tag{14.148}$$

$$Y_{23} = Y_{32} = -0.25(\frac{\sigma_{11d}^2}{E_{2311}}) \tag{14.149}$$

Hence the expanded form of the damage criterion equation (14.107) now reduces to,

$$P_{1111}Y_{11}^2 - 1 = 0 \tag{14.150}$$

$$\frac{1}{\lambda^2 v^4}\frac{\sigma_d^4}{E^2} = 1 \tag{14.151}$$

Thus it can be seen that

$$v = \frac{\sigma_d}{\sqrt{(\lambda+2\mu)\lambda}} \tag{14.152}$$

For the given material the elastic modulus is approximately equal to 225,000 MPa and the Poisson ratio is $v = 0.25$. This gives the range of values of v to be used. The tensor u represents the growth of damage in the damage evolution criterion.

Continuum Damage Model Results

The same loading that was studied in an earlier work by the authors [196], and used in the experimental work of Nigam et. al. [186] is also used in this work. The damage parameters found suitable for this material are $q = 1.0$ and $r = 7.0$. This effectively makes it dependent only on one parameter. Figures 14.10, 14.11, 14.12, and 14.13 show the results of this generation for the stress strain comparison in the transverse and shear directions. These curves compare the cyclic plasticity model with that of the damage-plasticity model. It can be seen that the strains predicted by the damage model are higher than that of the pure plasticity case. It can also be seen that during the unloading - reloading situation, when reloading takes place even in the elastic range, the damage criterion is exceeded, and hence the elastic-stiffness is reduced.

This can be clearly seen in the two lines of different inclinations in Figures 14.10 and 14.11.

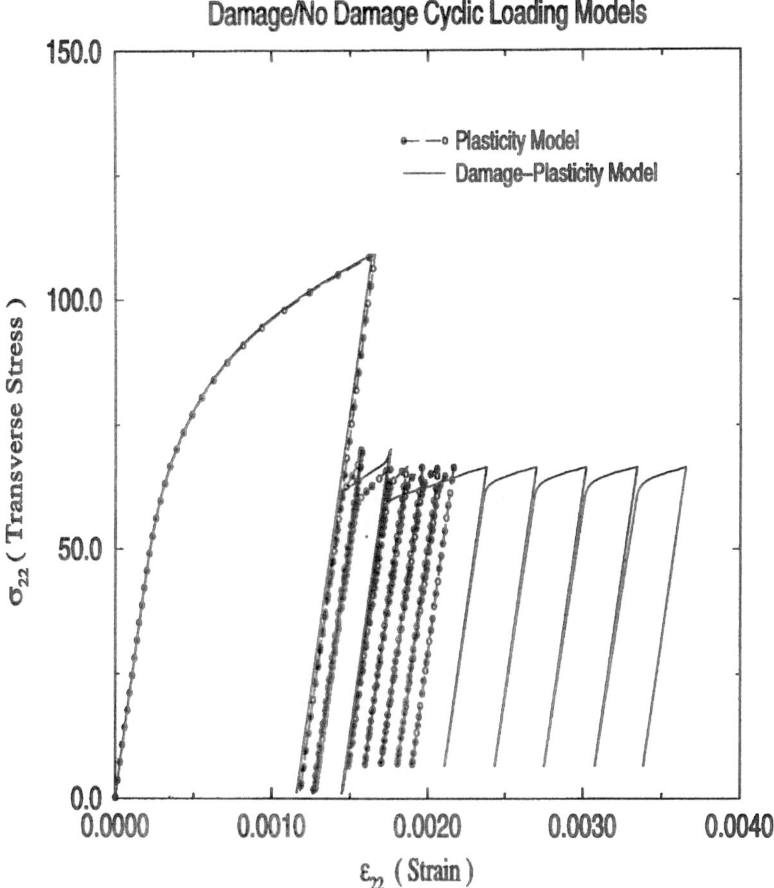

Fig. 14.10, Shear Stress-Strain Curves for Continuum Damage and Pure Plasticity Models

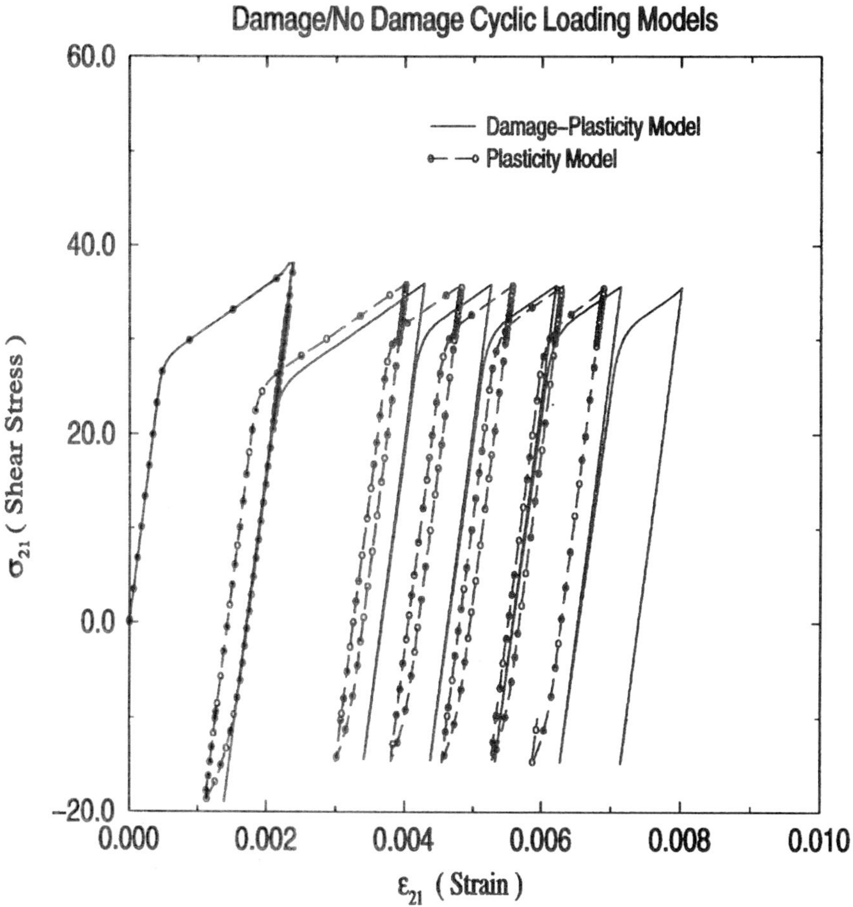

Fig. 14.11, Transverse Stress-Strain Curves for Continuum Damage and Pure Plasticity Models

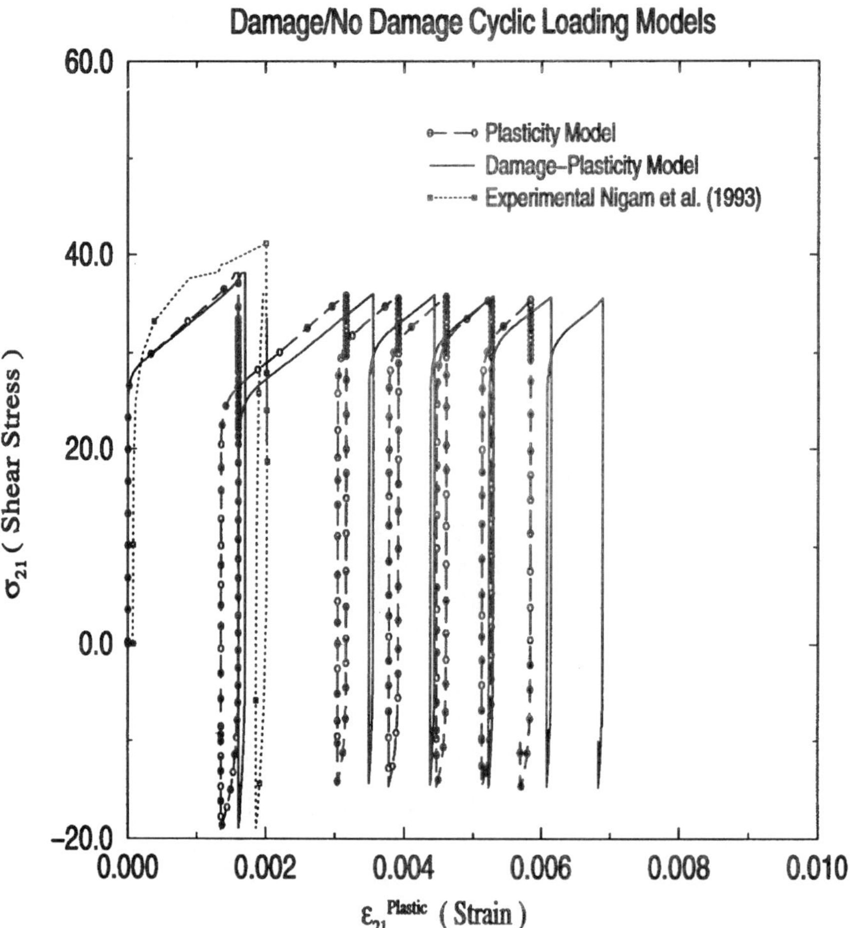

Fig. 14.12, Shear Stress Plastic Strain Curves for Continuum Damage and Pure Plasticity Models

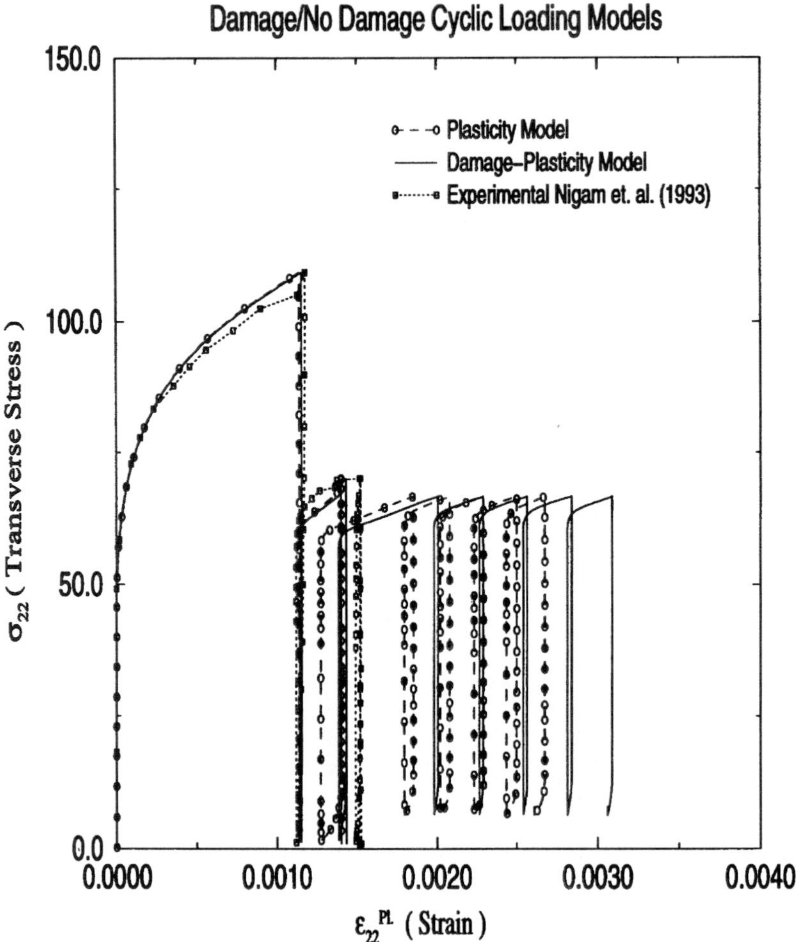

Fig. 14.13, Transverse Stress Plastic Strain Curve for Continuum Damage and Pure Plasticity Models

As seen in Figures 14.12 and 14.13 which depict the stress-plastic strain relationships in the shear and transverse directions respectively, due to successive reduction in the elastic stiffness, the plastic strains are also affected hence resulting in a higher prediction of plastic strain. Although this model assumes a decoupling between the damage and plasticity situations in modeling the behavior, there is an inherent coupling that is present.

Figure 14.14 shows the evolution of the damage parameter ϕ with stresss in the transverse direction under a cyclic loading type situation. One apparent behavior that is observed due to the nature of these curves is that as stress is increased, the same stress increment tends to produce a higher amount of damage. Upon unloading no significant change in damage is observed, and evolution of damage upon reloading takes place at a lower stress level for successive loading cases. Another behavior observed in that under constant load cycling, the amount of damage is higher as the number of times the load applied increases. These behaviors observed are reasonable with what one would expect the material to do in real life.

Micromechanical Damage Model Results

The flow of the program is modified and developed to model the micromechanical model as outlined here. The same loading data is used in order to evaluate and compare results. The major difficulty lies in evaluation of the parameters corresponding to the in-situ matrix properties. The in-situ elastic properties are obtained from the program by Aboudi [199] which can extract the in-situ properties of the matrix given the overall continuum elastic properties and the isotropic fiber properties. However, experiments need to be conducted to evaluate accurately the in-situ properties of the matrix. The observations pertaining to the micromechanical damage model are shown in Figures 14.15 and 14.16. The predictions are comparable to the experimental data. .

The micromechanical damage model gives a more accurate predictions of the reduction in the elastic stiffness during unloading. To improve the prediction of the micromechanical model on needs to measure the in-situ material parameters more accurately. The advantage of the micromechanical model over the continuum model is that it presents a much better picture of the behavior of the individual component and since most MMC's are applied in high temperature regions, this model is better suited for representing the behavior of the material.

Two approaches to model the damage-plasticity behavior in MMC's are presented in this chapter. The results are compared with experimental results. The continuum damage model results give a better prediction at this stage since the elastic properties in the continuum model are determined experimentally. In the micromechanical model the initial elastic stiffness of the in-situ matrix is derived mathematically rather than measured experimentally. Further research should focus on this aspect of better evaluation of in-situ properties of the matrix experimentally.

398

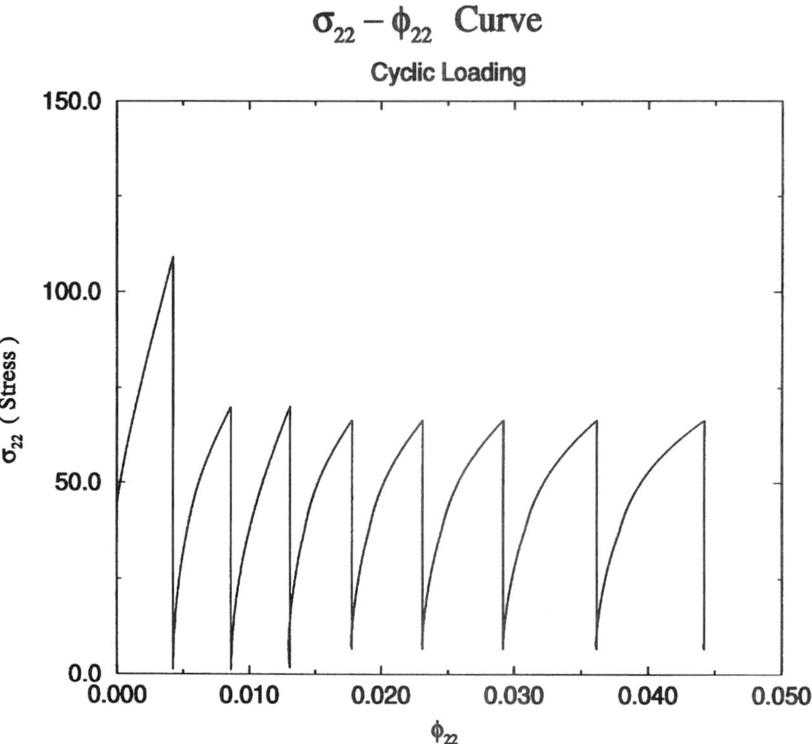

Fig. 14-14, Evolution of Damage ϕ_{22} with Transverse Stress

Transverse Stress – Strain

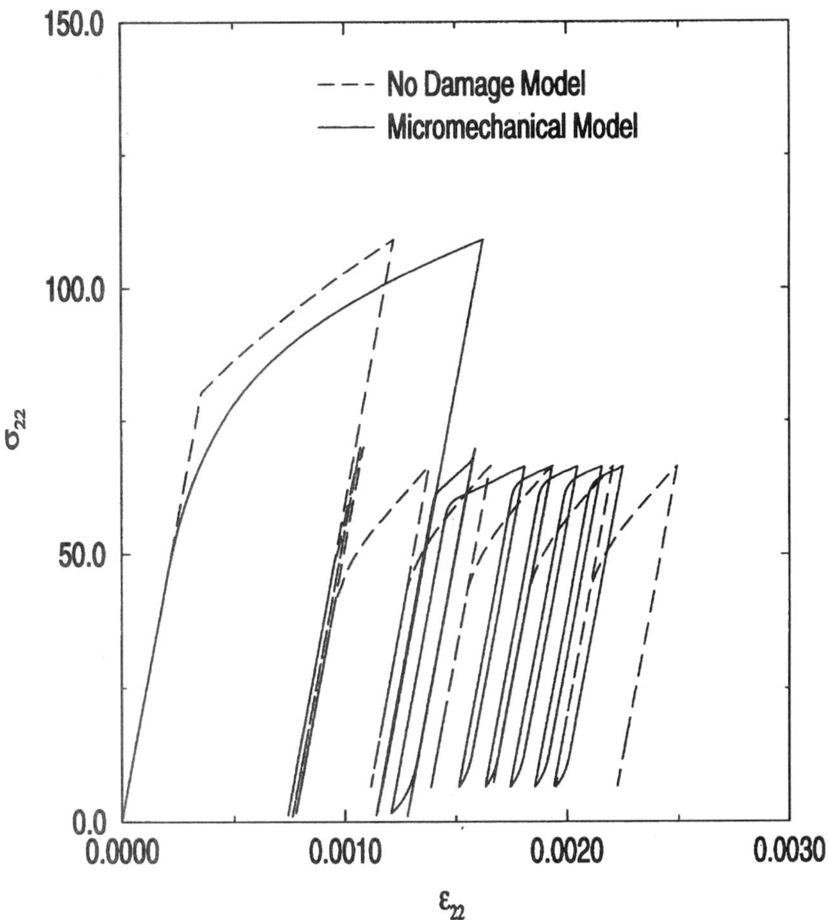

Fig. 14.15, Transverse Stress-Strain Curves for Micromechanical Damage and Pure Plasticity Models

Transverse Stress – Plastic Strain

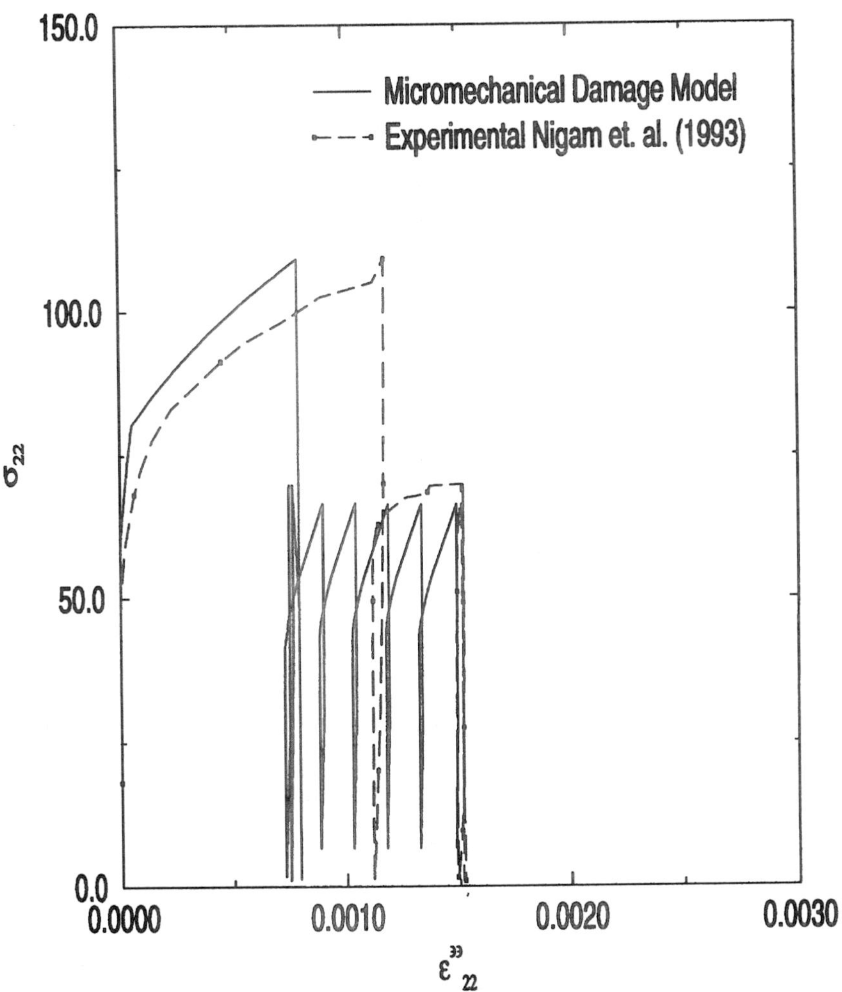

Fig. 14.16, Transverse Stress-Strain Curves for Micromechanical Damage and Pure Plasticity Models

<u>**PART III**</u>

ADVANCED TOPICS IN DAMAGE MECHANICS

CHAPTER 15

DAMAGE IN METAL MATRIX COMPOSITES USING THE GENARALIZED CELLS MODEL

In this Chapter damage is incorporated in the generalized cells model (GCM) (Paley and Aboudi [203]) and applied to metal matrix composites. The local incremental damage model of Voyiadjis and Park [176] is used here in order to account for damage in each subcell separately. The resulting micromechanical analysis establishes elasto-plastic constitutive equations which govern the overall behavior of the damaged composite. The elasto-plastic constitutive model is first derived in the undamaged configuration for each constituent of the metal matrix composite. The plasticity model used here is based on the existence of a yield surface and flow rule. The relations are then transformed for each constituent to the damaged configuration by applying the local incremental constituent damage tensors. The overall damaged quantities are then obtained by applying the local damage concentration factors that are obtained by employing the rate of displacement and traction continuity conditions at the interface between subcells and between neighboring repeating cells in the generalized cells model. Examples are solved numerically in order to explore the physical interpretation of the proposed theory for a unit cell composite element.

15.1 Theoretical Preliminaries

15.1.1 The Generalized Cells Model

The generalized cells model is the generalization of the method of cells (Aboudi [204]) by taking any number of subcells rather than four subcells and considering the rate dependent relations of the subcell for modeling the multiphase composite materials. This generalization is particularly advantageous when dealing with elasto-plastic composites, since yielding and plastic flow of the metallic phase may take place at different locations. The GCM is able to provide a more accurate representation of the actual microstructure.

This micromechanical analysis, based on the theory of the continuum media in which equilibrium is ensured, can be summarized essentially as follows. A repeating volume element of periodic multiphase composite is first identified. This is followed by defining the macroscopic average stresses and strains from the microscopic ones. Continuity of traction and displacement rates on the average basis are then imposed at the interfacesbetween the constituents. The micro equilibrium is guaranteed by the assumption that the velocity vector is linearly expanded in terms of the local coordinates of the subcell. This forms the relation between the microscopic strains, and the macroscopic strains through the relevant concentration tensors. In the final step the overall elasto-plastic behavior of multiphase inelastic composite is determined. This is expressed as a constitutive relation between the average stress, strain, and plastic strain, in conjunction with the effective elastic stiffness tensor of the composite. In this study the same steps are followed but in addition the damage mechanics is incorporated by using the micromechanical approach in order to obtain the damaged response of each constituent as well as overall instantaneous damaged behavior of the elasto-plastic composite.

A unidirectional fibrous composite is considered here in the method of cells. It is assumed that the composite has a periodic structure in which unidirectional fibers are extended in the x_1 direction. This representative volume element is shown in Figure 15.1a. The representative volume element (Figure 15.1b) consists of N_β by N_γ subcells such that

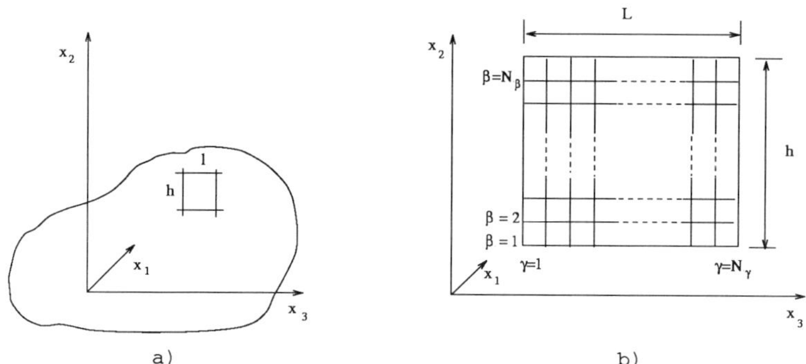

Figure 15.1   Representative Volume Element

the area of the cross section of each subcell is $h_\beta l_\gamma$ with $\beta = 1...N_\beta, \gamma = 1...N_\gamma$ and each subcell has its own local coordinate system $(x_1, x_2^{(\beta)}, x_3^{(\gamma)})$ with its origin located at the center of each subcell.

Unlike the method of cells, in this work the instantaneous behavior of the composite is considered. The displacement rate $\dot{u}_i^{(\beta\gamma)}$ (dot denotes time derivative) is expanded linearly in terms of the distance from the center of the subcell (Paley and Aboudi [203]). This leads the following first order expression

$$\dot{u}_i^{(\beta\gamma)} = \dot{w}_i^{(\beta\gamma)} + \bar{x}_2^{(\beta)}\dot{\theta}_i^{(\beta\gamma)} + \bar{x}_3^{(\gamma)}\dot{\psi}_i^{(\beta\gamma)} \tag{15.1}$$

where, $\dot{w}_i^{(\beta\gamma)}$ is the rate tensor of the displacement components at the center of the subcell, and $\dot{\theta}_i^{(\beta\gamma)}, \dot{\psi}_i^{(\beta\gamma)}$ are microvariables rates that characterize the linear dependence of the displacement rates on the local coordinates $\bar{x}_2^{(\beta)}, \bar{x}_3^{(\gamma)}$.

The small strain rate tensor and the constitutive law for the material that occupies the subcell $(\beta\gamma)$ are given by the following expressions respectively

$$\dot{\bar{\epsilon}}_{ij}^{(\beta\gamma)} = \frac{1}{2}\left(\frac{\partial \dot{u}_i^{(\beta\gamma)}}{\partial \bar{x}_j} + \frac{\partial \dot{u}_j^{(\beta\gamma)}}{\partial \bar{x}_i}\right) \tag{15.2}$$

$$\dot{\bar{\sigma}}_{ij}^{(\beta\gamma)} = \bar{D}_{ijkl}^{(\beta\gamma)}\dot{\bar{\epsilon}}_{ij}^{(\beta\gamma)} \tag{15.3}$$

The instantaneous stiffness tensor, $\bar{D}^{(\beta\gamma)}$. depends on the deformation history, loading path and applied loading rate. In this study for elasto-plastic materials, the von Mises yield criterion with an associated flow rule and the Ziegler-Prager kinematic hardening rule are used. This elasto-plastic tensor in the undamaged material is given by the following relation

$$\bar{D}_{ijkl}^{(\beta\gamma)} = \bar{E}_{ijkl}^{(\beta\gamma)} - \frac{1}{\bar{Q}} \frac{\partial \bar{f}^{(\beta\gamma)}}{\partial \bar{\sigma}_{pq}^{(\beta\gamma)}} \bar{E}_{pqij}^{(\beta\gamma)} \bar{E}_{klrs}^{(\beta\gamma)} \frac{\partial \bar{f}^{(\beta\gamma)}}{\partial \bar{\sigma}_{rs}^{(\beta\gamma)}} \tag{15.4}$$

where $\bar{Q}$ is given by

$$\bar{Q} = \frac{\partial \bar{f}^{(\beta\gamma)}}{\partial \bar{\sigma}_{ij}^{(\beta\gamma)}} \bar{E}_{ijkl}^{(\beta\gamma)} \frac{\partial \bar{f}^{(\beta\gamma)}}{\partial \bar{\sigma}_{kl}^{(\beta\gamma)}} - \frac{\partial \bar{f}^{(\beta\gamma)}}{\partial \bar{\alpha}_{ij}^{(\beta\gamma)}} (\bar{\sigma}_{ij}^{(\beta\gamma)} - \bar{\alpha}_{ij}^{(\beta\gamma)}) \frac{\frac{\partial \bar{f}^{(\beta\gamma)}}{\partial \bar{\sigma}_{pq}^{(\beta\gamma)}} \frac{\partial \bar{f}^{(\beta\gamma)}}{\partial \bar{\sigma}_{pq}^{(\beta\gamma)}}}{(\bar{\sigma}_{ij}^{(\beta\gamma)} - \bar{\alpha}_{ij}^{(\beta\gamma)}) \frac{\partial \bar{f}^{(\beta\gamma)}}{\partial \bar{\sigma}_{kl}^{(\beta\gamma)}}} b^{(\beta\gamma)} \tag{15.5}$$

In equation (15.5), $\bar{f}^{(\beta,\gamma)}$ is the von Mises yield criterion with kinematic hardening expressed in terms of the backstress tensor $\bar{\alpha}^{(\beta,\gamma)}$. The material parameter $b^{(\beta,\gamma)}$ pertains to the evolution behavior of the back stress (Voyiadjis and Kattan [99]). In the special case of perfectly elastic materials $\bar{D}^{(\beta\gamma)}$ is replaced by the standard elastic stiffness tensor $\bar{E}^{(\beta\gamma)}$ which characterizes the behavior of elastic materials in the subcells. More elaborate plasticity models for the in-situ characterization of metal matrix composites is given by the first author in other works(Voyiadjis and Thiagarajan [188]) . However, in this work a simple model is used.

The objective of the work outlined by Paley and Aboudi [203] is to solve the microvariables given in equation (15.1). This equation is substituted into the small strain tensor by employing the rate of displacement and traction continuity conditions at the interfaces between the subcells and between neighboring repeating cells in order to obtain the relation between the average subcell strain rate components and the average overall strain rate components via strain concentration tensors. The first step is to write a set of continuum equations in terms of the microvariables. These interface conditions are shown in Figures 15.2 and 15.3. Since it is ensured that at any instant the component of displacement rates should be continuous at the interfaces, the following relations can be obtained in terms of the micro variable rates using the continuity conditions of displacement rates at the interfaces between the subcells and the neighboring cell and these relations are given by

$$\dot{w}_i^{(\beta\gamma)} + \frac{1}{2} h_\beta \dot{\theta}_i^{(\beta\gamma)} = \dot{w}_i^{(\hat{\beta}\gamma)} - \frac{1}{2} h_{\hat{\beta}} \dot{\theta}_i^{(\hat{\beta}\gamma)} \tag{15.6}$$

and

$$\dot{w}_i^{(\beta\gamma)} + \frac{1}{2} \ell_\gamma \dot{\psi}_i^{(\beta\gamma)} = \dot{w}_i^{(\beta\hat{\gamma})} - \frac{1}{2} \ell_{\hat{\gamma}} \dot{\psi}_i^{(\beta\hat{\gamma})} \tag{15.7}$$

All the field variables in equations (15.6) and (15.7) are evaluated at the centerline $x_2^{(\beta)}$ for the subcell $(\beta\gamma)$ and $x_2^{(\hat{\beta})}$ for the subcell $(\hat{\beta}\gamma)$. As indicated in Figure 15.2 since the interface is along the x_3 direction one has $x_3^{(\gamma)}$ for the subcell $(\beta\gamma)$ and $x_3^{(\hat{\gamma})}$ for the subcell $(\beta\hat{\gamma})$ and the interface is along the x_2 direction. This relation can be expressed by

$$x_2^{(\beta)} = x^{(I)} - h_\beta/2 \quad \text{or} \quad x_2^{(\hat{\beta})} = x^{(I)} + h_{\hat{\beta}} \tag{15.8}$$

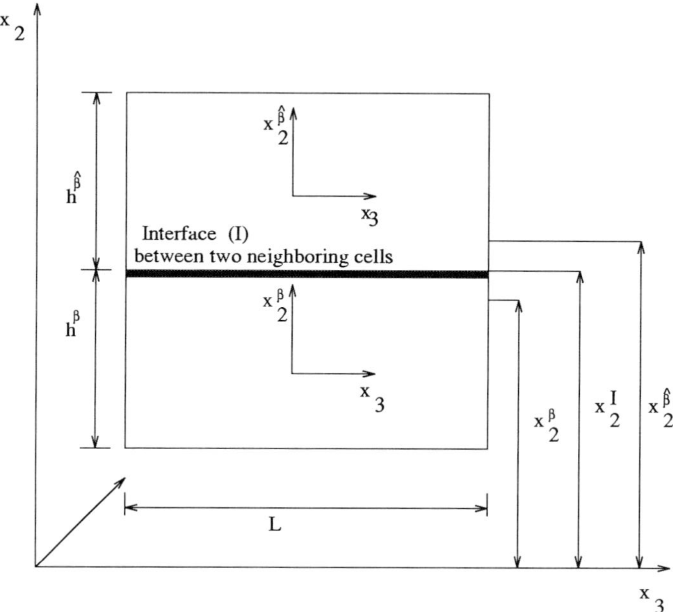

Figure 15.2 Interface Element in x_3 direction

and

$$x_3^{(\gamma)} = x^{(I)} - \ell_\gamma/2 \quad \text{or} \quad x_3^{(\hat{\gamma})} = x^{(I)} + \ell_{\hat{\gamma}}/2 \tag{15.9}$$

By employing a Taylor expansion of field variables in equation (15.6) together with equation (15.8) and omitting second and higher order terms, one obtains

$$\dot{w}_i^{(\beta\gamma)} - \frac{1}{2}h_{(\beta)}\left(\frac{\partial}{\partial x_2}\dot{w}_i^{(\beta\gamma)} - \dot{\theta}_i^{(\beta\gamma)}\right) = \dot{w}_i^{(\hat{\beta}\gamma)} + \frac{1}{2}h_{(\hat{\beta})}\left(\frac{\partial}{\partial x_2}\dot{w}_i^{(\hat{\beta}\gamma)} - \dot{\theta}_i^{(\hat{\beta}\gamma)}\right) \tag{15.10}$$

A similar expression can obtained by using equation (15.7) in equation (15.9) for the interface conditions along x_3 direction such that

$$\dot{w}_i^{(\beta\gamma)} - \frac{1}{2}\ell_{(\gamma)}\left(\frac{\partial}{\partial x_2}\dot{w}_i^{(\beta\gamma)} - \dot{\psi}_i^{(\beta\gamma)}\right) = \dot{w}_i^{(\beta\hat{\gamma})} + \frac{1}{2}\ell_{(\hat{\gamma})}\left(\frac{\partial}{\partial x_2}\dot{w}_i^{(\beta\hat{\gamma})} - \dot{\psi}_i^{(\beta\hat{\gamma})}\right) \tag{15.11}$$

These equations are valid in the equivalent continuum medium in which the repeating volume element can be defined by a point P. This mapping procedure of repeating volume elements at P within the equivalent homogeneous medium eliminates the discrete structure of the composite. Since a composite is subjected to homogeneous boundary conditions, the behavior of all repeating cells are identical, and a uniform field exits at the equivalent homogeneous medium. The governing constitutive laws of this equivalent continuum medium can be established by the generalized cells model.

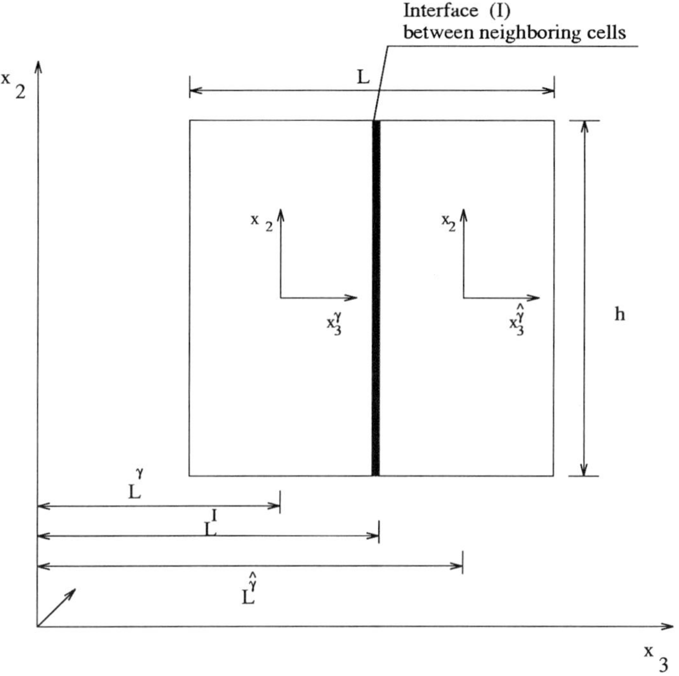

Figure 15.3 Interface Element in x_2 direction

From equations (15.10) and (15.11) the $N_\beta + N_\gamma$ continuum relations can be written in terms of the microvariables $\dot\theta_i^{(\beta\gamma)}$ and $\dot\psi_i^{(\beta\gamma)}$, and their explicit expression can be found in reference [203]. The composite standard average strain rate $\dot{\bar\epsilon}^{(\beta\gamma)}$ is given by

$$\dot{\bar\epsilon}_{ij} = \frac{1}{hl} \sum_{\beta=1}^{N\beta} \sum_{\gamma=1}^{N\gamma} h_\beta \, \ell_\gamma \dot{\bar\epsilon}_{ij}^{(\beta\gamma)} \tag{15.12}$$

It is possible to derive a $2(N_\beta + N_\gamma) + N_\beta \, N_\gamma + 1$ system of continuum relations expressed in terms of the subcell strain rate tensors $\dot\epsilon_{ij}^{(\beta\gamma)}$ by using the previous $N_\beta + N_\gamma$ continuum equations together with expression (15.12). After tedious mathematical manipulations, these relation can be given as follow

$$\dot{\bar{\epsilon}}_{11} = \dot{\bar{\epsilon}}_{11}^{(\beta\gamma)} \qquad\qquad \beta,\gamma = 1\cdots, N_\beta, N_\gamma \quad (N_\beta\, N_\gamma \quad \text{relations}) \tag{15.13}$$

$$\dot{\bar{\epsilon}}_{22} = \frac{1}{h}\sum_{\beta=1}^{N\beta} h_\beta \dot{\bar{\epsilon}}_{22}^{(\beta\gamma)} \qquad\qquad \gamma = 1\cdots, N_\gamma \quad (N_\gamma \quad \text{relations}) \tag{15.14}$$

$$\dot{\bar{\epsilon}}_{33} = \frac{1}{\ell}\sum_{\gamma=1}^{N\gamma} \ell_\gamma \dot{\bar{\epsilon}}_{33}^{(\beta\gamma)} \qquad\qquad \beta = 1\cdots, N_\beta \quad (N_\beta \quad \text{relations}) \tag{15.15}$$

$$2\dot{\bar{\epsilon}}_{23} = \frac{1}{h\ell}\sum_{\beta=1}^{N\beta}\sum_{\gamma=1}^{N\gamma} h_\beta\, \ell_\gamma \dot{\bar{\epsilon}}_{23}^{(\beta\gamma)} \qquad\qquad \text{(one relation)} \tag{15.16}$$

$$2\dot{\bar{\epsilon}}_{13} = \frac{1}{\ell}\sum_{\gamma=1}^{N\gamma} \ell_\gamma \dot{\bar{\epsilon}}_{13}^{(\beta\gamma)} \qquad\qquad \beta = 1\cdots, N_\beta \quad (N_\beta \quad \text{relations}) \tag{15.17}$$

$$2\dot{\bar{\epsilon}}_{12} = \frac{1}{h}\sum_{\beta=1}^{N\beta} h_\beta \dot{\bar{\epsilon}}_{12}^{(\beta\gamma)} \qquad\qquad \gamma = 1\cdots, N_\gamma \quad (N_\gamma \quad \text{relations}) \tag{15.18}$$

The above $2(N_\beta + N_\gamma) + N_\beta N_\gamma + 1$ continuum relations are expressed in matrix form by Paley and Aboudi [203] as follows

$$\boldsymbol{A_G}\dot{\bar{\boldsymbol{\epsilon}}}_{\boldsymbol{s}} = \boldsymbol{J}\dot{\bar{\boldsymbol{\epsilon}}} \tag{15.19}$$

where the 6-order average strain-rate vector is of the form

$$[\dot{\bar{\epsilon}}] = [\dot{\bar{\epsilon}}_{11}, \dot{\bar{\epsilon}}_{22}, \dot{\bar{\epsilon}}_{33}, \dot{\bar{\epsilon}}_{23}, \dot{\bar{\epsilon}}_{13}, \dot{\bar{\epsilon}}_{12}] \tag{15.20}$$

and the $6N_\beta N_\gamma$ order subcells strain-rate vector is defined as follows

$$[\dot{\bar{\boldsymbol{\epsilon}}}] = [\dot{\bar{\boldsymbol{\epsilon}}}^{(11)}, \dot{\bar{\boldsymbol{\epsilon}}}^{(12)}, \dot{\bar{\boldsymbol{\epsilon}}}^{(21)} \cdots \dot{\bar{\boldsymbol{\epsilon}}}^{(\beta\gamma)}] \tag{15.21}$$

The $\boldsymbol{A_G}$ is $2(N_\beta+N_\gamma)+N_\beta N_\gamma+1$ by $6N_\beta N_\gamma$ matrix and involves the geometrical properties of the repeating cells while $\boldsymbol{J}$ is a $2(N_\beta + N_\gamma) + N_\beta N_\gamma + 1$ by 6 matrix .

One now needs $5N_\beta N_\gamma - 2(N_\beta + N_\gamma) - 1$ continuum relations to complete the 6 $N_\beta N_\gamma$ set of continuum equations. They can be obtained by imposing the continuity of the rates of traction at the interfaces between the subcells and between neighboring repeating cells. The continuity of average stress rates at the interfaces can be expressed by the following relations

$$\dot{\bar{\sigma}}_{2j}^{(\beta\gamma)} = \dot{\bar{\sigma}}_{2j}^{(\beta\gamma)}, j = 1, 2, 3 \tag{15.22}$$

and

$$\dot{\bar{\sigma}}_{3j}^{(\beta\gamma)} = \dot{\bar{\sigma}}_{3j}^{(\beta\gamma)}, j = 1, 2, 3 \tag{15.23}$$

One can express the average stress rate $\dot{\bar{\sigma}}_{ij}^{(\beta\gamma)}$ in the subcells in terms of the average strain rate $\dot{\bar{\epsilon}}_{ij}^{(\beta\gamma)}$ by using the constitutive law of the material (equation (15.3)) in the subcells.

Using equation (15.22) and (15.23) the remaining continuum equations which can be written in the matrix form as follow

$$A_m \dot{\bar{\epsilon}}_s = 0 \tag{15.24}$$

A_m is $5N_\beta N_\gamma - 2(N_\beta + N_\gamma) - 1$ by $6N_\beta N_\gamma$ matrix. A_m involves the instantaneous properties of the material in the various subcells. The $6N_\beta N_\gamma$ continuum equation can be written in the following matrix form by combining equations (15.19) and (15.24)

$$A_m^* \dot{\bar{\epsilon}}_s = K \dot{\bar{\epsilon}} \tag{15.25}$$

where the $6N_\beta N_\gamma$ order square matrix A_m^* is given in the form

$$A_m^* = \begin{bmatrix} A_m \\ A_G \end{bmatrix} \quad \text{and} \quad K = \begin{bmatrix} 0 \\ J \end{bmatrix} \tag{15.26}$$

One can now solve the linear system of equations (15.25) in order to obtain the following expression

$$\dot{\bar{\epsilon}}_s = A_c \dot{\bar{\epsilon}} \tag{15.27}$$

where

$$A_c = \left[A_m^* \right]^{-1} K \tag{15.28}$$

A_c is the instantaneous strain concentration tensor that relates the average strain-rate tensor in the subcell to the average overall strain-rate tensor. The matrix A_c can be partitioned into a number of $N_\beta N_\gamma$ by 6x6 matrices as shown below

$$A_c = \begin{bmatrix} A_c^{(11)} \\ \vdots \\ A_c^{(N_\beta N_\gamma)} \end{bmatrix} \tag{15.29}$$

$A_c^{(\beta\gamma)}$ is the instantaneous strain concentration tensor for the subcell which relates the average strain rate tensor in the subcell $(\beta\gamma)$ to the average total strain rate tensor. One can now obtain the overall effective instantaneous stiffness tensor of the composite by using the strain concentration tensor of the subcell along with its respective subcell constitutive equations (Paley and Aboudi [203]).

15.1.2 Incremental Damage Model

In this study, the incremental damage model is used in order to characterize the damage using the fourth order incremental damage tensor $m^{(\beta\gamma)}$ where $(\beta\gamma)$ designates the subcell . The concept of effective stress as generalized by Murakami [58] is used here in order to introduce the damage for the $(N_\beta$ by $N_\gamma)$ constituents of the composite system. The $m^{(\beta\gamma)}$ is assumed to reflect all types of damages that corresponding subcells undergo such

410

as nucleation and coalescence of voids, and microcracks. This local damage response is linked to the overall damage response of the composite medium through the stress and strain concentration tensors. The elasto-plastic stiffness tensor is derived for the damaged composite using the subcell incremental damage tensors in the generalized cells model, and the relation between the subcell incremental damage tensor $m^{(\beta\gamma)}$ and the incremental overall damage tensor m.

Kachanov [1] introduced a simple scalar damage model for isotropic materials by using the concept of the effective stress. The incremental damage model was further developed subsequently on the base of the effective stress concept for anisotropic materials by Voyiadjis and Park [176] and Voyiadjis and Guelzim [205] In its formulation three configurations are assumed namely the initial undeformed and undamaged configuration C_0 , the deformed and damaged configuration C, and the state of the body after it has only deformed without damage $\bar{C}$, (Voyiadjis and Kattan [99]).

The rate (incremental) expression of damage tensor m can be written as follows

$$\dot{\bar{\sigma}} = M : \dot{\sigma} + \dot{M} : \sigma \tag{15.30}$$

The superposed dot implies the material time differentiation. In order for equation (15.30) to be homogeneous in time of order one (i.e stress-rate independent) $\dot{M}$ should be a linear function of $\dot{\sigma}$. This is demonstrated by the following expression

$$\dot{\phi} = X : \dot{\sigma} \tag{15.31}$$

Since $\dot{M}$ is a function of ϕ, one obtains therefore

$$\dot{M}_{ijkl} = \frac{\partial M_{ijkl}}{\partial \phi_{pq}} \dot{\phi}_{pq} \tag{15.32}$$

Consequently by substituting equations (15.31) and (15.32) into equation (15.30), the following relation may be written in the form

$$\dot{\bar{\sigma}} = m : \dot{\sigma} \tag{15.33}$$

where m represent the fourth order incremental damage tensor and is given by Voyiadjis and Guelzim [205]

$$m_{ijkl} = M_{ijkl} + G_{ijpqkl}\sigma_{pq} \tag{15.34}$$

where

$$G_{ijpqrs} = \frac{\partial M_{ijkl}}{\partial \phi_{pq}} X_{pqrs} \tag{15.35}$$

The explicit expression for the fourth order tensor X in equation (15.35) is given in Section 15.2.4. The proposed damage model was used successfully for both monotonic and cyclic loads Vojiadjis and Ganesh [196]

15.2 Theoretical Formulation

15.2.1 Basic Assumptions

In this work, C_0 denotes the initial undeformed and undamaged configuration of a single laminate while $C_0{}^{(\beta\gamma)}$ is the initial undeformed and undamaged subcell subconfiguration of a single laminate. The composite material is assumed to undergo elasto-plastic deformation and damage due to the applied loads. The resulting overall configuration for a single laminate is denoted by C. Damage is expressed by generalizing the concept proposed by Kachanov [1]. The fictitious configurations $\bar{C}^{(\beta\gamma)}$ is obtained from $C^{(\beta\gamma)}$ by removing the different types of damages that the corresponding subcell $(\beta\gamma)$ has undergone due to the applied stresses. The total or incremental subcell stress at configuration $C^{(\beta\gamma)}$ is converted to the respective total or incremental stress at the fictitious configuration $\bar{C}^{(\beta\gamma)}$ through the damage tensor $\boldsymbol{M}^{(\beta\gamma)}$ or $\boldsymbol{m}^{(\beta\gamma)}$ respectively. The incremental damage tensor $\boldsymbol{m}^{(\beta\gamma)}$ reflects the damage related that subcell only. Following this local damage description, local-overall relations are used to transfer the local damage effect to the whole composite system in configuration C. This is accomplished through the stress and strain concentration tensors of the subcells.

The coupled formulation of plastic flow and damage propagation is quite complex due to the presence of the two different dissipative mechanisms that influence each other. This could be indicated by the fact that the position of the slip planes affect the orientation of nucleated microcracks. A phenemological model of interaction can then be applied. In this work use is made of the concept of the effective stress (Lemaitre [93]). Making use of a fictitious undamaged system, the dissipation energy due to plastic flow in this undamaged system is assumed to be equal to the dissipation energy due to plastic flow in the damaged system. The damages at the single laminate level are described separately by the damage in the subcells according to the material in the subcells. The subcell incremental damage tensors, $\boldsymbol{m}^{(\beta\gamma)}$, is better suited for use in the formulation of the constitutive equation of the damaged material behavior due to the incremental nature of plasticity.

15.2.2 Local-Overall Relations of the Damage Tensors

In this section the relations between incremental damage tensor $\boldsymbol{m}^{(\beta\gamma)}$ of subcells $(\beta\gamma)$ and overall incremental damage tensor $\boldsymbol{m}$ of the composite medium are derived by using the fact that the average damaged stress rates $\dot{\boldsymbol{\sigma}}$ can be obtained as the average sum of the the damaged stress rates $\dot{\boldsymbol{\sigma}}^{(\beta\gamma)}$ of the subcells in the damaged configuration $C^{(\beta\gamma)}$ and is given by the following relation

$$\dot{\boldsymbol{\sigma}} = \frac{1}{V} \sum_{\beta=1}^{N_\beta} \sum_{\gamma=1}^{N_\gamma} v_{\beta\gamma} \dot{\boldsymbol{\sigma}}^{(\beta\gamma)} \tag{15.36}$$

In equation (15.36) V is the total area of the representative volume element while $v_{\beta\gamma}$ is the area of the individual subcell in the damaged configuration. Subcell incremental damage tensor $\boldsymbol{m}^{(\beta\gamma)}$ can be introduced in a similar form to equation (15.33) such that

$$\dot{\bar{\boldsymbol{\sigma}}}^{(\beta\gamma)} = \boldsymbol{m}^{(\beta\gamma)} : \dot{\boldsymbol{\sigma}}^{(\beta\gamma)} \tag{15.37}$$

where $m^{(\beta\gamma)}$ encompasses all the pertinent damages that the corresponding subcell undergoes. The effective subcell Cauchy stress rate $\dot{\bar{\sigma}}^{(\beta\gamma)}$ is related to the overall effective stress rate $\dot{\bar{\sigma}}$ in the composite through the stress concentration tensor $\bar{B}^{(\beta\gamma)}$ as follows

$$\dot{\bar{\sigma}}^{(\beta\gamma)} = \bar{B}^{(\beta\gamma)} : \dot{\bar{\sigma}} \tag{15.38}$$

where the effective stress concentration tensor $\bar{B}^{(\beta\gamma)}$ is given in the following expression by Paley and Aboudi [203]

$$\bar{B}^{(\beta\gamma)} = \bar{C}^{(\beta\gamma)} : \bar{A}^{(\beta\gamma)} : [\bar{C}]^{-1} \tag{15.39}$$

where $\bar{C}^{(\beta\gamma)}$ is the effective stiffness tensor for the subcell, $\bar{A}^{(\beta\gamma)}$ is the undamaged strain concentration tensor for the subcell and the $\bar{C}$ is the overall undamaged effective stiffness tensor for the composite. One can solve $\dot{\sigma}^{(\beta\gamma)}$ from equation (15.37) such that

$$\dot{\sigma}^{(\beta\gamma)} = [m^{(\beta\gamma)}]^{-1} : \dot{\bar{\sigma}}^{(\beta\gamma)} \tag{15.40}$$

Making use of relations (15.38) and (15.40) in (15.36), one obtains the following expression(Voyiadjis and Deliktas [202]

$$\dot{\sigma} = \frac{1}{V} \sum_{\beta=1}^{N_\beta} \sum_{\gamma=1}^{N_\gamma} v_{\beta\gamma} [m^{(\beta\gamma)}]^{-1} : \bar{B}^{(\beta\gamma)} : \dot{\bar{\sigma}} \tag{15.41}$$

This equation can be easily written in a similar form to equation (15.33) where m represents the overall incremental damage tensor which reflects all types of damages that the composite undergoes including that due to the interaction between the subcells. The resulting expression is given by

$$m = \left[\frac{1}{V} \sum_{\beta=1}^{N_\beta} \sum_{\gamma=1}^{N_\gamma} v_{\beta\gamma} [m^{(\beta\gamma)}]^{-1} : \bar{B}^{(\beta\gamma)} \right]^{-1} \tag{15.42}$$

This expression defines the cumulative incremental damage of the composite as a function of its subcell components.However, m may be expressed in terms of the fiber damage m^f, the matrix damage m^m, and the damage due to debonding m^d(Voyiadjis and Park [176]).

15.2.3 Damaged Strain and Stress Concentration Tensors

Concentration tensors do not remain constant as the composite undergoes damage. However, they are constant in the undamaged elastic domain. In this work undamaged concentration factors are modified for the incremental damage model in conjunction with the hypothesis of the equivalence of elastic strain energy [16]. The effective elastic strain and stress concentration tensors are obtained by using the generalized cells model. The subcell strain rate tensors can be related to the overall strain rate tensor in the following way

$$\dot{\bar{\epsilon}}^{(\beta\gamma)} = \bar{A}^{(\beta\gamma)} : \dot{\bar{\epsilon}} \tag{15.43}$$

where fourth order tensor, $\bar{\boldsymbol{A}}^{(\beta\gamma)}$ is the instantaneous strain concentration tensor for the subcell $(\beta\gamma)$ and is given by equation (15.29). The undamaged stress concentration tensors $\bar{\boldsymbol{B}}^{(\beta\gamma)}$ of the subcells are already defined in the previous section and their relations are given by equations (15.38) and (15.39).

The damaged concentration tensors can be obtained in terms of the undamaged concentration factors and incremental damage tensors in connection with the elastic energy equivalence, given by

$$d\bar{U}^{(\beta\gamma)} = dU^{(\beta\gamma)} \tag{15.44}$$

or

$$\frac{1}{2}d\dot{\bar{\boldsymbol{\sigma}}}^{(\beta\gamma)} : d\dot{\bar{\boldsymbol{\epsilon}}}^{(\beta\gamma)} = \frac{1}{2}d\dot{\boldsymbol{\sigma}}^{(\beta\gamma)} : d\dot{\boldsymbol{\epsilon}}^{(\beta\gamma)} \tag{15.45}$$

Substituting equation (15.37) into equation (15.45), one obtains the following relation

$$\dot{\bar{\boldsymbol{\epsilon}}}^{(\beta\gamma)} = [\boldsymbol{m}^{(\beta\gamma)}]^{-1} : \dot{\boldsymbol{\epsilon}}^{(\beta\gamma)} \tag{15.46}$$

The above equation can be written for the overall behavior in similar form as shown below

$$\dot{\bar{\boldsymbol{\epsilon}}} = [\boldsymbol{m}]^{-1} : \dot{\boldsymbol{\epsilon}} \tag{15.47}$$

Consequently by combining equations (15.46) and (15.47) with equation (15.43), the relation between the damaged strain rate $\dot{\boldsymbol{\epsilon}}^{(\beta\gamma)}$ of the subcell and the damaged strain rate $\dot{\boldsymbol{\epsilon}}$ can be obtained in the form shown below

$$\dot{\boldsymbol{\epsilon}}^{(\beta\gamma)} = \boldsymbol{A}^{(\beta\gamma)} : \dot{\boldsymbol{\epsilon}} \tag{15.48}$$

where $\boldsymbol{A}^{(\beta\gamma)}$ is the damaged stress concentration tensor for subcell $(\beta\gamma)$ and its expression is given by

$$\boldsymbol{A}^{(\beta\gamma)} = \boldsymbol{m}^{(\beta\gamma)} : \bar{\boldsymbol{A}}^{(\beta\gamma)} : [\boldsymbol{m}]^{-1} \tag{15.49}$$

Similarly by using equations (15.33) and (15.37) with (15.38), the damaged stress concentration tensor for the subcell can be given as follows

$$\dot{\boldsymbol{\sigma}}^{(\beta\gamma)} = \boldsymbol{B}^{(\beta\gamma)} : \dot{\boldsymbol{\sigma}} \tag{15.50}$$

where $\boldsymbol{B}^{(\beta\gamma)}$ is the damaged stress concentration tensor for subcell $(\beta\gamma)$ and its expression is given by

$$\boldsymbol{B}^{(\beta\gamma)} = [\boldsymbol{m}^{(\beta\gamma)}]^{-1} : \bar{\boldsymbol{B}}^{(\beta\gamma)} : \boldsymbol{m} \tag{15.51}$$

15.2.4 Damage Criterion

In order to study the evolution of damage in composite materials, one first needs to investigate the damage criterion. The anisotropic damage criterion used here is expressed in term of a tensorial parameter $\boldsymbol{h}$ (Voyiadjis and Park [176]). It is clear that the damage mechanism for each subcell of the composite materials should be accounted separately since each subcell can be occupied by a different type of material in addition their boundary and geometric conditions can be different for each subcell. Therefore one single damage mechanism cannot be considered for all subcells in the multiphase composite medium. The anisotropic damage criterion based on the Mroz model [139] is generalized by Voyiadjis and Park [176] as follows

$$g^{(\beta\gamma)} \equiv g^{(\beta\gamma)}(\boldsymbol{Y},\kappa) = 0 \tag{15.52}$$

or

$$g^{(\beta\gamma)} \equiv Y_{ij}^{(\beta\gamma)} P_{ijkl}^{(\beta\gamma)} Y_{kl}^{(\beta\gamma)} - 1 = 0 \tag{15.53}$$

where

$$P_{ijkl}^{(\beta\gamma)} = h_{ij}^{-(\beta\gamma)} h_{kl}^{-(\beta\gamma)} \tag{15.54}$$

$\boldsymbol{Y}^{(\beta\gamma)}$ is the generalized thermodynamic force conjugate to the damage tensor $\phi^{(\beta\gamma)}$. The hardening tensor $\boldsymbol{h}^{(\beta\gamma)}$ is expressed as follows

$$h_{ij}^{(\beta\gamma)} = U_{ij}^{(\beta\gamma)} + V_{ij}^{(\beta\gamma)} \tag{15.55}$$

where tensors $\boldsymbol{U}^{(\beta\gamma)}$ and $\boldsymbol{V}^{(\beta\gamma)}$ are defined for orthotropic materials in terms of the generalized Lame constant $\lambda_1^{(\beta\gamma)}$, $\lambda_2^{(\beta\gamma)}$, $\lambda_3^{(\beta\gamma)}$ and the material parameters $\nu_1^{(\beta\gamma)}$, $\nu_2^{(\beta\gamma)}$, $\nu_3^{(\beta\gamma)}$, $\xi_1^{(\beta\gamma)}$, $\xi_2^{(\beta\gamma)}$, $\xi_3^{(\beta\gamma)}$, and $\eta_1^{(\beta\gamma)}$, $\eta_2^{(\beta\gamma)}$, $\eta_3^{(\beta\gamma)}$ which are obtained by matching the theory with the experimental results. Voyiadjis and Park [176] used the following expressions for $U_{ij}^{(\beta\gamma)}$ and $V_{ij}^{(\beta\gamma)}$.

$$U_{ij}^{(\beta\gamma)} = \begin{pmatrix} \lambda_1\eta_1\left(\frac{\kappa}{\lambda_1}\right)^{\xi_1} & 0 & 0 \\ 0 & \lambda_2\eta_2\left(\frac{\kappa}{\lambda_2}\right)^{\xi_2} & 0 \\ 0 & 0 & \lambda_3\eta_3\left(\frac{\kappa}{\lambda_3}\right)^{\xi_3} \end{pmatrix}^{(\beta\gamma)} \tag{15.56}$$

and

$$V_{ij}^{(\beta\gamma)} = \begin{pmatrix} \lambda_1\nu_1^2 & 0 & 0 \\ 0 & \lambda_2\nu_2^2 & 0 \\ 0 & 0 & \lambda_3\nu_3^2 \end{pmatrix}^{(\beta\gamma)} \tag{15.57}$$

$\kappa^{(\beta\gamma)}$ is the scalar representing the total damage energy and is given by the following relation

$$\kappa^{(\beta\gamma)} = \int_0^t \boldsymbol{Y}^{(\beta\gamma)} : \dot{\phi}^{(\beta\gamma)} \, dt \tag{15.58}$$

or

$$\dot{\kappa}^{(\beta\gamma)} = \mathbf{Y}^{(\beta\gamma)} : \dot{\boldsymbol{\phi}}^{(\beta\gamma)} \tag{15.59}$$

The generalized Lame constants are defined as follow (Voyiadjis and Park [176])

$$\lambda_i^{(\beta\gamma)} = \bar{E}_i^{(\beta\gamma)}(1 - \phi_i^{(\beta\gamma)})^2 \tag{15.60}$$

$\bar{E}_i^{(\beta\gamma)}$ are the magnitudes of the effective moduli of elasticity along the principle axes defined along the direction of the fiber and transversely to them. In order to check the damaging state of the material, the following four steps are outlined below by (Stumvoll and Swoboda [140]) which are also indicated in equation (10.82)

$$g^{(\beta\gamma)} < 0, \qquad \text{(elastic unloading)} \tag{15.61}$$

$$g^{(\beta\gamma)} = 0, \qquad \frac{\partial g^{(\beta\gamma)}}{\partial \mathbf{Y}^{(\beta\gamma)}} : \dot{\mathbf{Y}}^{(\beta\gamma)} < 0, \quad \text{(elastic unloading)} \tag{15.62}$$

$$g^{(\beta\gamma)} = 0, \qquad \frac{\partial g^{(\beta\gamma)}}{\partial \mathbf{Y}^{(\beta\gamma)}} : \dot{\mathbf{Y}}^{(\beta\gamma)} = 0, \quad \text{(neutral loading)} \tag{15.63}$$

$$g^{(\beta\gamma)} = 0 \qquad \frac{\partial g^{(\beta\gamma)}}{\partial \mathbf{Y}^{(\beta\gamma)}} : \dot{\mathbf{Y}}^{(\beta\gamma)} > 0, \quad \text{(loading from a damaging state)} \tag{15.64}$$

The case corresponding to loading or unloading from an elastic state is given by relation (15.61). For elastic unloading it is represented by relation (15.62). In the case of neutral loading it is represented by relation (15.63). Finally for the loading case it is given by relation (15.64) from a damaging state. It is clear from the above outlined steps that, the damage criterion ($g^{(\beta\gamma} \equiv 0$) should be satisfied for the state of damage to occur. As mentioned before for the damage evolution of materials , different types of micro-mechanics damage are considered for each subcell depending on the material properties within the subcell. In this work, for an elasto-plastic matrix, the subcell is assumed to undergo ductile damage while the elastic fiber in the subcell undergoes brittle damage and their total energy dissipation is different from each other.

As mentioned previously in Chapters (10) and (13), one can similarly obtain the evolution expressions for damage and plastic strain(Voyiadjis and Deliktas [202])

$$\dot{\bar{\epsilon}}^{(\beta\gamma)} - \dot{\Lambda}_1^{(\beta\gamma)} \frac{\partial \bar{f}^{(\beta\gamma)}}{\partial \bar{\sigma}^{(\beta\gamma)}} = 0 \tag{15.65}$$

and

$$\dot{\boldsymbol{\phi}}^{(\beta\gamma)} - \dot{\Lambda}_2^{(\beta\gamma)} \frac{\partial g^{(\beta\gamma)}}{\partial \mathbf{Y}^{(\beta\gamma)}} = 0 \tag{15.66}$$

The Lagrange multiplier $\dot{\Lambda}_1^{(\beta\gamma)}$ in equation (15.65) can be obtained by using the consistency condition for the yield function for the elasto-plastic matrix in conjunction with the Ziegler-Prager kinematic hardening rule. The corresponding yield function is given by

$$\bar{f}^{(\beta\gamma)} = \frac{3}{2}(\bar{\sigma} - \bar{\alpha})^{(\beta\gamma)} : (\bar{\sigma} - \bar{\alpha})^{(\beta\gamma)} - \bar{\sigma}_o{}^2 \tag{15.67}$$

416

where $\boldsymbol{\alpha}^{(\beta\gamma)}$ is given by

$$\dot{\boldsymbol{\alpha}}^{(\beta\gamma)} = \dot{\bar{\mu}}^{(\beta\gamma)}(\bar{\boldsymbol{\sigma}} - \bar{\boldsymbol{\alpha}})^{(\beta\gamma)} \tag{15.68}$$

and $\dot{\bar{\mu}}^{(\beta\gamma)}$ is defined such that

$$\dot{\bar{\mu}}^{(\beta\gamma)} = 3b^{(\beta\gamma)}\dot{\Lambda}_1^{(\beta\gamma)} \tag{15.69}$$

In equation (15.69) $b^{(\beta\gamma)}$ is the kinematic hardening parameter for the elasto-plastic subcells. The consistency condition of the yield function in equation (15.67) can be written in the following form

$$\dot{\bar{f}}^{(\beta\gamma)} \equiv 0 \tag{15.70}$$

or

$$\frac{\partial \bar{f}^{(\beta\gamma)}}{\partial \bar{\boldsymbol{\sigma}}^{(\beta\gamma)}} : \dot{\bar{\boldsymbol{\sigma}}}^{(\beta\gamma)} + \frac{\partial \bar{f}^{(\beta\gamma)}}{\partial \bar{\boldsymbol{\alpha}}^{(\beta\gamma)}} : \dot{\boldsymbol{\alpha}}^{(\beta\gamma)} = 0 \tag{15.71}$$

This condition assures that in a plastic loading process the subsequent stress and deformation state remains on the subsequent yield surface. One can use this consistency condition together with the equations (15.68) and (15.69) in order to obtain $\dot{\Lambda}_1^{(\beta\gamma)}$ in the following form

$$\dot{\Lambda}_1^{(\beta\gamma)} = \frac{1}{H^{(\beta\gamma)}} \frac{\partial \bar{f}^{(\beta\gamma)}}{\partial \bar{\boldsymbol{\sigma}}^{(\beta\gamma)}} \dot{\bar{\boldsymbol{\sigma}}}^{(\beta\gamma)} \tag{15.72}$$

where the scalar quantity H is given by

$$H^{(\beta\gamma)} = 3b^{(\beta\gamma)} \frac{\partial \bar{f}^{(\beta\gamma)}}{\partial \bar{\boldsymbol{\alpha}}^{(\beta\gamma)}} : (\bar{\boldsymbol{\sigma}}^{(\beta\gamma)} - \bar{\boldsymbol{\alpha}}^{(\beta\gamma)}) \tag{15.73}$$

Equation (15.66) gives the incremental relation of the damage variable for each subcell. Similarly using the consistency condition of the damage potential $g^{(\beta\gamma)}$, one can obtain the parameter $\dot{\Lambda}_2^{(\beta\gamma)}$. The corresponding damage consistency relation can be given as follows

$$\dot{g}^{(\beta\gamma)} \equiv 0 \tag{15.74}$$

where $g^{(\beta\gamma)}$ can be defined as a function of $g^{(\beta\gamma)} \equiv g^{(\beta\gamma)}(\boldsymbol{Y}, \kappa)$ or $g^{(\beta\gamma)} \equiv g^{(\beta\gamma)}(\boldsymbol{\sigma}, \boldsymbol{\phi}, \kappa)$. Equation (15.74) can be written as follows

$$\frac{\partial g^{(\beta\gamma)}}{\partial \boldsymbol{\sigma}^{(\beta\gamma)}} : \dot{\boldsymbol{\sigma}}^{(\beta\gamma)} + \frac{\partial g^{(\beta\gamma)}}{\partial \boldsymbol{\phi}^{(\beta\gamma)}} : \dot{\boldsymbol{\phi}}^{(\beta\gamma)} + \frac{\partial g^{(\beta\gamma)}}{\partial \kappa^{(\beta\gamma)}} \dot{\kappa}^{(\beta\gamma)} = 0 \tag{15.75}$$

By substituting equation (15.59) and (15.66) into (15.75), the above equation can be expressed in terms of the parameter $\dot{\Lambda}_2^{(\beta\gamma)}$ where

$$\frac{\partial g^{(\beta\gamma)}}{\partial \boldsymbol{\sigma}^{(\beta\gamma)}} : \dot{\boldsymbol{\sigma}}^{(\beta\gamma)} + \Lambda_2^{(\beta\gamma)} \frac{\partial g^{(\beta\gamma)}}{\partial \boldsymbol{\phi}^{(\beta\gamma)}} : \frac{\partial g^{(\beta\gamma)}}{\partial \boldsymbol{Y}^{(\beta\gamma)}} + \frac{\partial g^{(\beta\gamma)}}{\partial \kappa^{(\beta\gamma)}} \boldsymbol{Y}^{(\beta\gamma)} : \dot{\boldsymbol{\phi}}^{(\beta\gamma)} = 0 \tag{15.76}$$

One can solve for the parameter $\dot{\Lambda}_2^{(\beta\gamma)}$ from equation (15.76) such that

$$\Lambda_2^{(\beta\gamma)} = \frac{\dfrac{\partial g^{(\beta\gamma)}}{\partial \boldsymbol{\sigma}^{(\beta\gamma)}} : \dot{\boldsymbol{\sigma}}^{(\beta\gamma)}}{\dfrac{\partial g^{(\beta\gamma)}}{\partial \boldsymbol{\phi}^{(\beta\gamma)}} : \dfrac{\partial g^{(\beta\gamma)}}{\partial \boldsymbol{Y}^{(\beta\gamma)}} + \dfrac{\partial g^{(\beta\gamma)}}{\partial \kappa^{(\beta\gamma)}} \boldsymbol{Y}^{(\beta\gamma)} : \dfrac{\partial g^{(\beta\gamma)}}{\partial \boldsymbol{Y}^{(\beta\gamma)}}} \tag{15.77}$$

Using relation (15.77) with (15.66), the incremental damage evolution equation for the subcell $(\beta\gamma)$ can be obtained in the following form

$$\dot{\boldsymbol{\phi}}^{(\beta\gamma)} = \boldsymbol{X}^{(\beta\gamma)} : \dot{\boldsymbol{\sigma}}^{(\beta\gamma)} \tag{15.78}$$

where $\boldsymbol{X}^{(\beta\gamma)}$ is the fourth order tensor such that

$$\boldsymbol{X}^{(\beta\gamma)} = c\frac{\dfrac{\partial g^{(\beta\gamma)}}{\partial \boldsymbol{Y}^{(\beta\gamma)}} \dfrac{\partial g^{(\beta\gamma)}}{\partial \boldsymbol{\sigma}^{(\beta\gamma)}}}{\dfrac{\partial g^{(\beta\gamma)}}{\partial \boldsymbol{\phi}^{(\beta\gamma)}} : \dfrac{\partial g^{(\beta\gamma)}}{\partial \boldsymbol{Y}^{(\beta\gamma)}} + \dfrac{\partial g^{(\beta\gamma)}}{\partial \kappa^{(\beta\gamma)}} \boldsymbol{Y}^{(\beta\gamma)} : \dfrac{\partial g^{(\beta\gamma)}}{\partial \boldsymbol{Y}^{(\beta\gamma)}}} \tag{15.79}$$

The thermodynamic force tensor $\boldsymbol{Y}^{(\beta\gamma)}$ associated with damage can be obtained by using the enthalpy of the damaged materials. This energy equation is given by

$$V^{(\beta\gamma)}(\boldsymbol{\sigma}, \boldsymbol{\phi}) = \frac{1}{2}\boldsymbol{\sigma}^{(\beta\gamma)} : \boldsymbol{E}^{-(\beta\gamma)}(\boldsymbol{\phi}) : \boldsymbol{\sigma}^{(\beta\gamma)} - \Phi^{(\beta\gamma)}(\alpha) \tag{15.80}$$

where $\Phi^{(\beta\gamma)}$ is the specific energy due to kinematic hardening. $\boldsymbol{E}^{-(\beta\gamma)}$ is the damaged elastic compliance tensor for the subcell. It can be expressed in terms of the undamaged compliance tensor $\bar{\boldsymbol{E}}^{(\beta\gamma)}$ and the damage tensor $\boldsymbol{M}^{(\beta\gamma)}$ such that

$$\boldsymbol{E}^{-(\beta\gamma)} = \boldsymbol{M}^{(\beta\gamma)} : \bar{\boldsymbol{E}}^{-(\beta\gamma)} : \boldsymbol{M}^{(\beta\gamma)} \tag{15.81}$$

The thermodynamic force $\boldsymbol{Y}^{(\beta\gamma)}$ of the subcell $(\beta\gamma)$ is given as the partial derivative of enthalpy of the damaged material equation (15.80) with respect to the second order damage tensor $\boldsymbol{\phi}^{(\beta\gamma)}$ in the following expression

$$\boldsymbol{Y}^{(\beta\gamma)} = \frac{\partial V^{(\beta\gamma)}}{\partial \boldsymbol{\phi}^{(\beta\gamma)}} \tag{15.82}$$

Making use of equations (15.80) and (15.81) in equation (15.82), one can write the thermodynamic force $\boldsymbol{Y}^{(\beta\gamma)}$ explicitly (Voyiadjis and Park[176]) as follows

$$Y_{ij}^{(\beta\gamma)} = \frac{1}{2}(\sigma_{cd}^{(\beta\gamma)} \bar{E}_{abpq}^{(\beta\gamma)} M_{pqkl}^{(\beta\gamma)} \sigma_{kl}^{(\beta\gamma)} + \sigma_{rs}^{(\beta\gamma)} M_{uvrs}^{(\beta\gamma)} \bar{E}_{uvab}^{(\beta\gamma)} \sigma_{cd}^{(\beta\gamma)}) \frac{\partial M_{abcd}^{(\beta\gamma)}}{\partial \phi_{ij}^{(\beta\gamma)}} \tag{15.83}$$

If the material in the subcell $(\beta\gamma)$ is elastic, one can easily see that the gradual degradation of the elastic material in the corresponding subcell is caused only through damage and consequently no plastic dissipation occurs in the material. A similar procedure is followed as outlined before to investigate the damage evolution for elastic materials.

418

15.2.5 Overall Damaged Stiffness Tensor for the Model

In this section, the elasto-plastic constitutive model for the damaged multiphase composite medium is obtained. The procedure can be outlined by the following steps. First one obtains the subcell (local) damage quantities in their respective damaged configuration $C^{(\beta\gamma)}$ from their undamaged relations such as, stress, strain concentration tensors, and undamaged effective stiffness of composite. These quantities can be obtained through the generalized cells model. This is followed by combining the (N_β by N_γ) subcell constitutive relations by using equation (15.36) in conjunction with the concentration factors in the damaged configuration $C^{(\beta\gamma)}$ in order to obtain the constitutive relation of the overall composite system in the damaged configuration C.

One can start with by substituting equation (15.40) in equation (15.36). The following relation is then obtained.

$$\dot{\sigma} = \frac{1}{V} \sum_{\beta=1}^{N_\beta} \sum_{\gamma=1}^{N_\gamma} v_{\beta\gamma} [m^{(\beta\gamma)}]^{-1} : \dot{\bar{\sigma}}^{(\beta\gamma)} \tag{15.84}$$

The term $\dot{\bar{\sigma}}^{(\beta\gamma)}$ in equation (15.84) is replaced with the relation in equation (15.3), where the fourth order effective tensor $\bar{C}^{(\beta\gamma)}$ in the effective configuration is to be replaced by the corresponding stiffness tensor depending on the properties of the material in the respective subcells. The resulting expression is written as follows

$$\dot{\sigma} = \frac{1}{V} \sum_{\beta=1}^{N_\beta} \sum_{\gamma=1}^{N_\gamma} v_{\beta\gamma} [m^{(\beta\gamma)}]^{-1} : \bar{C}^{(\beta\gamma)} : \dot{\bar{\epsilon}}^{(\beta\gamma)} \tag{15.85}$$

By substituting equation (15.46) and (15.48) into (15.85), finally one can obtain the following relation

$$\dot{\sigma} = \frac{1}{V} \sum_{\beta=1}^{N_\beta} \sum_{\gamma=1}^{N_\gamma} v_{\beta\gamma} \left\{ [m^{(\beta\gamma)}]^{-1} : \bar{C}^{(\beta\gamma)} : [m^{(\beta\gamma)}]^{-1} \right\} : A^{(\beta\gamma)} : \dot{\epsilon} \tag{15.86}$$

or

$$\dot{\sigma} = C : \dot{\epsilon} \tag{15.87}$$

where C represents the instantaneous overall stiffness tensor of the multiphase composite medium in the damaged configuration C, and is given by

$$C = \frac{1}{V} \sum_{\beta=1}^{N_\beta} \sum_{\gamma=1}^{N_\gamma} v_{\beta\gamma} [m^{(\beta\gamma)}]^{-1} : \bar{C}^{(\beta\gamma)} : [m^{(\beta\gamma)}]^{-1} : A^{(\beta\gamma)} \tag{15.88}$$

From equation (15.88), one concludes that the overall stiffness tensor in the damaged configuration C can be expressed through its subcell (local) stiffness tensors and strain concentration factors in the damaged configuration $C^{(\beta\gamma)}$.

15.3 Numerical Simulation of the Model

The numerical implementation of the proposed model is done for the special case of the unit cell model. The applicability of the incremental damage model is assessed herein by using the unidirectional metal matrix composite material. The damaged response of the subcells as well as for the overall composite system is obtained.

The unit cell model used here assumes that the unidirectional array of fibers(SiC) extending in the X_1 direction is elastic and isotropic while the matrix (Ti-14A1-21Nb) is elasto-plastic work-hardening material and constitutes the three subcell regions around the fiber. Table 15.1 gives the material properties of this composite.

Table 15.1 Material Properties

	Matrix(Ti-14A1-21Nb)	Fiber(SiC)
Modulus	$8 X 10^4$ MPa	$41 X 10^4$ MPa
Poisson's Ratio	0.30	0.22
Initial Volume Fraction	0.65	0.35

The loading is assumed applied incrementally along the fiber direction and damage is checked only for the elastic region. .The representative unit cell used here can be described using non-dimensional quantities and the subcell volume fractions can be given as a function of its non-dimensional quantities (h_1, h_2, ℓ_1, ℓ_2 and h, ℓ) such that

$$c^{(11)} = \frac{h_1\ell_1}{h\ell} \quad , \quad c^{(12)} = \frac{h_1\ell_2}{h\ell} \tag{15.89}$$

$$c^{(21)} = \frac{h_2\ell_1}{h\ell} \quad , \quad c^{(22)} = \frac{h_2\ell_2}{h\ell} \tag{15.90}$$

These non dimensional quantities can be related to the volume fractions of the fiber and matrix as follows

$$c^f = \frac{h_1\ell_1}{h\ell} \quad c^m = \frac{h_1\ell_2 + h_2\ell_1 + h_2\ell_2}{h\ell} \tag{15.91}$$

The relations between h_1 and ℓ_1 , h and ℓ are known. The above non-dimensional quantities can be easily calculated from the phase volume fractions.

In this work for simplicity, the fiber and the unit cell are assumed square i.e $h_1 = \ell_1$ and $h = \ell$. From this assumption one can find the non-dimensional quantities in terms of the phase volume fractions as follows

$$h_1 = \sqrt{c^f} \quad h_2 = 1 - \sqrt{c^f} \tag{15.92}$$

Once the non-dimensional quantities are determined, the next step is to follow the procedure outlined in Section 15.1 in order to obtain the strain concentration tensor $\bar{\boldsymbol{A}}^{(\beta\gamma)}$ of the subcells and the corresponding overall effective stiffness tensor $\bar{C}$ in the undamaged configuration. One can easily observe that in equation (15.27), the strain vector $\dot{\bar{\boldsymbol{\epsilon}}}_s$ is reduced from $(N_\beta N_\gamma$ by 1) to a (24 by 1) vector form. The matrix $\boldsymbol{A_c}$ becomes a (24 by 6) matrix and can be partitioned into four, (6 by 6) matrices where each one of them represents the strain concentration matrix of the corresponding subcell. This matrix is given as follows

$$\bar{\boldsymbol{A}}^{(\beta\gamma)} = \begin{bmatrix} 1 & 0 & 0 & 0 & 0 & 0 \\ A_{21} & A_{22} & A_{33} & 0 & 0 & 0 \\ A_{31} & A_{32} & A_{33} & 0 & 0 & 0 \\ 0 & 0 & 0 & A_{44} & 0 & 0 \\ 0 & 0 & 0 & 0 & A_{55} & 0 \\ 0 & 0 & 0 & 0 & 0 & A_{66} \end{bmatrix}^{(\beta\gamma)} \tag{15.93}$$

More elaborate information about the strain concentration matrix can be found in references [203,204].

The damage evolution for the subcell of the proposed model is performed by following the formulation in Section 15.2.4. The tensorial manipulation is preferred in the numerical solution in order to get more elaborate and consistent results. In the damage analysis of materials, the main objective is to satisfy the consistency condition ($g \equiv 0$) at any state of damage. This phenomenon can be explained as follows. Loading of the material by an increment of stress in the damaged state causes the stress tensor to move to the subsequent damage surface, which defines the boundary of the current undamaged region. A this state g is only a function of the three variables $\boldsymbol{\sigma}, \boldsymbol{\phi}$ and κ. If the stress point lies within the undamaged region, no damage takes places. i.e. $\boldsymbol{\phi} = \mathbf{0}$ and $\kappa = 0$. On the other hand if the state of stress at this point is increased by an increment of stress, the current state of stress will not be in equilibrium such that $g(\boldsymbol{\sigma} + d\boldsymbol{\sigma}, \boldsymbol{\phi}, \kappa) > 0$ which would mean that the current stress point has left the damage surface, which is impossible.

In order to bring the stress point back on the damage surface, an increment of damage $d\boldsymbol{\phi}$ and $d\kappa$ are induced by equation (15.59) and (15.78) respectively. The current damage surface $g(\boldsymbol{\sigma} + d\boldsymbol{\sigma}, \boldsymbol{\phi} + d\boldsymbol{\phi}, \kappa + d\kappa) = 0$ will then be satisfied.

The numerical solution investigates the damage evolution for each subcell separately by using different damage parameters for different constituents of the metal matrix composite. Small stress increments are applied along the fiber direction. These damage parameters for the matrix and fiber are given in Table 15.2.

In the one dimensional state of stress, the relation between the scalar value of the overall damage and the subcell damage can be obtained by assuming that the volume fractions of the material in the initial configuration configuration C_o and in the damaged configuration C to be the same. The volume fractions for these three configurations are given as follows

$$c_o = \frac{A_o^{(\beta\gamma)}}{A_o} \qquad c = \frac{A^{(\beta\gamma)}}{A} \qquad \bar{c} = \frac{\bar{A}^{(\beta\gamma)}}{\bar{A}} \tag{15.94}$$

where $c_o = c$ is assumed. One can express the total area of each configuration as a sum of the areas of the subcells such that

$$A_o = \sum_{\beta,\gamma=1}^{2} A_o \qquad A = \sum_{\beta,\gamma=1}^{2} A \qquad \bar{A} = \sum_{\beta,\gamma=1}^{2} \bar{A} \tag{15.95}$$

Table 15.2 Local Damage Parameters

	Matrix Damage	Fiber Damage
η_1	0.08	0.06
η_2	0.08	0.06
η_3	0.08	0.06
ξ_1	0.55	0.52
ξ_2	0.55	0.52
ξ_3	0.55	0.52
ν_1	0.0013	0.001
ν_2	0.0013	0.001
ν_3	0.0013	0.001

Making use of equations (15.94) and (15.95), one can obtain the following relation

$$\phi = \sum_{\beta,\gamma=1}^{2} \frac{A^{(\beta\gamma)}}{A} \phi^{(\beta\gamma)} \tag{15.96}$$

if the term $\frac{A^{(\beta\gamma)}}{A}$ is replaced by the initial volume fraction c_o, the above equation yields the following expression

$$\phi = \sum_{\beta,\gamma=1}^{2} c_o{}^{(\beta\gamma)} \phi^{(\beta\gamma)} \tag{15.97}$$

The program output gives the damage response of the material in each subcell as well as the overall. In Figure 15.4, different values for the parameter ν are used to plot the damage criterion, g, versus the stress in order to study the sensitivity and robustness of this parameter. For the range of values used here $1.8X10^{-3}$ to $8X10^{-4}$ the behavior of the parameter is quite robust. In Figure 15.5 different values for the parameter ν are used in order to plot the damage variable ϕ versus σ for the subcells (12) and (22). These subcells are chosen in order show how the damage can vary in each subcell even though both cells may have the same material properties. This implies that the boundary and geometry conditions are effective in analyzing the damage of the subcells. In Figures 15.6 and 15.7 the variation of parameters η and ξ is studied by plotting the damage versus the stress. It is observed that a 0.2 change between the different values of η is more sensitive to the damage behavior of the material than a difference in ξ values of 0.05.

The corresponding parameters η and ξ are evaluated for the fiber in subcell (11). It is observed that for η values between 0.1 and 0.06 and for ξ values between 0.48 and 0.52, the material is quite sensitive to damage, which is indicated in Figures 15.8 and 15.9 respectively.   In Figure 15.10 the damage versus the stress is plotted for the different subcells together with the overall damage in order to study the local versus the overall relation. The model gives the expected results such that the overall damage behavior is the average of the local ones.

Finally the stress strain curves for the subcells and for the overall composite are plotted and compared with their undamaged curves in Figure 15.11. It is clear that there is a reduction in the stiffnesses of the material with an accompanying non linear behavior after the damage is initiated in the material.

Damage Criterion for Different Damage Parameters
Subcell (12)

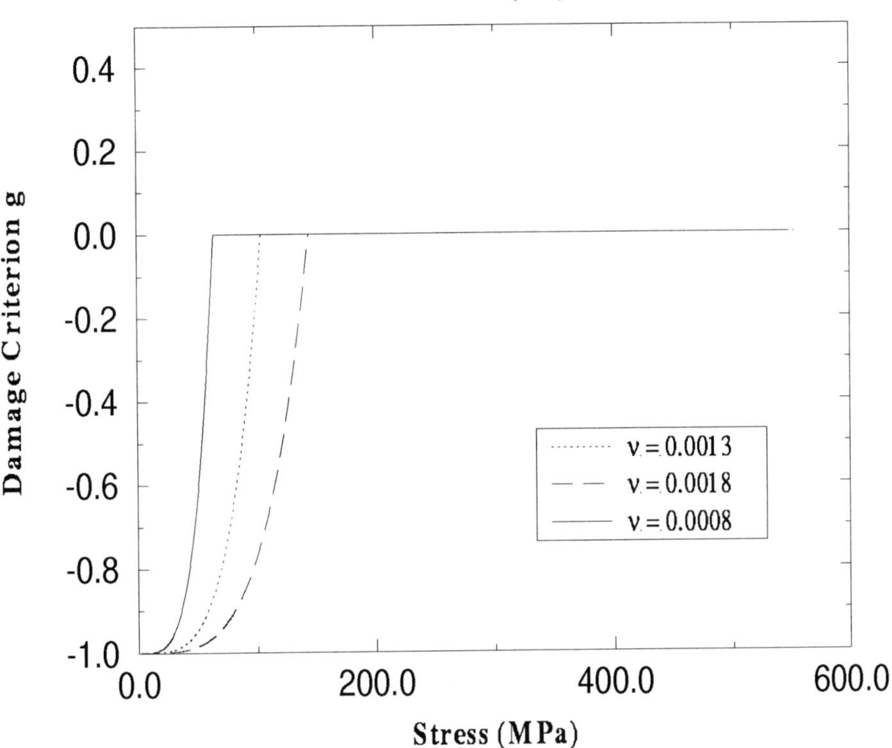

Figure 15.4 Damage Criterion for Subcell (12)

424

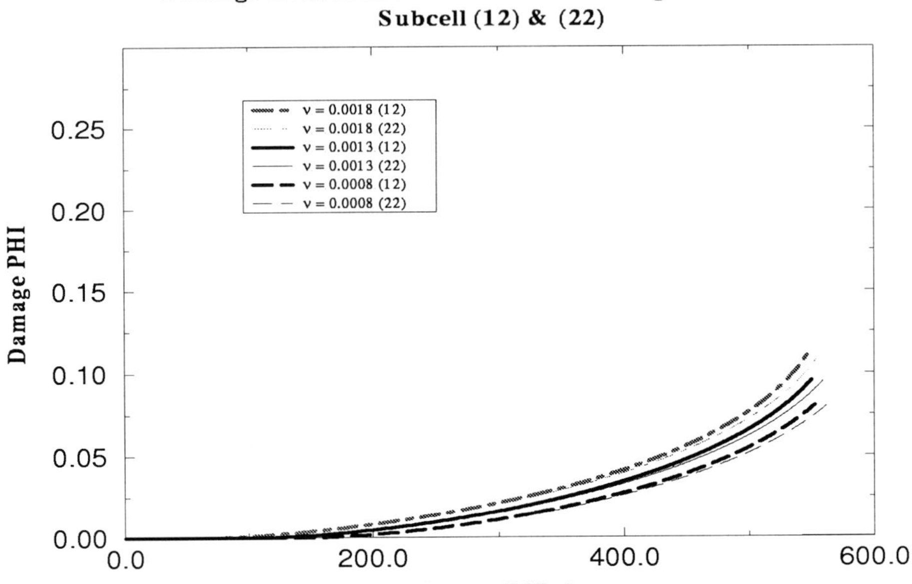

Figure 15.5 Damage Evolution for Different Values of ν in Subcell

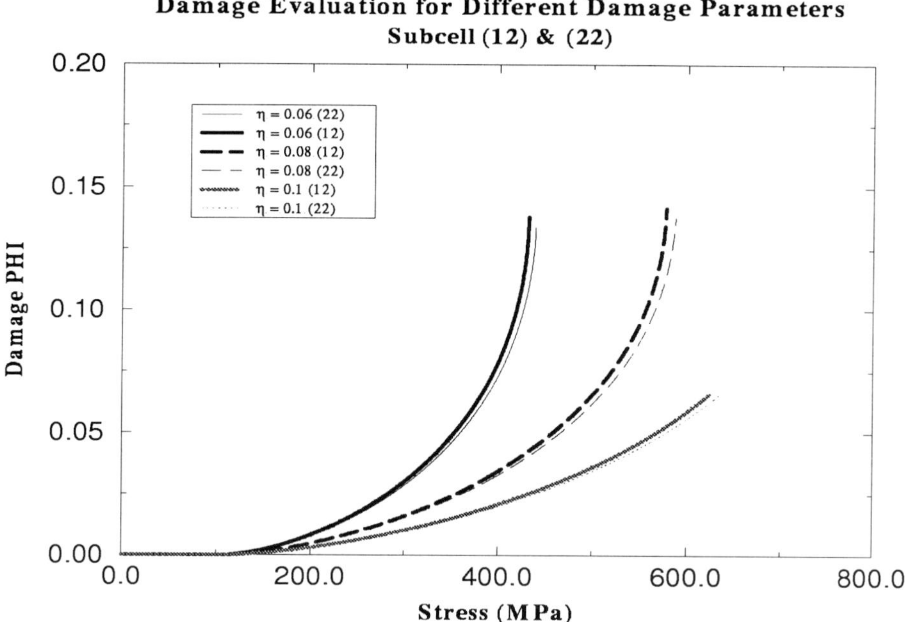

Figure 15.6 Damage Evolution for Different Values of η in Subcells

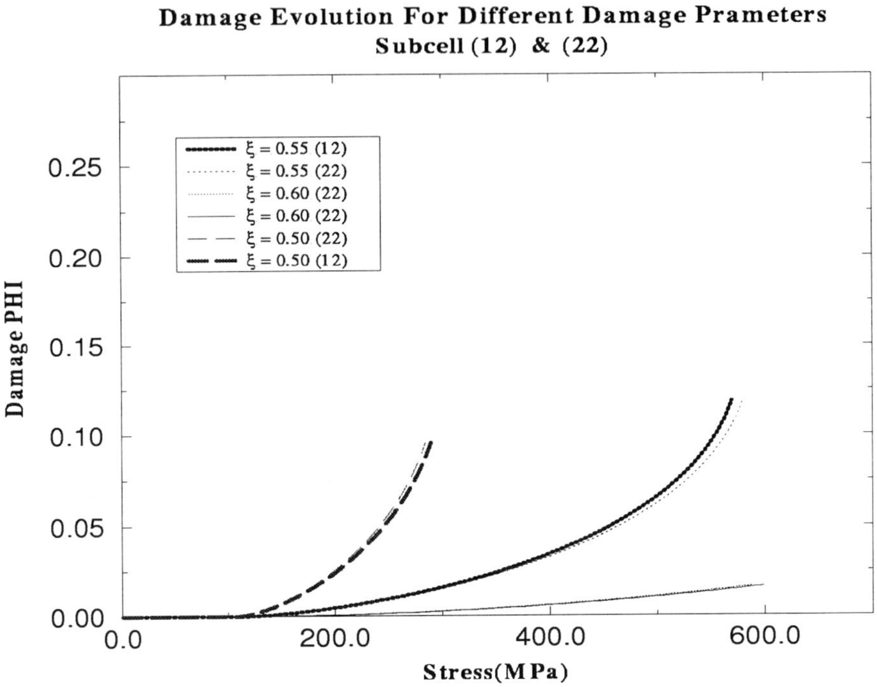

Figure 15.7 Damage Evolution for Different Values of ξ in Subcells

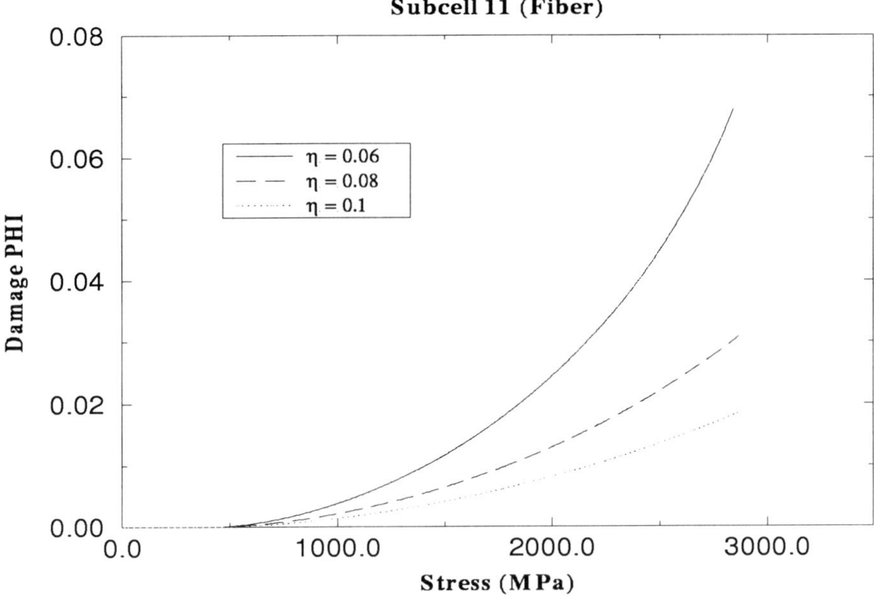

Figure 15.8 Damage Evolution for Different Values of η in Subcell (11)

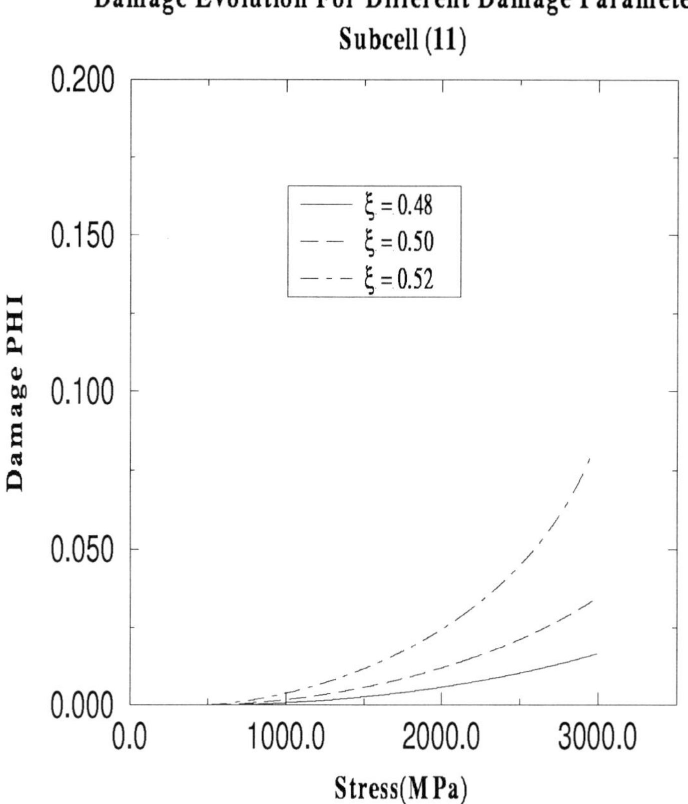

Figure 15.9 Damage Evolution for Different Values of ξ in Subcell

Damage Evolution for Unit Cell Element

(Subcells (11), (12) , (22)) and (Overall-1)

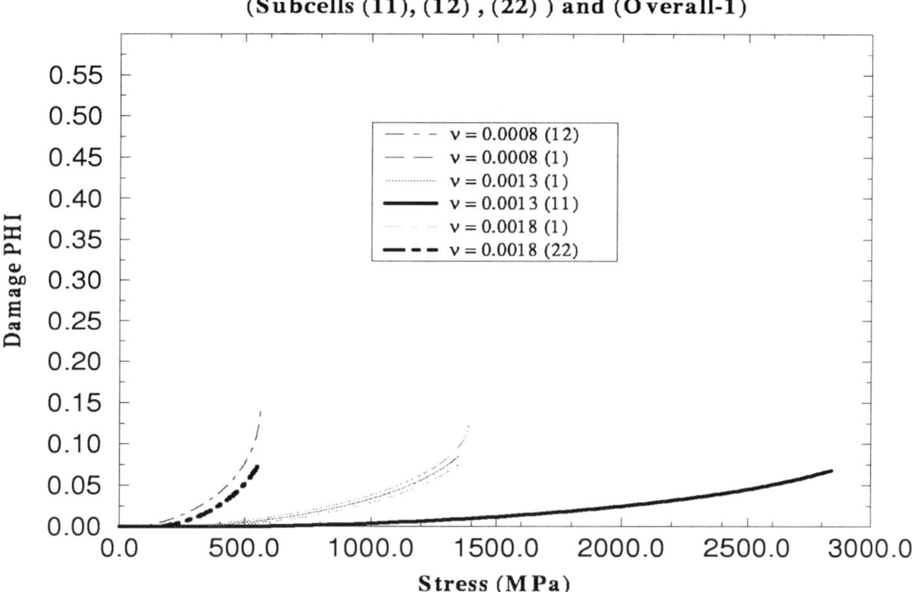

Figure 15.10 Damage Evolution for Different Subcells and Overall

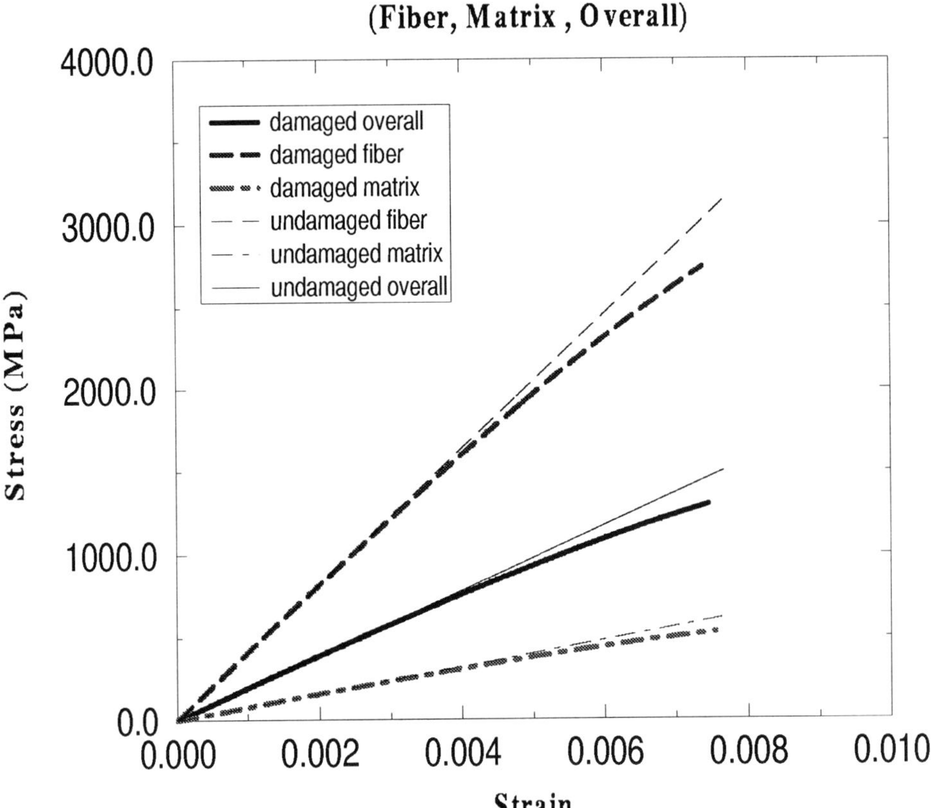

Figure 15.11   Material Stiffness

CHAPTER 16
THE KINEMATICS OF DAMAGE FOR FINITE-STRAIN ELASTO-PLASTIC SOLIDS

In this Chapter paper the kinematics of damage for finite strain, elasto-plastic deformation is introduced using the fourth-order damage effect tensor through the concept of the effective stress within the framework of continuum damage mechanics. In the absence of the kinematic description of damage deformation leads one to adopt one of the following two different hypotheses for the small deformation problems. One uses either the hypothesis of strain equivalence or the hyphothesis of energy equivalence in order to characterize the damage of the material. The proposed approach in this work provides a general description of kinematics of damage applicable to finite strains. This is accomplished by directly considering the kinematics of the deformation field and furthermore it is not confined to small strains as in the case of the strain equivalence or the strain energy equivalence approaches. In this work, the damage is described kinematically in both the elastic domain and plastic domain using the fourth order damage effect tensor which is a function of the second-order damage tensor. The damage effect tensor is explicitly characterized in terms of a kinematic measure of damage through a second-order damage tensor. Two kinds of second-order damage tensor representations are used in this work with respect to two reference configurations. The finite elasto-plastic deformation behavior with damage is also viewed here within the framework of thermodynamics with internal state variables. Using the consistent thermodynamic formulation one introduces seperately the strain due to damage and the associated dissipation energy due to this strain.

In continuum damage mechanics, the effective stress tensor is usually not symmetric. This leads to a complicated theory of damage mechanics involving micropolar media and the Cosserat continuum. Therefore, to avoid such a theory, symmetrization of the effective stress tensor is used to formulate a continuum damage theory in the classical sense (Lee et al. [130], Sidoroff [16], Cordebois and Sidoroff [111], and Murakami and Ohno [22], Betten [222]). Recently, Voyiadjis and Park [211] reviewed a linear transformation tensor, defined as a fourth-order damage effect tensor and focused on its geometric symmetrization method in order to describe the kinematics of damage using the second-order damage tensor. Voyiadjis and Park [211] introduced the kinematics of damage in the finite deformation field using the damage effect tensor which does not only symmetrize the effective stress tensor but can also be related to the deformation gradient of damage.

The kinematics of damage is described here using the second-order damage tensor. The deformation gradient of damage is defined using the second-order damage tensor. The Green deformation tensor of the damage elasto-plastic deformation is also derived.

For a detailed review of the principles of continuum damage machanics as used in this work, the reader is referred to the works of Kachanov [1], Lemaitre [11,31], Krajcinovic [89], Lubarda and Krajcinovic [227], Chaboche [35,36,87], Murakami [58], Sidoroff [16], and Voyiadjis and Kattan [99].

432

16.1 Theoretical Preliminaries

A continuous body in an initial undeformed configuration that consists of the material volume Ω° is denoted by C°, while the elasto-plastic damage deformed configuration at time t after the body is subjected to a set of external agencies is denoted by C^t. The corresponding material volume at time, t is denoted by Ω^t. Upon elastic unloading from the configuration C^t an intermediate stress free configuration is denoted by C^{dp}. In the framework of continuum damage mechanics a number of fictitious configurations, based on the effective stress concept, are assumed that are obtained by fictitiously removing all the damage that the body has undergone. Thus the fictitious configuration of the body denoted by $\bar{C}^t$ is obtained from C^t by fictitiously removing all the damage that the body has undergone at C^t. Also the fictitious configuration denoted by $\tilde{C}^p$ is assumed which is obtained from C^{dp} by fictitiously removing all the damage that the body has undergone at C^{dp}. While the configuration $\bar{C}^p$ is the intermediate configuration upon unloading from the configuration $\bar{C}^t$. The initial undeformed body may have a pre-existing damage state. The initial fictitious effective configuration denoted by $\bar{C}^\circ$ is defined by removing the initial damage from the initial undeformed configuration of the body. In the case of no initial damage existing in the undeformed body, the initial fictitious effective configuration is identical to the initial undeformed configuration. Cartesian tensors are used in this work and the tensorial index notation is employed in all equations. The tensors used in the text are denoted by boldface letters. However, superscripts in the notation do not indicate tensorial index but merely stand for corresponding deformation configurations such as "e" for elastic, "p" for plastic, and "d" for damage etc. The barred and tilded notations refer to the fictitious effective configurations.

16.2 Description of Damage State

The damage state can be described using an even order tensor (Leckie and Onat [26] and Betten [222]). Ju [228] pointed out that even for isotropic damage one should employ a damage tensor(not a scalar damage variable) to characterize the state of damage in materials. However, the damage generally is anisotropic due to the external agency condition or the material nature itself. Although the fourth-order damage tensor can be used directly as a linear transformation tensor to define the effective stress tensor, it is not easy to characterize physically the fourth-order damage tensor compared to the second-order damage tensor. In this work, the damage is considered as a symmetric second-order tensor. However, damage tensor for the finite elasto-plastic deformation can be defined in two reference systems [58]. The first one is the damage tensor denoted by ϕ representing the damage state with respect to the current damaged configuration, C^t. Another one is denoted by φ and is representing the damage state with respect to the elastically unloaded damage configuration, C^{dp}. Both are given by Murakami [23] as follows

$$\phi_{ij} = \sum_{k=1}^{3} \hat{\phi}_k \hat{n}_i^k \hat{n}_j^k \quad (no\ sum\ in\ k) \tag{16.1}$$

and

$$\varphi_{ij} = \sum_{k=1}^{3} \hat{\varphi}_k \hat{m}_i^k \hat{m}_j^k \quad (no\ sum\ in\ k) \tag{16.2}$$

where $\hat{\boldsymbol{n}}^k$ and $\hat{\boldsymbol{m}}^k$ are eigenvectors corresponding to the eigenvalues, $\hat{\phi}_k$ and $\hat{\varphi}_k$, of the damage tensors, $\boldsymbol{\phi}$ and $\boldsymbol{\varphi}$, respectively. Equations (16.1) and (16.2) can be written alternatively as follows

$$\phi_{ij} = b_{ir} b_{js} \hat{\phi}_{rs} \tag{16.3}$$

and

$$\varphi_{ij} = c_{ir} c_{js} \hat{\varphi}_{rs} \tag{16.4}$$

The damage tensors in the coordinate system that coincides with the three orthogonal principal directions of the damage tensors, $\hat{\phi}_{rs}$ and $\hat{\varphi}_{rs}$, in equations (16.3) and (16.4) are obviously of diagonal form and are given by

$$\hat{\phi}_{ij} = \begin{bmatrix} \hat{\phi}_1 & 0 & 0 \\ 0 & \hat{\phi}_2 & 0 \\ 0 & 0 & \hat{\phi}_3 \end{bmatrix} \tag{16.5}$$

$$\hat{\varphi}_{ij} = \begin{bmatrix} \hat{\varphi}_1 & 0 & 0 \\ 0 & \hat{\varphi}_2 & 0 \\ 0 & 0 & \hat{\varphi}_3 \end{bmatrix} \tag{16.6}$$

and the second order transformation tensors, $\boldsymbol{b}$ and $\boldsymbol{c}$ are given by

$$b_{ir} = \begin{bmatrix} n_1^1 & n_2^1 & n_3^1 \\ n_1^2 & n_2^2 & n_3^2 \\ n_1^3 & n_2^3 & n_3^3 \end{bmatrix} \tag{16.7}$$

$$c_{ir} = \begin{bmatrix} m_1^1 & m_2^1 & m_3^1 \\ m_1^2 & m_2^2 & m_3^2 \\ m_1^3 & m_2^3 & m_3^3 \end{bmatrix} \tag{16.8}$$

This proper orthogonal transformation tensor requires that

$$\begin{aligned} b_{ij} b_{kj} &= c_{ij} c_{kj} \\ &= \delta_{ik} \end{aligned} \tag{16.9}$$

where δ_{ik} is a kronecker delta and the determinants of the matrix $[b]$ and $[c]$ are given by

$$\begin{aligned} \|[b]\| &= \|[c]\| \\ &= 1 \end{aligned} \tag{16.10}$$

The relation between the damage tensors $\boldsymbol{\phi}$ and $\boldsymbol{\varphi}$ is shown in section 16.5.

16.3 Fourth-Order Anisotropic Damage Effect Tensor

In a general state of deformation and damage, the effective stress tensor $\bar{\sigma}$ is related to the Cauchy stress tensor σ by the following linear transformation (Murakani and Ohno [22])

$$\bar{\sigma}_{ij} = M_{ikjl}\sigma_{kl} \tag{16.11}$$

where M is a fourth-order linear transformation operator called the damage effect tensor. Depending on the form used for M, it is very clear from equation (17.109) that the effective stress tensor $\bar{\sigma}$ is generally nonsymmetric. Using a non-symmetric effective stress tensor as given by equation (17.109) to formulate a constitutive model will result in the introduction of the Cosserat and a micropolar continua. However, the use of such complicated mechanics can be easily avoided if the proper fourth-order linear transformation tensor is formulated in order to symmetrize the effective stress tensor. Such a linear transformation tensor called the damage effect tensor is obtained in the literature [16,130] using symmetrization methods. One of the symmetrization methods given by Cordebois and Sidoroff [17] and Lee et al. [130] is expressed as follows

$$\bar{\sigma}_{ij} = (\delta_{ik} - \phi_{ik})^{-1/2}\sigma_{kl}(\delta_{jl} - \phi_{jl})^{-1/2} \tag{16.12}$$

The fourth-order damage effect tensors corresponding to equations (16.12) is defined such that

$$M_{ikjl} = (\delta_{ik} - \phi_{ik})^{-1/2}(\delta_{jl} - \phi_{jl})^{-1/2} \tag{16.13}$$

In order to describe the kinematics of damage, the physical meaning of the fourth-order damage effect tensor should be interpreted and not merely given as the symmertrization of the effective stress. In this work, the fourth-order damage effect tensor given by equation (16.13) will be used because of its geometrical symmetrization of the effective stress [17]. However, it is very difficult to obtain the explicit representation of $(\delta_{ik} - \phi_{ik})^{-1/2}$. The explicit representation of the fourth-order damage effect tensor M using the second-order damage tensor ϕ is of particular importance in the implementation of the constitutive modeling of damage mechanics. Therefore, the damage effect tensor M of equation (16.13) should be obtained using the coordinate transformation of the principal damage direction coordinate system. Thus the fourth-order damage effect tensor given by equation (16.13) can be written as follows (Voyiadjis and Park [211])

$$M_{ikjl} = b_{mi}b_{nj}b_{pk}b_{ql}\hat{M}_{mnpq} \tag{16.14}$$

where $\hat{M}$ is a fourth-order damage effect tensor with reference to the principal damage direction coordinate system. The fourth-order damage effect tensor $\hat{M}$ can be written as follows (Voyiadjis and Park [210])

$$\hat{M}_{mpnq} = \hat{a}_{mp}\hat{a}_{nq} \tag{16.15}$$

where the second-order tensor $\boldsymbol{a}$ in the principal damage direction coordinate system is given by

$$\hat{a}_{mp} = [\delta_{mp} - \hat{\phi}_{mp}]^{-\frac{1}{2}}$$

$$= \begin{bmatrix} \dfrac{1}{\sqrt{1-\hat{\phi}_1}} & 0 & 0 \\[3ex] 0 & \dfrac{1}{\sqrt{1-\hat{\phi}_2}} & 0 \\[3ex] 0 & 0 & \dfrac{1}{\sqrt{1-\hat{\phi}_3}} \end{bmatrix} \tag{16.16}$$

Substituting equation (16.15) into equation (16.14), one obtains the following relation

$$\begin{aligned} M_{ikjl} &= b_{mi}b_{nj}b_{pk}b_{ql}\hat{a}_{mp}\hat{a}_{nq} \\ &= a_{ik}a_{jl} \end{aligned} \tag{16.17}$$

Using equation (16.17), a second-order tensor $\boldsymbol{a}$ is defined as follows

$$a_{ik} = b_{mi}b_{pk}\hat{a}_{mp} \tag{16.18}$$

The matrix form of equation (16.18) is as follows(Voyiadjis and Park [211])

$$[\,a\,] = [\,b\,]^T[\,\hat{a}\,][\,b\,]$$

$$= \begin{bmatrix} \dfrac{b_{11}b_{11}}{\sqrt{1-\hat{\phi}_1}} + \dfrac{b_{21}b_{21}}{\sqrt{1-\hat{\phi}_2}} + \dfrac{b_{31}b_{31}}{\sqrt{1-\hat{\phi}_3}} & \dfrac{b_{11}b_{12}}{\sqrt{1-\hat{\phi}_1}} + \dfrac{b_{21}b_{22}}{\sqrt{1-\hat{\phi}_2}} + \dfrac{b_{31}b_{32}}{\sqrt{1-\hat{\phi}_3}} & \dfrac{b_{11}b_{13}}{\sqrt{1-\hat{\phi}_1}} + \dfrac{b_{21}b_{23}}{\sqrt{1-\hat{\phi}_2}} + \dfrac{b_{31}b_{33}}{\sqrt{1-\hat{\phi}_3}} \\[4ex] \dfrac{b_{12}b_{11}}{\sqrt{1-\hat{\phi}_1}} + \dfrac{b_{22}b_{21}}{\sqrt{1-\hat{\phi}_2}} + \dfrac{b_{32}b_{31}}{\sqrt{1-\hat{\phi}_3}} & \dfrac{b_{12}b_{12}}{\sqrt{1-\hat{\phi}_1}} + \dfrac{b_{22}b_{22}}{\sqrt{1-\hat{\phi}_2}} + \dfrac{b_{32}b_{32}}{\sqrt{1-\hat{\phi}_3}} & \dfrac{b_{12}b_{13}}{\sqrt{1-\hat{\phi}_1}} + \dfrac{b_{22}b_{23}}{\sqrt{1-\hat{\phi}_2}} + \dfrac{b_{32}b_{33}}{\sqrt{1-\hat{\phi}_3}} \\[4ex] \dfrac{b_{13}b_{11}}{\sqrt{1-\hat{\phi}_1}} + \dfrac{b_{23}b_{21}}{\sqrt{1-\hat{\phi}_2}} + \dfrac{b_{33}b_{31}}{\sqrt{1-\hat{\phi}_3}} & \dfrac{b_{13}b_{12}}{\sqrt{1-\hat{\phi}_1}} + \dfrac{b_{23}b_{22}}{\sqrt{1-\hat{\phi}_2}} + \dfrac{b_{33}b_{32}}{\sqrt{1-\hat{\phi}_3}} & \dfrac{b_{13}b_{13}}{\sqrt{1-\hat{\phi}_1}} + \dfrac{b_{23}b_{23}}{\sqrt{1-\hat{\phi}_2}} + \dfrac{b_{33}b_{33}}{\sqrt{1-\hat{\phi}_3}} \end{bmatrix} \tag{16.19}$$

16.4   The Kinematics of Damage for Elasto-Plastic Behavior with Finite Strains

A position of a particle in C° at t° is denoted by $\boldsymbol{X}$ and can be defined at its corresponding position in C^t at t, denoted by $\boldsymbol{x}$. Futhermore, assuming that the deformation is smooth regardless of damage, one can assume a one-to-one mapping such that

$$x_k = x_k(\boldsymbol{X}, t) \tag{16.20}$$

or

$$X_k = X_k(\boldsymbol{x}, t) \tag{16.21}$$

The corresponding deformation gradient is expressed as follows

$$F_{ij} = \frac{\partial x_i}{\partial X_j} \tag{16.22}$$

and the change in the squared length of a material filament $d\boldsymbol{X}$ is used as a measure of deformation such that

$$\begin{aligned}
(ds)^2 - (dS)^2 &= dx_i\, dx_i - dX_i\, dX_i \\
&= 2\,\mathcal{E}_{ij}\, dX_i\, dX_j
\end{aligned} \tag{16.23}$$

or

$$(ds)^2 - (dS)^2 = 2\,\epsilon_{ij}\, dx_i\, dx_j \tag{16.24}$$

where $(ds)^2$ and $(dS)^2$ are the squared lengths of the material filaments in the deformed with damage configuration C^t, and the initial undeformed configuration C° respectively. $\mathcal{E}$ and ϵ are the Lagrngian and Eulerian strain tensors respectively and are given by

$$\begin{aligned}
\mathcal{E}_{ij} &= \frac{1}{2}[F_{ki}F_{kj} - \delta_{ij}] \\[2mm]
&= \frac{1}{2}(C_{ij} - \delta_{ij})
\end{aligned} \tag{16.25}$$

$$\begin{aligned}
\epsilon_{ij} &= \frac{1}{2}[\delta_{ij} - F_{ki}^{-1}F_{kj}^{-1}] \\[2mm]
&= \frac{1}{2}[\delta_{ij} - B_{ij}^{-1}]
\end{aligned} \tag{16.26}$$

where C and B are the right Cauchy-Green and the left Cauchy-Green tensors, respectively. The velocity vector field in the current configuration at time t is given by

$$v_i = \frac{dx_i}{dt} \tag{16.27}$$

The velocity gradient in the current configuration at time t is given by

$$\begin{aligned}
\mathcal{L}_{ij} &= \frac{\partial v_i}{\partial x_j} \\[2mm]
&= \dot{F}_{ik}F_{kj}^{-1} \\[2mm]
&= \mathcal{D}_{ij} + \mathcal{W}_{ij}
\end{aligned} \tag{16.28}$$

where the dot designates the material time derivative and where $\mathcal{D}$ and $\mathcal{W}$ are the rate of deformation (stretching) and the vorticity, respectively. The rate of deformation, $\mathcal{D}$ is equal to the symmetric part of the velocity gradient $\mathcal{L}$ while the vorcity, $\mathcal{W}$ is the antisymmetric part of the velocity gradient $\mathcal{L}$ such that

$$\mathcal{D}_{ij} = \frac{1}{2}(\mathcal{L}_{ij} + \mathcal{L}_{ji}) \qquad (16.29)$$

$$\mathcal{W}_{ij} = \frac{1}{2}(\mathcal{L}_{ij} - \mathcal{L}_{ji}) \qquad (16.30)$$

Strain rate measures are obtained by differentiating equations (16.23) and (16.24) such that

$$\frac{d}{dt}[(ds)^2 - (dS)^2] = 2\,dX_i\,\dot{\mathcal{E}}_{ij}\,dX_j \qquad (16.31a)$$
$$= 2\,dx_i\,\mathcal{D}_{ij}\,dx_j \qquad (16.31b)$$
$$= 2\,dX_i\,F_{ik}\mathcal{D}_{ij}F_{jm}\,dX_m \qquad (16.31c)$$
$$= 2\,dx_i\,[\dot{\epsilon}_{ij} + \epsilon_{ik}\mathcal{L}_{kj} + \mathcal{L}_{ik}\epsilon_{kj}]\,dx_j \qquad (16.31d)$$

By comparing equations (16.31a) and (16.31c) one obtains the rate of the Lagrangian strain that is the projection of $\mathcal{D}$ onto the reference frame as follows

$$\dot{\mathcal{E}}_{ij} = F_{ki}\mathcal{D}_{kl}F_{lj} \qquad (16.32)$$

while the deformation rate $\mathcal{D}$ is equal to the Cotter-Rivin convected rate of the Eulerian strain as follows

$$\mathcal{D}_{ij} = \dot{\epsilon}_{ij} + \epsilon_{ik}\mathcal{L}_{kj} + \mathcal{L}_{ik}\epsilon_{kj} \qquad (16.33)$$

The convected derivative shown in equation (16.33) can also be interpreted as the Lie derivative of the Eulerian strain.

16.4.1 A Multiplicative Decomposition

A schematic drawing representing the kinematics of elasto-plastic damage deformation is shown in Figure 16.1. C° is the initial undeformed configuration of the body which may have an initial damage in the material. C^t represents the current elasto-plastically deformed and damaged configuration of the body. The configuration $\bar{C}^\circ$ represents the initial configuration of the body that is obtained by fictitiously removing the initial damage from the C° configuration. If the initial configuration is undamaged consequently there is no difference between configurations C° and $\bar{C}^\circ$. Configuration $\bar{C}^t$ is obtained by fictitiously removing the damage from configuration C^t. Configuration C^{dp} is an intermediate configuration upon elastic unloading. In the most general case of large deformation processes, damage may be involved due to void and microcrack development because of external agencies. Although damage in the microlevel is a material discontinuity, damage can be considered as an irreversible deformation process in the framework of Continuum Damage Mechanics.

438

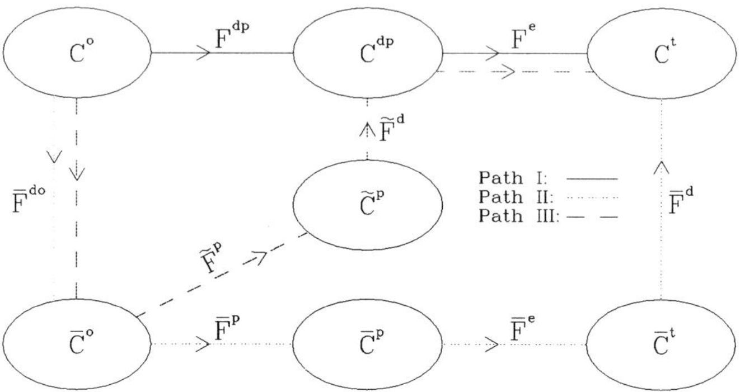

Figure 16.1: Schematic representation of elasto-plastic damage deformation configurations

Furthermore, one assumes that upon unloading from the elasto-palstic damage state, the elastic part of the deformation can be completely recovered while no additional plastic deformation and damage takes place. Thus upon unloading the elasto-plastic damage deformed body from the current configuration C^t will elastically unload to an intermediate stress free configuration denoted by C^{dp} as shown in Figure 16.1. Although the damage process is an irreversible deformation thermodynamically, however, deformation due to damage itself can be partially or completely recovered upon unloading due to closure of micro-cracks or contraction of micro-voids. Nevertheless, recovery of damage deformation does not mean the healing of damage. No naterials are brittle or ductile. The deformation gradient tensor and the Green deformation tensor of the elasto-plastic damage deformation can be obtained through Path I, Path II or Path III as shown in Figure 16.1. Considering Path I the deformation gradient referred to the undeformed configuration, C^o is denoted by $\boldsymbol{F}$ and is polarly decomposed into the elastic deformation gradient denoted by $\boldsymbol{F}^e$ and the damage-plastic deformation gradient denoted by $\boldsymbol{F}^{dp}$ such that

$$F_{ij} = F^{e}_{ik} F^{dp}_{kj} \tag{16.34}$$

The elastic deformation gradient is given by

$$F^{e}_{ij} = \frac{\partial x_i}{\partial x^{dp}_j} \tag{16.35}$$

The corresponding damage-plastic deformation gradient is given by

$$F^{dp}_{ij} = \frac{\partial x^{dp}_i}{\partial X_j} \tag{16.36}$$

The Right Cauchy Green deformation tensor, $\boldsymbol{C}$, is given by

$$C_{ij} = F^{dp}_{nk} F^{e}_{ki} F^{e}_{nm} F^{dp}_{mj} \tag{16.37}$$

The finite deformation damage models by Ju [228] and Zbib [229] emphasize that "added flexibility" due to the existence of microcracks or microvoids is already embedded in the deformation gradient implicity. Murakami [58] presented the kinematics of damage deformation using the second-order damage tensor. However, the lack of an explicit formulation for the kinematics of finite deformation with damage leads to the failure in obtaining an explicit derivation of the kinematics that directly consider the damage deformation. Although most finite strain elasto-plastic deformation processes involve damage such as micro-voids, nucleations and micro-crack development due to external agencies, however, only the elastic and plastic deformation processes are cosidered kinematically due to the complexity in the involvement of damage deformation. In this work, the kinematics of damage will be explicitly characterized based on continuum damage mechanics. The elastic deformation gradient corresponds to elastic stretching and rigid body rotations due to both internal and external constraints. The plastic deformation gradient is arising from purely irreversible processes due to dislocations in the material. Damage may be initiated and evolves in both the elastic and plastic deformation processes. Particularly, damage in the elastic deformation state is termed elastic damage which is the case for most brittle materials while damage in the plastic deformation state is termed plastic damage which is mainly for ductile materials. Additional deformation due to damage consists of damage itself with additional deformation due to elastic and plastic deformation. This causes loss of elastic and plastic stiffness. In this work, kinematics of damage deformation is completely described for both damage and the coupling of damage with elasto-plastic deformation. The total Lagrangian strain tensor is expressed as follows

$$\begin{aligned}
\mathcal{E}_{ij} &= \frac{1}{2}(F_{ki}^{dp} F_{kj}^{dp} - \delta_{ij}) + \frac{1}{2} F_{mi}^{dp}(F_{km}^{e} F_{kn}^{e} - \delta_{mn}) F_{nj}^{dp} \\
&= \mathcal{E}_{ij}^{dp} + F_{mi}^{dp} \varepsilon_{mn}^{e} F_{nj}^{dp} \\
&= \mathcal{E}_{ij}^{dp} + \mathcal{E}_{ij}^{e}
\end{aligned} \tag{16.38}$$

where $\mathcal{E}^{dp}$ and $\mathcal{E}^{e}$ are the Lagrangian damage-plastic strain tensor and the Lagrangian elastic strain tensor measured with respect to the reference configuration C^{o}, respectively. While ε^{e} is the Lagrangian elastic strain tensor measured with respect to the intermediate configuration C^{dp}. Similiarly, the Eulerian strains corresponding to deformation gradients F^{e} and F^{dp} are given by

$$e_{ij}^{dp} = \frac{1}{2}(\delta_{ij} - F_{ki}^{dp^{-1}} F_{kj}^{dp^{-1}}) \tag{16.39}$$

$$\epsilon_{ij}^{e} = \frac{1}{2}(\delta_{ij} - F_{ki}^{e^{-1}} F_{kj}^{e^{-1}}) \tag{16.40}$$

The total Eulerian strain tensor can be expressed as follows

$$\begin{aligned}
\epsilon_{ij} &= \epsilon_{ij}^{e} + F_{ki}^{e^{-1}} e_{km}^{dp} F_{mj}^{e^{-1}} \\
&= \epsilon_{ij}^{e} + \epsilon_{ij}^{dp}
\end{aligned} \tag{16.41}$$

440

The strain $\boldsymbol{e}^{dp}$ is refered to the intermediate configuration C^{dp}, while the strains $\boldsymbol{\epsilon}$, $\boldsymbol{\epsilon}^e$, and $\boldsymbol{\epsilon}^{dp}$ are defined relative to the current configuration as a reference. The relationship between the Lagrangian and Eulerian strains is obtained directly in the form

$$\mathcal{E}_{ij} = F_{ki}\epsilon_{kl}F_{lj} \tag{16.42}$$

The change in the squared length of a material filament deformed elastically from C^t to C^{dp} is given by

$$(ds)^2 - (ds^{dp})^2 = dx_i\, dx_i - dx_i^{dp}\, dx_i^{dp}$$

$$= 2\, dX_i\, \mathcal{E}_{ij}^e\, dX_j \tag{16.43}$$

However, the change in the squared length of a material filament deformed due to damage and plastic deformation from C^{dp} to C° is given by

$$(ds^{dp})^2 - (dS)^2 = 2\, dX_i\, \mathcal{E}_{ij}^{dp}\, dX_j \tag{16.44}$$

The kinematics of finite strain elasto-plastic deformation including damage is completely described in Path I. In order to describe the kinematics of damage and plastic deformation, the deformation gradient given by equation (16.34) may be further decomposed into

$$F_{ij} = F_{ik}^e F_{km}^d F_{mj}^p \tag{16.45}$$

However, it is very difficult to characterize physically only the kinematics of deformation due to damage inspite of its obvious physical phenomenon. The damage, however, may be defined through the effective stress concept. Similarly the kinematics of damage can be described using the effective kinematic configuration. Considering Path II the deformation gradient can be alternatively expressed as follows

$$F_{ij} = \bar{F}_{ik}^d \bar{F}_{km}^e \bar{F}_{mn}^p \bar{F}_{nj}^{do} \tag{16.46}$$

where $\bar{\boldsymbol{F}}^d$ is the fictitious damage deformation gradient from configuration $\bar{C}^t$ to C^t and is given by

$$\bar{F}_{ij}^d = \frac{\partial x_i}{\partial \bar{x}_j} \tag{16.47}$$

The elastic deformation gradient in the effective configuration is given by

$$\bar{F}_{ij}^e = \frac{\partial \bar{x}_i}{\partial \bar{x}_j^p} \tag{16.48}$$

The corresponding plastic deformation gradient in the effective configuration is given by

$$\bar{F}_{ij}^p = \frac{\partial \bar{x}_i^p}{\partial \bar{X}_j} \tag{16.49}$$

while the fictitious initial damage deformation gradient from configuration $\bar{C}^o$ to C^o is given by

$$F_{ij}^{do} = \frac{\partial \bar{X}_i}{\partial X_j} \tag{16.50}$$

Similar to Path I, the Right Cauchy Green deformation tensor, $\boldsymbol{C}$, is given by

$$C_{ij} = \bar{F}_{mk}^{do} \bar{F}_{kp}^{p} \bar{F}_{pq}^{e} \bar{F}_{qi}^{d} \bar{F}_{mn}^{d} \bar{F}_{nr}^{e} \bar{F}_{rs}^{p} \bar{F}_{sj}^{do} \tag{16.51}$$

The Lagrangian damage strain tensor measured with respect to the fictious configuration $\bar{C}^t$ is given by

$$\bar{\varepsilon}_{ij}^{d} = \frac{1}{2}(\bar{F}_{ki}^{d} \bar{F}_{kj}^{d} - \delta_{ij}) \tag{16.52}$$

and the corresponding Lagrangian effective elastic strain tensor measured with respect to the fictious configuration $\bar{C}^p$ is given by

$$\bar{\varepsilon}_{ij}^{e} = \frac{1}{2}(\bar{F}_{ki}^{e} \bar{F}_{kj}^{e} - \delta_{ij}) \tag{16.53}$$

The Lagrangian effective plastic strain tensor measured with respect to the fictious undamaged initial configuration $\bar{C}^o$ is given by

$$\bar{\varepsilon}_{ij}^{p} = \frac{1}{2}(\bar{F}_{ki}^{p} \bar{F}_{kj}^{p} - \delta_{ij}) \tag{16.54}$$

The total Lagrangian strain tensor is therefore expressed as follows

$$\bar{\mathcal{E}}_{ij} = \frac{1}{2}(\bar{F}_{ki}^{do} \bar{F}_{kj}^{do} - \delta_{ij}) + \frac{1}{2}\bar{F}_{mi}^{do}(\bar{F}_{km}^{p} \bar{F}_{kn}^{p} - \delta_{mn})\bar{F}_{nj}^{do} + \frac{1}{2}\bar{F}_{ni}^{do} \bar{F}_{rn}^{p}(\bar{F}_{qr}^{e} \bar{F}_{qs}^{e} - \delta_{rs})\bar{F}_{sm}^{p} \bar{F}_{mj}^{do}$$
$$+ \frac{1}{2}\bar{F}_{wi}^{do} \bar{F}_{nw}^{p} \bar{F}_{rn}^{e}(\bar{F}_{qr}^{d} \bar{F}_{qs}^{d} - \delta_{rs})\bar{F}_{sm}^{e} \bar{F}_{mk}^{p} \bar{F}_{kj}^{do} \tag{16.55}$$

The Lagrangian initial damage strain tensor measured with respect to the reference configuration C^o is denoted by

$$\bar{\mathcal{E}}_{ij}^{do} = \frac{1}{2}(\bar{F}_{ki}^{do} \bar{F}_{kj}^{do} - \delta_{ij}) \tag{16.56}$$

The Lagrangian plastic strain tensor measured with respect to the reference configuration C^o is denoted by

$$\bar{\mathcal{E}}_{ij}^{p} = \bar{F}_{ki}^{do} \bar{\varepsilon}_{km}^{p} \bar{F}_{mj}^{do} \tag{16.57}$$

One now defines the Lagrangian elastic strain tensor measured with respect to the reference configuration C^o as follows

$$\bar{\mathcal{E}}_{ij}^{e} = \bar{F}_{ni}^{do} \bar{F}_{nk}^{p} \bar{\varepsilon}_{km}^{e} \bar{F}_{mr}^{p} \bar{F}_{rj}^{do} \tag{16.58}$$

and the corresponding Lagrangian damage strain tensor measured with respect to the reference configuration C^o is given by

$$\bar{\mathcal{E}}^d_{ij} = \bar{F}^{do}_{wi} \bar{F}^p_{wn} \bar{F}^e_{nk} \bar{\mathcal{E}}^d_{km} \bar{F}^e_{mr} \bar{F}^p_{rs} \bar{F}^{do}_{sj} \tag{16.59}$$

The total Lagrangian strain is now given as follows through the additive decomposition of the corresponding strains

$$\mathcal{E}_{ij} = \bar{\mathcal{E}}^{do}_{ij} + \bar{\mathcal{E}}^p_{ij} + \bar{\mathcal{E}}^e_{ij} + \bar{\mathcal{E}}^d_{ij} \tag{16.60}$$

The change in the squared length of a material filament deformed due to fictitiously removing of damage from C^t to $\bar{C}^t$ is given by

$$(ds)^2 - (\bar{d}s)^2 = dx_i\, dx_i - d\bar{x}_i\, d\bar{x}_i$$

$$= 2\, dX_i\, \bar{\mathcal{E}}^d_{ij}\, dX_j \tag{16.61}$$

The change in the squared length of a material filament deformed elastically from $\bar{C}^t$ to $\bar{C}^p$ is given by

$$(\bar{d}s)^2 - (\bar{d}s^p)^2 = d\bar{x}_i\, d\bar{x}_i - d\bar{x}^p_i\, d\bar{x}^p_i$$

$$= 2\, dX_i\, \bar{\mathcal{E}}^e_{ij}\, dX_j \tag{16.62}$$

The change in the squared length of a material filament deformed plastically from $\bar{C}^o$ to $\bar{C}^p$ is then given by

$$(\bar{d}s^p)^2 - (\bar{d}S)^2 = d\bar{x}^p_i\, d\bar{x}^p_i - d\bar{X}_i\, d\bar{X}_i$$

$$= 2\, \bar{\mathcal{E}}^p_{ij}\, dX_i\, dX_j \tag{16.63}$$

while the change in the squared length of a material filament deformed due to fictitious removing of the initial damage from $\bar{C}^o$ to C^o is given by

$$(\bar{d}S)^2 - (dS)^2 = d\bar{X}_i\, d\bar{X}_i - dX_i\, dX_i$$

$$= 2\, dX_i\, \bar{\mathcal{E}}^{do}_{ij}\, dX_j \tag{16.64}$$

Finaly Path III gives the deformation gradient as follows

$$F_{ij} = F^e_{il} \tilde{F}^d_{lm} \tilde{F}^p_{mn} \bar{F}^{do}_{nj} \tag{16.65}$$

where $\tilde{\boldsymbol{F}}^d$ is the fictious damage deformation gradient from configuration $\tilde{C}^p$ to C^{dp} and is given by

$$\tilde{F}^d_{ij} = \frac{\partial x^{dp}_i}{\partial \tilde{x}^p_j} \tag{16.66}$$

and the corresponding plastic deformation gradient in the effective configuration is given by

$$\tilde{F}^{p}_{ij} = \frac{\partial \tilde{x}^{p}_{i}}{\partial \bar{\tilde{X}}_{j}} \tag{16.67}$$

Similar to Path II, the Right Cauchy Green deformation tensor $\mathcal{C}$ is given by

$$\mathcal{C}_{ij} = \bar{F}^{do}_{mk} \tilde{F}^{p}_{kp} \tilde{F}^{d}_{pq} F^{e}_{qi} F^{e}_{mn} \tilde{F}^{d}_{nr} \tilde{F}^{p}_{rs} \bar{F}^{do}_{sj} \tag{16.68}$$

The Lagrangian damage strain tensor measured with respect to the fictitious intermediate configuration $\tilde{C}^{p}$ is given by

$$\tilde{\epsilon}^{d}_{ij} = \frac{1}{2}(\tilde{F}^{d}_{ki} \tilde{F}^{d}_{kj} - \delta_{ij}) \tag{16.69}$$

The total Lagrangian strain tensor is expressed as follows

$$\bar{\mathcal{E}}_{ij} = \frac{1}{2}(\bar{F}^{do}_{ki}\bar{F}^{do}_{kj} - \delta_{ij}) + \frac{1}{2}\bar{F}^{do}_{mi}(\tilde{F}^{p}_{km}\tilde{F}^{p}_{kn} - \delta_{mn})\bar{F}^{do}_{nj} + \frac{1}{2}\bar{F}^{do}_{ni}\tilde{F}^{p}_{rn}(\tilde{F}^{d}_{qr}\tilde{F}^{d}_{qs} - \delta_{rs})\tilde{F}^{p}_{sm}\bar{F}^{do}_{mj} +$$
$$\frac{1}{2}\bar{F}^{do}_{wi}\tilde{F}^{p}_{nw}\tilde{F}^{d}_{rn}(F^{e}_{qr}F^{e}_{qs} - \delta_{rs})\tilde{F}^{d}_{sm}\tilde{F}^{p}_{mk}\bar{F}^{do}_{kj} \tag{16.70}$$

The Lagrangian damage strain tensor measured with respect to the reference configuration C^{o} is denoted by

$$\tilde{\mathcal{E}}^{d}_{ij} = \bar{F}^{do}_{ki} \tilde{F}^{p}_{mk} \tilde{\epsilon}^{d}_{mn} \tilde{F}^{p}_{nq} \bar{F}^{do}_{qj} \tag{16.71}$$

The Lagrangian elastic strain tensor measured with respect to the reference configuration C^{o} is denoted by

$$\mathcal{E}^{e}_{ij} = \bar{F}^{do}_{li} \tilde{F}^{p}_{kl} \tilde{F}^{d}_{mk} \epsilon^{e}_{mn} \tilde{F}^{d}_{nq} \tilde{F}^{p}_{qr} \bar{F}^{do}_{rj} \tag{16.72}$$

The corresponding total Lagrangian strain is now given by

$$\mathcal{E}_{ij} = \bar{\mathcal{E}}^{do}_{ij} + \tilde{\mathcal{E}}^{p}_{ij} + \tilde{\mathcal{E}}^{d}_{ij} + \mathcal{E}^{e}_{ij} \tag{16.73}$$

The change in the squared length of a material filament deformed due to fictitious removal of damage from C^{dp} to $\tilde{C}^{p}$ is given by

$$(ds^{dp})^{2} - (d\tilde{s}^{p})^{2} = dx^{dp}_{i} dx^{dp}_{i} - d\tilde{x}^{p}_{i} d\tilde{x}^{p}_{i}$$

$$= 2\, dX_{i}\, \tilde{\mathcal{E}}^{d}_{ij}\, dX_{j} \tag{16.74}$$

The change in the squared length of a material filament deformed plastically from $\bar{C}^{o}$ to $\tilde{C}^{p}$ is then given by

$$(d\tilde{s}^{p})^{2} - (d\bar{S})^{2} = d\tilde{x}^{p}_{i} d\tilde{x}^{p}_{i} - d\bar{X}_{i} d\bar{X}_{i}$$

$$= 2\, dX_{i}\, \tilde{\mathcal{E}}^{p}_{ij}\, dX_{j} \tag{16.75}$$

444

The total Lagrangian strain tensors obtained by considering the three paths are given by equations (16.38), (16.60) and (16.73). From the equivalency of these total strains, one obtains the explicit presentations of the kinematics of damage as follows. With the assumption of the equivalence between the elastic strain tensors given by equations (16.38) and (16.73), the damage-plastic deformation gradient given by (16.36) and the Lagrangian damage plastic strain tensor can be expressed as follows

$$F_{ij}^{dp} = \bar{F}_{ik}^{do} \tilde{F}_{kl}^{p} \tilde{F}_{lj}^{d} \tag{16.76}$$

and

$$\mathcal{E}_{ij}^{dp} = \bar{\mathcal{E}}_{ij}^{do} + \tilde{\mathcal{E}}_{ij}^{p} + \tilde{\mathcal{E}}_{ij}^{d} \tag{16.77}$$

Furthermore one obtains the following expression from equations (16.60) and (16.73) as follows

$$\bar{\mathcal{E}}_{ij}^{e} + \bar{\mathcal{E}}_{ij}^{d} = \tilde{\mathcal{E}}_{ij}^{d} + \mathcal{E}_{ij}^{e} \tag{16.78}$$

which concludes that $\tilde{C}^p$ and $\bar{C}^p$ are the same. Substituting equations (16.59), (16.71) and (16.72) into equation (16.78), one obtains the effective Lagrangian elastic strain tensor as follows

$$\bar{\mathcal{E}}_{ij}^{e} = \bar{F}_{ki}^{do} \bar{F}_{mk}^{p} [\tilde{\epsilon}_{mn}^{d} - \bar{F}_{qm}^{e} \bar{\epsilon}_{qr}^{d} \bar{F}_{rn}^{e} + \tilde{F}_{qm}^{d} \epsilon_{qr}^{e} \tilde{F}_{rn}^{d}] \bar{F}_{ns}^{p} \bar{F}_{sj}^{do} \tag{16.79}$$

Using equations (16.58) and (16.79) one can now express $\bar{\epsilon}$ as follows

$$\bar{\epsilon}_{ij}^{e} = \tilde{\epsilon}_{ij}^{d} - \bar{F}_{mi}^{e} \bar{\epsilon}_{mn}^{d} \bar{F}_{nj}^{e} + \tilde{F}_{mi}^{d} \epsilon_{mn}^{e} \tilde{F}_{nj}^{d} \tag{16.80}$$

This expression gives a general relation of the effective elastic strain for finite strains of elasto-plasic damage deformation. For the special case when one assumes that

$$\tilde{\epsilon}_{ij}^{d} - \bar{F}_{mi}^{e} \bar{\epsilon}_{mn}^{d} \bar{F}_{nj}^{e} = 0 \tag{16.81}$$

equation (16.80) can be reduced to the following expression

$$\bar{\epsilon}_{ij}^{e} = \tilde{F}_{ki}^{d} \epsilon_{kl}^{e} \tilde{F}_{lj}^{d} \tag{16.82}$$

This relation is similar to that obtained without the consideration of the kinematics of damage and only utilizing the hypothesis of elastic energy equivalence. However, equation (16.82) for the case of finite strains is given by relation (16.80) which cannot be obtained through the hypothesis of elastic energy equivalence. Equation (16.81) maybe valid only for some special cases of the small strain theory.

16.4.2 Fictitious Damage Deformation Gradients

The two fictitious deformation gradients given by equations (16.47) and (16.66) may be used to define the damage tensor in order to describe the damage behavior of solids. Since the fictitious effective deformed cofiguration denoted by $\bar{C}^t$ is obtained by removing the damages from the real deformed configuration denoted by C^t, therefore the differential volume of the fictitious effective deformed volumes denoted by $d\bar{\Omega}^t$ is obtained as follows Park and Voyiadjis [211]

$$d\bar{\Omega}^t = d\Omega^t - d\Omega^d$$

$$= \sqrt{(1 - \hat{\phi}_1)(1 - \hat{\phi}_2)(1 - \hat{\phi}_3)}\, d\Omega^t \tag{16.83}$$

or

$$d\Omega^t = \bar{J}^d d\bar{\Omega}^t \tag{16.84}$$

where Ω^d is the volume of damage in the configuration C^t and $\bar{J}^d$ is termed the Jacobian of the damage deformation which is the determinant of the fictitious damage deformation gradient. Thus the Jacobian of the damage deformation can be written as follows

$$\bar{J}^d = |\bar{F}^d_{ij}|$$

$$= \frac{1}{\sqrt{(1 - \hat{\phi}_1)(1 - \hat{\phi}_2)(1 - \hat{\phi}_3)}} \tag{16.85}$$

The determinant of the matrix $[\,a\,]$ in equation (16.19) is given by

$$|[\,a\,]| = |[\,b\,]^T|\,|[\,\hat{a}\,]|\,|[\,b\,]|$$

$$= |[\,\hat{a}\,]|$$

$$= \frac{1}{\sqrt{(1 - \hat{\phi}_1)(1 - \hat{\phi}_2)(1 - \hat{\phi}_3)}} \tag{16.86}$$

Thus one assumes the following relation without loss of generality

$$\bar{F}^d_{ij} = [\delta_{ij} - \phi_{ij}]^{-\frac{1}{2}} \tag{16.87}$$

Although the identity is established between $\bar{J}^d$ and $|a|$, however, this is not sufficient to demonstrate the validity of equation (16.87). This relation is assumed here based on the physics of the geometrically symmetrized effective stress concept [210]. Similiarly, the fictitious damage deformation gradient $\tilde{F}^d$ can be written as follows

$$\tilde{F}^d_{ij} = [\delta_{ij} - \varphi_{ij}]^{-\frac{1}{2}} \tag{16.88}$$

Finally, assuming that $\bar{x} = \tilde{x}$ based on equation (16.78) the relations between $\tilde{F}^d$ and $\bar{F}^d$, and φ and ϕ are given by

$$\tilde{F}^d_{ij} = F^e_{ki}\bar{F}^d_{kl}F^{e^{-1}}_{lj} \tag{16.89}$$

and

$$\varphi_{ij} = F^e_{ki}\phi_{kl}F^{e^{-1}}_{lj} \tag{16.90}$$

16.4.3   An Additive Decomposition

The kinematics of finite deformation is described here based on the polar decomposition by considering three paths as indicated in the previous section. In order to proceed further, one assumes a homogeneous state of deformation such that the completely unloaded stress free configuration C^{dp} has open cracks and micro-cavities. Furthermore one assumes that these cracks and micro-cavities can be completely closed by subjecting them to certain additional stress. The configuration that is subjected to the additional sresses is denoted by $\tilde{C}^p$ and is assumed that this configuration has deformed only plastically. The additional stress which can close all micro cracks and micro cavities is assumed as follows

$$\sigma_{ij}^* = \sigma_{ij} - \bar{\sigma}_{ij} \tag{16.91}$$

If no initial damage is assumed in the configuration C°, it can be assumed such that $C^p = \tilde{C}^p$. The total displacement vector $\boldsymbol{u}(\boldsymbol{X}, t)$ can be decomposed in the Cartesian reference frame in the absence of rigid body displacement such that

$$u_i = u_i^e + u_i^d + u_i^p \tag{16.92}$$

$$u_i = x_i - X_i \tag{16.93}$$

$$u_i^e = x_i - x_i^d \tag{16.94}$$

$$u_i^d = x_i^d - x_i^p \tag{16.95}$$

$$u_i^p = x_i^p - X_i \tag{16.96}$$

where $\boldsymbol{x}^d$ is a point in the intermediate unloaded configuration C^{dp} and $\boldsymbol{x}^p$ is a point in the configuration $\tilde{C}^p$. Recalling that $\boldsymbol{u} = \boldsymbol{x} - \boldsymbol{X}$ and using the notation $u_{i,j} = \frac{\partial x_i}{\partial X_j}$, the corresponding total Lagrangian strain tensor $\boldsymbol{\mathcal{E}}$ given by equation (16.25) can be written in the usual form as follows

$$\mathcal{E}_{ij} = \frac{1}{2}(u_{i,j} + u_{j,i} + u_{k,i}u_{k,j}) \tag{16.97}$$

Substituting equation (16.92) in to equation (16.97), one obtains

$$\mathcal{E}_{ij} = \varepsilon_{ij}^p + \varepsilon_{ij}^d + \varepsilon_{ij}^e + \varepsilon_{ij}^{de} + \varepsilon_{ij}^{pe} + \varepsilon_{ij}^{pd} \tag{16.98}$$

where ε^p termed the pure plastic strain is given by

$$\varepsilon_{ij}^p = \frac{1}{2}(u_{i,j}^p + u_{j,i}^p + u_{k,i}^p u_{k,j}^p) \tag{16.99}$$

ε^d termed the pure damage strain is given by

$$\varepsilon_{ij}^d = \frac{1}{2}(u_{i,j}^d + u_{j,i}^d + u_{k,i}^d u_{k,j}^d) \tag{16.100}$$

ε^e termed the pure elastic strain is given by

$$\varepsilon_{ij}^e = \frac{1}{2}(u_{i,j}^e + u_{j,i}^e + u_{k,i}^e u_{k,j}^e) \tag{16.101}$$

ε^{de} termed the coupled elastic-damage strain is given by

$$\varepsilon_{ij}^{de} = \frac{1}{2}(u_{k,i}^e u_{k,j}^d + u_{k,i}^d u_{k,j}^e) \tag{16.102}$$

ε^{pe} termed the coupled elastic-plastic strain is given by

$$\varepsilon_{ij}^{ep} = \frac{1}{2}(u_{k,i}^e u_{k,j}^p + u_{k,i}^p u_{k,j}^e) \tag{16.103}$$

and ε^{pd} termed the coupled plastic-damage strain is given by

$$\varepsilon_{ij}^{pd} = \frac{1}{2}(u_{k,i}^p u_{k,j}^d + u_{k,i}^d u_{k,j}^p) \tag{16.104}$$

One defines the Lagrangian elastic strain as follows

$$\mathcal{E}_{ij}^e = \varepsilon_{ij}^e + \varepsilon_{ij}^{de} + \varepsilon_{ij}^{pe} \tag{16.105}$$

the Lagrangian damage strain as follows

$$\mathcal{E}_{ij}^d = \varepsilon_{ij}^d \tag{16.106}$$

and the Lagrangian plastic strain as follows

$$\mathcal{E}_{ij}^p = \varepsilon_{ij}^p + \varepsilon_{ij}^{pd} \tag{16.107}$$

The coupled term of elastic-damage and plastic-damage strains are linked respectively with the elastic and plastic strains since they directly influence the stresses acting on the body. Consequently the total Lagrangian strain can be written as follows

$$\mathcal{E}_{ij} = \mathcal{E}_{ij}^p + \mathcal{E}_{ij}^d + \mathcal{E}_{ij}^e \tag{16.108}$$

The differential displacement is given by

$$du_i = x_i^{t+dt} - x_i^t \tag{16.109}$$

Then, the corresponding the differential total displacement can be decomposed into an elastic, plastic and damage part as follows

$$du_i = du_i^e + du_i^d + du_i^p \tag{16.110}$$

Equivalently one obtains the following decomposition of the velocity tensor field $\boldsymbol{v}(\boldsymbol{x}, t)$:

$$v_i(x_i, t) = v_i^e(x_i, t) + v_i^d(x_i, t) + v_i^p(x_i, t) \tag{16.111}$$

where $\boldsymbol{v}^e$ is the velocity vector field due to elastic stretching and rigid body rotations and $\boldsymbol{v}^d$ is the velocity vector field due to damage process and $\boldsymbol{v}^p$ is the velocity vector field arising from the plastic deformations due to dislocation motion. The gradient of the velocity vector with respect to the current frame $\boldsymbol{x}$ is given by the following relation

$$\mathcal{L}_{ij} = \mathcal{L}_{ij}^e + \mathcal{L}_{ij}^d + \mathcal{L}_{ij}^p \tag{16.112}$$

$$\mathcal{D}_{ij} = \mathcal{D}_{ij}^e + \mathcal{D}_{ij}^d + \mathcal{D}_{ij}^p \tag{16.113}$$

$$\mathcal{W}_{ij} = \mathcal{W}_{ij}^e + \mathcal{W}_{ij}^d + \mathcal{W}_{ij}^p \tag{16.114}$$

16.5   Irreversible Thermodynamics

The finite elasto-plastic deformation behavior with damage can be viewed within the framework of thermodynamics with internal state variables.

The Helmholtz free energy per unit mass in an isothermal deformation process at the current state of the deformation and material damage is assumed as follows:

$$\Psi = \psi + \Upsilon \tag{16.115}$$

where ψ is the strain energy which is a purely reversible stored energy, while Υ is the energy associated with specific microstructural changes produced by damage and plastic yielding. Conceptionally, the energy Υ is assumed to be an irreversible energy. In generall, an explicit presentation of the energy Υ and its rate $\dot{\Upsilon}$ is limited by the complexities of the internal microstructural changes, however, only two internal variables which are associated with damage and plastic hardening, respectively are considered in this work(Voyiadjis and Park [211]). In the equation that follows $\bar{E}$ is the initial undamaged Young's modulus, E is the damaged Young's modulus, S is the second Piola Kirchhoff stress, and $\mathcal{E}$ is the Lagrangian strain. The total Lagrangian strain tensor $\mathcal{E}$ is given by

$$\mathcal{E}_{ij} = \mathcal{E}_{ij}^{p} + \mathcal{E}_{ij}^{e} + \mathcal{E}_{ij}^{d} \tag{16.116}$$

where $\mathcal{E}^{p}$ is the plastic strain tensor, $\mathcal{E}^{e}$ is the elastic strain tensor, and $\mathcal{E}^{d}$ is the additional strain tensor due to damage. Comparing equations (16.38) and (16.116) one notes that

$$\mathcal{E}_{ij}^{dp} = \mathcal{E}_{ij}^{p} + \mathcal{E}_{ij}^{d} \tag{16.117}$$

Furthermore the additional strain tensor due to damage can be decomposed as follows

$$\mathcal{E}_{ij}^{d} = \mathcal{E}_{ij}^{d'} + \mathcal{E}_{ij}^{d''} \tag{16.118}$$

where $\mathcal{E}^{d''}$ is the irrecoverable damage strain tensor due to lack of closure of the microcracks and microvoids during unloading, while $\mathcal{E}^{d'}$ is the elastic damage strain due to reduction of the elastic stiffness tensor. Thus the purely reversible strain tensor,$\mathcal{E}^{E}$ due to unloading can be obtained by

$$\mathcal{E}_{ij}^{E} = \mathcal{E}_{ij}^{e} + \mathcal{E}_{ij}^{d'} \tag{16.119}$$

The srain energy ψ is assumed as follows

$$\psi = \frac{1}{2\rho}\mathcal{E}_{ij}^{E} E_{ijkl}\mathcal{E}_{kl}^{E} \tag{16.120}$$

where ρ is the specific density. Furthermore this strain energy can be decomposed into the elastic strain energy ψ^{e} and the damage strain energy ψ^{d} as follows

$$\psi = \psi^{e} + \psi^{d} \tag{16.121}$$

The elastic strain energy, ψ^e is given by

$$\psi^e = \frac{1}{2\rho}\mathcal{E}^e_{ij}\bar{E}_{ijkl}\mathcal{E}^e_{kl} \qquad (16.122)$$

and the corresponding damage strain energy ψ^d is given by

$$\psi^d = \frac{1}{2\rho}\mathcal{E}^E_{ij}E_{ijkl}\mathcal{E}^E_{kl} - \frac{1}{2\rho}\mathcal{E}^e_{ij}\bar{E}_{ijkl}\mathcal{E}^e_{kl} \qquad (16.123)$$

where $\bar{E}$ and E are the initial undamaged elastic stiffness and the damaged elastic stiffness, respectively. These stiffnesses are defined such that

$$\bar{E}_{ijkl} = \frac{\partial^2 \Psi}{\partial \mathcal{E}^e_{ij} \partial \mathcal{E}^e_{kl}} \qquad (16.124)$$

and

$$E_{ijkl} = \frac{\partial^2 \Psi}{\partial \mathcal{E}^E_{ij} \partial \mathcal{E}^E_{kl}} \qquad (16.125)$$

The damaged elastic stifness in the case of finite deformation is given by Voyiadjis and Park [211] as follows

$$E_{ijrs} = N_{ikjl}\bar{E}_{klpq}N_{prqs} \qquad (16.126)$$

where

$$\begin{aligned} N_{ikjl} &= M^{-1}_{ikjl} \\ &= a^{-1}_{ik}a^{-1}_{jl} \end{aligned} \qquad (16.127)$$

The elastic damage stiffness given by equation (16.126) is symmetric. This is in line with the classic sense of continnum mechanics which is violated by using the hypothesis of strain equivalence. Using the similar relation between the Lagrangian and the Eulerian strain tensors given by equation (16.42), the corresponding strain energy given by equation(16.120) can be written as follows

$$\begin{aligned} \psi &= \frac{1}{2\rho}\epsilon^E_{mn}F_{mi}F_{nj}E_{ijkl}F_{rk}F_{sl}\epsilon^E_{rs} \\ &= \frac{1}{2\rho}\epsilon^E_{mn}\Lambda_{mnrs}\epsilon^E_{rs} \end{aligned} \qquad (16.128)$$

where ϵ^E is the Eulerian strain corresponding to the Lagrangian strain shown in equation (16.119), and Λ is termed the Eulerian elastic stiffness which is given by

$$\Lambda_{mnrs} = F_{mi}F_{nj}E_{ijkl}F_{rk}F_{sl} \qquad (16.129)$$

The second Piola-Kirchhoff stress tensor, $\boldsymbol{S}$ is defined as follows

$$\begin{aligned}
\mathcal{S}_{ij} &= \rho \frac{\partial \psi}{\partial \mathcal{E}_{ij}^E} \\
&= \rho \frac{\partial \psi^e}{\partial \mathcal{E}_{ij}^e}
\end{aligned} \qquad (16.130)$$

The second Piola-Kirchhoff stress tensor, $\boldsymbol{S}$ is related to the Cauchy stress tensor, $\boldsymbol{\sigma}$ by the following relation

$$\mathcal{S}_{ij} = J F_{ik}^{-1} \sigma_{km} F_{jm} \qquad (16.131)$$

The Kirchhoff stress tensor $\boldsymbol{T}$ is related to the Cauchy stress tensor by

$$\mathcal{T}_{ij} = J \sigma_{ij} \qquad (16.132)$$

The rate of the Helmholtz free energy is then given as follows

$$\dot{\Psi} = \dot{\psi} + \dot{\Upsilon} \qquad (16.133)$$

where $\dot{\Upsilon}$ is the rate of Υ associated with the two neighboring constrained equilibrium states with two different sets of internal variables, $\boldsymbol{\phi}$ and $\boldsymbol{\alpha}$. Using equations (16.120) or (16.121) the rate form of the strain energy can be given as follows since $\dot{\bar{\boldsymbol{E}}} = 0$

$$\rho \dot{\psi} = \frac{1}{2} \dot{E}_{ijkl} \mathcal{E}_{ij}^E \mathcal{E}_{kl}^E + E_{ijkl} \mathcal{E}_{ij}^E \dot{\mathcal{E}}_{kl}^E - \frac{\dot{\rho}}{2\rho} \mathcal{E}_{ij}^E E_{ijkl} \mathcal{E}_{kl}^E \qquad (16.134)$$

or

$$\rho \dot{\psi}^e = \dot{\mathcal{E}}_{ij}^e \bar{E}_{ijkl} \mathcal{E}_{kl}^e - \frac{\dot{\rho}}{2\rho} \mathcal{E}_{ij}^e \bar{E}_{ijkl} \mathcal{E}_{kl}^e \qquad (16.135)$$

and

$$\rho \dot{\psi}^d = \dot{\mathcal{E}}_{ij}^E E_{ijkl} \mathcal{E}_{kl}^E + \frac{1}{2} \mathcal{E}_{ij}^E \dot{E}_{ijkl} \mathcal{E}_{kl}^E - \dot{\mathcal{E}}_{ij}^e \bar{E}_{ijkl} \mathcal{E}_{kl}^e - \frac{\dot{\rho}}{2\rho} (\mathcal{E}_{ij}^E E_{ijkl} \mathcal{E}_{kl}^E - \mathcal{E}_{ij}^e \bar{E}_{ijkl} \mathcal{E}_{kl}^e) \qquad (16.136)$$

If the deformation process is assumed to be isothermal with negligible temperature non-uniformoties, the rate of the Helmholtz free energy can be written using the first law of thermodynamics (balance of energy) as follows

$$\dot{\Psi} = \mathcal{T}_{ij} \mathcal{D}_{ij} - T\eta \qquad (16.137)$$

where T is the temperature and η is the irreversible entropy production rate. The product $T\eta$ represents the energy dissipation rate associated with both the damage and plastic deformation processes. The energy of the dissipation rate is given as follows

$$T\eta = \mathcal{S}_{ij} \dot{\mathcal{E}}_{ij}^{d''} + \mathcal{S}_{ij} \dot{\mathcal{E}}_{ij}^p - \dot{\Upsilon} \qquad (16.138)$$

The first two terms on the right-hand side of equation (16.137) represent a macroscopically non-recoverable rate of work expanded on damage and plastic processes, respectively. Furthermore the rate of the additional strain tensor due to damage is given by

$$\dot{\mathcal{E}}_{ij}^{d} = \dot{\mathcal{E}}_{ij}^{d''} + \dot{\mathcal{E}}_{ij}^{d'} \tag{16.139}$$

If we assume that a fraction of the additional strain tensor can be recoverd during unloading, then the elastic damage tensor due to the reduction of the elastic stiffness is given by

$$\dot{\mathcal{E}}_{ij}^{d'} = c\,\dot{\mathcal{E}}_{kj}^{d} \tag{16.140}$$

where c is a fraction which ranges from 0 to 1. Then the permenant damage strain due to lack of closure of micro-cracks and micro-cavities is given by

$$\dot{\mathcal{E}}_{ij}^{d''} = (1-c)\dot{\mathcal{E}}_{kj}^{d} \tag{16.141}$$

Thus the energy of the dissipation rate given by equation (16.137) can be written as follows

$$\begin{aligned}
T\eta &= (1-c)\mathcal{S}_{ij}\dot{\mathcal{E}}_{ij}^{d} + \mathcal{S}_{ij}\dot{\mathcal{E}}_{ij}^{p} - \dot{\Upsilon} \\
&= (1-c)\mathcal{T}_{ij}\mathcal{D}_{ij}^{d} + \mathcal{T}_{ij}\mathcal{D}_{ij}^{p} - \dot{\Upsilon}
\end{aligned} \tag{16.142}$$

The rate of energy associated with a specific microstructural change due to both the damage and the plastic processes can be decomposed as follows

$$\dot{\Upsilon} = \dot{\Upsilon}^{d} + \dot{\Upsilon}^{p} \tag{16.143}$$

where one defines that

$$\rho\dot{\Upsilon}^{d} = \mathcal{Y}_{ij}\dot{\phi}_{ij} \tag{16.144}$$

and

$$\rho\dot{\Upsilon}^{p} = \mathcal{A}_{ij}\dot{\alpha}_{ij} \tag{16.145}$$

where $\mathcal{Y}$ and $\mathcal{A}$ are the general forces conjugated by damage and plastic yielding, respectively. They are defined as follows

$$\begin{aligned}
\mathcal{Y}_{ij} &= \rho\frac{\partial\Psi}{\partial\phi_{ij}} \\
&= \rho\frac{\partial\psi^{ed}}{\partial\phi_{ij}} + \rho\frac{\partial\Upsilon}{\partial\phi_{ij}}
\end{aligned} \tag{16.146}$$

$$\mathcal{A}_{ij} = \rho\frac{\partial\Psi}{\partial\alpha_{ij}} \tag{16.147}$$

In view of equation (16.142) one notes that it is equivalent to the work by Lubarda and Krajcinovic [227] when $(1-c) = \frac{1}{2}$.

16.6 Constitutive Equation for Finite Elasto-Plastic Deformation with Damage Behavior

The kinematics and the thermodynamics discussed in the previous sections provide the basis for a finite deformation damage elasto-plasticity. In this section the basic structure of the constitutive equations are reviewed based on the generalized Hooke's law, originally obtained for small elastic strains such that the second Piola-Kirchoff stress tensor $\boldsymbol{S}$ is the gradient of free energy Ψ with respect to the Lagrangian elastic strain tensor $\boldsymbol{\mathcal{E}}^E$ given by equation (16.130). The following relation is between the three dimensional state of stress and strain

$$\begin{aligned}
\mathcal{S}_{ij} &= \bar{E}_{ijkl}(\mathcal{E}_{kl} - \mathcal{E}_{kl}^p - \mathcal{E}_{kl}^d) & \text{(16.148a)}\\
&= \bar{E}_{ijkl}\mathcal{E}_{kl}^e & \text{(16.148b)}\\
&= E_{ijkl}(\mathcal{E}_{kl}^e + \mathcal{E}_{kl}^{d'}) & \text{(16.148c)}\\
&= E_{ijkl}(\mathcal{E}_{kl} - \mathcal{E}_{kl}^{d''} - \mathcal{E}_{kl}^p) & \text{(16.148d)}
\end{aligned}$$

From the incremental analysis one obtains the following rate form of the constitutive equation

$$\dot{\mathcal{S}}_{ij} = \bar{E}_{ijkl}(\dot{\mathcal{E}}_{kl} - \dot{\mathcal{E}}_{kl}^p - \dot{\mathcal{E}}_{kl}^d) \tag{16.149}$$

Consiquently the constitutive equation of the elasto-pastic damage behavior can be written as follows

$$\dot{\mathcal{S}}_{ij} = E_{ikjl}^{DP}\dot{\mathcal{E}}_{kl} \tag{16.150}$$

where $\boldsymbol{E}^{DP}$ is the damage elasto-plastic stifness and is expressed as follows

$$E_{ijkl}^{DP} = \bar{E}_{ikjl} - E_{ikjl}^p - E_{ikjl}^d \tag{16.151}$$

where $\boldsymbol{E}^p$ is the plastic stiffness and $\boldsymbol{E}^d$ is the damage stiffness. Both $\boldsymbol{E}^p$ and $\boldsymbol{E}^d$ are the reduction in stiffness due to the plastic and damage deteriorations, respectively. The plastic stiffness and the damage stiffness can be obtaind by using the flow rule and damage evolution law, respectively. By assuming that the reference state coincides with the current configuration, the second Piola-Kirchoff stress rate, $\dot{\boldsymbol{S}}$ can be replaced by the corotational rate of the Cauchy stress tensor $\boldsymbol{\sigma}$ and the rate of Lagrangian strain tensor $\dot{\boldsymbol{\mathcal{E}}}$ by the deformation rate $\boldsymbol{D}$ as follows

$$\sigma_{ij} = E_{ijkl}^{DP}\mathcal{D}_{kl} \tag{16.152}$$

The corotational rate of the Cauchy stress tensor, $\boldsymbol{\sigma}$ is related to the rate of the rate of the Cauchy stress tensor, $\dot{\boldsymbol{\sigma}}$ as follows

$$\sigma_{ij} = \dot{\sigma}_{ij} - \mathcal{W}_{ik}^*\sigma_{kj} + \sigma_{ik}\mathcal{W}_{kj}^* \tag{16.153}$$

where

$$W^* = W - W^p - W^d \tag{16.154}$$

The details of the complete constitutive models using the proposed kinematics and the evolution laws of damage will be stated in the forthcoming paper.

16.7 Application to Metals

In this application the authors show the feasibility of the proposed work to metals. The constitutive model developed in sections 6 and 7 is demonstrated here through the degradation of the material stiffness due to both the plastic deformation and damage. The metal investigated here is the aluminum alloy 2024-T3. The case of uniaxially loaded specimens is shown in Figure 16.2 (Chow and Wang [20]. The proposed model using explicitly the kinematics of damage presented in this paper shows good agreement with the experimental results for the case of large deformations. This model shows a more accurate prediction than the previous model of the authors using the energy equivalence hypothesis. The constitutive models using the energy equivalence hypothesis by Voyiadjis and Park [176] and Chow and Wang [20] lose the explicit material degradation when plastic deprmation and damage are coupled.

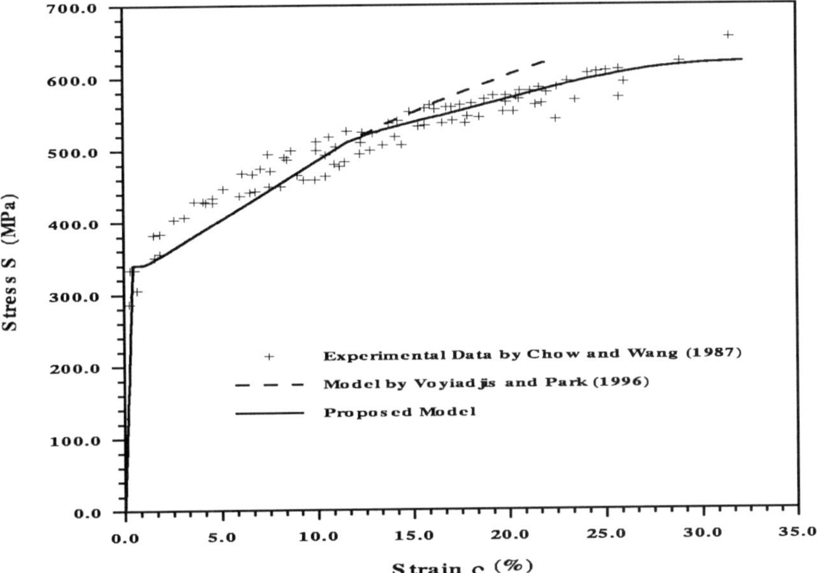

Figure 16.2 Uniaxial stress-strain curve for aluminum alloy 2024-T3.

<u>CHAPTER 17</u>

A COUPLED ANISOTROPIC DAMAGE MODEL FOR THE INELASTIC RESPONSE OF COMPOSITE MATERIALS

A coupled incremental damage and plasticity theory for rate independent and rate dependent composite materials is introduced here. This allows damage to be path dependent either on the stress history or thermodynamic force conjugate to damage. This is achieved through the use of an incremental damage tensor. Damage and inelastic deformations are incorporated in the proposed model that is used for the analysis of fiber-reinforced metal matrix composite materials. The damage is described kinematically in both the elastic and inelastic domains using the fourth order damage effect tensor which is a function of the second-order damage tensor. A physical interpretation of the second order damage tensor is given in this work which relates to the microcrack porosity within the unit cell. The inelastic deformation behavior with damage is viewed here within the frame work of thermodynamics with internal state variables.

Computational aspects of both the rate independent and rate dependent models are also discussed in this Chapter. The Newthon Rapson iterative scheme is used for the overall laminate system. The constitute equations of both the rate independent and the rate dependent plasticity coupled with damage models are additively decomposed into the elastic, inelastic and damage deformations by using the three step split operator algorithm [206]. The main framework return maping algorithm [207] is used for the correction of the elasto-plastic and viscoplastic states. However, for the case of the damage model these algorithms are redefined according to the governed damage constitutive relations.

In order to test the validity of the model, a series of laminated systems $(0_{(8s)})$, $(90_{(8s)})$, $(0/90)_{(4s)}$, $(-45/45)_{(2s)}$ are investigated at both room and elevated temperatures of $538°C$ and $649°C$. The results obtained from the special purpose developed computer program, DVP-CALSET(Damage and Viscoplastic Coupled Analysis of Laminate Systems at Elevated Temperatures), are then compared with the available experimental results and other existing theoretical material models obtained from the work of Majumdar and Newaz [208] and Voyiadjis and Venson [143].

The coupling of damage and inelastic deformation in materials have been studied only recently [205,206,209,211]. Both Ju[206] and Johansson and Runesson[209] made use of the effective stress utilizing a scalar measure of isotropic damage. Voyiadjis and Park [210], made use of the effective configuration by invoking the kinematics of damage through the use of a second order damage tensor. Recently Voyiadjis and Park [211] reviewed a linear transformation tensor, defined as a fourth order damage effect tensor and focused on its geometric symmetrization method in order to describe the kinematics of damage using the second order damage tensor. Voyiadjis and Park [211] utilized the polar decomposition of the deformation gradient and introduced the kinematics of damage using the damage effect tensor which does not only symmetrize the effective stress tensor but can also be related to the deformation gradient of damage. Using the consistent thermodynamic formulation one introduces separately the strain due to damage and the associated dissipation energy of this strain. Bammann and Aifantis [212] proposed a similar but less general and elaborate decomposition of the deformation gradient to include the damage contribution through a scalar damage variable.

The damage model presented here is within the thermodynamics constitutive framework making use of internal state variables. The second order damage tensor presented here is physically based and is a measurable quantity that is defined within the context of the damage internal variables. The damage definition is related to the microcrack porosity(concentration ratio) within the unit cell. Both rate independent and rate dependent plasticity constitutive models for damaged composite materials are presented here.

For the numerical simulation of boundary value problems involving damage Simo and Ju [133] assumed an additive split of the stress tensor. Ju [206] in his work assumed an additive split of the strain tensor into the elastic-damage and plastic-damage parts from the outset. This is more appealing since it is analogous to the J integral in nonlinear fracture mechanics. It also results in more robust tangent moduli than the "stress-split" formulation. In the present work a three step split operator algorithm is used in order to additively decompose the set of differential equations into the elastic, inelastic, and damage deformations. This is accomplished by making use of the effective undamaged configuration of the material [210,211]. The elastic and inelastic deformations are additively split through the strain tensor in the effective undamaged configuration of the material. Although, Ju [206]also used the effective configuration in his numerical analysis, however, this was only applied to the stress tensor. This is because the kinematics of damage is not introduced in this work but is accounted indirectly through the reduction in stiffness. Ju [206] emphasizes that " added flexibility" due to the existence of microcrack is already embedded in the deformation gradient implicitly. The damage deformation in the work presented here is characterized separately using the kinematics of damage [211]. This introduces a distinct kinematic measure of damage which is complimentary to the deformation kinematic measure of strain. Voyiadjis and Venson [143] quantified the physical values of the eigenvalues of the second-order damage tensor for the unidirectional fibrous composite by measuring the crack density.

17.1 Theoretical Formulation

Damage variables can be presented through the internal state variables of thermodynamics for irreversible processes in order to describe the effects of damage and its microscopic growth on the macromechanical properties of the materials. One of the most crucial aspect of continuum damage mechanics is the appropriate choice of the damage variable since the accuracy and reliability of the developed damage model is mostly dependent on the suitable selection of the type and numbers of these variables. For the current application a second order symmetric damage tensor is selected as an internal state variable. It characterizes the anisotropic phenomenon of micro-crack distribution in the material and it can be interpreted as the effective area reduction caused by distributed micro-cracks and cavities due to the material damage. Since the elasto-plastic response of the damaged material is considered here, both hardening effects due to plasticity and damage together with the above description of micro-crack distribution can be introduced as hidden internal state variables in the thermodynamic state potential. The free energy is considered here as the thermodynamic state potential depending on both observable and internal state variables. The form of this potential in terms of observable and internal state variables can be given as follows

$$\Psi = \Psi(\epsilon^e, T, \phi, p, \alpha, \kappa, \Upsilon) \tag{17.1}$$

where p and α variables characterize the isotropic and kinematic hardening in plasticity/ viscoplasticity respectively and κ and Υ variables characterize respectively the isotropic and kinematic hardening in damage. In equation (17.1) T characterizes the temperature, and ϵ^e is the elastic component of the strain tensor. Superscripts do not indicate a tensorial character but only a particular state of the variable such as elastic (e), plastic (p), viscoplastic (vp), damage (d), viscoplastic-damage (vpd), etc. In some particular cases, wherever indicated explicitly, superscripts will be used as exponents. The time derivative of equation (17.1) with respect to its state variables is given by :

$$\dot{\Psi} = \frac{\partial \Psi}{\partial \epsilon^e} : \dot{\epsilon}^e + \frac{\partial \Psi}{\partial \phi} : \dot{\phi} + \frac{\partial \Psi}{\partial p}\dot{p} + \frac{\partial \Psi}{\partial \alpha} : \dot{\alpha} + \frac{\partial \Psi}{\partial \kappa}\dot{\kappa} + \frac{\partial \Psi}{\partial \Upsilon} : \dot{\Upsilon} + \frac{\partial \Psi}{\partial T}\dot{T} \tag{17.2}$$

where superdot implies time differentiation, " : " denotes $\boldsymbol{A} : \boldsymbol{A} = A_{ij}A_{ij}$ for second order tensors, and "·" denotes $\boldsymbol{A} \cdot \boldsymbol{A} = A_{ik}A_{kj}$. The isotropic hardening variable of plasticity is a scalar quantity denoted by p and expressed as follows

$$\dot{p} = \sqrt{\frac{2}{3}\dot{\epsilon}^p : \dot{\epsilon}^p} \tag{17.3}$$

where $\dot{\epsilon}^p$ is a second order tensor describing the plastic strain rate. The corresponding isotropic hardening variable of damage is also a scalar variable denoted by κ. An expression for this scalar variable can be taken as follows

$$\dot{\kappa} = \sqrt{\frac{2}{3}\dot{\phi} : \dot{\phi}} \tag{17.4}$$

An alternative definition is suggested by Chow and Lu [179] and Voyiadjis and Park [176] such as

$$\dot{\kappa} = -\boldsymbol{Y} : \dot{\phi} \tag{17.5}$$

From the second law of thermodynamics [91], Clausius-Duhem inequality can be written as follows

$$\boldsymbol{\sigma} : \dot{\epsilon} - \rho(\dot{\Psi} - s\dot{T}) - q_i\frac{T_{,i}}{T} \geq 0 \tag{17.6}$$

where $\boldsymbol{\sigma}$ is the Cauchy stress, ρ is the material density, $T_{,i}$ is the temperature gradient, q_i is the heat flux, and s is known as the specific entropy per unit mass. Substituting equation (17.2) into equation (17.6) results into the following expression

$$\left(\boldsymbol{\sigma} - \rho\frac{\partial \Psi}{\partial \epsilon}\right) : \dot{\epsilon}^e + \boldsymbol{\sigma} : \dot{\epsilon}^p - \left(s - \rho\frac{\partial \Psi}{\partial T}\right)\dot{T} - \rho\frac{\partial \Psi}{\partial \phi} : \dot{\phi} - \rho\frac{\partial \Psi}{\partial p}\dot{p} - \rho\frac{\partial \Psi}{\partial \alpha} : \dot{\alpha} - \rho\frac{\partial \Psi}{\partial \kappa}\dot{\kappa} -$$

$$-\rho\frac{\partial \Psi}{\partial \Upsilon} : \dot{\Upsilon} - \frac{\vec{q}}{T}\overrightarrow{grad}T \geq 0 \tag{17.7}$$

458

From this equation the following thermodynamic state laws can be obtained

$$\sigma = \rho \frac{\partial \Psi}{\partial \epsilon^e} \tag{17.8}$$

$$s = -\rho \frac{\partial \Psi}{\partial T} \tag{17.9}$$

$$A_k = \rho \frac{\partial \Psi}{\partial V_k} \tag{17.10}$$

where equation (17.10) describes the relations between the internal state variables ($V_k = \epsilon^e, \phi, p, \alpha, \kappa, \Upsilon$) and their associated thermodynamic conjugate forces ($A_k = \sigma, Y, R, X, K, \Gamma, s$). These thermodynamic state variables with their associated force variables are summarized in Table 17.1. The associated variable Y is the thermodynamic conjugate

Table 17.1 Thermodynamic Variables

State Variables		
Observable	Internal	Associated Variables
ϵ^e		σ
T		s
	ϵ^p	$-\sigma$
	p	R
	α	X
	ϕ	Y
	κ	K
	Υ	Γ

force associated with the damage variable ϕ. The damage associated tensor Γ is analogous to the back stress term in plasticity. It represents the translation of the damage surface as loading progresses akin to the kinematic hardening of plasticity.

Since the internal state variables are selected independently from one another, it is possible to decouple the Helmholtz free energy Ψ into a potential function for each corresponding internal state variable. Therefore an analytical expression for the thermodynamic potential can be given as a quadratic from of its internal state variables as follows [213,214]

$$\rho \Psi = \frac{1}{2}(\epsilon - \epsilon^p) : E(\phi) : (\epsilon - \epsilon^p) + \frac{1}{2}k_1 \alpha : \alpha + \frac{1}{2}k_2 p^2 + \frac{1}{2}k_3 \Upsilon : \Upsilon + \frac{1}{2}k_4 \kappa^2 \tag{17.11}$$

Numeral superscripts in equation (17.11) indicate exponents. In equation (17.11) ρ denotes the constant material density, $E(\phi)$ is the fourth order damaged elastic stiffness tensor, and k_1, k_2, k_3, and k_4 are the material dependent constants. The material dependent constants maybe expressed in terms of scaling parameters such as the spacing and radius of the fibers and other microstructural parameters such as dislocation densities, etc.

The state laws can be written from the thermodynamic potential equation (17.11) in the following form

$$\boldsymbol{\sigma} = \rho\frac{\partial\Psi}{\partial\epsilon^e} = \boldsymbol{E}(\boldsymbol{\phi}) : \epsilon^e \tag{17.12a}$$

$$s = -\rho\frac{\partial\Psi}{\partial T} \tag{17.12b}$$

$$\boldsymbol{Y} = \rho\frac{\partial\Psi}{\partial\boldsymbol{\phi}} \tag{17.12c}$$

$$R = \rho\frac{\partial\Psi}{\partial p} = k_2 r \tag{17.12d}$$

$$\boldsymbol{X} = \rho\frac{\partial\Psi}{\partial\boldsymbol{\alpha}} = k_1\boldsymbol{\alpha} \tag{17.12e}$$

$$K = \rho\frac{\partial\Psi}{\partial\kappa} = k_4\kappa \tag{17.12f}$$

$$\boldsymbol{\Gamma} = \rho\frac{\partial\Psi}{\partial\boldsymbol{\Upsilon}} = k_3\boldsymbol{\Upsilon} \tag{17.12g}$$

In equation (17.12a) $\boldsymbol{E}(\boldsymbol{\phi}) : \epsilon^e$ implies the following product between fourth and second order tensors $E_{ijkl}\epsilon^e_{kl}$. As it is clearly seen from equations (17.12), the only explicit expressions can be obtained from the relations between the observable state variables and the associated variables in the thermodynamic potential. However, internal variables give only the definition of their associated variables. This implies the necessity of the complementary formalism of the dissipation processes in order to describe the evolution of the internal variables. This dissipation processes can be expressed as the sum of the product of the associated variables with the respective flux variables in the following form

$$\Pi = \boldsymbol{\sigma} : \dot{\epsilon}^p - \boldsymbol{Y} : \dot{\boldsymbol{\phi}} - R\dot{p} - K\dot{\kappa} - \boldsymbol{X} : \dot{\boldsymbol{\alpha}} - \boldsymbol{\Gamma} : \dot{\boldsymbol{\Upsilon}} \geq 0 \tag{17.13}$$

Based on the previous assumption of decoupling, the dissipation energy can be written as the summation of dissipation energies due to plasticity and damage:

$$\Pi = \Pi^p + \Pi^d \tag{17.14}$$

where

$$\Pi^p = \boldsymbol{\sigma} : \dot{\epsilon}^p - R\dot{p} - \boldsymbol{X} : \dot{\boldsymbol{\alpha}} \geq 0 \tag{17.15a}$$
$$\Pi^d = -\boldsymbol{Y} : \dot{\boldsymbol{\phi}} - K\dot{\kappa} - \boldsymbol{\Gamma} : \dot{\boldsymbol{\Upsilon}} \geq 0 \tag{17.15b}$$

Two energy dissipative mechanisms for plasticity and damage are exhibited by the material. These two energy dissipative behaviors influence each other. As will be outlined later the plastic strain rate and the damage rate are each functions of the stress and the conjugate force to damage. Consequently, the energies dissipated due to damage and that due to plasticity are interdependent through the stress and the conjugate force to damage. In equation (17.15a) $\boldsymbol{X}$ is the backstress associated with kinematic hardening in plasticity. In equation (17.15b) $\boldsymbol{Y}$ is the force conjugate of the damage tensor $\boldsymbol{\phi}$. Coupling does occur in the plastic potential given by equation (17.15a) between plasticity and damage since the plastic strain is expressed in the current deformed and damaged configuration of the material. Complementary laws can be defined related to the dissipation processes given by equations (17.15a) and (17.15b). This implies the existence of the dissipation potential expressed as a continuous and convex scalar valued function of the flux variables

$$\Theta = \Theta(\dot{\epsilon}^p, \dot{\phi}, \dot{p}, \dot{\kappa}, \dot{\alpha}, \dot{\Upsilon}) \tag{17.16}$$

By using the Legendre-Fenchel transformation of the dissipation potential (Θ), one can obtain complementary laws in the form of the evolution laws of flux variables as function of the dual variables

$$\Theta^* = \Theta^*(\sigma, Y, R, K, X, \Gamma) \tag{17.17}$$

It is possible to decouple the potential Θ^* into the plastic and the damage dissipation potential parts as follows

$$\Theta^* = F(\sigma, R, X) + G(Y, K, \Gamma) \tag{17.18}$$

However, one can notice that there is no exact explicit decoupling of the potentials in the above equation. There is an implicit coupling between these two potentials through the force conjugate which is a function of both σ and ϕ. Keeping this in mind, evolution laws for the plastic strain rate $\dot{\epsilon}^p$, and the damage rate $\dot{\phi}$ can be obtained now by utilizing the calculus of function of several variables with the Lagrange multipliers $\dot{\lambda}^p$ and $\dot{\lambda}^d$. This function Ω can be written in the following form

$$\Omega = \Pi^p + \Pi^d - \dot{\lambda}^p F - \dot{\lambda}^d G \tag{17.19}$$

In order to extrimize the function Ω, one uses the necessary conditions such that

$$\frac{\partial \Omega}{\partial \sigma} = 0 \tag{17.20}$$

and

$$\frac{\partial \Omega}{\partial Y} = 0 \tag{17.21}$$

The two equation (17.20) and (17.21) yield the corresponding plastic strain rate and damage rate evolution equations respectively, which are coupled as shown below. For the case when $F \geq 0$ and $G \geq 0$, one obtains the following expressions:

$$\dot{\epsilon}^p = \dot{\lambda}^p \frac{\partial F}{\partial \sigma} + \dot{\lambda}^d \frac{\partial G}{\partial \sigma} \tag{17.22}$$

and

$$\dot{\phi} = \dot{\lambda}^p \frac{\partial F}{\partial Y} + \dot{\lambda}^d \frac{\partial G}{\partial Y} \tag{17.23}$$

Equations (17.22) and (17.23) indicate non-associativity in both the $\dot{\boldsymbol{\epsilon}}^p$ and $\dot{\boldsymbol{\phi}}$ for the case when coupling occurs between damage and plasticity. In the case of the individual constituents of a composite such as metals in metal matrix composites one notices that non-associativity occurs in the metal primarily because of its in-situ behavior when surrounded by fibers [188,196,215].

Equations (17.22) and (17.23) give respectively the increments of the plastic strain and damage from the damage potential G and the yield function F. Coupling therefore, exists between the plastic strain rate and the damage rate in the material. In the case when either F or G is less than zero decoupling occurs between $\dot{\boldsymbol{\epsilon}}^p$ and $\dot{\boldsymbol{\phi}}$. Complementary laws for the evolution of the other internal variables can be obtained directly from the generalized normality rules. They can be written as follows

$$\dot{p} = -\dot{\lambda}^p \frac{\partial F}{\partial R} \tag{17.24}$$

$$\dot{\boldsymbol{\alpha}} = -\dot{\lambda}^p \frac{\partial F}{\partial \boldsymbol{X}} \tag{17.25}$$

$$\dot{\kappa} = -\dot{\lambda}^d \frac{\partial G}{\partial K} \tag{17.26}$$

$$\dot{\boldsymbol{\Upsilon}} = -\dot{\lambda}^d \frac{\partial G}{\partial \boldsymbol{\Gamma}} \tag{17.27}$$

The next important step is the selection of the appropriate form for the dissipation potentials for both the plastic potential F and the damage potential G in order to establish the desired constitutive equations that describe the mechanical behavior of the material.

17.1.1 Plastic Potential and Yield Criterion

A non-linear kinematic hardening model is selected in this work. In the case of composite materials both the associated flow rule and the von Mises type yield criterion are for the individual constituents of the composite that deform plastically. However, due to the pressence of damage, coupling does occur between plasticity and damage and a non-associative plastic flow results as indicated by equation (17.22). In this work the plastic potential takes the same expression as the yield criterion $(F = f)$. The yield function f can be given as a function of $f(\boldsymbol{\sigma}, \boldsymbol{\phi}, R, \boldsymbol{X})$. For the case of a von Mises type , f is given as follows

$$f = \left(\frac{3}{2}(\boldsymbol{\sigma} - \boldsymbol{X}) : (\boldsymbol{\sigma} - \boldsymbol{X}) \right)^{\frac{1}{2}} - R(p) - \sigma_y \leq 0 \tag{17.28}$$

σ_y is the initial yield threshold value. The suitable form for the back stress $\boldsymbol{X}$ is given by the Armstrong and Frederic model as follows [216]

$$\dot{\boldsymbol{X}} = \frac{2}{3}C^p \dot{\boldsymbol{\epsilon}}^p - \gamma^p \boldsymbol{X} \dot{p} \tag{17.29}$$

C^p and γ^p are the material dependent kinematic hardening parameters. In order to solve for the plastic multiplier, the consistency condition $(\dot{f} = 0)$ is used

$$\dot{f} = \frac{\partial f}{\partial \boldsymbol{\sigma}} : \dot{\boldsymbol{\sigma}} + \frac{\partial f}{\partial \boldsymbol{\phi}} : \dot{\boldsymbol{\phi}} + \frac{\partial f}{\partial \boldsymbol{X}} : \dot{\boldsymbol{X}} + \frac{\partial f}{\partial R} \dot{R} = 0 \tag{17.30}$$

By defining $\dot{R}$ as follows

$$\dot{R} = \frac{\partial R}{\partial p}\dot{p} \quad \text{and} \quad \dot{p} = \dot{\lambda}^p \quad \text{(for von Mises)} \tag{17.31}$$

and making use of equations (17.29) for $\dot{\boldsymbol{X}}$ and (17.23) for $\dot{\phi}$ into equation (17.30), one obtains the following expression

$$\dot{f} = \frac{\partial f}{\partial \boldsymbol{\sigma}} : \dot{\boldsymbol{\sigma}} + \left((\frac{2}{3}C^p\frac{\partial f}{\partial \boldsymbol{\sigma}} - \gamma^p \boldsymbol{X}) + \frac{\partial f}{\partial \boldsymbol{\phi}} : \frac{\partial f}{\partial \boldsymbol{Y}} + \frac{\partial R}{\partial p}\frac{\partial f}{\partial R} \right)\dot{\lambda}^p + \frac{\partial f}{\partial \boldsymbol{\phi}} : \frac{\partial G}{\partial \boldsymbol{Y}}\dot{\lambda}^d \tag{17.32}$$

By defining the following relations

$$b_1 = \frac{\partial f}{\partial \boldsymbol{\sigma}} : \dot{\boldsymbol{\sigma}} \tag{17.33}$$

$$a_{11} = \left((\frac{2}{3}C^p\frac{\partial f}{\partial \boldsymbol{\sigma}} - \gamma^p \boldsymbol{X}) + \frac{\partial f}{\partial \boldsymbol{\phi}} : \frac{\partial f}{\partial \boldsymbol{Y}} + \frac{\partial R}{\partial p}\frac{\partial f}{\partial R} \right) \tag{17.34}$$

$$a_{12} = \frac{\partial f}{\partial \boldsymbol{\phi}} : \frac{\partial G}{\partial \boldsymbol{Y}} \tag{17.35}$$

equation (17.32) may be rewritten in the following linear form

$$a_{11}\dot{\lambda}^p + a_{12}\dot{\lambda}^d = -b_1 \tag{17.36}$$

The second linear equation required for the solution of $\dot{\lambda}^p$ and $\dot{\lambda}^d$ can be obtained from the damage criterion and the corresponding damage consistency condition which will be presented in the next section.

17.1.2 Rate Independent Damage

The anisotropic damage criterion model of Voyiadjis and Park [176] for metal matrix composites is used here. However, unlike that model which is restricted to the isotropic hardening growth of damage, the current model includes translation of the damage surface akin to kinematic hardening. Non-linearity makes the damage potential non-associative to the damage criterion($G \neq g$) but allows one for a better modeling of the randomly distributed micro-cracks and cavities in the material especially under multi axial loading conditions.

Based on the additional effect of non-linear kinematic hardening, the new damage criterion can be written in terms of the tensorial hardening parameters $\boldsymbol{h}$ and $\boldsymbol{\Gamma}$ by satisfying the requirement that g is an isotropic function of its tensorial arguments ($\boldsymbol{Y}$, $\boldsymbol{h}$, and $\boldsymbol{\Gamma}$). This requirement is a necessary condition to make the analytical expression for g not to be dependent on the orientation of the employed coordinate system [140,176]. Therefore one can write an analytical expression for the damage criterion g in terms of the invariants of its tensorial variables in the following form for each individual constituent of the composite

$$g = (Y_{ij} - \Gamma_{ij})P_{ijkl}(Y_{kl} - \Gamma_{kl}) - 1 = 0 \tag{17.37}$$

where the fourth order tensor $\boldsymbol{P}$ describes the anisotropic nature of the damage growth and the initiation of damage. Its form is given as a function of the hardening tensor $\boldsymbol{h}$

$$P_{ijkl} = h_{ij}^{-1} h_{kl}^{-1} \tag{17.38}$$

where $\boldsymbol{h}^{-1}$ is the inverse of the tensor $\boldsymbol{h}$,

$$h_{ij} = \left(\lambda \eta \left(\frac{\kappa}{\lambda} \right)^{\xi} \delta_{ik} \phi_{kj} + \delta_{ij} \lambda \nu^2 \right) \tag{17.39}$$

and ξ indicates an exponent in the above equation. In equation (17.39) the first term is associated with the anisotropic growth of damage. As it is clearly seen, anisotropy is introduced through the anisotropic nature of the second order damage tensor while hardening is introduced in the equation through the multiplier $\lambda \eta (\kappa/\lambda)^{\xi}$. The parameter λ is the Lamé constant in the damaged configuration as given by Voyiadjis and Park [176]. The tensor $\boldsymbol{h}$ maybe expressed in terms of scaling parameters such as the radius and spacing of fibers.

In order to incorporate the non-linear kinematic hardening rule within the framework of the generalized normality hypothesis of equations (17.27), one selects the damage flow potential to be non associative and it should be different than the loading surface g. Here one can take the damage potential as follows(Voyiadjis and Deliktas [223])

$$G = g + \frac{k_5}{2k_3} \boldsymbol{\Gamma} : \boldsymbol{\Gamma} \tag{17.40}$$

where k_5 is the material dependent constant. The following relations can be obtained from equations (17.37) and (17.40) for use in the derivations that follow

$$\frac{\partial g}{\partial \boldsymbol{\Gamma}} = -\frac{\partial g}{\partial \boldsymbol{Y}} \tag{17.41}$$

$$\frac{\partial G}{\partial \boldsymbol{\Gamma}} = \frac{\partial g}{\partial \boldsymbol{\Gamma}} + \frac{k_5}{k_3} \boldsymbol{\Gamma} \tag{17.42}$$

$$\frac{\partial G}{\partial \boldsymbol{Y}} = \frac{\partial g}{\partial \boldsymbol{Y}} \tag{17.43}$$

The second linear relation can be derived from the consistency condition of the damage criterion ($\dot{g} = 0$)

$$\dot{g} = \frac{\partial g}{\partial \boldsymbol{\sigma}} : \dot{\boldsymbol{\sigma}} + \frac{\partial g}{\partial \boldsymbol{\phi}} : \dot{\boldsymbol{\phi}} + \frac{\partial g}{\partial \kappa} \dot{\kappa} + \frac{\partial g}{\partial \boldsymbol{\Gamma}} : \dot{\boldsymbol{\Gamma}} = 0 \tag{17.44}$$

where $\dot{\boldsymbol{\Gamma}}$ can be obtained by making use of the state laws in equations (17.12). The time derivative of equation (17.12g) is given by

$$\dot{\boldsymbol{\Gamma}} = k_3 \dot{\boldsymbol{\Upsilon}} \tag{17.45}$$

By substituting for $\dot{\boldsymbol{\Upsilon}}$ from the normality rule in equation (17.27) into the above relation one can obtain

$$\dot{\mathbf{\Gamma}} = -k_3 \dot{\lambda}^d \frac{\partial G}{\partial \mathbf{\Gamma}} \tag{17.46}$$

Now making use of relation (17.43) into equation (17.46), the final relation for $\dot{\mathbf{\Gamma}}$ can be written as follows

$$\dot{\mathbf{\Gamma}} = \dot{\lambda}^d \left(k_3 \frac{\partial g}{\partial \mathbf{Y}} - k_5 \mathbf{\Gamma} \right) \tag{17.47}$$

where the coefficients k_3 and k_5 are defined as $k_3 = \frac{2}{3} C^d$ and $k_5 = \gamma^d$. Substituting $\dot{\mathbf{\Gamma}}$ from equation (17.47), $\dot{\phi}$ from equation (17.23), and $\dot{\kappa}$ from relation (17.5) into equation (17.44) one obtains the following relations

$$\dot{g} = \frac{\partial g}{\partial \boldsymbol{\sigma}} : \dot{\boldsymbol{\sigma}} + \dot{\lambda}^p \left(\frac{\partial g}{\partial \phi} : \frac{\partial f}{\partial \mathbf{Y}} + \frac{\partial g}{\partial \kappa} \sqrt{\frac{2}{3} \frac{\partial g}{\partial \mathbf{Y}} : \frac{\partial g}{\partial \mathbf{Y}}} \right) +$$
$$+ \dot{\lambda}^d \left(\frac{\partial g}{\partial \phi} : \frac{\partial f}{\partial \mathbf{Y}} + \frac{\partial g}{\partial \kappa} \sqrt{\frac{2}{3} \frac{\partial g}{\partial \mathbf{Y}} : \frac{\partial g}{\partial \mathbf{Y}}} - k_3 \frac{\partial g}{\partial \mathbf{Y}} : \frac{\partial g}{\partial \mathbf{Y}} + k_5 \frac{\partial g}{\partial \mathbf{Y}} : \mathbf{\Gamma} \right) \tag{17.48}$$

By defining

$$b_2 = \frac{\partial g}{\partial \boldsymbol{\sigma}} : \dot{\boldsymbol{\sigma}} \tag{17.49}$$

$$a_{21} = \left(\frac{\partial g}{\partial \phi} : \frac{\partial f}{\partial \mathbf{Y}} + \frac{\partial g}{\partial \kappa} \sqrt{\frac{2}{3} \frac{\partial g}{\partial \mathbf{Y}} : \frac{\partial g}{\partial \mathbf{Y}}} \right) \tag{17.50}$$

$$a_{22} = \left(\frac{\partial g}{\partial \phi} : \frac{\partial g}{\partial \mathbf{Y}} + \frac{\partial g}{\partial \kappa} \sqrt{\frac{2}{3} \frac{\partial g}{\partial \mathbf{Y}} : \frac{\partial g}{\partial \mathbf{Y}}} - k_3 \frac{\partial g}{\partial \mathbf{Y}} : \frac{\partial g}{\partial \mathbf{Y}} + k_5 \frac{\partial g}{\partial \mathbf{Y}} : \mathbf{\Gamma} \right) \tag{17.51}$$

equation (17.48) may be re written in the following form

$$a_{21} \dot{\lambda}^p + a_{22} \dot{\lambda}^d = -b_2 \tag{17.52}$$

The plastic multiplier $\dot{\lambda}^p$ and the damage multiplier $\dot{\lambda}^d$ can be solved from the linear system of equations given by (17.36) and (17.52) such that

$$\begin{pmatrix} \dot{\lambda}^p \\ \dot{\lambda}^d \end{pmatrix} = \frac{1}{\Delta} \begin{bmatrix} a_{22} & -a_{12} \\ -a_{21} & a_{11} \end{bmatrix} \begin{pmatrix} -b_1 \\ -b_2 \end{pmatrix} \tag{17.53}$$

where

$$\Delta = a_{11} a_{22} - a_{12} a_{21} \tag{17.54}$$

Substituting $\dot{\lambda}^p$ and $\dot{\lambda}^d$ from equations (17.53) into equation (17.22), the evolution equation for the plastic strain rate $\dot{\boldsymbol{\epsilon}}^p$ can be written in the following form

$$\dot{\boldsymbol{\epsilon}}^p = \boldsymbol{\chi}^p : \dot{\boldsymbol{\sigma}} \tag{17.55}$$

where the fourth order tensor χ^p is defined as the sum of the fourth order tensors $\boldsymbol{K}$ and $\boldsymbol{L}$ which are expressed respectively as follows

$$\boldsymbol{K} = \frac{1}{\Delta}\frac{\partial f}{\partial \boldsymbol{\sigma}} \otimes \left(a_{12}\frac{\partial f}{\partial \boldsymbol{\sigma}} - a_{22}\frac{\partial g}{\partial \boldsymbol{\sigma}}\right) \tag{17.56}$$

$$\boldsymbol{L} = \frac{1}{\Delta}\frac{\partial g}{\partial \boldsymbol{\sigma}} \otimes \left(a_{21}\frac{\partial f}{\partial \boldsymbol{\sigma}} - a_{11}\frac{\partial g}{\partial \boldsymbol{\sigma}}\right) \tag{17.57}$$

Similarly the evolution equation for damage can be expressed as follows

$$\dot{\boldsymbol{\phi}} = \chi^d : \dot{\boldsymbol{\sigma}} \tag{17.58}$$

where the fourth order tensor χ^d is now defined as the sum of the fourth order tensors $\boldsymbol{P}$ and $\boldsymbol{Q}$ which are expressed respectively as follows

$$\boldsymbol{P} = \frac{1}{\Delta}\frac{\partial f}{\partial \boldsymbol{Y}} \otimes \left(a_{12}\frac{\partial f}{\partial \boldsymbol{\sigma}} - a_{22}\frac{\partial g}{\partial \boldsymbol{\sigma}}\right) \tag{17.59}$$

$$\boldsymbol{Q} = \frac{1}{\Delta}\frac{\partial g}{\partial \boldsymbol{Y}} \otimes \left(a_{21}\frac{\partial f}{\partial \boldsymbol{\sigma}} - a_{11}\frac{\partial g}{\partial \boldsymbol{\sigma}}\right) \tag{17.60}$$

and

$$\chi^d = \boldsymbol{P} + \boldsymbol{Q} \tag{17.61}$$

17.1.3   Rate Dependent Damage Coupled with Rate Dependent Plasticity

In order to account for both loading rate dependency and regularizing the localization problems a viscous anisotropic damage mechanism needs to be implemented. Such a model accounts for retardation of the micro-crack growth at higher strain rates. The proposed rate dependent damage model is based on the mathematical formulation of the overstress type modeling of rate dependent plasticity. For rate dependent damage an overstress conjugate force type damage function is postulated. However rate dependency of damage is considered only after inelastic deformation occurs. In the elastic region, damage is considered as rate independent and the formulation made in the previous section is used for the damage response in the elastic region. This is because the proposed theory does not encompasses a viscoelastic behavior in the elastic region. The more accurate response of the damaged materials may be possible by replacing the elastic domain with a corresponding viscoelastic domain coupled with damage. This is beyond the scope of this work and therefore it is not considered here. An extension of equations (17.22) and (17.23) leads to the rate dependent plastic strain rate [217,218] and the damage rate given as follows

$$\dot{\boldsymbol{\epsilon}}^{vp} = \|\dot{\boldsymbol{\epsilon}}^{vp}\|\boldsymbol{n}^{vp} + \|\dot{\boldsymbol{\phi}}\|\boldsymbol{n}^{vpd} \tag{17.62a}$$

$$\dot{\boldsymbol{\epsilon}}^{vp} = \dot{\boldsymbol{\epsilon}}^{vp(1)} + \dot{\boldsymbol{\epsilon}}^{vp(2)} \tag{17.62b}$$

and

$$\dot{\boldsymbol{\phi}} = \|\dot{\boldsymbol{\epsilon}}^{vp}\|\boldsymbol{n}^{dvp} + \|\dot{\boldsymbol{\phi}}\|\boldsymbol{n}^d \tag{17.63a}$$

$$\dot{\boldsymbol{\phi}} = \dot{\boldsymbol{\phi}}^{(1)} + \dot{\boldsymbol{\phi}}^{(2)} \tag{17.63b}$$

Superscripts in this work do not imply tensorial indices but only describe the type of material inelasticity. In equations (17.62) and (17.63) $\|\dot{\boldsymbol{\epsilon}}^{vp}\|$ and $\|\dot{\boldsymbol{\phi}}\|$ are the magnitudes of the plastic strain rate and damage rate which can be decomposed into a product of two functions respectively [219] using the Zener parameters such that

$$\|\dot{\boldsymbol{\epsilon}}^{vp}\| = \vartheta^{vp}(T)Z^{vp} \geq 0 \tag{17.64a}$$
$$\|\dot{\boldsymbol{\phi}}\| = \vartheta^{d}(T)Z^{d} \geq 0 \tag{17.64b}$$

The unit tensors $\boldsymbol{n}^{vp}, \boldsymbol{n}^{vpd}, \boldsymbol{n}^{dvp}$ and $\boldsymbol{n}^{d}$ are used to identify the direction of flow of the plastic strain and damage and are expressed as follows, respectively

$$\boldsymbol{n}^{vp} = \frac{\dfrac{\partial F^{vp}}{\partial \boldsymbol{\sigma}}}{\left\|\dfrac{\partial F^{vp}}{\partial \boldsymbol{\sigma}}\right\|}, \quad \boldsymbol{n}^{vpd} = \frac{\dfrac{\partial G^{d}}{\partial \boldsymbol{\sigma}}}{\left\|\dfrac{\partial G^{d}}{\partial \boldsymbol{\sigma}}\right\|} \tag{17.65}$$

$$\boldsymbol{n}^{dvp} = \frac{\dfrac{\partial F^{vp}}{\partial \boldsymbol{Y}}}{\left\|\dfrac{\partial F^{vp}}{\partial \boldsymbol{Y}}\right\|}, \quad \boldsymbol{n}^{d} = \frac{\dfrac{\partial G^{d}}{\partial \boldsymbol{Y}}}{\left\|\dfrac{\partial G^{d}}{\partial \boldsymbol{Y}}\right\|} \tag{17.66}$$

where F^{vp} and G^{d} are the dynamic potentials for viscoplasticity and damage and their expression are given respectively as follows [220]

$$F^{vp} = \left\{\frac{f^{*}}{[R(r) + \sigma_{y}(\dot{\epsilon}_{o})]} + 1\right\}^{\frac{1}{2}} - 1 \tag{17.67}$$

and

$$G^{d} = \left\{\frac{g^{*}}{[K(\kappa) + \sigma_{d}(\dot{\epsilon}_{o})]} + 1\right\}^{\frac{1}{2}} - 1 \tag{17.68}$$

where f^{*} and g^{*} represent the equilibrium surfaces of viscoplasticity and damage and are strain rate dependent. Therefore, equations (17.28) and (17.37) are modified here in order to describe the equilibrium surfaces

$$f^{*} = \left(\frac{3}{2}(\boldsymbol{\sigma}^{*} - \boldsymbol{X}) : (\boldsymbol{\sigma}^{*} - \boldsymbol{X})\right)^{\frac{1}{2}} - \left[R(p) + \sigma_{y}(\dot{\epsilon})\right] \leq 0 \tag{17.69}$$

and

$$g^{*} = (Y_{ij}^{*} - \Gamma_{ij})P_{ijkl}(Y_{kl}^{*} - \Gamma_{kl}) - 1 = 0 \tag{17.70}$$

The functional dependency of the initial threshold values of plasticity and damage on the strain rate, is obtained through the function $Q(z)$ [221] such that

$$\sigma_{y}(z) = A \tanh z$$

and

$$z = C \ln\left[1 + \frac{1}{2} B_{abcd} \dot{e}_{ab} \dot{e}_{cd}\right]$$

where A, $\boldsymbol{B}$, and C are appropriate material parameters. $\boldsymbol{\sigma}^*$ and $\boldsymbol{Y}^*$ are the stresses and conjugate forces respectively on the equilibrium surfaces. It is postulated that $\boldsymbol{\sigma}^*$ lies on the line joining the current state of stress and the center of the equilibrium surface [215]. The same applies for the conjugate equilibrium force. The equilibrium stresses are given in Figure 17.1 and can be written as follows

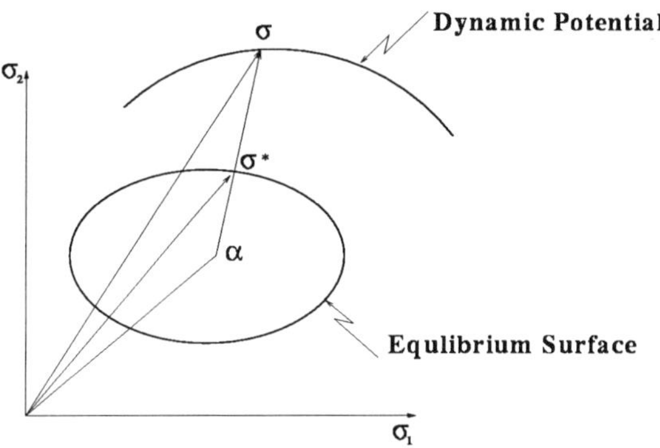

Figure 17.1 Equilibrium Surface and Viscoplastic Potential

$$\boldsymbol{\sigma}^* = \boldsymbol{X} + c^{vp}(\boldsymbol{\sigma} - \boldsymbol{X}) \tag{17.71}$$

A similar expression is obtained for the conjugate equilibrium force

$$\boldsymbol{Y}^* = \boldsymbol{\Gamma} + c^{d}(\boldsymbol{Y} - \boldsymbol{\Gamma}) \tag{17.72}$$

c^{vp} and c^{d} can be obtained by using equations (17.71) and (17.72) in equations (17.69) and (17.70) respectively and the corresponding expressions are given below

$$c^{vp} = \sqrt{\frac{[R(p) + \sigma_y(z)]}{\frac{3}{2}(\boldsymbol{\sigma} - \boldsymbol{X}) : (\boldsymbol{\sigma} - \boldsymbol{X})}} \tag{17.73}$$

and

$$c^{d} = \sqrt{\frac{1}{(Y_{ij} - \Gamma_{ij})P_{ijkl}(Y_{kl} - \Gamma_{kl})}} \tag{17.74}$$

The simpler form of the Zener parameters, in equation (17.64a) and (17.64b), can be expressed by substituting the expression for F^{vp} and G^{d} in equations (17.67) and (17.68) such that

$$Z^{vp} = Z\left(\frac{< \|\boldsymbol{\sigma} - \boldsymbol{X}\| - \sigma_y^{*vp} >}{D^{vp}}\right) = \frac{< \sigma^{vp} >}{D^{vp}} \tag{17.75}$$

and

$$Z^d = Z\left(\frac{< \|\boldsymbol{Y} - \boldsymbol{\Gamma}\| - Y^{*d} >}{D^d}\right) = \frac{< Y^d >}{D^d} \tag{17.76}$$

The terms, σ^{vp} and Y^d, are the overstress of viscoplasticity and damage respectively. σ_y^{*vp} is defined as $\sigma_y^{*vp} = [R(r) + \sigma_y(\dot{\epsilon})]$. Similarly $Y^{*d} = [K(\kappa) + Y_d(\dot{\epsilon})]$. The terms, D^{vp} and D^d are the drug forces which represent the isotropic hardening effects. They can be considered as internal variables and their evolution equations can be derived. However, in this paper they are treated as constant parameters. The final form of the viscoplastic strain rate and damage rate can be rewritten in the uncoupled form as follows

$$\dot{\boldsymbol{\epsilon}}^{vp} = \left(\frac{< \|\boldsymbol{\sigma} - \boldsymbol{X}\| - \sigma_y^{*vp} >}{D^{vp}}\right)^{n_1} \boldsymbol{n}^{vp} = \frac{F^{n_1}}{\eta^{vp}} \boldsymbol{n}^{vp} \tag{17.77}$$

and

$$\dot{\boldsymbol{\phi}} = \left(\frac{< \|\boldsymbol{Y} - \boldsymbol{\Gamma}\| - Y^{*d} >}{D^d}\right)^{n_2} \boldsymbol{n}^d = \frac{G^{n_2}}{\eta^d} \boldsymbol{n}^d \tag{17.78}$$

where η^{vp} and η^d are defined as $(\frac{1}{D^{vp}})^{n_1}$ and $(\frac{1}{D^d})^{n_2}$ respectively and "n_1", and "n_2" are the exponents for the potential functions of viscoplasticity and damage respectively. Superscripts imply exponents only in the case of the bracketed terms.

17.1.4 Characterizing Internal State Variables of the Rate Dependent Models

As pointed out earlier the internal state variables are introduced in the material model to represent the true response of the material due to the variation of the microstructure when subjected to external forces. The anisotropic structure of the material is usually defined in two forms either as material inherited or deformation induced. The anisotropic nature of the composite material is material inherited anisotropy. However, at the local level its constituents are isotropic materials. Therefore, the use of a micromechanical model to analyze the composite material deals with deformation induced anisotropy. This deformation induced anisotropy is considered here due to both the plasticity and damage in the material. This phenomenon is characterized in the theory by using internal variables for the hardening terms and through the use of the second order tensorial form of the damage variable.

In the case of the rate independent models, internal variables are defined in section 17.1.1. However, the ones used in the rate dependent model need to be redefined in order to characterize the time and thermal recovery effects due to the rate and temperature dependency of the material. For this reason the general form of the internal variables can be defined as follows [219]

$$\dot{A}_k = \textbf{hardening} - \textbf{dynamic recovery} - \textbf{static recovery} \tag{17.79}$$

The hardening terms represent the strengthening mechanism, while the recovery terms represents softening mechanism. The hardening and dynamic recovery terms evolves with the deformation due to either plasticity or damage or both. The static recovery term evolves with time. The evolution equation of the internal variables for the rate dependent behavior are described below

$$\dot{\boldsymbol{X}} = \frac{3}{2}H^{vp}\left(\dot{\boldsymbol{\epsilon}}^{vp} - \frac{3}{2}\frac{\|\boldsymbol{X}\|}{L^{vp}}\|\dot{\boldsymbol{\epsilon}}^{vp}\|\boldsymbol{d}^{vp}\right) - \vartheta^{vp}\boldsymbol{X}\cdot\boldsymbol{b}^{vp} \tag{17.80a}$$

$$\dot{R} = Q^{vp}(1 - \frac{R}{Q^{vp}})\|\dot{\boldsymbol{\epsilon}}^{vp}\| - \vartheta^{vp}\Upsilon(r) \tag{17.80b}$$

$$\boldsymbol{b}^{vp} = \frac{\boldsymbol{X}}{\|\boldsymbol{X}\|} \tag{17.80c}$$

$$\boldsymbol{d}^{vp} = (1 - \rho^{vp})\boldsymbol{b}^{vp} + \rho^{vp}\frac{\boldsymbol{n}^{vp}\cdot\boldsymbol{X}\cdot\boldsymbol{n}^{vp}}{\|\boldsymbol{X}\|} \tag{17.80d}$$

Similarly the evolution equations for the hardening variables of damage can be written analogously to that of plasticity as follows

$$\dot{\boldsymbol{\Gamma}} = \frac{3}{2}H^d\left(\dot{\boldsymbol{\phi}} - \frac{3}{2}\frac{\|\boldsymbol{\Gamma}\|}{L^d}\|\dot{\boldsymbol{\phi}}\|\boldsymbol{d}^d\right) - \vartheta^d\boldsymbol{\Gamma}\cdot\boldsymbol{b}^d \tag{17.81a}$$

$$\dot{K} = Q^d(1 - \frac{K}{Q^d})\|\dot{\boldsymbol{\phi}}\| - \vartheta^d\Upsilon(\kappa) \tag{17.81b}$$

$$\boldsymbol{b}^d = \frac{\boldsymbol{\Gamma}}{\|\boldsymbol{\Gamma}\|} \tag{17.81c}$$

$$\boldsymbol{d}^d = (1 - \rho^d)\boldsymbol{b}^d + \rho^d\frac{\boldsymbol{n}^d\cdot\boldsymbol{\Gamma}\cdot\boldsymbol{n}^d}{\|\boldsymbol{\Gamma}\|} \tag{17.81d}$$

where ρ^{vp} in the above equations defines the non-proportionality condition. In the case of $\rho^{vp} = 0$ equations (17.80a) and (17.81a) reduce to the Armstrong and Frederic type backstress evolution which is used in this work. The other extreme case can be obtained by taking $\rho^{vp} = 1$ which gives a relation similar to the non-linear Prager model for the backstress definition. In the above equations H^{vp}, L^{vp}, Q^{vp}, H^d, L^d, and Q^d are the model parameters.

17.1.5 A Physical Interpretation of the Damage Tensor ϕ

Damage in this work is characterized as the net area decrease due to a three-dimensional distribution of micro-cracks or micro-voids [58]. A differential tetrahedron is considered at point "O" in an undamaged continuum in the initial configuration, C_o, as indicated in Figure 17.2a [222]. An element PQR of an arbitrary orientation is shown in Figure 17.2b for the deformed damaged material in the current configuration C. The line elements OP, OQ, OR and area PQR in the current configuration, C, are represented respectively by the differential lengths dx_1, dx_2, dx_3 and the vector $d\boldsymbol{S}$ in the three-dimensional vector

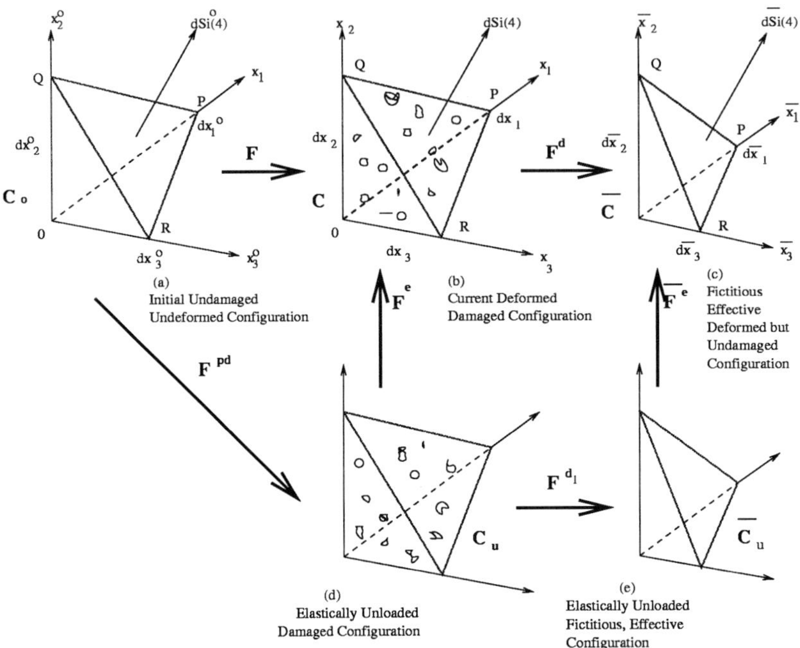

Figure 17.2: Schematic Representation of the Deformation Gradients for Micro-cracks

space where the x_i axes coincide with the principal damage axes. Figure 17.2a shows the corresponding differential lengths dx_1^o, dx_2^o, dx_3^o and the vector dS^o in the initial undamaged configuration, C_o. The deformation gradient from C_o to C is represented by $\boldsymbol{F}$. A fictitious effective undamaged configuration, $\bar{C}$, is postulated as shown in Figure 17.2c with an area reduction due to the damage brought about by the micro-cracks and the micro-cavities. The deformation gradient from C to $\bar{C}$ is represented by $\boldsymbol{F}^d$. The direction of vectors dS and $d\bar{S}$ are not necessarily coincident since the reduction due to damage is not only confined in the PQR plane but in other planes with other orientations. In Figures 17.2d and 17.2e the elastically unloaded configurations C_u and $\bar{C}_u$ are respectively postulated. C_u represents the elastically unloaded damaged configuration with the deformation gradient from C to C_u being represented by $\boldsymbol{F}^e$. However, $\bar{C}_u$, represents the fictitious, effective elastically unloaded configuration with the deformation gradient from $\bar{C}$ to $\bar{C}_u$ being represented by $\bar{\boldsymbol{F}}^e$. The two deformation gradient $\boldsymbol{F}^e$ and $\bar{\boldsymbol{F}}^e$ are not equal since $\boldsymbol{F}^e$ incorporates some elastic recovered damage. This does not imply the healing of the material.

The work of Betten [222] is followed here in characterizing the anisotropic damage tensor. In three dimensional space a parallelogram formed by the vectors $\boldsymbol{V}$ and $\boldsymbol{W}$ can be represented as follows:

$$S_i = \epsilon_{ijk} V_j W_k \tag{17.82}$$

or in dual form as follows

$$S_{ij} = \epsilon_{ijk} S_k \qquad (17.83)$$

such that

$$S_i = \frac{1}{2} \epsilon_{ijk} S_{jk} \qquad (17.84)$$

where ϵ_{ijk} is the permutation tensor. In a rectilinear three-dimensional space the absolute values of the components S_{12}, S_{23}, S_{31} are projections of the area of the parallelogram on the coordinate planes. The S_{ij} represents the area vector in a three-dimensional space and has an orientation fixed by the cross product shown in equation(17.82).

The deformation gradient $\boldsymbol{F}^d$ is used to express the differential lengths $d\bar{x}_i$ in the effective configuration $\bar{C}$ in terms of the differential lengths dx_i in the current deformed damaged configuration C such that

$$d\bar{x}_i = F_{ij}^d dx_j \qquad (17.85)$$

The components of the area vector $d\boldsymbol{S}$ in the three-dimensional space of the current configuration ,C, are given by

$$dS_1 = -\frac{1}{2} dx_2 dx_3, \qquad dS_2 = -\frac{1}{2} dx_1 dx_3, \qquad dS_3 = -\frac{1}{2} dx_1 dx_2 \qquad (17.86)$$

The corresponding area vector components of $d\bar{\boldsymbol{S}}$ in the effective configuration, $\bar{C}$, are given by

$$d\bar{S}_1 = -\frac{1}{2} d\bar{x}_2 d\bar{x}_3, \qquad d\bar{S}_2 = -\frac{1}{2} d\bar{x}_1 d\bar{x}_3, \qquad d\bar{S}_3 = -\frac{1}{2} d\bar{x}_1 d\bar{x}_2 \qquad (17.87)$$

The reduction in area between the current, C, and effective, $\bar{C}$, configurations may be described in terms of the eigenvalues of the second order tensor ϕ such that

$$d\bar{S}_1 = (1 - \hat{\phi}_1) dS_1, \qquad d\bar{S}_2 = (1 - \hat{\phi}_2) dS_2, \qquad d\bar{S}_3 = (1 - \hat{\phi}_3) dS_3 \qquad (17.88)$$

Making use of equations (17.85) through (17.88) one obtains the eigenvalues of $\boldsymbol{F}^d$ in terms of the eigenvalues of ϕ such that

$$\hat{F}_{11}^d = \sqrt{\frac{(1 - \hat{\phi}_2)(1 - \hat{\phi}_3)}{(1 - \hat{\phi}_1)}} \qquad (17.89)$$

$$\hat{F}_{22}^d = \sqrt{\frac{(1 - \hat{\phi}_1)(1 - \hat{\phi}_3)}{(1 - \hat{\phi}_2)}} \qquad (17.90)$$

$$\hat{F}_{33}^d = \sqrt{\frac{(1 - \hat{\phi}_1)(1 - \hat{\phi}_2)}{(1 - \hat{\phi}_3)}} \qquad (17.91)$$

The resulting Jacobian of the damage deformation gradient is expressed as follows

$$J^d = \hat{F}^d_{11}\hat{F}^d_{22}\hat{F}^d_{33} = \sqrt{(1 - \hat{\phi}_1)(1 - \hat{\phi}_2)(1 - \hat{\phi}_3)} \tag{17.92}$$

This similar to the one derived in Chapter 16(Voyiadjis and Park [211]) Since the fictitious effective deformed configuration denoted by $\bar{C}$ is obtained by removing the damages from the real deformed configuration denoted by C, therefore the differential volume of the fictitious effective deformed volume denoted by $d\bar{V}$ is obtained as follows(Voyiadjis and Deliktas [223])

$$d\bar{V} = dV - dV^d \tag{17.93}$$

$$d\bar{V} = \left(\sqrt{(1 - \hat{\phi}_1)(1 - \hat{\phi}_2)(1 - \hat{\phi}_3)}\right)dV \tag{17.94}$$

or

$$d\bar{V} = J^d dV \tag{17.95}$$

dV^d is the volume of damage in configuration C. Equation (17.94) may be expressed alternatively as

$$\left(1 - d\right)^2 \equiv \left(\frac{dV - dV^d}{dV}\right)^2 = (1 - \hat{\phi}_1)(1 - \hat{\phi}_2)(1 - \hat{\phi}_3) \tag{17.96}$$

where d is a measure of volume reduction due to the presence of micro-cavities and micro-cracks caused by damage. Rearranging the terms in equation (17.96) one obtains

$$(1 - d)^2 = 1 - (\hat{\phi}_1 + \hat{\phi}_2 + \hat{\phi}_3) + (\hat{\phi}_1\hat{\phi}_2 + \hat{\phi}_1\hat{\phi}_3 + \hat{\phi}_2\hat{\phi}_3) - (\hat{\phi}_1\hat{\phi}_2\hat{\phi}_3)$$
$$= \left(\frac{dV - dV^d}{dV}\right)^2 \tag{17.97}$$

In the case when the volume reduction is infinitesimal (that is when $\hat{\phi}_1\hat{\phi}_2$, $\hat{\phi}_1\hat{\phi}_3$, $\hat{\phi}_2\hat{\phi}_3$, and $\hat{\phi}_1\hat{\phi}_2\hat{\phi}_3$ can be considered negligible when compared to $\hat{\phi}_i$), then equation (17.97) reduces to the following by ignoring higher order terms in ϕ

$$(1 - d^i)^2 = 1 - (\hat{\phi}_1 + \hat{\phi}_2 + \hat{\phi}_3) \tag{17.98}$$

Infinitesimal damage does not reflect necessarily small strain theory. In equations (17.97) and (17.98) d and d^i are measures of volume reduction due to damage. The measures d and d^i are equal to $(\frac{a^3}{dV})$ assuming only one single micro-crack where 'a" is the radius of an assumed single spherical micro-crack and dV is the volume of a representative unit cell in the mesostructure [21,206,224]. The measure $(\frac{a^3}{dV})$ relates to the microcrack porosity(concentration ratio) within the unit cell . A fourth order damage tensor representation is a generalization of this measure. However,in this work a second order damage tensor representation is used through ϕ which is in turn used to describe the fourth order damage effect tensor M defined in the next section.

In the formulation of Budiansky and O'Conell [224], the distributed energy due to microcracks was explicitly related to the fracture mechanics released energy for similar non-interacting cracks. However, that was limited to a homogeneous state of uniform pressure or for an axial load. In their interpretation the volume stress and strain are assumed to be identical in both the cracked and uncracked states. In the present formulation the effective space is used mainly as an interpretation of the damage deformation gradient to allow one to obtain the kinematics of damage. In the formulation presented here, the stress and strain fields differ in the effective and cracked configurations and the dissipated energy is accounted for in plasticity and damage through the theory of thermodynamics. However, the dissipated energy is not expressed through fracture mechanics and does not identify different modes of fracture. Instead it provides a damage strengthening criterion through internal variables to account for the interaction of cracks and the corresponding arresting of cracks.

Both d and d^i are isotropic measures of damage. It is clear that although the damage distribution may not be isotropic, however, d and d^i are simplified kinematic measures of damage that are scalar valued. In the special case when damage is indeed isotropic and the volume reduction is infinitesimal then one obtains

$$\phi = \hat{\phi}_1 = \hat{\phi}_2 = \hat{\phi}_3 \tag{17.99}$$

Using equation (17.99) in equation (17.98), one obtains d^i in terms of ϕ as follows

$$d^i = 1 - \sqrt{1 - 3\phi} \tag{17.100}$$

or

$$d^i \cong \frac{3}{2}\phi \tag{17.101}$$

In the general damage case given by equation (17.97) but under the constraint of isotropic damage given by equation (17.99) one obtains d in terms of ϕ as follows

$$d = 1 - (1 - \phi)^{\frac{3}{2}} \tag{17.102}$$

In the absence of damage, $\phi = 0$, and consequently d^i and d are both zero.

The scalar measure of $"d"$, is obtained from the second order damage tensor ϕ. This measure ϕ does not compromise an anisotropic damage distribution by interpreting it as isotropic. Bammann and Aifantis [212] introduced a polar decomposition for the deformation gradient utilizing the kinematics of plastic materials with voids such that(as shown in Figure 17.3

$$F = F^e F^v F^p \tag{17.103}$$

where $\boldsymbol{F}^v$ is the deformation gradient in terms of a continuous variable related directly to the void dissipation. B_{pd} of Figure 17.3 is equivalent to the state, C_u given in Figure 17.2d. $B(t)$ of Figure 17.3 is equivalent to the state, $C(t)$ given in Figure 17.2b. B_p of Figure 17.3 is equivalent to the state, $\bar{C}_u$, in Figure 17.2e The representation of the deformation gradient in the proposed work presented here is given by (as shown in Figure 17.4)

$$\boldsymbol{F} = \boldsymbol{F}^{ed}\boldsymbol{F}^{pd} \tag{17.104}$$

whereby part of the damage occurs in the elastic state and the other part in the inelastic state. However, damage here is not continuous and is subject to a damage criterion and the Kuhn Tucker restrictions.

In the work of Bammann and Aifantis [212] dV_v denotes the portion of the elementary volume due to voids and dV_o is the elementary volume in the initial configuration. The elementary volume due to the plastic deformation gradient $\boldsymbol{F}^p$, is given by dV_1 and the elementary volume in the final configuration is dV_2. Bammann and Aifantis [212] assumed $dV_o = dV_1$ due to the assumption of incompressible plastic flow. The determinant of $\boldsymbol{F}^v$ is given by

$$\det \boldsymbol{F}^v \;=\; \frac{dV_2}{dV_o} = \frac{dV_2}{(dV_o + dV_v) - dV_v} \tag{17.105a}$$

$$\det \boldsymbol{F}^v \;=\; \frac{1}{1 - \bar{d}} \tag{17.105b}$$

where

$$\bar{d} \;=\; \frac{dV_v}{dV_2} \tag{17.106a}$$

$$dV_2 \;=\; dV_o + dV_v = dV_1 + dV_v \tag{17.106b}$$

This definition of $"\bar{d}"$ coincides with that of equation (17.96) for the reduction in volume due to the effective configuration given by the symbol $"d"$. In this formulation on the basis that void nucleation and growth results in volumetric changes only, one can express[212]

$$\boldsymbol{F}^v = \frac{1}{(1 - \bar{d})^{\frac{1}{3}}} \boldsymbol{I} \tag{17.107}$$

with the corresponding velocity gradient given by

$$\boldsymbol{L}^v = \frac{\dot{\bar{d}}}{3(1 - \bar{d})} \boldsymbol{I} \tag{17.108}$$

It is clear from equations (17.106a) and (17.101) that for isotropic behavior and infinitesimal volume reduction the parameters $"\bar{d}"$ and ϕ are similar and describe the reduction in volume due to cracks or voids. In the more general case of anisotropy and damage the parameter $"d"$ is given by equation (17.96) in terms of the second order tensor ϕ. In the general case of damage occurring in both the elastic and plastic domain the polar decomposition of the

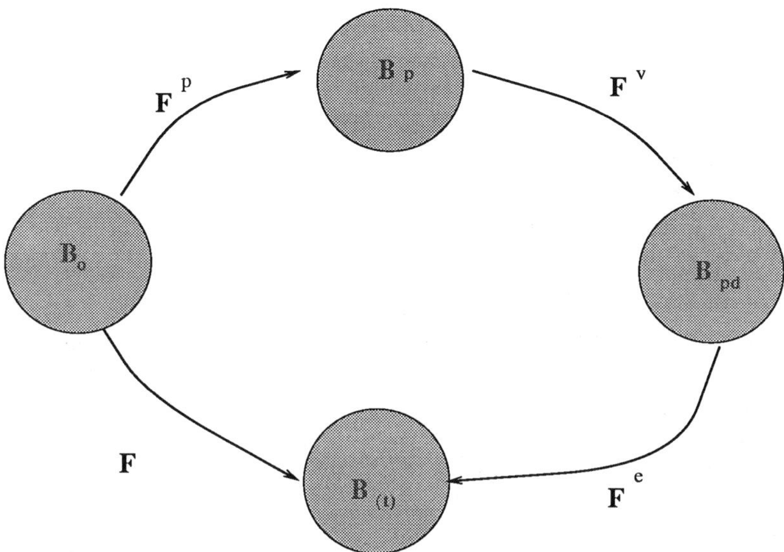

Figure 17.3 Schematic Representation of Elasto-Plastic Damage Deformation with Voids

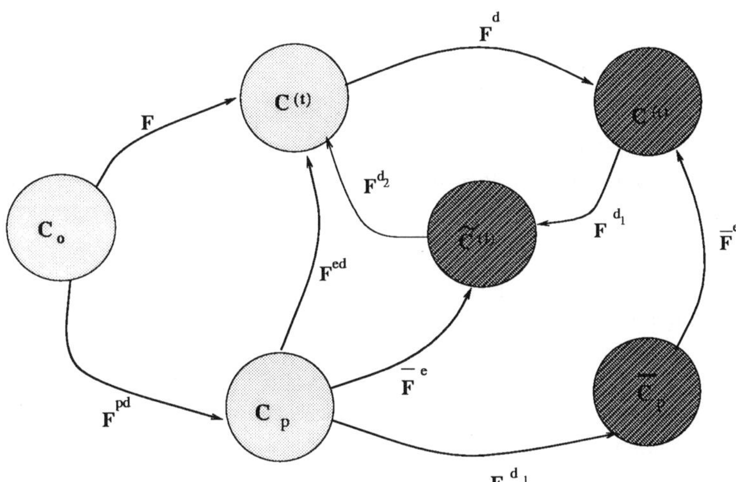

Figure 17.4 Schematic Representation of Elasto-Plastic Damage Deformation for the Proposed Model

deformation gradient is given in Figure 4 (see also [211]). Damage closure in the elastic domain does not imply healing of damage during unloading.

In Figure 17.4 a schematic drawing is representing the kinematics of elasto-plastic damage deformation. C_o is the initial undeformed configuration of the body which may have an initial damage in the material. However, for simplicity this is ignored in this work. The configuration $C(t)$ represents the current elasto-plastic with damage state of the body. The state C_p is an intermediate configuration upon elastic unloading which incorporates part of the damage. This damage is in the deformation gradient $\boldsymbol{F}^{ed}$ and does not constitute healing of the material upon its removal but merely closure of some voids and cracks.

Damage in the microlevel is a material discontinuity. Damage can be considered as an irreversible deformation process in the framework of the continuum damage mechanics. One assumes that upon unloading from the elasto-plastic damage state, the elastic part of the deformation can be completely recovered while no additional plastic deformation and damage takes place. Upon unloading the elasto-plastic damage deformed body from the current configuration $C(t)$ will elastically unload to an intermediate stress free configuration denoted by C_p as shown in Figure 4. Damage process is an irreversible deformation thermodynamically, however, deformation due to damage itself can be partially or completely recovered upon unloading due to closure of microcracks or micro-voids. Nevertheless, recovery of damage deformation does not mean the healing of the damage. No materials are brittle or ductile.

Configuration $\bar{C}(t)$ is the effective configuration of $C(t)$ with all damage removed that occurred during both the elastic and plastic deformation gradients. Damage associated with the elastic deformation gradient is $\boldsymbol{F}^{d_2}$, while that associated with the plastic deformation gradient is $\boldsymbol{F}^{d_1}$. Configuration $\bar{C}_p$ is the effective plastic configuration obtained from configuration C_p by removing the damage that occurred during the plastic deformation, $\boldsymbol{F}^{d_1}$.

17.1.6 Incremental Expression for the Damage Tensor

In a general state of deformation and damage, the effective stress tensor $\bar{\boldsymbol{\sigma}}$ is related to the Cauchy stress tensor $\boldsymbol{\sigma}$ by the following linear transformation Murakami and Ohno [22]

$$\bar{\boldsymbol{\sigma}} = \boldsymbol{M} : \boldsymbol{\sigma} \tag{17.109}$$

where $\boldsymbol{M}$ is a fourth-order linear transformation operator called the damage effect tensor. Depending on the form used for $\boldsymbol{M}$, it is very clear from equation (17.109) that the effective stress tensor $\bar{\boldsymbol{\sigma}}$ is generally non-symmetric. Using a non-symmetric effective stress tensor as given by equation (17.109) to formulate a constitutive model will result in the introduction of the Cosserat and micropolar continua. However, the use of such complicated mechanics can be easily avoided by symmetrizing the effective stress. One of the symmetrization methods is given by Cordebois and Sidoroff [17]

By defining the incremental damage tensor $\boldsymbol{m}$, the above equation can be written as Voyiadjis and Guelzim [205] follows

$$\dot{\bar{\boldsymbol{\sigma}}} = \boldsymbol{m} : \dot{\boldsymbol{\sigma}} \tag{17.110}$$

The next step is to obtain the incremental effective strain by using the hypothesis of elastic strain energy equivalence Cordebois and Sidoroff [17]:

$$\frac{1}{2}\dot{\bar{\sigma}} : \dot{\bar{\epsilon}}^{e} = \frac{1}{2}\dot{\sigma} : \dot{\epsilon}^{e} \tag{17.111}$$

Voyiadjis and Park [211], through the use of the kinematics of damage, obtained a general relation of the effective elastic strains for the case of finite strains of elasto-plastic deformations. For the special case of small strains with small rotations this equations reduces to the hypothesis of elastic energy equivalence. Using equation(17.110) in equation (17.111), the effective elastic strain rate can be given as follows

$$\dot{\bar{\epsilon}}^{e} = m^{-1} : \dot{\epsilon}^{e} \tag{17.112}$$

However, the hypothesis of strain energy equivalence is not sufficient to obtain the expression for the total strain rate tensor, $\dot{\bar{\epsilon}}$, therefore one needs to postulate the hypothesis of plastic dissipation equivalence [15,215]. This provides the following relation

$$\dot{\bar{\epsilon}}^{p} = M^{-1} : \dot{\epsilon}^{p} \tag{17.113}$$

The effective plastic strain rate can be also obtained from classical plasticity in the following form

$$\dot{\bar{\epsilon}}^{p} = \bar{H} : \dot{\bar{\epsilon}} \tag{17.114}$$

$\bar{H}$ is a fourth order plastic hardening tensor. The additive decomposition of total strain rate in the effective and damage configurations are given respectively in the following form

$$\dot{\bar{\epsilon}} = \dot{\bar{\epsilon}}^{e} + \dot{\bar{\epsilon}}^{p} \tag{17.115}$$

$$\dot{\epsilon} = \dot{\epsilon}^{e} + \dot{\epsilon}^{p} \tag{17.116}$$

Making use of equations (17.112) through (17.116) one is able to obtain the relation between the total effective strain rate and the total damage strain rate in the following form

$$\dot{\bar{\epsilon}} = \left(m + \bar{H} : (M - m) \right)^{-1} : \dot{\epsilon} \tag{17.117}$$

The total strain rate damage transformation tensor m^{*} can be defined from the above equation as follows(Voyiadjis and Deliktas [223])

$$m^{*} = \left(m + \bar{H} : (M - m) \right)^{-1} \tag{17.118}$$

This tensor m^{*} is different than the stress incremental damage tensor m given in equation (17.110). However, it can be easily seen that if there is no plastic deformation ($H = 0$), both the incremental damage transformation tensors m and m^{*} become identical. Tensor m^{*} will be refered to for clarity as the strain incremental damage tensor as opposed to m the stress incremental damage tensor.

478

17.2 Constitutive Equations

The elasto-plastic stiffness for the damaged material can be obtained using the incremental relation of Hooke's Law in the effective stress space as follows

$$\dot{\bar{\sigma}} = \bar{E} : \dot{\bar{\epsilon}}^{e} \tag{17.119}$$

The effective stress rate, $\dot{\bar{\sigma}}$, and the elastic component of the effective strain rate, $\dot{\bar{\epsilon}}^{e}$, in equation (17.119) can be transformed into the damage configuration by using equations (17.110) and (17.112) respectively. The resulting relation can be given as follows

$$\dot{\sigma} = m^{-1} : \bar{E} : m^{-1} : \dot{\epsilon}^{e} \tag{17.120}$$

Making use of the both equations (17.116) and (17.55) in equation (17.120) one obtains the following

$$\begin{aligned}
\dot{\sigma} &= E^{d} : (\dot{\epsilon} - \chi^{p} : \dot{\sigma}) \\
\dot{\sigma} &= (I + E^{d} : \chi^{p})^{-1} : E^{d} : \dot{\epsilon} \\
&= D : \dot{\epsilon}
\end{aligned} \tag{17.121}$$

where E^{d} represents the elastic damaged stiffness and is defined from equation (17.120) as follows

$$E^{d} = m^{-1} : \bar{E} : m^{-1} \tag{17.122}$$

D in equation (17.121)represents the elasto-plastic damaged stiffness and is given by

$$D = (I + E^{d} : \chi^{p})^{-1} : E^{d} \tag{17.123}$$

17.2.1 Constitutive Equations of the Composite Material

The inelastic response of damaged composite materials can be analyzed by using the coupled anisotropic damage model proposed in the previous sections. In the analysis of the composite materials, there are two approaches which are commonly used the phenomenological based approach and the micromechanical based approach. The advantages of the micromechanical model over the strictly phenomenological continuum model are discussed by Paley and Aboudi [203] , Voyaidjis and Kattan [105], and Boyd et. al. [225]. The micromechanical models enable one to investigate damage that occurs in the composite material by incorporating the physics of damage. In addition the micromechanical model can account for different types of damage within the composite such as matrix cracks, matrix/fiber debonding and fiber cracks. It is obvious that this distinction for different types of damage at the local level and their effects on the macromechanical properties of the composite can not be accounted for by using the phenomenological continuum approach. Therefore, in this work the micromechanical model based on the Mori-Tanaka averaging scheme [14,108]is selected for incorporation into the developed damage model in order to investigate the inelastic response of the metal matrix composite.

In the micromechanical models, the information obtained from the individual properties of the different materials at the local level can be linked to the overall properties by using a certain homogenization procedure [108,176]. The objective of this section is to obtain overall constitutive relations for composite materials in terms of the developed constitutive relations of damaged materials presented in the previous sections.

The derived stress rate damage operator tensor, $\boldsymbol{m}^r$ for each constituent of the composite given by equation (17.110) can be linked to the overall stress rate damage operator tensor, $\boldsymbol{m}$ by making use of the micromechanical model. The superscript r represents the different constituents of the composite material at the local level such as the matrix material($r = m$) and the fiber material($r = f$). This model postulates the relation between the effective Cauchy stress rate, $\dot{\bar{\boldsymbol{\sigma}}}$, for the overall composite and the effective Cauchy stress rate, $\dot{\bar{\boldsymbol{\sigma}}}^r$, for the constituents of the composite in the following form

$$\dot{\bar{\boldsymbol{\sigma}}} = \sum \bar{c}^r \dot{\bar{\boldsymbol{\sigma}}}^r \qquad \text{where} \quad r = f, m \tag{17.124}$$

Through the same analogy, the above equation can be written for the damaged configuration as follows

$$\dot{\boldsymbol{\sigma}} = \sum c^r \dot{\boldsymbol{\sigma}}^r \tag{17.125}$$

Similarly the relation between the effective strain rate, $\dot{\bar{\boldsymbol{\epsilon}}}$, for the overall composite and the effective strain rate, $\dot{\bar{\boldsymbol{\epsilon}}}^r$, for the constituents can be given in the following form

$$\dot{\bar{\boldsymbol{\epsilon}}} = \sum \bar{c}^r \dot{\bar{\boldsymbol{\epsilon}}}^r \tag{17.126}$$

A similar relation can be obtained in the damaged configuration as follows

$$\dot{\boldsymbol{\epsilon}} = \sum c^r \dot{\boldsymbol{\epsilon}}^r \tag{17.127}$$

$\bar{c}^r$ and c^r in the above equations represent the volume fractions of the constituents in the composite material in the effective and damage configurations respectively. The effective stress rate of the constituents, $\dot{\bar{\boldsymbol{\sigma}}}^r$, in equation (17.124) and the effective strain rate of the constituents, $\dot{\bar{\boldsymbol{\epsilon}}}^r$, in equation (17.126) can be obtained respectively from the overall effective stress rate, $\dot{\bar{\boldsymbol{\sigma}}}$, and from the overall effective strain rate, $\dot{\bar{\boldsymbol{\epsilon}}}$, respectively by using the stress and strain concentration tensors as shown in the following relations

$$\dot{\bar{\boldsymbol{\sigma}}}^r = \bar{\boldsymbol{B}}^r : \dot{\bar{\boldsymbol{\sigma}}} \tag{17.128}$$

and

$$\dot{\bar{\boldsymbol{\epsilon}}}^r = \bar{\boldsymbol{A}}^r : \dot{\bar{\boldsymbol{\epsilon}}} \tag{17.129}$$

In the case of elastic deformation without damage the effective stress concentration tensor, $\bar{\boldsymbol{B}}^r$, and the effective strain concentration tensor, $\bar{\boldsymbol{A}}^r$, are constant and can be obtained using the Mori-Tanaka method. However, for the case of inelastic deformation, they may be obtained using the numerical method by Gavazzi amd Lagoudas [107]. In this case these

480

tensors are not constant and their evolutions are dependent on the evolution of the internal state variables of thermodynamics.

One may obtain the overall incremental stress damage operator, $\boldsymbol{m}$, in terms of the local incremental stress damage operators, $\boldsymbol{m}^r$, by using equation (17.110) in equation (17.125). The resulting expression becomes

$$\dot{\boldsymbol{\sigma}} = \sum c^r (\boldsymbol{m}^r)^{-1} : \dot{\boldsymbol{\sigma}}^r \tag{17.130}$$

Using equation(17.128) into equation (17.130) one obtains the following expression

$$\dot{\boldsymbol{\sigma}} = \left\{ \sum c^r \boldsymbol{m}^{-r} : \bar{\boldsymbol{B}}^r \right\} : \dot{\bar{\boldsymbol{\sigma}}}$$
$$\dot{\bar{\boldsymbol{\sigma}}} = \boldsymbol{m} : \dot{\boldsymbol{\sigma}} \tag{17.131}$$

where $\boldsymbol{m}$ represents the overall stress incremental damage operator tensor for the composite material and its expression can be given from equation (17.131) in the following form

$$\boldsymbol{m} = \left\{ \sum c^r \boldsymbol{m}^{-r} : \bar{\boldsymbol{B}}^r \right\}^{-1} \tag{17.132}$$

Similarly the expression for the overall strain damage operator, $\boldsymbol{m}^*$ can be derived by making use of equation (17.117) in conjunction with equation (17.129) in equation (17.127). The resulting relation for $\boldsymbol{m}^*$ can be written in the following form

$$\boldsymbol{m}^* = \left\{ \sum c^r \boldsymbol{m}^{*-r} : \bar{\boldsymbol{A}}^r \right\}^{-1} \tag{17.133}$$

The damaged stress concentration tensor , $\boldsymbol{B}^r$, and the damaged strain concentration tensor, $\boldsymbol{A}^r$, can be obtained by using equations (17.110) and (17.132) in conjunction with equation (17.128), and using equations (17.117) and (17.133) in conjunction with equation (17.129) respectively. The resulting forms for both concentration tensors are given respectively as follows

$$\boldsymbol{B}^r = \boldsymbol{m}^{-r} : \bar{\boldsymbol{B}}^r : \boldsymbol{m} \tag{17.134}$$

and

$$\boldsymbol{A}^r = \boldsymbol{m}^{*-r} : \bar{\boldsymbol{A}}^r : \boldsymbol{m}^* \tag{17.135}$$

Finally, the overall composite damaged stiffness tensor can be obtained by making use of equation (17.121) in conjunction with equation (17.135) in equation (17.125) such that

$$\dot{\boldsymbol{\sigma}} = \sum c^r \dot{\boldsymbol{\sigma}}^r$$
$$= \sum c^r \boldsymbol{D}^r : \dot{\boldsymbol{\epsilon}}^r$$
$$= \left\{ \sum c^r \boldsymbol{D}^r : \boldsymbol{A}^r \right\} : \dot{\boldsymbol{\epsilon}} \tag{17.136}$$
$$= \boldsymbol{D} : \dot{\boldsymbol{\epsilon}}$$

In equation (17.136) $\boldsymbol{D}$ is the resulting overall elasto-plastic damaged stiffness tensor for the composite.

17.2.2 Laminate Analysis

Laminate analyses is performed by using the classical lamination theory. It is considered in this work that a lamina layup of $2n$ layers is placed with their different in plane orientations symmetrically with respect to the mid plane. The global coordinates for the plate are denoted by (x_1, x_2, x_3) and the local coordinates for the lamina are denoted by (x'_1, x'_2, x'_2). These are shown in Figure 17.5.

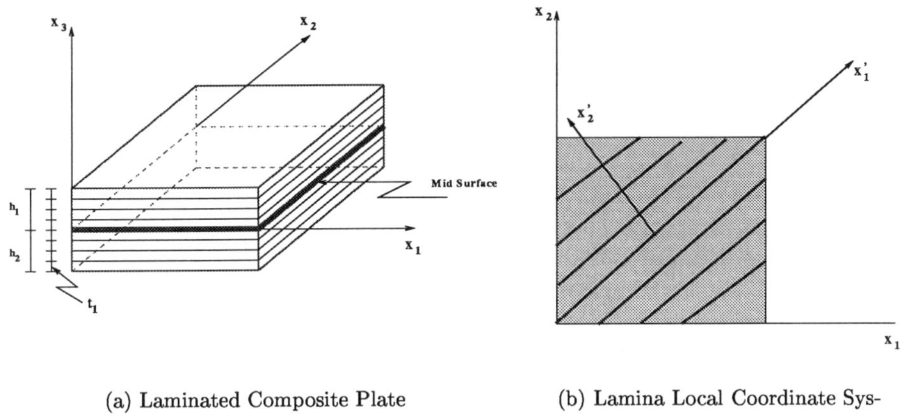

(a) Laminated Composite Plate

(b) Lamina Local Coordinate System

Figure 17.5 Laminated Composite

The laminate in plane forces and out of plane moments can be related to the deformation and the curvature of the laminate by the following expression

$$\begin{bmatrix} N \\ M \end{bmatrix} = \begin{bmatrix} A & B \\ B & D \end{bmatrix}^{-1} \begin{bmatrix} \epsilon \\ \kappa \end{bmatrix} \tag{17.137}$$

where

$$A = \int_{-h_2}^{h_1} D_k^{-1} dz, \quad B = \int_{-h_2}^{h_1} z D_k^{-1} dz, \quad D = \int_{-h_2}^{h_1} z^2 D_k^{-1} dz \tag{17.138}$$

where D is the overall stiffness matrix of the kth ply . In the case of pure elastic behavior, D should be replaced by the elastic stiffness tensor, E. The local stresses in each constituent can be obtained from the applied or known increment of loading by using the assumption of the lamination theory that at any instant of loading, the bond between the laminates is assumed to remain perfect. This strain constraint has to be satisfied and may be expressed by the following relations

$$\dot{\epsilon} = \dot{\epsilon}_{(1)} = \dot{\epsilon}_{(2)} = \cdots = \dot{\epsilon}_{(n)} \tag{17.139}$$

where $\dot{\boldsymbol{\epsilon}}$ is the strain rate of the laminate and $\dot{\boldsymbol{\epsilon}}_{(1)}, \dot{\boldsymbol{\epsilon}}_{(2)}, \cdots, \dot{\boldsymbol{\epsilon}}_{(n)}$ are the strain rates for each layer. However, the strain rate vector of each layer in equation (17.139) is in the global coordinate system and needs to be transfered into the local coordinate system such as

$$\dot{\boldsymbol{\epsilon}}_{(k)} = \boldsymbol{T}\dot{\boldsymbol{\epsilon}}'_{(k)} \tag{17.140}$$

where $\boldsymbol{T}$ is the orthogonal transformation matrix. The strain rate for each phase can be obtained by using the strain concentration matrix

$$\dot{\boldsymbol{\epsilon}}^r_{(k)} = \boldsymbol{A}^r_{(k)}\boldsymbol{T}_k\dot{\boldsymbol{\epsilon}}'_{(k)} \tag{17.141}$$

Making use of equation (17.139) into equation (17.141), the following expression can be written

$$\dot{\boldsymbol{\epsilon}}^r_{(k)} = \boldsymbol{A}^r_{(k)}\boldsymbol{T}_k[\boldsymbol{A}]^{-1}\dot{\boldsymbol{N}} \tag{17.142}$$

Although the classical lamination theory is the easiest and fastest to implement, and to solve the problem numerically, however, it may cause inaccurate predictions of the overall response of the composite laminates. This is because the local deformation could not be coupled with the overall response of the material. This effect may be resolved by using the functionally graded concept which enables one to couple directly microstructural variation in the local materials on the overall response of the laminated material. However, the damage criterion in the current formulation needs to include gradient terms which may be attributed to the nonlocal damage approach. This is an ongoing research by the first author to develop a new gradient damage model which can be incorporated into functionally graded materials.

17.3　Computational Aspects of the Model

The developed elasto-plastic and viscoplastic damage models are used here to numerically predict the inelastic response of composite materials. For this reason the following laminate systems, $0_{(8s)}$, $90_{(8s)}$, $(0/90)_{(4s)}$, and $(-45/45)_{(2s)}$, are analyzed. The composite laminate investigated here is made of a titanium alimunide matrix ($T_i - 15 - 3$) reinforced with continuous (SCS-6) fibers. The typical properties of the materials are obtained from the pertinent literature [143,208] and presented in Tables 17.2, and 17.3.

The three step split algorithm [206] is adopted here in order to additively decompose the constitutive equations into elastic, inelastic, and damage behaviors. The effective space is used for the elasto-plastic and viscoplastic analyses. The damage variables in this space are assumed to be removed fictitiously so that the elasto-plastic and viscoplastic solutions can be performed in the absence of damage. However, for the case of damage a new algorithm needs to be developed for the damage model. In the work of Ju [206], the damaged state is obtained by linear multiplication of the defined damage variables with the undamaged virgin state. This simplification allows the damage correction not to require any iterations for the correction of the damage state. However, in this work, damage state is characterized using the kinematics of the deformed body along with the concept of the effective space. This description of damage in turn gives the relation between the effective stress and Cauchy

Table 17.2 Typical Properties of the MMCs[143]

Materials	Fiber (SiC)	Matrix (Ti-14Al-21Nb)
Tensile Strength	3440.0Gpa	448.0MPa.
Young's modulus	414.0GPa.	84.1GPa.
Poisson ratio	0.22	0.30
Volume fractions	0.4	0.6

Table 17.3 Typical Properties of the MMCs[208]

Materials	Fiber (SiC)	Matrix (Ti-15-3)
Tensile Strength	3440.0Gpa	689.5MPa.
Young's modulus	400.0GPa.	91.8GPa.
Poisson ratio	0.25	0.36
Volume Fractions	0.34	0.66

stress through the fourth order damage operator tensor. Therefore, the resulting non-linear relation does require iteration steps to correct the damage state. A fully implicit algorithm is used here to achieve this correction.

The Newton Raphson iterative scheme is applied here to correct the final stage of the strain. It gives faster convergence, however, it requires more computational effort in order to update both the local strain concentration tensors as well as the overall tangent modulus. The program flow followed in this work is outlined in the following steps, and the inelastic and damage correction algorithms are presented accordingly.

17.3.1 Program Flow for Elasto-Plastic and Damage Model

The program flow for the elastic-plastic damage behavior of the material is outlined below (Voyiadjis and Deliktas [223])

1. **Compute** the strain increments $\Delta\epsilon$ by using the following relation

$$\left[\Delta\epsilon\right] = \left[A^{-1}\right]\left[\Delta N\right]$$

2. **Compute** the strain increment of each lamina by using the obove equation through the transformation matrix T_k

$$\left[\Delta\epsilon_k\right] = \left[T_k\right] : \left[A^{-1}\right]\left[\Delta N\right]$$

3. **Compute** the phase strain increment using the constituent strain concentration matrix, $\boldsymbol{A}^r$

$$\left[\Delta\boldsymbol{\epsilon}_k^r\right] = \left[\boldsymbol{A}^r\right]\left[\boldsymbol{T}_k\right]\left[\boldsymbol{A}^{-1}\right]\left[\Delta\boldsymbol{N}\right]$$

4. **Split** the constitutive equation into the Elastic, Plastic, and Damage parts:

Elasticity	Plasticity	Damage
$\Delta\boldsymbol{\epsilon}$	$\Delta\boldsymbol{\epsilon}=0$	$\Delta\boldsymbol{\epsilon}=0$

$$\dot{\boldsymbol{\phi}}=0 \qquad\qquad \dot{\boldsymbol{\phi}}=0 \qquad\qquad \dot{\boldsymbol{\phi}}=\begin{cases}0 & \text{if } g\le 0,\\[2mm] \dot{\lambda}^d\dfrac{\partial g}{\partial \boldsymbol{Y}} & \text{if } \dfrac{\partial g}{\partial \boldsymbol{Y}}\dot{\boldsymbol{Y}}\ge 0\end{cases}$$

$$\dot{\kappa}=0 \qquad\qquad \dot{\kappa}=0 \qquad\qquad \dot{\kappa}=\sqrt{\tfrac{2}{3}\dot{\boldsymbol{\phi}}:\dot{\boldsymbol{\phi}}}$$

$$\dot{\boldsymbol{\Gamma}}=0 \qquad\qquad \dot{\boldsymbol{\Gamma}}=0 \qquad\qquad \dot{\boldsymbol{\Gamma}}=Equation\ (17.47)$$

$$\dot{\boldsymbol{\sigma}}=\boldsymbol{E}(\boldsymbol{\phi}):\dot{\boldsymbol{\epsilon}} \qquad \dot{\boldsymbol{\sigma}}=-\boldsymbol{E}(\boldsymbol{\phi}):\dot{\boldsymbol{\epsilon}}^p \qquad \dot{\boldsymbol{\sigma}}=\dfrac{\partial \boldsymbol{M}^{-1}}{\partial\boldsymbol{\phi}}:\dot{\boldsymbol{\phi}}:\boldsymbol{E}:\boldsymbol{\epsilon}^e$$

$$\dot{\boldsymbol{\epsilon}}^p=0 \qquad \dot{\boldsymbol{\epsilon}}^p=\begin{cases}0 & \text{if } f\le 0,\\[2mm] \dot{\lambda}^p\dfrac{\partial f}{\partial\boldsymbol{\sigma}} & \text{if } \dfrac{\partial f}{\partial\boldsymbol{\sigma}}\dot{\boldsymbol{\sigma}}\ge 0\end{cases} \qquad \dot{\boldsymbol{\epsilon}}^p=0$$

$$\dot{\boldsymbol{\alpha}}=0 \qquad\qquad \dot{\boldsymbol{\alpha}}=Equation\ (17.29) \qquad \dot{\boldsymbol{\alpha}}=0$$

$$\dot{r}=0 \qquad\qquad \dot{r}=\dot{p}=\dot{\lambda}^p \qquad\qquad \dot{r}=0$$

5. **Check** the plasticity condition. **If** yielding occurs **then** perform the plastic return algorithm

6. **Check** the damage condition. **If** damage occurs **then** perform the damage return algorithm

7. **Compute** the load $\boldsymbol{N}$ at the current updated stress $\boldsymbol{\sigma}_k$ by using the relation

$$\boldsymbol{N}=\int_{-h_2}^{h_1}\boldsymbol{\sigma}_k dz$$

8. **Check** the condition **if** $(\boldsymbol{N}_{n+1}^{(i+1)}-\boldsymbol{N}_n^{(i)}\le)$TOL **then** go to the next loading case **else** goto next iteration

17.3.2 Plastic Corrector Algorithm

The evolution equations of the plastic and the damage parameters, obtained in the previous sections, require the solution of a set of differential equations. One needs to use a numerical procedure to obtain the approximate solution of these equations. It is imperative that one ensures that the state of stress does not lie outside the yield surface. For the return path, the relaxation equation or elastic predictor using the plastic corrector can be carried out by iterative steps at each increment. The relaxation relation can be given as follows

$$\dot{\boldsymbol{\sigma}}=-\bar{\boldsymbol{E}}:\dot{\boldsymbol{\epsilon}}^p$$

$$=-\dot{\lambda}^p\bar{\boldsymbol{E}}:\dfrac{\partial f}{\partial\bar{\boldsymbol{\sigma}}} \tag{17.143}$$

If one discretizes the above equation and the plastic hardening equation (17.29) around the current value of the state variables, then one obtains the following relations

$$\bar{\sigma}_{n+1}^{(i+1)} - \bar{\sigma}_{n+1}^{(i)} = -\Delta\lambda^p \bar{E} : \left.\frac{\partial f}{\partial\bar{\sigma}}\right._{n+1}^{(i)} \tag{17.144a}$$

$$\bar{X}_{n+1}^{(i+1)} - \bar{X}_{n+1}^{(i)} = \Delta\lambda^p(\frac{2}{3}C^p\dot{\bar{\epsilon}}^p - \gamma^p X\dot{p})_{n+1}^{(i)} \tag{17.144b}$$

where the superscript $(i+1)$ and (i) indicate the respective iterations and the subscript $(n+1)$ represents the corresponding load step. The plastic multiplier $\Delta\lambda^p$ can be obtained from the linearized yield function f around the current values of the state variables such that

$$f = f_{n+1}^{(i)} + \left.\frac{\partial f}{\partial\bar{\sigma}}\right|_{n+1}^{(i)} : (\bar{\sigma} - \bar{\sigma})_{n+1}^{(i)} + \left.\frac{\partial f}{\partial\bar{X}}\right|_{n+1}^{(i)} : (\bar{X} - \bar{X})_{n+1}^{(i)} \tag{17.145}$$

By substituting equations (17.144a) and (17.144b) into equation (17.145), one can solve for the plastic multiplier $\Delta\lambda^p$ such that (Voyiadjis and Deliktas [223]])

$$\Delta\lambda^p = \frac{f_{n+1}^{(i)}}{\frac{\partial f}{\partial\bar{\sigma}} : \left(\bar{E} : \frac{\partial f}{\partial\bar{\sigma}} + \frac{2}{3}C\dot{\bar{\epsilon}}^p + \gamma X\dot{p} \right)} \tag{17.146}$$

Equations (17.144a) and (17.144b) are updated at each iteration step until the convergence criterion is satisfied with the given tolerance. The final stage for the plastic analysis with the frozen values of damage variables in the effective space is given as

$$f = f(\bar{\sigma}_{n+1}, \bar{X}_{n+1}, \phi_n, \kappa_n, \Gamma_n) \tag{17.147}$$

The final stage of plasticity is taken as the initial condition for the damage equations, while the plastic variables are frozen in the damage analysis. This defines a damage corrector whereby the plastically predicted stress values are corrected and the corresponding damage variables are updated at each step of iteration until the consistency condition is satisfied.

17.3.3 Damage Corrector Algorithm

The damage corrector process is carried out by updating the damage variables in an iterative fashion at each increment. Therefore, one can obtain the relaxation stress due to damage through the use of the effective stress definition such as

$$\sigma = [M(\phi)]^{-1} : \bar{\sigma} \tag{17.148}$$

By taking the time derivative of equation (17.148), the following expression can be written

$$\dot{\sigma} = \dot{M}^{-1} : \bar{\sigma} + M^{-1} : \dot{\bar{\sigma}} \tag{17.149}$$

However, since the $\bar{\sigma}$ is obtained from the previous elasto-plastic analysis, it is set to a fixed value ($\dot{\bar{\sigma}} = 0$). Hence, equation (17.149) reduces to the following

$$\dot{\sigma} = \dot{\boldsymbol{M}}^{-1} : \bar{\sigma} \tag{17.150}$$

where

$$\dot{\boldsymbol{M}}^{-1} = \frac{\partial \boldsymbol{M}^{-1}}{\partial \phi} : \dot{\phi} \tag{17.151}$$

One can descritize equation(17.150) as well as the damage hardening relations such that (Voyiadjis and Deliktas [223])

$$\sigma_{n+1}^{(i+1)} - \sigma_{n+1}^{(i)} = \Delta\lambda^d \left[\frac{\partial \boldsymbol{M}^{-1}}{\partial \phi} : \frac{\partial g}{\partial \boldsymbol{Y}} \right]_{n+1}^{(i+1)} : \bar{\sigma} \tag{17.152a}$$

$$\phi_{n+1}^{(i+1)} - \phi_{n+1}^{(i)} = \Delta\lambda^d \left. \frac{\partial g}{\partial \boldsymbol{Y}} \right|_{n+1}^{(i)} \tag{17.152b}$$

$$\kappa_{n+1}^{(i+1)} - \kappa_{n+1}^{(i)} = \Delta\lambda^d \left. \sqrt{\frac{2}{3}\left(\frac{\partial g}{\partial \boldsymbol{Y}} : \frac{\partial g}{\partial \boldsymbol{Y}} \right)} \right|_{n+1}^{(i)} \tag{17.152c}$$

$$\boldsymbol{\Gamma}_{n+1}^{(i+1)} - \boldsymbol{\Gamma}_{n+1}^{(i)} = \Delta\lambda^d \left(C^d \frac{\partial g}{\partial \boldsymbol{Y}} + \gamma^d \boldsymbol{\Gamma} \sqrt{\frac{2}{3}\left(\frac{\partial g}{\partial \boldsymbol{Y}} : \frac{\partial g}{\partial \boldsymbol{Y}} \right)} \right)_{n+1}^{(i)} \tag{17.152d}$$

The damage multiplier in equations (17.152) can be solved for by linearizing the function g around the current values of the damage variables at each iteration step,

$$g = g_{n+1}^{(i)} + \left. \frac{\partial g}{\partial \sigma} \right|_{n+1}^{(i)} : (\sigma - \sigma_{n+1}^{(i)}) + \left. \frac{\partial g}{\partial \phi} \right|_{n+1}^{(i)} : (\phi - \phi_{n+1}^{(i)}) + \left. \frac{\partial g}{\partial \kappa} \right|_{n+1}^{(i)} (\kappa - \kappa_{n+1}^{(i)}) +$$
$$\left. \frac{\partial g}{\partial \boldsymbol{\Gamma}} \right|_{n+1}^{(i)} : (\boldsymbol{\Gamma} - \boldsymbol{\Gamma}_{n+1}^{(i)}) + \tag{17.153}$$

Back substituting equations (17.152) into equation (17.153) one can obtain the damage multiplier $\Delta\lambda^d$ as follows

$$\Delta\lambda^d = \frac{-g_{n+1}^{(i)}}{H^d} \tag{17.154}$$

where H_d is given as follows

$$H^d = \left[\frac{\partial g}{\partial \sigma} : \frac{\partial \boldsymbol{M}^{-1}}{\partial \phi} : \frac{\partial g}{\partial \boldsymbol{Y}} : \bar{\sigma} \right]_{n+1}^{(i+1)} + \left[\frac{\partial g}{\partial \phi} : \frac{\partial g}{\partial \boldsymbol{Y}} \right]_{n+1}^{(i)} + \left[\frac{\partial g}{\partial \kappa} \sqrt{\frac{2}{3}} \frac{\partial g}{\partial \boldsymbol{Y}} : \frac{\partial g}{\partial \boldsymbol{Y}} \right]_{n+1}^{(i)} +$$
$$\frac{\partial g}{\partial \boldsymbol{Y}} : \left[C^d \frac{\partial g}{\partial \boldsymbol{Y}} + \gamma^d \boldsymbol{\Gamma} \sqrt{\frac{2}{3}} \frac{\partial g}{\partial \boldsymbol{Y}} : \frac{\partial g}{\partial \boldsymbol{Y}} \right]_{n+1}^{(i)}$$
$$\tag{17.155}$$

One can now update the damage equations (17.51) by using the damage multiplier from the above equation. The iteration process continues until the consistency condition of the damage is satisfied within a prescribed tolerance and the final stage for the overall behavior can be described by the following variables $(\sigma_{n+1}, X_{n+1}, \phi_{n+1}, \kappa_{n+1}, \Gamma_{n+1})$ at the $(n+1)$ configuration.

17.3.4 Discussion of the Results for the Elasto-Plastic Damage Analysis

The capability of the developed elasto-plastic damage model to predict the inelastic response of the laminated composite material is discussed here. For this purpose the program generated curves using the proposed theory are presented in this section.

In order to show the effect of damage, the computational algorithm is first run without the damage model and with only the elasto-plastic behavior. It is then re-run again including the damage model for the laminated system $(0/90)_{4s}$. The stress strain curves are generated for the undamaged(UD) and damaged(D) cases and are compared with each other, which are shown in Figures 17.6 through 17.8. In Figure 17.6, the undamaged and

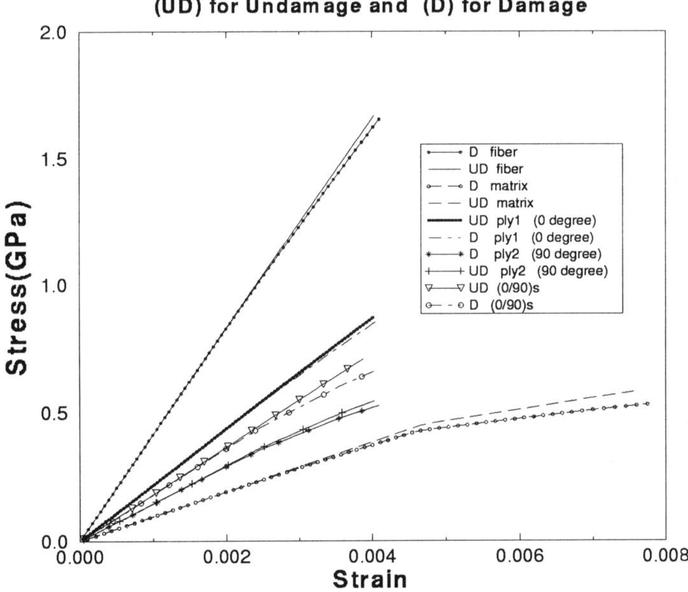

Figure 17.6 Stress Strain Curves for the Damaged Elasto-Plastic Composite

damaged curves for the (0/90) laminated system with its components the fiber and the matrix, for the 0° ply, and the 90° ply are first presented. As it is expected, less damage is obtained for the case of the 0° ply than the 90° ply. Figures 17.7 and 17.8 show separately the individual stress-strain curves for the damaged and undamaged 90° ply, and (0/90)s laminates respectively.

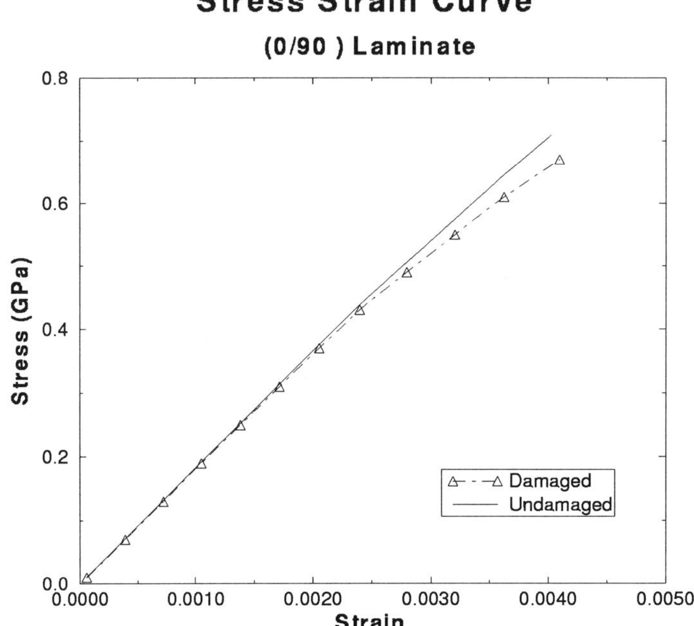

Figure 17.7 Stress Strain Curves for Damaged Elasto-Plastic 0/90 Layup

For different stacking sequences of the laminated composite systems, the numerical analyses are performed. The results obtained by these analyses are then compared with the available experimental results [208,143]. As it is pointed out by Majumdar and Newaz [208], the material elasto-plastic models are adequate to predict the overall response of certain laminated sytems where plasticity and not the damage is the dominant deformation mechanism on the overall inelastic behavior of the material. However, in the case of damage dominant deformation mechanisms on the overall inelastic response of the material, most of the plasticity models are not capable of predicting accurately the experimentally observed behavior of the material. Hence this dictates the necessity for the plasticity models to be coupled with damage.

Experimental studies [208,143] for the laminate system $(0)_{8s}$ indicate that the unloading curves from the various stages of load are parallel to the initial elastic curve. This implies that the inelastic response of the material is due to plasticity , therefore, plasticity models such as that by Dvorak and Bahei-El-Din [69], and Voyiadjis and Thiagarajan [196] show good agreement with the experimental results as shown in the references [143,197,208]. Therefore, the elasto-plastic model presented in Section 17.1.1 is used without including the damage model in the computation. The proposed elasto-plastic

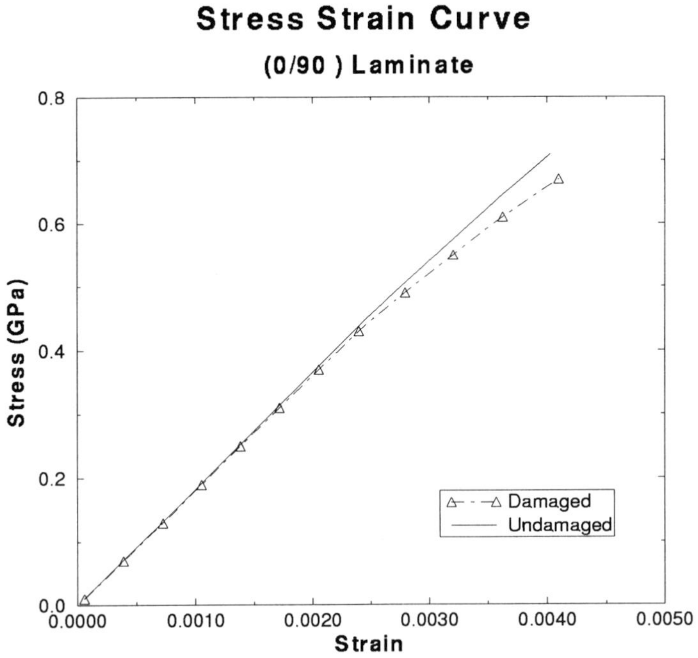

Figure 17.8 Stress Strain Curves for Damaged Elasto–Plastic 0/90 Layup

490

model also showed very good agreement with the experimental results as indicated in Figure 17.9.

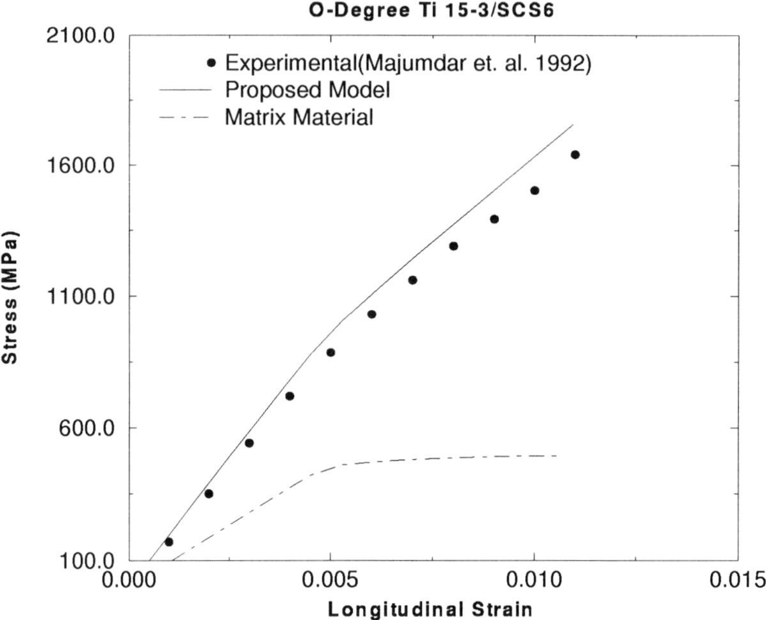

Figure 17.9 Comparison of the Elasto-Plastic Model with Experimental Results [208] of the $(0)_{8s}$ Layup (Without Damage)

However, in the case of the laminated systems $(90)_{8s}$ and $(45/-45)_{8s}$, as shown in Figures 17.10 and 17.11 respectively the plasticity models overestimate the inelastic response of the material. This is because damage itself or coupling of damage with plasticity plays a crucial role in the inelastic deformation of the material. Therefore, in the analysis of these laminated systems, one should include damage into the elasto-plastic model.   At this stage it is not possible to differentiate directly the effect of damage or plasticity on the overall inelastic deformation response of the material, however, one can see the priority of the occurrence of the damage or plasticity in the materials. The numerical results using the proposed formulation indicate that damage occurs before plasticity and the resulting numerical curves show very good agreement with the experimental results. Excellent corelation between the proposed model and the experimental results [143] is also obtained for the $(0/90)_{4s}$ lamaninated system as shown in Figure 17.12.

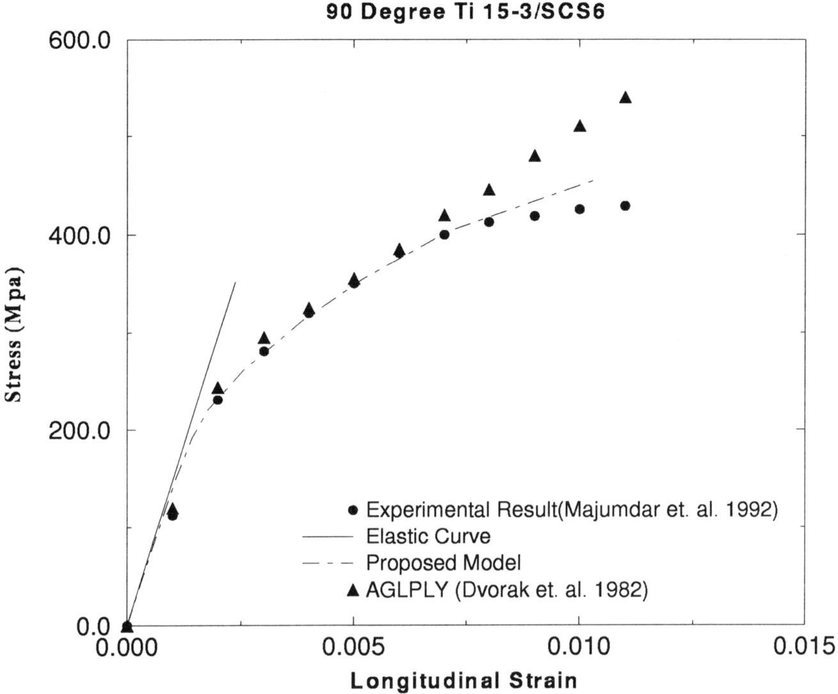

Figure 17.10 Comparison of the Elasto-Plastic Damage Model with Experimental Results [208] of the $(90)_{8s}$ Layup

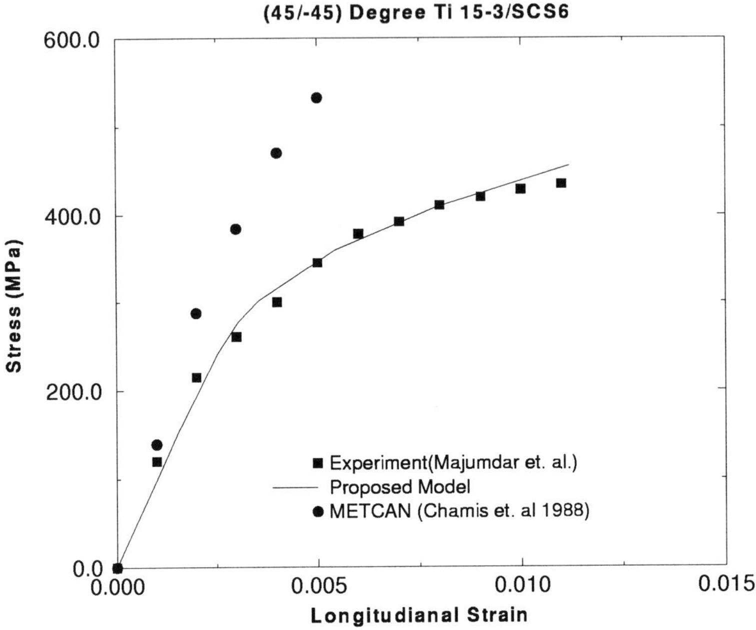

Figure 17.11   Comparison of the Elasto-Plastic Damage Model with Experimental Results [208] of the $(45/-45)_{2s}$

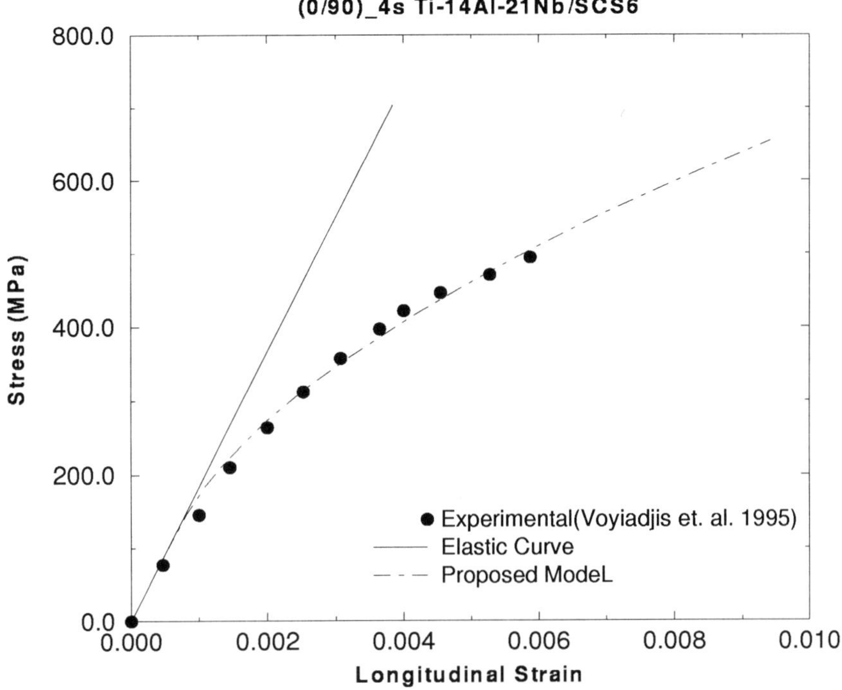

Figure 17.12   Comparison of the Elasto-Plastic Damage Model with Experimental Results [143] of the $(0/90)_{4s}$ Layup

494

17.4 Implementation of the Viscoplastic Damage Model

The general procedure for the implementation of the viscoplastic damage model is similar to the one defined for the elasto-plastic damage model in Section 17.3.1. However, unlike the elasto-plastic damage analysis, there may be three different scenarios which may occur. Some cases will be coupled or uncoupled due to inelastic behavior of the material. These cases may be uncoupled due to the occurrence of damage in the elastic region followed by the one coupled with viscoplasticity. The three possible cases for analysis are using the rate independent damage model, or the viscoplasticity without damage, or the viscoplasticy with damage. The last case may be analyzed using the viscoplastic and rate dependent damage models. The procedure of the program flow is summarized in the next section. In the numerical implementation the thermal recovery term is ignored in order to simplify the algorithm.

17.4.1 Flow of the Program

The above outlined procedure is summarized in the following steps:

1. **Load** Increments $n = 1, 2, 3, \cdots$

$$\dot{\epsilon} = \text{constant}$$
$$\Delta N = \text{Applied Load Increment}$$
$$N_{n+1} = N_n + \Delta N_n$$

2. **Newthon Raphson** Iteration $i = 1, 2, 3, \cdots$

$$\left[\Delta\epsilon\right]_{n+1}^{(i+1)} = \left[A(\dot{\epsilon})^{-1}\right]_{n+1}^{(i)} \left[\Delta N\right]_{n+1}^{(i)}$$
$$\epsilon_{n+1}^{i+1} = \epsilon_{n+1}^{i} + \Delta\epsilon_{n+1}^{i+1}$$

3. **Loop 1** over the number of plies $k = 1, 2, 3, \cdots$

$$\left[\Delta\epsilon\right]_{n+1}^{(i+1,k)} = \left[T_k\right] : \left[A^{-1}\right]_{n+1}^{(i)} \left[\Delta N\right]_{n+1}^{(i)}$$
$$\epsilon_{n+1}^{(i+1,k)} = \epsilon_{n+1}^{(i,k)} + \Delta\epsilon_{n+1}^{(i+1,k)}$$

4. **Loop** over the number of the phases $r = m, f$

$$\left[\Delta\epsilon\right]_{n+1}^{(i+1,k,r)} = \left[A\right]_{n+1}^{(i,k,r)} \left[T_k\right] \left[A\right]_{n+1}^{(i+1,k,r)} \left[\Delta N\right]_{n+1}^{(i+1,k,r)}$$

5. **Split** the constitutive equation into Elastic, Viscoplastic, and Damage parts

6. **Check** the viscoplasticity condition **If** the case is viscoplastic **then** perform the viscoplastic correction algorithm

7. **Check** the damage condition **If** the case is damage **and if** the case in Step 6 is viscoplastic **then** perform the rate dependent damage correction algorithm **goto** Step 8 **else if** the case in Step 6 is elastic **then** perform only rate independent damage correction algorithm in Section

8. **Compute** the load N at the current updated stress σ_k by using the relation

$$N = \int_{-h_2}^{h_1} \sigma_k dz$$

9. **Check** the condition **if** $(N^{(i+1)} - N^{(i)} \leq)$TOL **then** next loading **else** goto next iteration

Since the constitutive equations of viscoplastic and rate dependent damage are both rate dependent and temperature dependent, consequently, the correction algorithm defined for the elasto-plastic damage should be modified accordingly.

17.4.2 Viscoplastic Corrector Algorithm

The set of the constitutive equations for viscoplasticity are summarized here

$$\dot{\epsilon}^{vp} = -\frac{(F^{vp})^{n_1}}{\eta^{vp}} \frac{\partial F^{vp}}{\partial \sigma} \tag{17.156}$$

$$\dot{\sigma} = -E : \dot{\epsilon}^{vp} \tag{17.157}$$

$$\dot{R} = Q^{vp}(1 - \frac{R}{Q^{vp}})\|\dot{\epsilon}^{vp}\| \tag{17.158}$$

$$\dot{X} = \frac{3}{2}H^{vp}\left(\dot{\epsilon}^{vp} - \frac{2}{3}\frac{X}{L^{vp}}\|\epsilon^{vp}\|\right) \tag{17.159}$$

where

$$\|\dot{\epsilon}^{vp}\| = \sqrt{\frac{2}{3}\dot{\epsilon}^{vp} : \dot{\epsilon}^{vp}} \tag{17.160}$$

The rate of change in the overstress, $(F^{vp})^{n_1}$ which is defined here as β^{vp}, during the relaxation process is written as follows

$$\dot{\beta}^{vp} = \frac{\partial \beta^{vp}}{\partial \sigma} : \dot{\sigma} + \frac{\partial \beta^{vp}}{\partial X} : \dot{X} + \frac{\partial \beta^{vp}}{\partial R} \dot{R} \tag{17.161}$$

Rearranging the above equation and substituting equation (17.159), one obtains the following relation

$$\dot{\beta}^{vp} = \frac{\beta^{vp}}{\eta^{vp}}\left(-\frac{\partial \beta^{vp}}{\partial \sigma} : E : \frac{\partial F^{vp}}{\partial \sigma} + \frac{2}{3}H^{vp}\frac{\partial \beta^{vp}}{\partial \sigma} : \frac{\partial F^{vp}}{\partial X} - \sqrt{\frac{2}{3}}\frac{H^{vp}}{L^{vp}}\right) \tag{17.162}$$

or

$$\dot{\beta}^{vp} = \frac{\beta^{vp}}{\bar{t}^{vp}} \qquad (17.163)$$

where the instantaneous relaxation time, $\bar{t}^{vp}$ is given by

$$\bar{t}^{vp} = \frac{\eta^{vp}}{\left(-\dfrac{\partial \beta^{vp}}{\partial \boldsymbol{\sigma}} : \boldsymbol{E} : \dfrac{\partial F^{vp}}{\partial \boldsymbol{\sigma}} + \dfrac{2}{3} H^{vp} \dfrac{\partial \beta^{vp}}{\partial \boldsymbol{\sigma}} : \dfrac{\partial F^{vp}}{\partial \boldsymbol{X}} - \sqrt{\dfrac{2}{3} \dfrac{H^{vp}}{L^{vp}}} \right)} \qquad (17.164)$$

An iterative process can now be adopted using an algorithm similar to the return path method that is defined for rate independent elasto-plastic behavior. However, in viscoplasticity the stress point may not be on the yield surface due to the relaxation in stress. The procedure for determining the final location of the stress point within the return path is outlined by Ortiz and Simo [207] and Voyiadjis and Mohammad [226]. This is summarized below

$$\Delta \lambda^{vp} = \frac{\beta^{vp} \bar{t}^{vp}}{\eta^{vp}}$$

$$\boldsymbol{\sigma}_{n+1}^{(i+1)} = \boldsymbol{\sigma}_{n+1}^{(i)} - \Delta \lambda^{vp} \boldsymbol{E} : \left. \frac{\partial F^{vp}}{\partial \boldsymbol{\sigma}} \right|_{n+1}^{(i)}$$

$$\boldsymbol{X}_{n+1}^{(i+1)} = \Delta \lambda^{vp} \frac{3}{2} H^{vp} \left(\frac{\partial F^{vp}}{\partial \boldsymbol{\sigma}} - \frac{2}{3} \frac{\boldsymbol{X}}{L^{vp}} \|\boldsymbol{\epsilon}^{vp}\| \right)_{n+1}^{(i)}$$

$$t_{n+1}^{(i+1)} = t_{n+1}^{(i)} + \bar{t}_{n+1}^{(i)} \log \left(\frac{F_{n+1}^{(i)}}{F_{n+1}^{(i+1)}} \right)$$

Check the following relaxation condition

$$t_{n+1}^{(i+1)} \geq h = \sum_i \bar{t}^i \log \frac{F_{n+1}^{(i)}}{F_{n+1}^{(i+1)}}$$

If the above condition is satisfied, then we have

$$\Delta \lambda^{vp} = \frac{\beta^i \bar{t}^i}{\eta^{vp}} [1 - \exp (h - \frac{t^i}{\bar{t}^i})]$$

$$\boldsymbol{\sigma}_{n+1}^{(i+1)} = \boldsymbol{\sigma}_{n+1}^{(i)} - \Delta \lambda^{vp} \boldsymbol{E} : \left. \frac{\partial F^{vp}}{\partial \boldsymbol{\sigma}} \right|_{n+1}^{(i)}$$

$$\boldsymbol{X}_{n+1}^{(i+1)} = \Delta \lambda^{vp} \frac{3}{2} H^{vp} \left. \left(\frac{\partial F^{vp}}{\partial \boldsymbol{\sigma}} - \frac{2}{3} \frac{\boldsymbol{X}}{L^{vp}} \|\boldsymbol{\epsilon}^{vp}\| \right) \right|_{n+1}^{(i)}$$

else $i = i + 1$ and **goto** the next iteration

17.4.3   Damage Corrector Algorithm

For the case of rate independent damage the algorithm defined in section 17.5.1 is used. However, once the material exceeds the elastic range, the rate dependent damage model is used instead of the rate independent damage model. The correction algorithm for the rate dependent damage model is similar to the one described for the viscoplastic model in the previous section. The results generated from the viscoplastic analysis is discussed in the next section.

17.4.4   Discussion for the Results of Viscoplastic Damage Analysis

The computational analysis of the viscoplastic damage model is performed for the laminate systems of $(90)_{8s}$ at elevated temperatures of 538°C and 649°C, and for $(45/-45)_{2s}$ at the temperature of the 538°C . The viscoplastic model parameters are given in Table 17.4 They are obtained by best fit of the the viscoplasticy model with the available experimental

Table 17.4: Viscoplastic Model Parameters

Properties	$T = 21$C	$T = 482$C	$T = 649$C
$E^f =$	400 Gpa.	393 Gpa	370 GPa
$E^m =$	92.4 GPa	72.2 Gpa	55 GPa
$\sigma_y =$	689 MPa	45 MPa	15.5 MPa
$H^{vp} =$	5000 MPa	50,000 MPa	50,000 MPa
$L^{vp} =$	100 MPa	85 MPa	75 MPa
$D^{vp} =$	840 MPa	450 Mpa	85 MPa
$n_1 =$	5.4	1.55	1.3

results(Bahei-El-Din et. al 230]. This is indicated in Figure 17.13. The viscoplastic model parameters obtained from this analysis are then used in the viscoplastic damage analysis.

In Figure 17.14, the viscoplastic damage model predictions for uniaxially loading of the $(90)_{8s}$ system at 538°C are compared with the experimental results and other viscoplastic models, and finite element analyses which are obtained by Majumadar and Newaz [208]. As seen clearly from the plots, the proposed model provides better predictions for the response of the material at elevated temperature than the other models. However, at the initial stage of the deformation such as 0.15 to 0.30 percent of strain, none of these theoretical models show good agreement with the experimental results. A better response is observed for the case of 649°C which is illustrated in Figures 17.15 and 17.16.

498

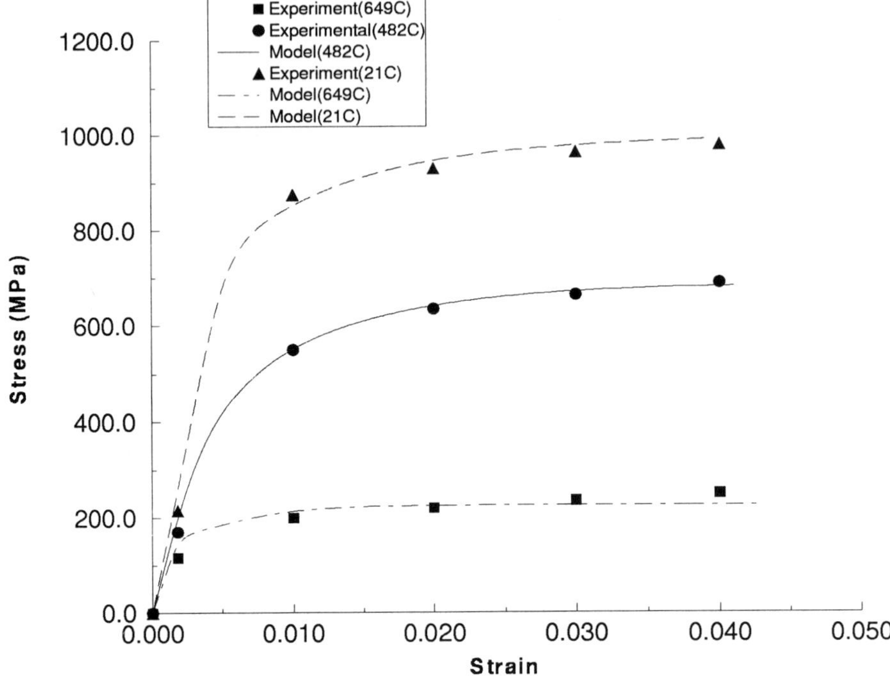

Figure 17.13 Comparision of the Proposed Viscoplasticity Model with Experimental Results [230] at Elevated Temperatures

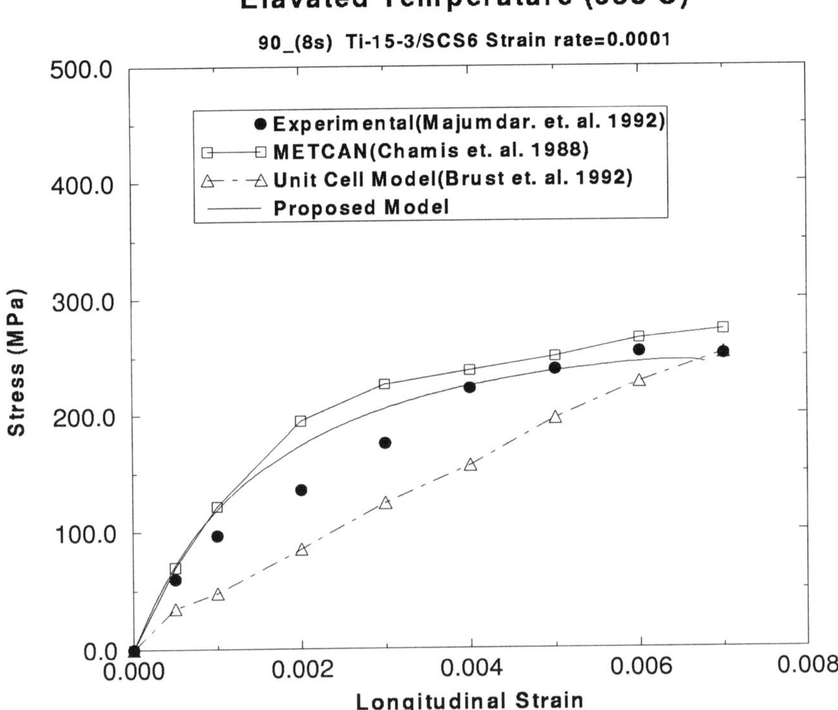

Figure 17.14   Comparison of the Viscoplastic Damage Model with Experimental Results [208] of the $(90)_{8s}$ Layup at Elevated Temperature of $538C$

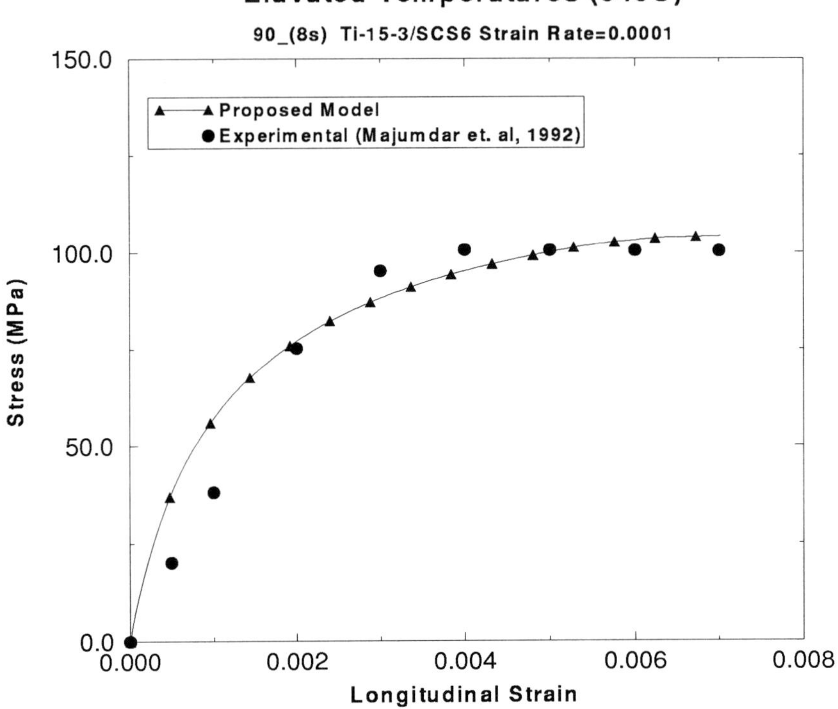

Figure 17.15   Comparison of the Viscoplastic Damage Model with Experimental Results [208] of the $(90)_{8s}$ at an Elevated Temperature of $649C$ Layup

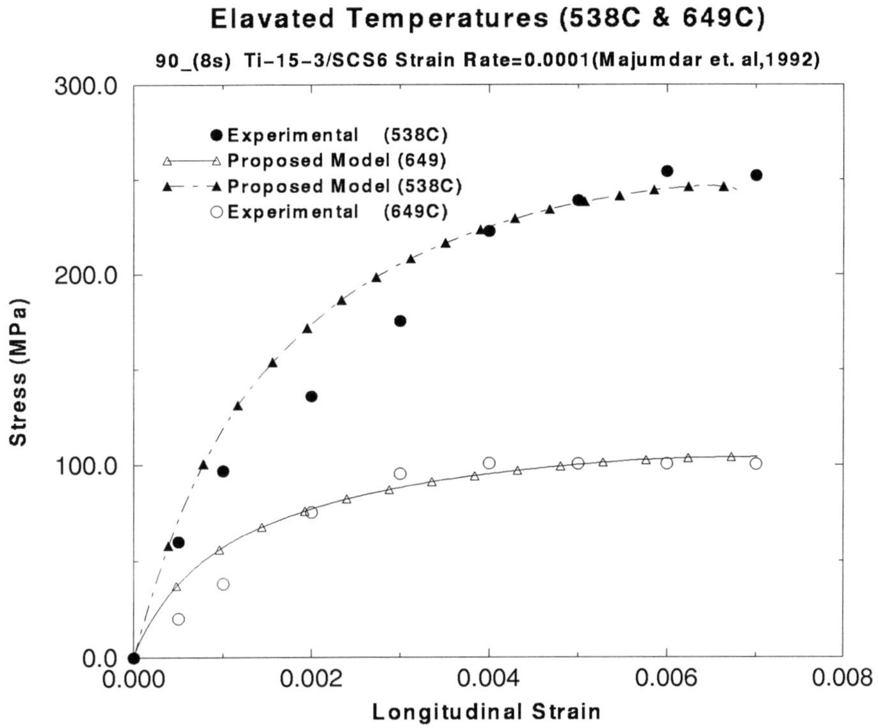

Figure 17.16   Comparison of the Viscoplastic Damage Model with Experimental Results [208] of the $90_{(8s)}$ Layup at differenet Elevated Temperatures of 538°C and 649°C

It is clear that in order to obtain a better prediction of the model with the experimental results at very low strains one needs to show a dramatic initial increase of damage at the early stage of loading. This is indicated by the reduction in stifness at lower strains. However, at higher strains plasticity is predominant with a decrease in the rate of damage. However, for the $(90)_{(8s)}$ at room temperature the model predictions show good correlation with the experimental results for both the initial and final stages of loading (Figure 10). Damage evolution for both systems is presented in Figures 17.17 and 17.18 for 538°C and 649°C respectively.

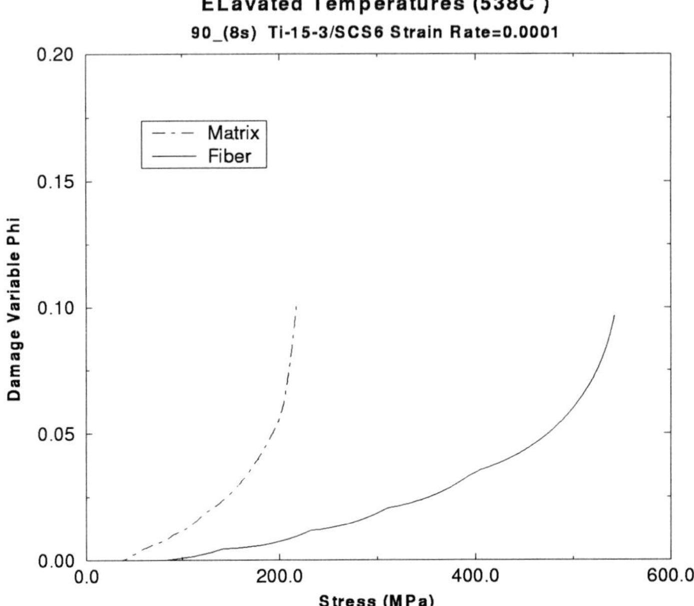

Figure 17.17 Evolution of the Damage Variable ϕ for $(90_{8_s}$ at an Elevated Temperature of 538°C

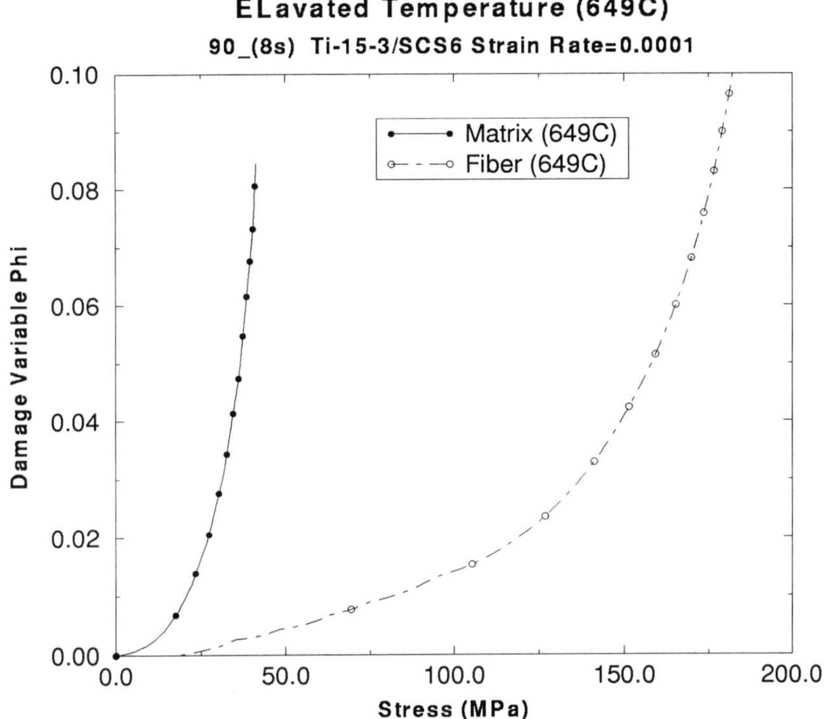

Figure 17.18 Evolution of the Damage Variable ϕ for $(90)_{8s}$ Layup at an Elevated Temperature of 649°C

At elevated temperatures the material becomes more ductile which may cause retardation of the damage in the material. This is because of the possibility of increase in the bond strength and the reaction zone. Yielding occurs at low stress values for elevated temperatures. However, the debonding may require higher stress levels. This temperature effect is investigated, as well as, the response of the evolution of damage versus stress. The theoretical model shows similar behavior as the experimental observations which is shown in Figure 17.19.

The strain rate effect on the evolution of the damage variable is also studied here. As indicated by Ju [208], higher strain rates cause retardation of the growth of damage in the materials. This characteristic behavior of the material is also validated by the proposed theory. For this reason different strain rates are used and the corresponding damage evolution curves are generated. As expected with the strain rate increase less damage occurs due to the hardening and consequently less damage is obtained at the same stress level. This is shown for both phases, matrix and fiber in Figure 17.20

504

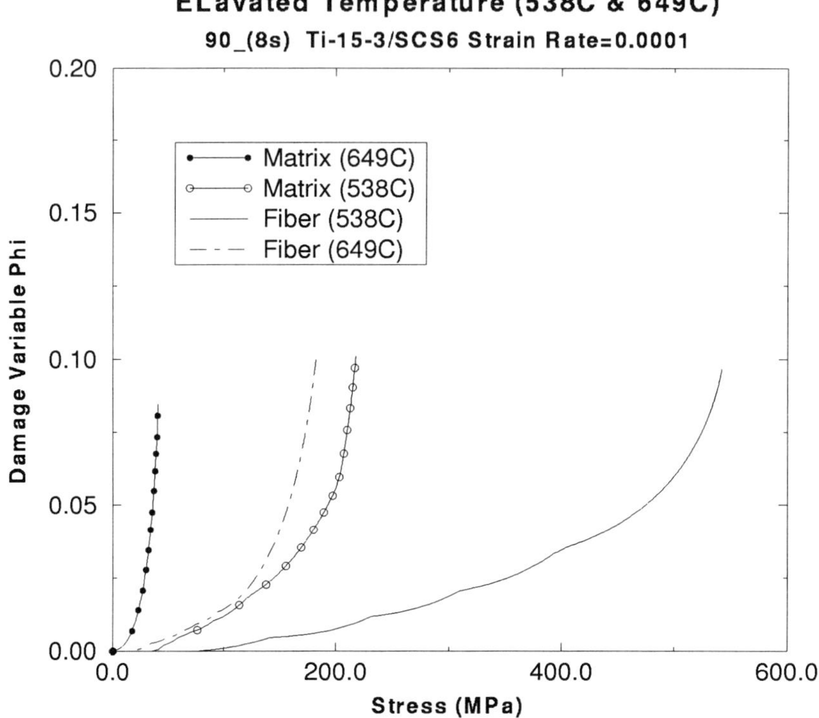

Figure 17.19 Temperature Effect on the Damage Variable ϕ for $90_{(8s)}$ Layup

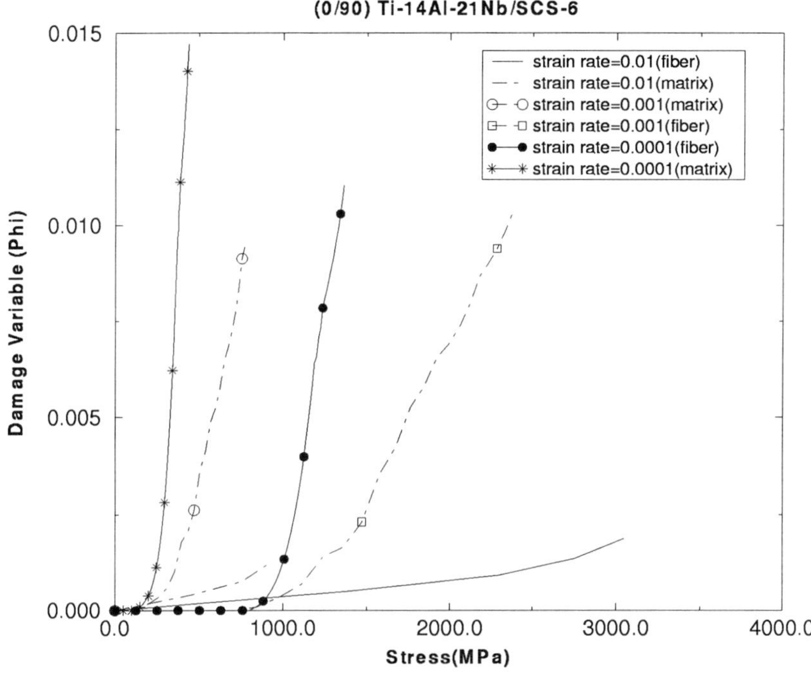

Figure 17.20 Strain Rate Effect on the Damage Variable ϕ of the $90_{(8s)}$ Layup

References

[1] L. M. Kachanov, "On the Creep Fracture Time," Izv Akad. Nauk USSR Otd. Tekh., Vol. 8, pp. 26-31, 1958 (in Russian).

[2] Y. Wang, D. Mukjerji, W. Chen, T. Kuttner, R. Prasad Wahi and H. Wever, "The Cyclic Behavior of Nickel Base Superalloy, IN738LC," Zeitschrift für Metallkunde, 86(5), pp.365-370, 1995.

[3] D. Bettge, W. Osterle, and J. Ziebs, "Temperature Dependence of Yield Strength and Elongation of the Nickel-base Superalloy IN738LC and the Corresponding Microstructural Evolution," Zeitschrift für Metallkunde, 86(5), pp.190-197, 1995.

[4] G.Z. Voyiadjis, A. Venson and P.I. Kattan,"Experimental Determination of Damage Parameters in Uniaxially-Loaded Metal-Matrix Composites Using The Overall Approach, " International Journal of Plasticity, Vo. 11, No. 8, pp. 895 - 926, 1995.

[5] J. Lemaitre, A course on Damage Mechanics, Springler-Verlag, New York, 1992

[6] J. L. Chaboche, "Une Loi Differentielle d'Endommagement de Fatigue avec Cumulation Non-Lineare, " Rev. Francaise Mecanique, No. 50-51, 1974 (in French).

[7] F. A. Leckie and D. Hayhurst, "Creep Rupture of Structures," Proc. Royal Society, London, Vol. A340, pp. 323-347, 1974.

[8] J. Hult, "Creep in Continua and Structures," in Topics in Applied Continuum Mechanics (Edited by Zeman and Ziegler), pp. 137, Springer, N.Y., 1974.

[9] J. Lemaitre and J. L. Chaboche, "A Nonlinear Model of Creep Fatigue Cumulation and Interaction," Proc. IUTAM, Symposium on Mechanics of Viscoelastic Media and Bodies (Edited by Hult), pp. 291-301, 1975.

[10] J. Lemaitre and J. Dufailly, "Modelisation et Identification de l'Endommagement Plastique de Mataux," Zeme Congres Francaise de Mecanique, Grenoble, 1977.

[11] J. Lemaitre, "A Continuous Damage Mechanics Model for Ductile Fracture," Journal of Engineering Materials and Technology, Vol. 107, pp. 83-89, 1985.

[12] J. Lemaitre, "How to Use Damage Mechanics," Nuclear Engineering and Design, Vol. 80, pp. 233-245, 1984.

508

[13]   D. R. Hayhurst, "Creep Rupture under Multiaxial States of Stress," Journal of the Mechanics and Physics of Solids, Vol. 20, pp. 381-390, 1972.

[14]   C. L. Chow and J. Wang, "An Anisotropic Theory of Elasticity for Continuum Damage Mechanics," International Journal of Fracture, Vol. 33, pp. 3-16, 1987.

[15]   H. Lee, K. Peng and J. Wang, "An Anisotropic Damage Criterion for Deformation Instability and its Application to Forming Limit Analysis of Metal Plates," Engineering Fracture Mechanics, Vol. 21, pp. 1031-1054, 1985.

[16]   F. Sidoroff, "Description of Anisotropic Damage Application to Elasticity," in IUTAM Colloqium on Physical Nonlinearities in Structural Analysis, pp. 237-244, Springer-Verlag, Berlin, 1981.

[17]   J. P. Cordebois and F. Sidoroff, "Damage Induced Elastic Anisotropy," Colloque Euromech, 115, Villard de Lans, 1979.

[18]   J. P. Cordebois, "Criteres d'Instabilite Plastique et Endommagement Ductile en Grandes Deformations," These de Doctorat, Presente a l'Universite Pierre et Marie Curie, 1983.

[19]   C. L. Chow and J. Wang," An Anisotropic Theory of Continuum Damage Mechanics for Ductile Fracture," Engineering Fracture Mechanics, Vol. 27, pp. 547-558, 1987.

[20]   C.L. Chow and J. Wang, "Ductile Fracture Characterization with an Anisotropic Continuum Damage Theory," Engineering Fracture Mechanics, Vol. 30, pp. 547-563, 1988.

[21]   D. Krajcinovic and G.U. Foneska, "The Continuum Damage Theory for Brittle Materials," Journal of Applied Mechanics, Vol. 48, pp. 809-824, 1981.

[22]   S. Murakami and N. Ohno, "A Continuum Theory of Creep and Creep Damage," in Proc. 3rd IUTAM Symposium on Creep in Structures, pp. 422-444, Springer, Berlin, 1981.

[23]   S. Murakami, "Notion of Continuum Damage Mechanics and its Application to Anisotropic Creep Damage Theory," Journal of Engineering Materials and Technology, Vol. 105, pp. 99-105, 1983.

[24]   D. Krajcinovic," Constitutive Equations for Damaging Materials," Journal of Applied Mechanics, Vol. 50, pp. 355-360, 1983.

[25]   E. Krempl, "On the Identification Problem in Materials Deformation Modeling," Euromech, 147, on Damage Mechanics, Cachan, France, 1981.

[26]   F. A. Leckie and E. T. Onat, "Tensorial Nature of Damage Measuring Internal Variables," in IUTAM Colloquim on Physical Nonlinearities in Structural Analysis, pp. 140-155, Springer-Verlag, Berlin, 1981.

[27]   E. T. Onat, "Representation of Mechanical Behavior in the Presence of Internal Damage," Engineering Fracture Mechanics, Vol. 25, pp. 605-614, 1986.

[28]   E.T. Onat and F. A. Leckie, "Representation of Mechanical Behavior in the Presence of Changing Internal Structure," Journal of Applied Mechanics, Vol. 55, pp. 1-10, 1988.

[29]   J. Betten, "Damage Tensors in Continuum Mechanics," J. Mecanique Theorique et Appliquees, Vol. 2, pp. 13-32. (Presented at Euromech Colloqium 147 on Damage Mechanics, Paris-VI, Cachan, 22 September, 1981).

[30]   J. Betten, "Applications of Tensor Functions to the Formulation of Continutive Equations Involving Damage and Initial Anisotropy," Engineering Fracture Mechanics, Vol. 25, pp. 573-584, 1986.

[31]   J. Lemaitre, "Local Approach of Fracture," Engineering Fracture Mechanics, Vol. 25, Nos. 5/6, pp. 523 -537, 1986.

[32]   J. Lemaitre and J. Dufailly, "Damage Measurements," Engineering Fracture Mechanics, Vol. 28, Nos. 5/6, pp. 643-661, 1987.

[33]   L. M. Kachanov, Introduction to Continuum Damage Mechanics, Martinus Nijhoff Publishers, the Netherlands, 1986.

[34]   J. L. Chaboche, "Continuum Damage Mechanics: Present State and Future Trends," International Seminar on Local Approach of Fracture, Moret-sur-Loing, France, 1986.

[35]   J. L. Chaboche, "Continuum Damage Mechanics: Part I - General Concepts," Journal of Applied Mechanics, Vol. 55, pp. 59-64, 1988.

[36]   J. L. Chaboche, "Continuum Damage Mechanics: Part II - Damage Growth, Crack Initiation, and Crack Growth," Journal of Applied Mechanics, Vol. 55, pp. 65-72, 1988.

[37]   E. H. Lee, R. L. Mallet and T. B. Wertheimer, "Stress Analysis for Anisotropic Hardening in Finite-Deformation Plasticity," Transactions of the ASME, Journal of Applied Mechanics, Vol. 50, pp. 554-560, 1983.

[38]   Y. F. Dafalias, "Corotational Rates for Kinematic Hardening at Large Plastic Deformations," Transactions of the ASME, Journal of Applied Mechanics, Vol. 50, pp. 561-565, 1983.

[39]   E.T. Onat, "Shear Flow of Kinematically Hardening Rigid-Plastic Materials," Mechanics of Materials Behavior (Edited by G. J. Dvorak and R. T. Shield), Elsevier, Amsterdam-Oxford, NY - Tokyo, pp. 311-324, 1984.

[40]   E. T. Onat, "Representation of Inelastic Behavior in the Presence of Anisotropy and of Finite

510

Deformations," Recent Advances in Creep and Fracture of Engineering Materials and Structures (Edited by B. Wilshire and D. R. J. Owen), Pineridge Press, Swansea, U. K., pp. 231-264, 1982.

[41]   Y. F. Dafalias, "A Missing Link in the Macroscopic Constitutive Formulation of Large Plastic Deformations," Plasticity Today: Modeling, Methods and Applications, Proceedings of the International Symposium on Current Trends and Results in Plasticity, CISM, Udine, June 27-30, 1983, (Edited by a Sawczuk and G. Bianchi), Elsevier, London, New York, pp. 135-151, 1985.

[42]   B. Loret, "On the Effects of Plastic Rotation in the Deformation of Anisotropic Elasto-Plastic Materials," Mechanics of Materials Journal, Vol. 2, pp. 287-304, 1983.

[43]   J. Mandel, "Relations de Comportement des Milieux Elastiques - Plastiques et Elastiques - Viscoplastiques, Notion de Repere Directeur," Foundations of Plasticity, (Edited by A. Sawcauk), Noordhoff, Leyden, pp. 387-400, 1973.

[44]   J. Mandel, "Director Vectors and Constitutive Equations for Plastic and Viscoplastic Media," Problems of Plasticity, Noordhof, Leyden, pp. 135-143, 1973.

[45]   F. Fardshisheh and E. T. Onat, " Representation of Elasto-Plastic Behavior by Means of State Variables," Problems of Plasticity, Noordhoff, Leyden, pp. 89-115., 1973

[46]   J. A. Geary and E. T. Onat, "Representation of Nonlinear Hereditary Mechanical Behavior," Oak Ridge National Laboratory Report, ORNL-TM-4525, 1974.

[47]   J. L. Chaboche, "Le Concept de Constrainte Effective Applique a L'Elasticite et a la Viscoplasticite en Presence d'un Endommagement Anisotrope, "Colloque Internationaux du CNRS, No. 295, pp. 738-759, 1982 (in French).

[48]   R. B. Pecherski, "Discussion of Sufficient Condition of Plastic Flow Localization," Proceedings of the International Symposium on Current Trends and Results in Plasticity, CISM, Udine, June 27-30, 1983, Engineering Fracture Mechanics, the Olszak Memorial Volume, Vol. 21, pp. 767-779, 1985.

[49]   J. E. Paulun and R. B. Pecherski, "Study of Corotational Rates for Kinematic Hardening in Finite Deformation Plasticity," Archives of Mechanics, Vol. 37, No. 6, pp. 661-677, 1985.

[50]   S. N. Atturi, "On Constitutive Relations at Finite Strain: Hypo-Elasticity and Elasto-Plasticity with Isotropic and Kinematic Hardening, "Computer Methods in Applied Mechanics and Engineering, Vol. 43, pp. 137-171, 1984.

[51]   G. C. Johnson and D. J. Bammann, "A Discussion of Stress Rates in Finite Deformation Problems," International Journal of Solids and Structures, Vol. 20, No. 8, pp. 725-737, 1984.

[52]   C. Fressengeas and A. Molinari, "Representations de Comportement Plastiques Anisotrope aux Grandes Deformations," Archives of Mechanics, Vol. 36, 1984.

[53]   W. C. Moss, "On Instabilities in Large Deformation Simple Shear Loading," Computer Methods in Applied Mechanics and Engineering, Vol. 46, pp. 329-338, 1984.

[54]   J. C. Simo and K. S. Pister, "Remarks on Rate Constitutive Equations for Finite Deformation Problems: Computational Implications," Computer Methods in Applied Mechanics and Engineering, Vol. 46, pp. 201-215, 1984.

[55]   G. Z. Voyiadjis, "Experimental Determination of the Material Parameters of Elasto-Plastic Work-Hardening Metal Alloys," Material Science and Engineering Journal, Vol. 62, No. 1, pp. 99 - 107, 1984.

[56]   G. Z. Voyiadjis, and P. D. Kiousis, "Stress Rate and the Lagrangian Formulation of the Finite-Strain Plasticity for a von Mises Kinematic Hardening Model," International Journal of Solids and Structures, Vol. 23, No. 1, pp. 95-109, 1987.

[57]   K. J. Williams (editor), "Constitutive Equations: Macro and Computational Aspects," ASME Publications, New York, 1984 (Book No. G00274, 111).

[58]   S. Murakami," Mechanical Modeling of Material Damage," Journal of Applied Mechanics, Vol. 55, pp. 280-286, 1988.

[59]   R. Talreja, "A Continuum Mechanics Characterization of Damage in Composite Materials," Proceedings of the Royal Society, London, Vol A399, pp.195-216, 1985.

[60]   R. Talreja, "Stiffness Properties of Composite Laminates with Matrix Cracking and Interior Delamination," Engineering Fracture Mechanics, Vol. 25, Nos. 5/6, p. 751-762, 1986.

[61]   R. M. Christensen, "Tensor Transformations and Failure Criteria for the Analysis of Fiber Composite Materials," Journal of Composite Materials, Vol. 22, pp. 874-897, 1988.

[62]   R. M. Christensen, "Tensor Transformations and Failure Criteria for the Analysis of Fiber Composite Materials. Part II: Necessary and Sufficient Conditions for Laminate Failure," Journal of Composite Materials, Vol. 24, pp. 796-800, 1990.

[63]   W. Shen, B. Raio and H. Lee, "A Crack-Damage Mechanics Model for Composite Laminates," Engineering Fracture Mechanics, Vol. 21, No. 5, pp. 1019-1029, 1985.

[64]   F. Lene, "Damage Constitutive Relations for Composite Materials," Engineering Fracture Mechanics, Vol. 25, Nos. 5/6, pp. 713-728, 1986.

512

[65]   R. Badaliance, G. C. Sih and E. P. Chen, "Through Cracks in Multilayered Plates," in Mechanics of Fracture, Vol. 3, Plates and Shells with Cracks (Edited by G.C. Sih), pp. 85- 115, Noordhoff, Leyden, 1977.

[66]   R. Hill, "A Self-Consistent Mechanics of Composite Materials," Journal of the Mechanics and Physics of Solids, Vol. 13, pp. 213-222, 1965.

[67]   R. Hill, "On Constitutive Macro-Variables for Heterogeneous Solids at Finite Strain," Proceedings of the Royal Society, London, Vol. A326, pp. 131-147, 1972.

[68]   G. J. Dvorak and Y. A. Bahei-El-Din," Elastic-Plastic Behavior of Fibrous Composites," Journal of the Mechanics and Physics of Solids, Vol. 27, pp. 51-72, 1979.

[69]   G. J. Dvorak and Y. A. Bahei-El-Din, "Plasticity Analysis of Fibrous Composites," Journal of Applied Mechanics, Vol. 49, pp. 327-335, 1982.

[70]   G. J. Dvorak and Y. A. Bahei-El-Din, "A Bimodal Plasticity Theory of Fibrous Composite Materials," Acta Mechanica, Vol. 69, 219-244, 1987.

[71]   J. Aboudi, "Micromechanical Prediction of Initial and Subsequent Yield Surfaces of Metal Matrix Composites," International Journal of Plasticity, Vol. 6, pp. 471-484, 1990.

[72]   D. H. Allen and C. E. Harris, "A Thermomechanical Constitutive Theory for Elastic Composites with Distributed Damage - I. Theoretical Formulation," International Journal of Solids and Structures, Vol. 23, pp. 1301-1318, 1987.

[73]   D. H. Allen, C. E. Harris and S. E. Groves, "A Thermomechanemical Constitutive Theory for Elastic Composites with Distributed Damage - II. Application to Matrix Cracking in Laminated Composites," International Journal of Solids and Structures, Vol. 23, pp.1319-1338, 1987.

[74]   G. J. Dvorak, N. Laws and M. Hejazi, "Analysis of Progressive Matrix Cracking in Composite Laminates - I. Thermoelastic Properties of a Ply with Cracks," Journal of Composite Materials, Vol. 19, pp. 216-234, 1985.

[75]   G. J. Dvorak and N. Laws, "Analysis of Progressive Matrix Cracking in Composite Laminates - II. First Ply Failure," Journal of Composite Materials, Vol. 21, pp. 309-329, 1987.

[76]   N. Laws and G. J. Dvorak, "The Effect of Fiber Breaks and Aligned Penny - Shaped Cracks on the Stiffness and Energy Release Rates in Unidirectional Composites," International Journal of Solids and Structure, Vol. 23, No. 9, pp. 1269-1283, 1987.

[77]   D. H. Allen, C. E. Harris, S. E. Groves and R. G. Norvel, "Characteristics of Stiffness Loss in Crossply Laminates with Curved Matrix Cracks," Journal of Composite Materials, Vol. 22, pp.

71-80, 1988.

[78] J. W. Lee, D. H. Allen and C. E. Harris," Internal State Variable Approach for Predicting Stiffness Reduction in Fibrous Laminated Composites with Matrix Cracks," Journal of Composite Materials, Vol. 23, pp. 1273-1291, 1989.

[79] O. P. Allix, P. Ladeveze, D. Gilleta and R. Ohayon, "A Damage Prediction Method for Composite Structure," International Journal of Numerical Methods in Engineering, Vol. 27, No. 2, pp. 271-283, 1989.

[80] A. Poursatip, M. F. Ashby and P. W. R. Beaumont, "Damage Accumulation During Fatigue of Composites," in Progress in Science and Engineering of Composites. Proceedings of the Fourth International Conference on Composite Materials (Edited by Tsuyoshi Hayashi, Kozo Kawata and Sokichi Umekawa), Vol. 1, Japan Society for Composite Materials, pp. 693-700, 1982.

[81] P. Ladeveze, M. Poss and L. Proslier, "Damage and Fracture of Tridirectional Composites," in Progress in Science and Engineering of Composites. Proceedings of the Fourth International Conference on Composite Materials, Japan Society for Composite Materials, Vol. 1, pp. 649-658, 1982.

[82] S. S. Wang, H. Suemasu and E. S. M. Chim, "Analysis of Fatigue Damage Evolution and Associated Anisotropic Elastic Property Degradation in Random Short-Fiber Composites," Journal of Composite Materials, December, Vol. 21, pp. 1084-1105, 1987.

[83] B. S. Majumdar, G. M. Newaz and J. R. Ellis, "Evolution of Damage and Plasticity in Titanium Based Fiber Reinforced Composites," Metallurgical Transactions, Series A, July, Vol. 24A, pp. 1597-1610, 1993.

[84] T. M. Breunig, S. R. Stock, J. H. Kinney, A. Guvenilir and M. C. Nichols, "Impact of X-ray Tomographics Microscopy on Deformation Studies on a Si/Al MMC," in Material Research Society Symposium Proceedings, Materials Research Society, pp. 135-141, 1991.

[85] K. J. Baumann, W. H. Kennedy and D. L. Hebert "Computed Tomography X-ray Scanning of Graphite/Epoxy Coupons," Journal of Composite Materials, Vol. 18, pp. 537-544, 1984.

[86] J. E. Benci and D. P. Pope, "Measuring Creep Damage Using Microadiography, Metallurgical Transactions, Series A, April, Vol. 19A, pp. 837-847, 1988.

[87] J. L. Chaboche, "Continuous Damage Mechanics - A Tool to Describe Phenomena Before Crack Initiation," Nuclear Engineering and Design, Vol. 64, pp. 233-247, 1981.

[88] J. Hult, "CDM- Capabilities, Limitations and Promises," in Mechanisms of Deformation and Fracture, Edited by K. E. Easterling, Pergamon, Oxford, pp. 233-247, 1979.

514

[89]   D. Krajcinovic, "Continuum Damage Mechanics, Applied Mechanics Reviews, Vol. 37, pp. 1-6, 1984.

[90]   J. Lemaitre and J. L. Chaboche, "Aspect Phenomenologique de la Rupture par Endommagement," Journal de Mecanique Applique, Vol. 2, pp. 317-365, 1978.

[91]   J. Lemaitre and J. L. Chaboche, <u>Mecanique de Materiaux Solides</u>, Dunod, Paris, 1985.

[92]   Yu. N. Rabotnov, <u>Creep Problems of Structural Members</u>, North Holland, Amsterdam, 1969.

[93]   J. Lemaitre, "Evaluation of Dissipation and Damage in Metals Subjected to Dynamic Loading," Proceedings of I. C. M. 1 Kyoto, Japan, 1971.

[94]   P. Suquet, "Plasticite' et Homogeneisation," These d' Etat, Universite' Paris 6, 1982.

[95]   P. I. Kattan and G. Z. Voyiadjis, "Separation of Voids and Cracks in Continuum Damage Mechanics," Submitted for Publication, 1998.

[96]   P. I. Kattan and G. Z. Voyiadjis, "Bending of Damaged Beams," Submitted for Publication, 1997.

[97]   Y . A. Bahei-El-Din and G. J. Dvorak, "A Review of Plasticity Theory of Fibrous Composite Materials," In Metal Matrix Composites: Testing, Analysis and Failure Modes (Edited by W. S. Johnson), ASTM STP 1032, pp. 103-129, 1989.

[98]   G. Z. Voyiadjis and T. Park, "Local and Interfacial Damage Analysis of Metal Matrix Cpomosites Using the Finite Element Method," Engineering Fracture Mechanics, Vol.56, No.4, pp.483-511, 1997

[99]   G. Z. Voyiadjis and P. I. Kattan, "A Plasticity-Damage Theory for Large Deformation of Solids - Part I: Theoretical Formulation," International Journal of Engineering Science, Vol. 30, No. 9, pp. 1089-1108, 1992.

[100]   P. I. Kattan and G. Z. Voyiadjis, "Micromechanical Modeling of Damage in Uniaxially Loaded Unidirectional Fiber-Reinforced Composite Laminae," International Journal of Solids and Structures, Vol. 30, No. 1, pp. 19-36, 1993.

[101]   P. I. Kattan and G. Z. Voyiadjis, "Overall Damage and Elasto-Plastic Deformation in Fibrous Metal Matrix Composites," International Journal of Plasticity, Vol. 9, pp. 931-949, 1993.

[102]   P. I. Kattan and G. Z. Voyiadjis, "Damage - Plasticity in a Uniaxially Loaded Composite Lamina: Overall Analysis," International Journal of Solids and Structures, Vol. 33, No. 4, pp. 555-576, 1996.

[103] C. Stolz, "General Relationships Between Micro and Macro Scales for the Non-Linear Behavior of Heterogeneous Media," in Modeling Small Deformations of Polycrystals (Edited by J. Gittus and J. Zarka), pp. 89-115, Elsevier, 1986.

[104] G. Z. Voyiadjis and P. I. Kattan, "Damage of Fiber-Reinforced Composite Materials with Micromechanical Characterization, "International Journal of Solids and Structures, Vol. 30, pp. 2757-2778, 1993.

[105] G. Z. Voyadjis and P. I. Kattan, "Anisotropic Damage Mechanics Modeling in Metal Matrix Composites," Technical Report, Final Report Submitted to the Air Force Office of Scientific Research, 1993.

[106] T. Mori and K. Tanaka, "Average Stress in Matrix and Average Elastic Energy of Materials with Misfitting Inclusions,' Acta Metallurgica, Vol. 21, pp. 571-574, 1973.

[107] A. C. Gavazzi and D. C. Lagoudas, "On The Numerical Evaluation of Eshelby's Tensor and its Applications to Elasto-Plastic Fibrous Composites," Computational Mechanics, Vol. 7, pp. 13-19, 1990.

[108] D. C. Lagoudas, A. C. Gavazzi and H. Nigam, "Elasto-Plastic Behavior of Metal Matrix Composites Based on Incremental Plasticity and the Mori-Tanaka Averaging Scheme," Computational Mechanics, Vol. 8, pp. 193-203, 1991.

[109] G. Z. Voyiadjis and P. I. Kattan, "A Coupled Theory of Damage Mechanics and Finite Strain Elasto-Plasticity- Part II: Damage and Finite Strain Plasticity," International Journal of Engineering Science, Vol. 28, No. 6, pp. 505-524, 1990.

[110] P. I. Kattan and G. Z. Voyiadjis, "A Plasticity-Damage Theory for Large Deformation of Solids - Part II: Applications to Finite Simple Shear," International Journal of Engineering Science, Vol. 31, No. 1, pp. 183-199, 1993.

[111] J. P. Cordebois and F. Sidoroff, "Anisotropic Damage in Elasticity and Plasticity," J. Mec. Theor. Appl. (Numerous Special), pp. 45-60, 1982 (in French).

[112] O. C. Zienkiewicz and K. Morgan, Finite Elements and Approximation, Wiley, New York, 1983.

[113] K. J. Bathe, Finite Element Procedures, Prentice-Hall, 1996.

[114] S. Cescotto, F. Frey and G. Fonder, "Total and Updated Lagrangian Descriptions in Nonlinear Structural Analysis: A Unified Approach," in Energy Methods in Finite Element Analysis (Edited by R. Glowinski, E. Y. Rodin and O. C. Zienkiewicz), pp. 283-296, Wiley, New York, 1979.

[115] O. C. Zienkiewicz, The Finite Element Method, Third Edition, McGraw-Hill, New York, 1977.

516

[116] J. T. Oden, <u>The Mathematical Theory of Finite Elements,</u> Wiley-Interscience, New York, 1976.

[117] G. Tsamasphyros and A. E. Giannakopoulos, "The Optimum Finite Element Grids Around Crack Singularities in Bilinear Elasto-Plastic Materials, "Engineering Fracture Mechanics, Vol. 32, No. 4, pp. 515-522, 1989

[118] R. D. Henshell and K. G. Shaw, "Crack Tip Finite Elements Are Unnecessary," International Journal of Numerical Methods in Engineering, Vol. 9, pp. 495-507, 1975.

[119] R. S. Barsoum, "On the Use of Isoparametric Finite Elements in Linear Fracture Mechanics," International Journal of Numerical Methods in Engineering, Vol. 10, pp. 25-37, 1976.

[120] C. L. Chow and J. Wang, "A Finite Element Analysis of Continuum Damage Mechanics for Ductile Fracture," International Journal of Fracture, Vol. 38, pp. 83-102, 1988.

[121] A. L. Gurson, "Plastic Flow and Fracture Behavior of Ductile Materials Incorporating Void Nucleation, Growth, and Interaction," Ph.D. Thesis, Brown University.

[122] A. L. Gurson, "Continuum Theory of Ductile Rupture by Void Nucleation and Growth. Part I: Yield Criteria and Flow Rules for Porous Ductile Media," Journal of Engineering Materials and Technology, Vol. 99, No. 2, pp. 2-15, 1977

[123] V. Tvergaard, "Material Failure by Void Coalescence in Localized Shear Bands," International Journal of Solids and Structures, Vol. 18, pp. 659-672, 1982.

[124] V. Tvergaard and A. Needleman," Analysis of Cup-cone Fracture in a Round Tensile Bar," Acta Metallurgica, Vol. 32, p. 157, 1984.

[125] S. Nemat-Nasser, "Decomposition of Strain Measures and Their Rates in Finite Deformation Elastoplasticity," International Journal of Solids and Structures, Vol. 15, pp. 155-166, 1979.

[126] S. Nemat-Nasser, "On Finite Plastic Flow of Crystalline Solids and Geomaterials," Journal of Applied Mechanics, Vol. 50, pp. 1114-1126, 1983.

[127] E. H. Lee, "Some Comments on Elastic-Plastic Analysis," International Journal of Solids and Structures, Vol. 17, pp. 859-872, 1981.

[128] R. J. Asaro, "Micromechanics of Crystals and Polycrystals," Advances in Applied Mechanics, Vol. 23, pp. 1-115, 1983.

[129] H. Ziegler, "A Modification of Prager's Hardening Rule," Quarterly of Applied Mathematics, Vol. 17, pp. 55-65, 1959.

[130] H. Lee, G. Li, and S. Lee, "The Influence of Anisotropic Damage on the Elastic Behavior of Materials," International Seminar on Local Approach of Fracture, Moret-sur-Loing, France, pp. 79-90, 1986.

[131] G. Z. Voyiadjis and P. I. Kattan, "Eulerian Constitutive Model for Finite Strain Plasticity with Anisotropic Hardening," Mechanics of Materials Journal, Vol. 7, No. 4, pp. 279-293, 1989.

[132] J. G. Oldroyd, "On the Formulation of Rheological Equations of State, "Proceedings of the Royal Society, London, Vol. A 200, pp. 523-541, 1950.

[133] J. C. Simo and J. W. Ju, "Strain and Stress-based Continuum Damage Models. Part I: Formulation," International Journal of Solids and Structures, Vol. 23, No. 7, pp. 821-840, 1987.

[134] G. Z. Voyiadjis, "Degradation of Elastic Modulus in Elasto-Plastic Coupling with Finite Strains," International Journal of Plasticity, Vol. 4, pp. 335-353, 1988.

[135] C. L. Chow and J. Wang, "A Finite Element Analysis of Continuum Damage Mechanics for Ductile Fracture," International Journal of Fracture, Vol. 38, pp. 83-102, 1988.

[136] G. Z. Voyiadjis and P.I. Kattan, "A Continuum-Micromechanics Damage Model for Metal Matrix Composites," in Composite Material Technology 1992, ASME, Proceedings of the Composite Material, Symposium of the Energy Technology Conference and Exposition, Houston, Texas, Vol. 48, pp. 83-95, 1992.

[137] G. Weng, "Some Elastic Properties of Reinforced Solids with Special Reference to Isotropic Ones Containing Spherical Inclusions," International Journal of Engineering Science, Vol. 22, No. 7, pp. 845- 856, 1984.

[138] G. Z. Voyiadjis and P. I. Kattan, "Local Approach to Damage in Elasto-Plastic Metal Matrix Composites," International Journal of Damage Mechanics, Vol. 2, No. 1, pp. 92-114, 1993.

[139] Z. Mroz, Mathematical Models of Inelastic Material Behavior, University of Waterloo, pp. 120-146, 1973.

[140] Stumvoll and Swoboda, "Deformation Behabior of Ductile Solids Containing Anisotropic Damage," Journal of Engineering Mechanics, ASCE, Vol. 119, No. 7, pp. 169-192 , 1993.

[141] J. P. Cordebois and F. Sidoroff, Journal de Mecanique Theorique et Appliques, pp. 45 - 60, 1982.

[142] A. J. Levy, "Decohesion at a Circular Interface," in Studies in Applied Mechanics, Vol. 35, Mechanics of Materials and Structures (edited by G. Z. Voyiadjis, L. C. Bank and L. J. Jacobs), pp. 173-192, Elsevier, Amsterdam, 1994.

[143] G. Z. Voyiadjis and A. R. Venson, "Experimental Damage Investigation of a SiC-Ti Aluminide Metal Matrix Composite," International Journal of Damage Mechanics, Vo. 4, No. 4, pp. 338 - 361, 1995.

[144] G. Z. Voyiadjis, "Large Elasto-Plastic Deformation of Solids," Ph.D. Dissertation, Department of Civil Engineering and Engineering Mechanics, Columbia University, New York, NY. U.S.A., 1973

[145] J. Lemaitre and J. L. Chaboche, <u>Mechanics of Solids</u>, pp. 69 - 120, pp. 346 - 450, Cambridge University Press, 1990.

[146] L. A. Carlsson and R. B. Pipers, <u>Experimental Characterization of Advanced Composites Materials,</u> Prentice Hall, 1987.

[147] M. E. Tuttle and H. F. Brinson, "Resistance Foil Strain - Gage Technology as Applied to Composite Materials," Experimental Mechanics, March, Vol. 24, No. 1, pp 54-65, 1984 (Errata: Vol. 26, No. 2, June, 1986, pp. 153-154).

[148] P. K. Brindley, "SiC Reinforced Aluminide Composites" In Symposia Proceedings: High Temperature Ordered Intermetallic Alloys II (edited by N. S. Stoloff, C. C. Koch, C.T. Liu, and O. Imuzi), Vol. 81, Materials Research Society, 1987.

[149] R. A. Mackay, P. K. Brindley, and F. H. Froes, "Continuous Fiber-Reinforced Titanium Aluminide Composites," Journal of Minerals, Metals and Materials Society, May, Vol. 43, No. 5, pp. 23-29, 1991.

[150] G. Z. Voyiadjis, A. R. Venson, and P. I. Kattan, "Experimental Determination of Damage Parameters in Uniaxially - Loaded Metal Matrix Composites Using the Overall Approach," International Journal of Plasticity, Vol. 11, No. 8, pp. 895 - 926, 1995.

[151] R. M. Jones, <u>Mechanics of Composite Materials</u>, Hemisphere Publishing Co., 1975.

[152] G. Z. Voyiadjis, P. I. Kattan, and A. R. Venson, "Evolution of a Damage Tensor for Metal Matrix Composites," in MECAMAT 93: International Seminar on Micromechanics of Materials, Vol. 84, pp. 406 - 417, Moret-sur-Loing, France, July, 1993.

[153] G. Z. Voyiadjis, P. I. Kattan, A. R. Venson, and T. Park, "Anisotropic Damage Mechanics Modeling in Metal Matrix Composites," Technical Report, Final Report Submitted to Air Force Office of Scientific Research, 141 pages, 1993.

[154] M.J. Owen and R.J. Howe. The Accumulation of Damage in a Glass-reinforced Plastic under Tensile and Fatigue Loading. Journal of Physics, D:5:1637-1649, 1972.

[155] S. Subramanyan. A Cumulative Damage Rule Based on the Knee Point of the S-N-Curve.

Journal of Engineering Mechanics and Technology, pp.316-321, 1976.

[156] P. Srivatsavan and S. Subramanyan. A Cummulative Damage Rule Based on Successive Reduction in Fatigue Limit. Journal of Engineering Mechanics and Technology, 100:212-214, 1978.

[157] J. Lemaitre and A. Plumtree. Application of Damage Concepts to Predict Creep-Fatigue Failures. Journal of Engineering Mechanics and Technology, 101:248-292. 1979.

[158] J.T. Fong. What is Fatigue Damage? In K.L. Reifsnider, editor, Damage in Composite Materials, pp. 243-266. American Society for Testing and Materials, Philadelphia, PA, 1982.

[159] Z. Hashin. Cumulative Damage Theory for Composite Materials, Residual Life and Residual Strength Methods. Composite Science and Technology, 23:1-19, 1985.

[160] W. Hwang and K.S. Han. Commutative Fatigue Damage Models and Multi-Stress Fatigue Life Prediction. Journal of Composite Materials, 20:125-153, 1986a.

[161] W. Hwang and K. S. Han. Fatigue of Composites - Fatigue Modulus Concept and Life Prediction. Journal of Composite Materials, 20:154-165, 1986b.

[162] H.A. Whitworth. Cumulative Damage in Composites. Transactions of the ASME, 112:358-361, 1990.

[163] S.M. Arnold and S. Kruch. Differential Continuum Damage Mechanics Models for Creep and Fatigue of Unidirectional Metal Matrix Composites. Technical Memorandum, 105213, NASA, 1991a.

[164] S. M. Arnold and S. Kruch. A Differential CDM Model for Fatigue of Unidirectional Metal Matrix Composites. Technical Memorandum 105726, NASA, 1991b.

[165] J. L. Chaboche and P. M. Lesne. A Non-Linear Continuous Fatigue Damage Model. Fatigue and Fracture of Engineering Materials and Structures, 11(1):1-17, 1988.

[166] J. L. Chaboche. Fracture Mechanics and Damage Mechanics: Complementarity of Approaches. In Proceedings of the 4th International Conference on "NUMBERICAL METHODS IN FRACTURE MECHANICS", pp. 309-324, 1987.

[167] P. M. Lesne and S. Savalle. A differential Damage Rule with Microinitiation and Micropropagation. La. Recherche Aerospatiale, 1987(2):33-47, 1987.

[168] P.M. Lesne and G. Cailletaud. Creep-Fatigue Interaction under High Frequency Loading. Int. Conf. on Mechanical Behavior of Materials, Beijing, China, 1987.

[169] D. N. Robinson, S. F. Duffy, and J.R. Ellis. A Viscoplastic Constitutive Theory for Metal Matrix Composites at High Temperature. pp. 49-56,1987.

[170] D. N. Robinson and S. F. Duffy. Continuum Deformation Theory for High-Temperature Metallic Composites. Journal of Engineering Mechanics, 116(4):832-844, 1990.

[171] T. E. Wilt and S. M. Arnold. A Coupled/Uncoupled Deformation and Fatigue Damage Algorithm Utilizing the Finite Element Method. NASA TM 106526, NASA, Lewis Research Center, Cleveland, OH, 1994.

[172] T. Nicholas. Fatigue Life Prediction in Titanium Matrix Composites. Journal of Engineering Materials and Technology, 117:440-447, 1995.

[173] R. W. Neu. A Mechanistic-Based Thermomechanical Fatigue Life Prediction Model For Metal Matrix Composites. Fatigue and Fracture of Engineering Materials and Structures, 16(8):811-828, 1993.

[174] R. Talreja. Fatigue of Composite Materials. Technomic Publishing Co., Lancaster, PA, 1987.

[175] G.Z. Voyiadjis and P.I. Kattan. Micromechanical Characterization of Damage-Plasticity in Metal Matrix Composites. In G. Z. Voyiadjis, editor, Studies in Applied Mechanics, Vol. 34: Damage in Composite Materials, pp. 67-102. 1993c.

[176] G. Z. Voyiadjis and T. Park. Anisotropic Damage of Fiber Reinforced MMC Using An Overall Damage Analysis. Journal of Engineering Mechanics, 121(11):1209-1217, 1995.

[177] YU. N. Rabotnov. Creep Rupture. In Proceedings of the *XII International Congress on Applied Mechanics,* pp. 342-349. (Stanford-Springer, 1969), 1968.

[178] T. Chen, George J. Dvorak, and Y. Benveniste. Mori-Tanaka Estimates of the Overall Elastic Moduli of Certain Composite Materials. Journal of Applied Mechanics, 59:539-546, 1992.

[179] C. L. Chow and T. J .Lu. On Evolution Laws of Anisotropic Damage. Engineering Fracture Mechanics, 34:(3):679-701, 1989.

[180] W. S. Johnson, S. J. Lubowinski, and A. L. Highsmith. Mechanical Characterization of Unnotched SCS_6/Ti-15-3 Metal Matrix Composites at Room Temperature. In J. M. Kennedy, H. H. Moeller, and W. S. Johnson, editors, Thermal and Mechanical Behavior of Metal Matrix and Cermaic Matrix Composites, ASTM STP 1080, pp. 193-218. ASTM, Philadelphia, PA, 1990.

[181] W. S. Johnson. Fatigue Testing and Damage Development in Continuous Fiber Reinforced Metal Matrix Composites. In W. S. Johnson, editor, Metal Matrix Composites: Testing, Analysis and Failure Modes, pp. 194-221. 1989.

[182]   G. Z. Voyiadjis and R. Echle, "A Micro-Mechanical Fatigue Damage Model for Uni-Directional Metal Matrix Composites," ASTM, accepted for publication, to appear in 1998.

[183]   R. Hill, "A Theory of the Yielding and Plastic Flow of Anisotropic Metals," Proceeding of Royal Society of London, A193, pp.281-297, 1948

[184]   J, F. Mulhern, T. G. Rogers and A. J. M. Spencer, "A Continuum Model for a Fiber Reinforced Plastic Material," Proceedings of Royal Society of London," 7, pp. 129-152, 1967

[185]   G. J. Dvorak, Y. A. Bahei-El-Din, Y. Macheret and C. H. Liu, "An Experimental Study of Elastic-Plastic Behavior of Fibrous Boron-Aluminum Composite," Journal of the Mechanics and Physics of Solids, 38, 3, pp.419-441, 1988

[186]   H. Nigam, G. J. Dvorak, Y. A. Bahei-El-Din, "An Experimental Investigation of Elastic-Plastic Behavior of Fibrous Boron-Aluminum Composite, I. Matrix-Dominated Mode," International Journal of Plasticity, In Press. 1993

[187]   Y. H. Zhao, G. J. Weng, eds: G. J. Weng, M. Taya and H. Abe, "Theory of Plasticity for a Class of Inclusion and Fiber-Reinforced Composites," Micromechanics and Inhomogeneities: The Toshio Mura 65th Anniversary Volume, Springer-Verlar, New York, 1990

[188]   G. Z. Voyiadjis and G. Thiagarajan, "An Anisotropic Yield Surface Model for Directionally Reinforce Metal Matrix Composites," International Journal of Plasticity, 110, pp.151-172, 1995

[189]   G. Z. Voyiadjis and M. Foroozesh, "An Anisotropic Distortional Yield Model," ASME, Journal of Appliedd Mechanics, 57, pp537-547, 1990

[190]   M. A. Eisenberg and C. F. Chen, "The Anisotropic Deformation of yield Surfaces," ASME Journal of Engineering Material Technology, 106, pp.355-360, 1984

[191]   G. Z. Voyiadjis, G. Thigarajan and E. Petrakis, "Constitutive Modelling for Granular Media Using an Anisotropic Distortional Yield Model," Acta Mechanica, 107. pp.

[192]   H. C. Drucker, "A More Fundamental Approach to Plastic Stress-Strain Relations, Proceedings of First U.S. National Congress in Applied Mechanics, ASME, New York, pp.487-491, 1948

[193]   H. C. Drucker, "A Definition of Stable Inelastic Material," Journal of Applied Mechanics, Transactions ASME, 26. 101, 1959

[194]   S. W. Tsai and E. M. Wu, "A General Theory of Strength for Anisotropic Materials," Journal of Composite Materials, 5, pp.58-80, 1971

[195]   A. J. M. Spencer, "Plasticity Theory for Fiber-reinforced Composites," Journal of Engineering

Mathematics, 26, 107-118, 1992

[196] G. Z. Voyiadjis, and G. Thiagarajan, "A Cyclic Anisotropic-Plasticity Model for Metal Matrix Composites," International Journal of Plasticity, Vol.12, No.1, pp.69-91, 1996

[197] G. Z. Voyiasjis and G. Thiagarajan, "Micro and Macro Anisotropic Cyclic Damage-Plasticity Models for MMC's," Journal of Engineering Science, Vol.35, No.5, pp.467-484, 1997

[198] G. J. Weng, "The Overall Elastoplastic Stress-Strain Relations of Dual Phase Metals, Journal of Mechanics and Phisics of Solids, 36, pp.655-687

[199] J. Aboudi, Mechanics of Composite Materials : A Unified Micromechanical Approach, Elsevier, 1991

[200] A. J. M. Spencer, Deformations of Fiber-reinforced Materials, Clarendon Press. Oxford, 1972

[201] L. J. Walpole, "On the Overall Elastic Moduli of Composite Materials," Journal of the Mechanics of Physics and Solids, 17, pp.235-251

[202] G. Z. Voyiadjis and B. Deliktas, "Damage in MMCs using the GMC: theoretical formulation," Composite Part B, Vol.28B, pp.597-611, 1997

[203] M. Paley, and J. Aboudi, "Micromechanical Analysis of Composites by the Generalized Cells Model," Mechanics of Materials, Vol. 14, pp. 127-139, 1992.

[204] J. Aboudi, "Micromechanical Analysis of Composites by the Method of Cells ," Applied Mechanics of Review, Vol. 42, pp. 193-221, 1989.

[205] G.Z. Voyiadjis, and Z. Guelzim, "A Coupled Inceremntal Damage and Plasticity Theory for Metal Matrix Composites ," Journal of Mechanical Behavior of Materials, Vol. 6, pp. 193-219, 1996.

[206] J.W. Ju, "Energy Based Coupled Elastoplastic Damage Theories Constitutive Modeling and Computational Aspects ,"International Journal of Solids and Structures," Vol. 25, pp. 803-833, 1989.

[207] M. Ortiz, and C. Simo, "An Analysis of a New Class of Integration Algorithms for Elastoplastic Constitutive Relations ,"International Journal for Numerical Method in Engineering," Vol. 23, pp. 353-366, 1986.

[208] B.S. Majumdar, and G.M. Newaz, "Inelastic Deformation of Metal Matrix Composite Part I Plastic and Damage Mechanisms ," CR-1890095, NASA, 1992.

[209] M. Johansson, and K. Runesson, "Viscoplastic with Dynamic Yield Surface Coupled with Damage

," Computational Mechanics Vol. 20, pp. 53-59, 1997.

[210] G.Z. Voyiadjis, and T Park, "Anisotropic Damage Effect Tensor for the Symmetrization of the Effective Stress Tensor ," Journal of Applied Mechanics," Vol. 64, pp. 106-110, 1997.

[211] G.Z. Voyiadjis, and T. Park, "Kinematic of Damage for Finite Strain Plasticity ,"International Journal of Engineering Science, Vol. 0, pp. 1-28, 1999

[212] D.J. Bammann and E.C. Aifantis, "A Damage Model for Ductile Metals ," Neclear Engineering and Design, Vol. 116, pp. 355-362, 1989.

[213] I. Doghri, "Fully Implicit Integration and Consistent Tangent Modules in Elasto-Plasticity ," Mechanics of Solid Materials, Vol. 55, pp. 59-64, 1988.

[214] J.L. Chaboche, "Cyclic Viscoplastic Equations Part I Thermodynamically Consistent Formulation ," Journal of Applied Mechanics, Vol. 60, pp.81 3-821, 1993.

[215] G.Z. Voyiadjis, and G. Thiagarajan, "Micro and Macro Anisotropic Cyclic Damage Plasticity Models for MMCS ," International Journal of Engineering Science," Vol.35, pp. 467-484, 1997.

[216] G.Z. Voyiadjis, and I.N. Basuroychowdury, "A Plasticity Model for Multiaxial Cyclic Loading and Ratcheting," Acta Mechanica, Vol.126, pp.19-35 1998

[217] P. Perzyna,"The Constitutive Equations for Rate Sensitive Plastic Material," Applied Mathematics, Vol.20, pp.321-332, 1963

[218] P. Perzyna,"Thermodynamic Theory of Viscoplasticity,"Advances in Applied Mathematics, Vol.11, pp.313-345, 1971

[219] A.D. Freed, J.L. Chaboche and, K.P Walker ," A Viscoplasticity Theory with Thermodynamic Consideration," Acta Mechanica, Vol.90, pp.219-241, 1991

[220]   G.Z. Voyiadjis, and S.M. Sivakumar, "A Finite Strain and Rate Dependent Cyclic p Plasticity Model for Metals," In C. Teddosiu, J.L. Raphanel, and F. Sidoroff editors. Proceedings of the International Seminar MECAMAT91, Fontainnebleau, France on LARGE PLASTIC DEFORMATION, Fundamental Aspects and Application to Metal Forming, pp.353-360, 1992

[221] M.B. Rubin,"A Thermoelastic Viscoplastic Model with a Rate Dependent Yield Strength," Journal of Applied Mechanics, Vol.49, pp.305-311, 1982

[222] J. Betten,"Damage Tensor in Continuum Mechanics," Journal de Mechanique Theorique et Appliquee, Vol.2, pp.13-32, 1983

[223] G.Z. Voyiadjis, and B. Deliktas, " A Coupled Anisotropic Damage Models for the Inelastic response of Composite Materials," International Journal of Engineering Science," Vol.35, pp. 467-484, 1999.

[224] B. Budiansky, and R.J. O'Connell, "Elastic Moduli of Cracked Solids," International Journal of Solids and Structures," Vol.12, pp.81-97, 1976.

[225] J.G. Boyd, F. Costanza, and D.H. Allen, "A Micromechanics Approach for Constructing Locally Averaged Damage Dependent Constitutive Equations in Inelastic Composite," International Journal of Damage Mechanics, Vol.2, pp.209-228, 1993

[226] G. Z. Voyiadjis and L,N Mohammad,"Rate Equations for Viscoplastic Materials Subjected toi Finte Strain," International Journal of Solids and Structures," Vol.12, pp.81-97, 1976.

[227] V. A. Lubarda and D. Krojcinovic, "Some Fundamental Issues in Rate Theory of Damage Elastoplasticity," International Journal of Plasticity, Vol.11, pp.763-797, 1995

[228] J. W. Ju, "Isotropic and Aisotropic Damage Variables in Continuum Damage Mechanics," Journal of Engineering Mechanics, Vol.116, pp.2764-2770, 1990

[229] H. M. Zbib, "On the Mechanics of Large Inelastic Deformations: Kinematics and Constitutive Modeling," Acta Mechanica, Vol.96, pp.119-138, 1993

[230] Y. A. Bahei-El-Din, R. S. Shah, and G. J. Dvorak, "Numerical Analysis of the Rate Dependent Behavior of High Temperature Fibrous Composites," *Mechanics of Composite at Elevated Temperature*, The American Society of Mechanical Engineers, 118:67-78, 1991

[231] G. J. Voyiadjis, and P. I. Kattan, "On the Symmeterization of the Effective Stress Tensor in Continuum Damage Mechanics," Journal of the Mechanical Behavior of Materials, Vol.7, No.2, pp.139-165, 1996

Appendices

Listing of Damage Formulas

Appendix 1
Formulas for Chapter 7

The scalar functions of equations (7.20) are given by (see similar derivation in Chapter 6):

$$a_1 = \overline{Q} - \overline{E}_{klmn} \frac{\partial \overline{f}}{\partial \overline{\tau}_{kl}} \frac{\partial \overline{f}}{\partial \overline{\sigma}_{mn}} \tag{A1}$$

$$a_2 = Q - E_{klmn} \frac{\partial f}{\partial \tau_{kl}} \frac{\partial f}{\partial \tau_{mn}} \tag{A2}$$

$$a_3 = \frac{\partial f}{\partial \overline{\tau}_{kl}} \overline{E}_{klmn} dM_{mnpq}^{-T} E_{ijpq}^{-1} \sigma_{ij} \tag{A3}$$

where Q and $\overline{Q}$ are given by:

$$Q = E_{abcd} \frac{\partial f}{\partial \tau_{ab}} \frac{\partial f}{\partial \tau_{cd}} - b \frac{\partial f}{\partial \alpha_{ef}} (\tau_{ef} - \alpha_{ef}) \frac{\dfrac{\partial f}{\partial \sigma_{ij}} \dfrac{\partial f}{\partial \sigma_{ij}}}{(\tau_{gh} - \alpha_{gh}) \dfrac{\partial f}{\partial \sigma_{gh}}} \tag{A4}$$

$$\overline{Q} = \overline{E}_{abcd} \frac{\partial \overline{f}}{\partial \overline{\tau}_{ab}} \frac{\partial \overline{f}}{\partial \overline{\tau}_{cd}} - b \frac{\partial \overline{f}}{\partial \overline{\alpha}_{ef}} (\overline{\tau}_{ef} - \overline{\alpha}_{ef}) \frac{\dfrac{\partial \overline{f}}{\partial \overline{\sigma}_{ij}} \dfrac{\partial \overline{f}}{\partial \overline{\sigma}_{ij}}}{(\overline{\tau}_{gh} - \overline{\alpha}_{gh}) \dfrac{\partial \overline{f}}{\partial \overline{\sigma}_{gh}}} \tag{A5}$$

and b is a material constant assoiated with kinematic hardening as defined by equation (7.57).

The material time derivative $d\mathbf{M}^{-T}$ appearing in equation (7.19a) is given by:

$$dM_{klmn}^{-T} = - M_{ijkl}^{-T} dM_{ijpq}^{T} M_{pqmn}^{-T} \tag{A6}$$

where

$$dM_{ijpq} = \frac{\partial M_{ijpq}}{\partial \phi_{mn}} d\phi_{mn} \tag{A7}$$

Appendix   2
Formulas for Chapter 8

The damage transformation equations for the elastic and plastic parts of the overall strain rate tensor $d\overline{\varepsilon}$ are given by:

$$d\overline{\varepsilon}' = dM^{-T}:\varepsilon' + M^{-T}:d\varepsilon' \tag{A8}$$

$$d\overline{\varepsilon}'' = X:d\varepsilon'' + Z \tag{A9}$$

where the overall tensors X and Z are given by:

$$X = \frac{a_2}{a_1} M^{-1} \tag{A10}$$

$$Z = 3\frac{a_3}{a_2} M^{-1}:N:N:(\sigma - \beta) \tag{A11}$$

and the scalar coefficients a_1, a_2, a_3 are given by:

$$a_1 = -b\frac{\partial \overline{f}}{\partial \overline{\alpha}}:(\overline{\tau} - \overline{\alpha})\frac{\dfrac{\partial \overline{f}}{\partial \overline{\sigma}}:\dfrac{\partial \overline{f}}{\partial \overline{\sigma}}}{(\overline{\tau} - \overline{\alpha}):\dfrac{\partial \overline{f}}{\partial \overline{\sigma}}} \tag{A12}$$

$$a_2 = -b\frac{\partial f}{\partial \alpha}(\tau - \alpha):\frac{\dfrac{\partial f}{\partial \sigma}:\dfrac{\partial f}{\partial \sigma}}{(\tau - \alpha):\dfrac{\partial f}{\partial \sigma}} \tag{A13}$$

$$a_3 = \frac{\partial \overline{f}}{\partial \overline{\tau}}:\overline{E}:dM^{-T}:\sigma.E^{-1} \tag{A14}$$

In equations (A12) and (A13), b is a material constant associated with kinematic hardening. The material time derivative dM^{-T} appearing in equations (A8) and (A14) is given by:

$$dM^{-T} = -M^{-T}:dM^T:M^{-T} \tag{A15}$$

where

$$dM = \frac{\partial M}{\partial \phi} : d\phi \tag{A16}$$

The tensor equation $\dfrac{\partial M}{\partial \phi}$ is of the sixth rank and its components are given by $\dfrac{\partial M_{ijkl}}{\partial \phi_{mn}}$. The operation given in equation (A16) is defined by $a{:}b=\alpha_{ijklmmn}b_{mn}$ where a is a sixth-rank tensor and b is a second rank tensor.

In a way similar to the derivation of equation (A9), one can postulate the following damage transformation equation for the matrix plastic strain rate tensor $d\overline{\varepsilon}^{M''}$

$$d\overline{\varepsilon}^{\,M''} = X^{\,M} {:} d\varepsilon^{\,M''} + Z^{\,M} \tag{A17}$$

where the fourth-rank tensor X^M and the second-rank tensor Z^M can be defined in a way similar to the tensors X and Z, respectively. The remaining part of Appendix A-2 is devoted to deriving a relation between the matrix tensors X^M, Z^M and the overall tensors X and Z. Substituting equations (8.9a) and (8.11a) into equation (A17), simplifying and comparing the result with equation (A9), one obtains the desired transformation relations:

$$X = A^{\,M^{-1}} {:} X^{\,M} {:} A^{\,M} \tag{A18}$$

$$Z = Z^{\,M} {:} A^{\,M^{-1}} \tag{A19}$$

Alternatively, equations (A18) and (A19) can be written in terms of the matrix tensors X^M and Z^M, as follows:

$$X^{\,M} = A^{\,M} \cdot X {:} \overline{A}^{\,M^{-1}} \tag{A20}$$

$$Z^{\,M} = A^{\,M} {:} Z \tag{A21}$$

Appendix 3
Formulas for Chapter 11

The elements of the matrix [M] are listed below according to the implicit symmetrization method where $\psi_{ij} = \delta_{ij} - \phi_{ij}$. No attempt is made to simplify the expressions that follow.

$$
\begin{aligned}
M_{11} = (&\psi_{33}^3\psi_{22}^2 + \psi_{33}^3\psi_{22}\psi_{11} - \psi_{33}^3\phi_{12}^2 - 2\psi_{33}^2\phi_{23}^2\psi_{22} - \psi_{33}^2\phi_{23}^2\psi_{11} - 2\psi_{33}^2\phi_{23}\phi_{12}\phi_{13} \\
&+ \psi_{33}^2\psi_{22}^3 + 2\psi_{33}^2\psi_{22}^2\psi_{11} - \psi_{33}^2\psi_{22}\phi_{12}^2 - \psi_{33}^2\psi_{22}\phi_{13}^2 + \psi_{33}^2\psi_{22}\psi_{11}^2 - \psi_{33}^2\phi_{12}^2\psi_{11} \\
&+ \psi_{33}\phi_{23}^4 - 2\psi_{33}\phi_{23}^2\psi_{22}^2 - 2\psi_{33}\phi_{23}^2\psi_{22}\psi_{11} - \psi_{33}\phi_{23}^2\phi_{12}^2 - \psi_{33}\phi_{23}^2\psi_{11}^2 - 2\psi_{33}\phi_{23}\psi_{22}\phi_{12}\phi_{13} \\
&- 2\psi_{33}\phi_{23}\phi_{12}\phi_{13}\psi_{11} + \psi_{33}\psi_{22}^3\psi_{11} - \psi_{33}\psi_{22}^2\phi_{12}^2 - \psi_{33}\psi_{22}^2\phi_{13}^2 + \psi_{33}\psi_{22}^2\psi_{11}^2 - 2\psi_{33}\psi_{22}\phi_{12}^2\psi_{11} \\
&- 2\psi_{33}\psi_{22}\phi_{13}^2\psi_{11} + \psi_{33}\phi_{12}^4 + \psi_{33}\phi_{12}^2\phi_{13}^2 + \phi_{23}^4\psi_{22} - 2\phi_{23}^3\phi_{12}\phi_{13} - \phi_{23}^2\psi_{22}^2\psi_{11} - \phi_{23}^2\psi_{22}\phi_{13}^2 \\
&- \phi_{23}^2\psi_{22}\psi_{11}^2 + \phi_{23}^2\phi_{12}^2\psi_{11} + \phi_{23}^2\phi_{13}^2\psi_{11} - 2\phi_{23}\psi_{22}^2\phi_{12}\phi_{13} - 2\phi_{23}\psi_{22}\phi_{12}\phi_{13}\psi_{11} + 2\phi_{23}\phi_{12}^3\phi_{13} \\
&+ 2\phi_{23}\phi_{12}\phi_{13}^3 - \psi_{22}^3\phi_{13}^2 - \psi_{22}^2\phi_{13}^2\psi_{11} + \psi_{22}\phi_{12}^2\phi_{13}^2 + \psi_{22}\phi_{13}^4)/8\Delta
\end{aligned}
$$

$$
\begin{aligned}
M_{12} = (&\psi_{33}^3\phi_{12}^2 + 2\psi_{33}^2\phi_{23}\phi_{12}\phi_{13} + \psi_{33}^2\psi_{22}\phi_{12}^2 + \psi_{33}^2\phi_{12}^2\psi_{11} + \psi_{33}\phi_{23}^2\phi_{13}^2 + 2\psi_{33}\phi_{23}\psi_{22}\phi_{12}\phi_{13} \\
&+ 2\psi_{33}\phi_{23}\phi_{12}\phi_{13}\psi_{11} + \psi_{33}\psi_{22}\phi_{12}^2\psi_{11} - \psi_{33}\phi_{12}^4 + \phi_{23}^2\psi_{22}\phi_{13}^2 - \phi_{23}^2\phi_{12}^2\psi_{11} + \phi_{23}^2\phi_{13}^2\psi_{11} \\
&- 2\phi_{23}\phi_{12}^3\phi_{13} - \psi_{22}\phi_{12}^2\phi_{13}^2)/8\Delta
\end{aligned}
$$

$$
\begin{aligned}
M_{13} = (&\psi_{33}\phi_{23}^2\phi_{12}^2 + 2\psi_{33}\phi_{23}\psi_{22}\phi_{12}\phi_{13} + \psi_{33}\psi_{22}^2\phi_{13}^2 + \psi_{33}\psi_{22}\phi_{13}^2\psi_{11} - \psi_{33}\phi_{12}^2\phi_{13}^2 + \phi_{23}^2\psi_{22}\phi_{12}^2 \\
&+ \phi_{23}^2\phi_{12}^2\psi_{11} - \phi_{23}^2\phi_{13}^2\psi_{11} + 2\phi_{23}\psi_{22}^2\phi_{12}\phi_{13} + 2\phi_{23}\psi_{22}\phi_{12}\phi_{13}\psi_{11} - 2\phi_{23}\phi_{12}\phi_{13}^3 + \psi_{22}^3\phi_{13}^2 \\
&+ \psi_{22}^2\phi_{13}^2\psi_{11} - \psi_{22}\phi_{13}^4)/8\Delta
\end{aligned}
$$

$$
\begin{aligned}
M_{14} = (&\psi_{33}^2\phi_{23}\phi_{12}^2 + \psi_{33}^2\psi_{22}\phi_{12}\phi_{13} + \psi_{33}\phi_{23}^2\phi_{12}\phi_{13} + \psi_{33}\phi_{23}\psi_{22}\phi_{12}^2 + \psi_{33}\phi_{23}\psi_{22}\phi_{13}^2 + \psi_{33}\phi_{23}\phi_{12}^2\psi_{11} \\
&+ \psi_{33}\psi_{22}^2\phi_{12}\phi_{13} + 2\psi_{33}\psi_{22}\phi_{12}\phi_{13}\psi_{11} - \psi_{33}\phi_{12}^3\phi_{13} + \phi_{23}^2\psi_{22}\phi_{12}\phi_{13} + \phi_{23}\psi_{22}^2\phi_{13}^2 + \phi_{23}\psi_{22}\phi_{13}^2\psi_{11} \\
&- 2\phi_{23}\phi_{12}^2\phi_{13}^2 - \psi_{22}\phi_{12}\phi_{13}^3)/4\Delta
\end{aligned}
$$

$$
\begin{aligned}
M_{15} = (&\psi_{33}^2\phi_{23}\psi_{22}\phi_{12} + \psi_{33}^2\psi_{22}^2\phi_{13} + \psi_{33}^2\psi_{22}\phi_{13}\psi_{11} - \psi_{33}^2\phi_{12}^2\phi_{13} - \psi_{33}\phi_{23}^3\phi_{12} - \psi_{33}\phi_{23}^2\psi_{22}\phi_{13} \\
&- \psi_{33}\phi_{23}^2\phi_{13}\psi_{11} + \psi_{33}\phi_{23}\psi_{22}^2\phi_{12} + \psi_{33}\phi_{23}\phi_{12}^3 - 2\psi_{33}\phi_{23}\phi_{12}\phi_{13}^2 + \psi_{33}\psi_{22}^3\phi_{13} + \psi_{33}\psi_{22}^2\phi_{13}\psi_{11} \\
&- \psi_{33}\psi_{22}\phi_{13}^3 - \phi_{23}^3\psi_{22}\phi_{12} - \phi_{23}^2\psi_{22}^2\phi_{13} - \phi_{23}^2\psi_{22}\phi_{13}\psi_{11} + 2\phi_{23}^2\phi_{12}^2\phi_{13} + \phi_{23}\psi_{22}\phi_{12}\phi_{13}^2)/4\Delta
\end{aligned}
$$

$$
\begin{aligned}
M_{16} = (&\psi_{33}^3\psi_{22}\phi_{12} - \psi_{33}^2\phi_{23}^2\phi_{12} + \psi_{33}^2\phi_{23}\psi_{22}\phi_{13} + \psi_{33}^2\psi_{22}^2\phi_{12} + \psi_{33}^2\psi_{22}\phi_{12}\psi_{11} - \psi_{33}\phi_{23}^3\phi_{13} \\
&- \psi_{33}\phi_{23}^2\psi_{22}\phi_{12} - \psi_{33}\phi_{23}^2\phi_{12}\psi_{11} + \psi_{33}\phi_{23}\psi_{22}^2\phi_{13} + \psi_{33}\phi_{23}\phi_{12}^2\phi_{13} + \psi_{33}\psi_{22}^2\phi_{12}\psi_{11} \\
&- \psi_{33}\psi_{22}\phi_{12}^3 - \phi_{23}^3\psi_{22}\phi_{13} - \phi_{23}^2\psi_{22}\phi_{12}\psi_{11} + 2\phi_{23}^2\phi_{12}\phi_{13}^2 - 2\phi_{23}\psi_{22}\phi_{12}^2\phi_{13} + \phi_{23}\psi_{22}\phi_{13}^3 \\
&- \psi_{22}^2\phi_{12}\phi_{13}^2)/4\Delta
\end{aligned}
$$

$$M_{21} = M_{12}$$

$$\begin{aligned}
M_{22} = (&\psi_{33}^3\psi_{22}\psi_{11} - \psi_{33}^3\phi_{12}^2 + \psi_{33}^3\psi_{11}^2 - \psi_{33}^2\phi_{23}^2\psi_{11} - 2\psi_{33}^2\phi_{23}\phi_{12}\phi_{13} + \psi_{33}^2\psi_{22}^2\psi_{11} \\
&- \psi_{33}^2\psi_{22}\phi_{12}^2 - \psi_{33}^2\psi_{22}\phi_{13}^2 + 2\psi_{33}^2\psi_{22}\psi_{11}^2 - \psi_{33}^2\phi_{12}^2\psi_{11} - 2\psi_{33}^2\phi_{13}^2\psi_{11} + \psi_{33}^2\psi_{11}^3 \\
&- 2\psi_{33}\phi_{23}^2\psi_{22}\psi_{11} + \psi_{33}\phi_{23}^2\phi_{12}^2 - \psi_{33}\phi_{23}^2\psi_{11}^2 - 2\psi_{33}\phi_{23}\psi_{22}\phi_{12}\phi_{13} - 2\psi_{33}\phi_{23}\phi_{12}\phi_{13}\psi_{11} \\
&- \psi_{33}\psi_{22}^2\phi_{13}^2 + \psi_{33}\psi_{22}^2\psi_{11}^2 - 2\psi_{33}\psi_{22}\phi_{12}^2\psi_{11} - 2\psi_{33}\psi_{22}\phi_{13}^2\psi_{11} + \psi_{33}\psi_{22}\psi_{11}^3 + \psi_{33}\phi_{12}^4 \\
&- \psi_{33}\phi_{12}^2\phi_{13}^2 - \psi_{33}\phi_{12}^2\psi_{11}^2 + \psi_{33}\phi_{13}^4 - 2\psi_{33}\phi_{13}^2\psi_{11}^2 + \phi_{23}^4\psi_{11} + 2\phi_{23}^3\phi_{12}\phi_{13} + \phi_{23}^2\psi_{22}\phi_{13}^2 \\
&- \phi_{23}^2\psi_{22}\psi_{11}^2 + \phi_{23}^2\phi_{12}^2\psi_{11} - \phi_{23}^2\phi_{13}^2\psi_{11} - \phi_{23}^2\psi_{11}^3 - 2\phi_{23}\psi_{22}\phi_{12}\phi_{13}\psi_{11} + 2\phi_{23}\phi_{12}^3\phi_{13} \\
&- 2\phi_{23}\phi_{12}\phi_{13}^3 - 2\phi_{23}\phi_{12}\phi_{13}\psi_{11}^2 - \psi_{22}^2\phi_{13}^2\psi_{11} + \psi_{22}\phi_{12}^2\phi_{13}^2 - \psi_{22}\phi_{13}^2\psi_{11}^2 \\
&+ \phi_{13}^4\psi_{11})/8\Delta
\end{aligned}$$

$$\begin{aligned}
M_{23} = (&\psi_{33}\phi_{23}^2\psi_{22}\psi_{11} - \psi_{33}\phi_{23}^2\phi_{12}^2 + \psi_{33}\phi_{23}^2\psi_{11}^2 + 2\psi_{33}\phi_{23}\phi_{12}\phi_{13}\psi_{11} + \psi_{33}\phi_{12}^2\phi_{13}^2 - \phi_{23}^4\psi_{11} \\
&- 2\phi_{23}^3\phi_{12}\phi_{13} - \phi_{23}^2\psi_{22}\phi_{13}^2 + \phi_{23}^2\psi_{22}\psi_{11}^2 + \phi_{23}^2\psi_{11}^3 + 2\phi_{23}\psi_{22}\phi_{12}\phi_{13}\psi_{11} + 2\phi_{23}\phi_{12}\phi_{13}\psi_{11}^2 \\
&+ \psi_{22}\phi_{12}^2\phi_{13}^2 + \phi_{12}^2\phi_{13}^2\psi_{11})/8\Delta
\end{aligned}$$

$$\begin{aligned}
M_{24} = (&\psi_{33}^2\phi_{23}\psi_{22}\psi_{11} - \psi_{33}^2\phi_{23}\phi_{12}^2 + \psi_{33}^2\phi_{23}\psi_{11}^2 + \psi_{33}^2\phi_{12}\phi_{13}\psi_{11} - \psi_{33}\phi_{23}^3\psi_{11} - 2\psi_{33}\phi_{23}^2\phi_{12}\phi_{13} \\
&- \psi_{33}\phi_{23}\psi_{22}\phi_{13}^2 + \psi_{33}\phi_{23}\psi_{22}\psi_{11}^2 - \psi_{33}\phi_{23}\phi_{13}^2\psi_{11} + \psi_{33}\phi_{23}\psi_{11}^3 + \psi_{33}\phi_{12}^3\phi_{13} - \psi_{33}\phi_{12}\phi_{13}^3 \\
&+ \psi_{33}\phi_{12}\phi_{13}\psi_{11}^2 + \phi_{23}^2\phi_{12}\phi_{13}\psi_{11} - \phi_{23}\psi_{22}\phi_{13}^2\psi_{11} + 2\phi_{23}\phi_{12}^2\phi_{13}^2 - \phi_{23}\phi_{13}^2\psi_{11}^2 - \phi_{12}\phi_{13}^3\psi_{11})/4\Delta
\end{aligned}$$

$$\begin{aligned}
M_{25} = (&\psi_{33}^2\phi_{23}\phi_{12}\psi_{11} + \psi_{33}^2\phi_{12}^2\phi_{13} + \psi_{33}\phi_{23}^2\phi_{13}\psi_{11} + 2\psi_{33}\phi_{23}\psi_{22}\phi_{12}\psi_{11} - \psi_{33}\phi_{23}\phi_{12}^3 + \psi_{33}\phi_{23}\phi_{12}\phi_{13}^2 \\
&+ \psi_{33}\phi_{23}\phi_{12}\psi_{11}^2 + \psi_{33}\psi_{22}\phi_{12}^2\phi_{13} + \psi_{33}\phi_{12}^2\phi_{13}\psi_{11} - \phi_{23}^3\phi_{12}\psi_{11} + \phi_{23}^2\psi_{22}\phi_{13}\psi_{11} - 2\phi_{23}^2\phi_{12}^2\phi_{13} \\
&+ \phi_{23}^2\phi_{13}\psi_{11}^2 + \phi_{23}\phi_{12}\phi_{13}^2\psi_{11})/4\Delta
\end{aligned}$$

$$\begin{aligned}
M_{26} = (&\psi_{33}^3\phi_{12}\psi_{11} + \psi_{33}^2\phi_{23}\phi_{13}\psi_{11} + \psi_{33}^2\psi_{22}\phi_{12}\psi_{11} - \psi_{33}^2\phi_{12}\phi_{13}^2 + \psi_{33}^2\phi_{12}\psi_{11}^2 + \psi_{33}\phi_{23}\phi_{12}^2\phi_{13} \\
&- \psi_{33}\phi_{23}\phi_{13}^3 + \psi_{33}\phi_{23}\phi_{13}\psi_{11}^2 - \psi_{33}\psi_{22}\phi_{12}\phi_{13}^2 + \psi_{33}\psi_{22}\phi_{12}\psi_{11}^2 - \psi_{33}\phi_{12}^3\psi_{11} - \psi_{33}\phi_{12}\phi_{13}^2\psi_{11} \\
&+ \phi_{23}^3\phi_{13}\psi_{11} + 2\phi_{23}^2\phi_{12}\phi_{13}^2 - \phi_{23}^2\phi_{12}\psi_{11}^2 - 2\phi_{23}\phi_{12}^2\phi_{13}\psi_{11} - \phi_{23}\phi_{13}^3\psi_{11} - \psi_{22}\phi_{12}\phi_{13}^2\psi_{11})/4\Delta
\end{aligned}$$

$$M_{31} = M_{13}$$

$$M_{32} = M_{23}$$

$$\begin{aligned}
M_{33} = (&\psi_{33}^2\psi_{22}^2\psi_{11} - \psi_{33}^2\psi_{22}\phi_{12}^2 + \psi_{33}^2\psi_{22}\psi_{11}^2 - \psi_{33}^2\phi_{12}^2\psi_{11} - 2\psi_{33}\phi_{23}^2\psi_{22}\psi_{11} + \psi_{33}\phi_{23}^2\phi_{12}^2 \\
&- \psi_{33}\phi_{23}^2\psi_{11}^2 - 2\psi_{33}\phi_{23}\psi_{22}\phi_{12}\phi_{13} - 2\psi_{33}\phi_{23}\phi_{12}\phi_{13}\psi_{11} + \psi_{33}\psi_{22}^3\psi_{11} - \psi_{33}\psi_{22}^2\phi_{12}^2 - \psi_{33}\psi_{22}^2\phi_{13}^2 \\
&+ 2\psi_{33}\psi_{22}^2\psi_{11}^2 - 2\psi_{33}\psi_{22}\phi_{12}^2\psi_{11} - 2\psi_{33}\psi_{22}\phi_{13}^2\psi_{11} + \psi_{33}\psi_{22}\psi_{11}^3 + \psi_{33}\phi_{12}^2\phi_{13}^2 - \psi_{33}\phi_{12}^2\psi_{11}^2 \\
&+ \phi_{23}^4\psi_{11} + 2\phi_{23}^3\phi_{12}\phi_{13} - \phi_{23}^2\psi_{22}^2\psi_{11} + \phi_{23}^2\psi_{22}\phi_{13}^2 - \phi_{23}^2\psi_{22}\psi_{11}^2 - \phi_{23}^2\phi_{12}^2\psi_{11} \\
&+ \phi_{23}^2\phi_{13}^2\psi_{11} - \phi_{23}^2\psi_{11}^3 - 2\phi_{23}\psi_{22}^2\phi_{12}\phi_{13} - 2\phi_{23}\psi_{22}\phi_{12}\phi_{13}\psi_{11} - 2\phi_{23}\phi_{12}^3\phi_{13} + 2\phi_{23}\phi_{12}\phi_{13}^3 \\
&- 2\phi_{23}\phi_{12}\phi_{13}\psi_{11}^2 - \psi_{22}^3\phi_{13}^2 + \psi_{22}^3\psi_{11}^2 - 2\psi_{22}^2\phi_{12}^2\psi_{11} - \psi_{22}^2\phi_{13}^2\psi_{11} + \psi_{22}^2\psi_{11}^3 \\
&+ \psi_{22}\phi_{12}^4 - \psi_{22}\phi_{12}^2\phi_{13}^2 - 2\psi_{22}\phi_{12}^2\psi_{11}^2 + \psi_{22}\phi_{13}^4 - \psi_{22}\phi_{13}^2\psi_{11}^2 \\
&+ \phi_{12}^4\psi_{11})/8\Delta
\end{aligned}$$

$$M_{34} = (\psi_{33}\phi_{23}\psi_{22}^2\psi_{11} - \psi_{33}\phi_{23}\psi_{22}\phi_{12}^2 + \psi_{33}\phi_{23}\psi_{22}\psi_{11}^2 - \psi_{33}\phi_{23}\phi_{12}^2\psi_{11} - \phi_{23}^3\psi_{22}\psi_{11} - 2\phi_{23}^2\psi_{22}\phi_{12}\phi_{13}$$
$$+ \phi_{23}^2\phi_{12}\phi_{13}\psi_{11} - \phi_{23}\psi_{22}^2\phi_{13}^2 + \phi_{23}\psi_{22}^2\psi_{11}^2 - \phi_{23}\psi_{22}\phi_{12}^2\psi_{11} + \phi_{23}\psi_{22}\psi_{11}^3 + 2\phi_{23}\phi_{12}^2\phi_{13}^2$$
$$- \phi_{23}\phi_{12}^2\psi_{11}^2 + \psi_{22}^2\phi_{12}\phi_{13}\psi_{11} - \psi_{22}\phi_{12}^3\phi_{13} + \psi_{22}\phi_{12}\phi_{13}^3 + \psi_{22}\phi_{12}\phi_{13}\psi_{11}^2 - \phi_{12}^3\phi_{13}\psi_{11})/4\Delta$$

$$M_{35} = (\psi_{33}\psi_{22}^2\phi_{13}\psi_{11} - \psi_{33}\psi_{22}\phi_{12}^2\phi_{13} + \psi_{33}\psi_{22}\phi_{13}\psi_{11}^2 - \psi_{33}\phi_{12}^2\phi_{13}\psi_{11} + \phi_{23}^3\phi_{12}\psi_{11} + 2\phi_{23}^2\phi_{12}^2\phi_{13}$$
$$- \phi_{23}^2\phi_{13}\psi_{11}^2 + \phi_{23}\psi_{22}^2\phi_{12}\psi_{11} - \phi_{23}\psi_{22}\phi_{12}^3 + \phi_{23}\psi_{22}\phi_{12}\phi_{13}^2 + \phi_{23}\psi_{22}\phi_{12}\psi_{11}^2 - \phi_{23}\phi_{12}^3\psi_{11}$$
$$- 2\phi_{23}\phi_{12}\phi_{13}^2\psi_{11} + \psi_{22}^3\phi_{13}\psi_{11} - \psi_{22}^2\phi_{12}^2\phi_{13} + \psi_{22}^2\phi_{13}\psi_{11}^2 - \psi_{22}\phi_{12}^2\phi_{13}\psi_{11} - \psi_{22}\phi_{13}^3\psi_{11})/4\Delta$$

$$M_{36} = (\psi_{33}\phi_{23}^2\phi_{12}\psi_{11} + 2\psi_{33}\phi_{23}\psi_{22}\phi_{13}\psi_{11} + \psi_{33}\psi_{22}\phi_{12}\phi_{13}^2 - \phi_{23}^3\phi_{13}\psi_{11} + \phi_{23}^2\psi_{22}\phi_{12}\psi_{11} - 2\phi_{23}^2\phi_{12}\phi_{13}^2$$
$$+ \phi_{23}^2\phi_{12}\psi_{11}^2 + \phi_{23}\psi_{22}^2\phi_{13}\psi_{11} + \phi_{23}\psi_{22}\phi_{12}^2\phi_{13} - \phi_{23}\psi_{22}\phi_{13}^3 + \phi_{23}\psi_{22}\phi_{13}\psi_{11}^2 + \phi_{23}\phi_{12}^2\phi_{13}\psi_{11}$$
$$+ \psi_{22}^2\phi_{12}\phi_{13}^2 + \psi_{22}\phi_{12}\phi_{13}^2\psi_{11})/4\Delta$$

$$M_{41} = M_{14}/2$$

$$M_{42} = M_{24}/2$$

$$M_{43} = M_{34}/2$$

$$M_{44} = (\psi_{33}^2\psi_{22}^2\psi_{11} - \psi_{33}^2\psi_{22}\phi_{12}^2 + \psi_{33}^2\psi_{22}\psi_{11}^2 - \psi_{33}^2\phi_{12}^2\psi_{11} - \psi_{33}\phi_{23}^2\psi_{22}\psi_{11} - 2\psi_{33}\phi_{23}\psi_{22}\phi_{12}\phi_{13}$$
$$- \psi_{33}\psi_{22}^2\phi_{13}^2 + \psi_{33}\psi_{22}^2\psi_{11}^2 - \psi_{33}\psi_{22}\phi_{12}^2\psi_{11} - \psi_{33}\psi_{22}\phi_{13}^2\psi_{11} + \psi_{33}\psi_{22}\psi_{11}^3 + \psi_{33}\phi_{12}^2\phi_{13}^2$$
$$- \psi_{33}\phi_{12}^2\psi_{11}^2 - \psi_{22}^2\phi_{13}^2\psi_{11} + \psi_{22}\phi_{12}^2\phi_{13}^2 - \psi_{22}\phi_{13}^2\psi_{11}^2 + \phi_{12}^2\phi_{13}^2\psi_{11})/4\Delta$$

$$M_{45} = (\psi_{33}\phi_{23}^2\phi_{12}\psi_{11} + \psi_{33}\phi_{23}\psi_{22}\phi_{13}\psi_{11} + \psi_{33}\phi_{23}\phi_{12}^2\phi_{13} + \psi_{33}\psi_{22}^2\phi_{12}\psi_{11} - \psi_{33}\psi_{22}\phi_{12}^3$$
$$+ \psi_{33}\psi_{22}\phi_{12}\phi_{13}^2 + \psi_{33}\psi_{22}\phi_{12}\psi_{11}^2 - \psi_{33}\phi_{12}^3\psi_{11} + \phi_{23}\psi_{22}^2\phi_{13}\psi_{11} - \phi_{23}\psi_{22}\phi_{12}^2\phi_{13}$$
$$+ \phi_{23}\psi_{22}\phi_{13}\psi_{11}^2 - \phi_{23}\phi_{12}^2\phi_{13}\psi_{11})/4\Delta$$

$$M_{46} = (\psi_{33}^2\phi_{23}\phi_{12}\psi_{11} + \psi_{33}^2\psi_{22}\phi_{13}\psi_{11} + \psi_{33}\phi_{23}\psi_{22}\phi_{12}\psi_{11} - \psi_{33}\phi_{23}\phi_{12}\phi_{13}^2 + \psi_{33}\phi_{23}\phi_{12}\psi_{11}^2$$
$$+ \psi_{33}\psi_{22}\phi_{12}^2\phi_{13} - \psi_{33}\psi_{22}\phi_{13}^3 + \psi_{33}\psi_{22}\phi_{13}\psi_{11}^2 + \phi_{23}^2\psi_{22}\phi_{13}\psi_{11} + \phi_{23}\psi_{22}\phi_{12}\phi_{13}^2$$
$$- \phi_{23}\phi_{12}\phi_{13}^2\psi_{11} - \psi_{22}\phi_{13}^3\psi_{11})/4\Delta$$

$$M_{51} = M_{15}/2$$

$$M_{52} = M_{25}/2$$

$$M_{53} = M_{35}/2$$

$$M_{54} = M_{45}$$

$$M_{55} = (\psi_{33}^2\psi_{22}^2\psi_{11} - \psi_{33}^2\psi_{22}\phi_{12}^2 + \psi_{33}^2\psi_{22}\psi_{11}^2 - \psi_{33}^2\phi_{12}^2\psi_{11} - \psi_{33}\phi_{23}^2\psi_{22}\psi_{11} + \psi_{33}\phi_{23}^2\phi_{12}^2$$
$$- \psi_{33}\phi_{23}^2\psi_{11}^2 - 2\psi_{33}\phi_{23}\phi_{12}\phi_{13}\psi_{11} + \psi_{33}\psi_{22}^3\psi_{11} - \psi_{33}\psi_{22}^2\phi_{12}^2 + \psi_{33}\psi_{22}^2\psi_{11}^2 - \psi_{33}\psi_{22}\phi_{12}^2\psi_{11}$$
$$- \psi_{33}\psi_{22}\phi_{13}^2\psi_{11} - \phi_{23}^2\psi_{22}^2\psi_{11} + \phi_{23}^2\psi_{22}\phi_{12}^2 - \phi_{23}^2\psi_{22}\psi_{11}^2 + \phi_{23}^2\phi_{12}^2\psi_{11})/4\Delta$$

$$M_{56} = (\psi_{33}^2\phi_{23}\psi_{22}\psi_{11} + \psi_{33}^2\psi_{22}\phi_{12}\phi_{13} - \psi_{33}\phi_{23}^3\psi_{11} - \psi_{33}\phi_{23}^2\phi_{12}\phi_{13} + \psi_{33}\phi_{23}\psi_{22}^2\psi_{11} + \psi_{33}\phi_{23}\phi_{12}^2\psi_{11}$$
$$+ \psi_{33}\psi_{22}^2\phi_{12}\phi_{13} + \psi_{33}\psi_{22}\phi_{12}\phi_{13}\psi_{11} - \phi_{23}^3\psi_{22}\psi_{11} - \phi_{23}^2\psi_{22}\phi_{12}\phi_{13} + \phi_{23}^2\phi_{12}\phi_{13}\psi_{11}$$
$$+ \phi_{23}\psi_{22}\phi_{13}^2\psi_{11})/4\Delta$$

$$M_{61} = M_{16}/2$$

$$M_{62} = M_{26}/2$$

$$M_{63} = M_{36}/2$$

$$M_{64} = M_{46}$$

$$M_{65} = M_{56}$$

$$M_{66} = (\psi_{33}^3\psi_{22}\psi_{11} - \psi_{33}^2\phi_{23}^2\psi_{11} + \psi_{33}^2\psi_{22}^2\psi_{11} - \psi_{33}^2\psi_{22}\phi_{13}^2 + \psi_{33}^2\psi_{22}\psi_{11}^2 - \psi_{33}\phi_{23}^2\psi_{22}\psi_{11}$$
$$+ \psi_{33}\phi_{23}^2\phi_{13}^2 - \psi_{33}\phi_{23}^2\psi_{11}^2 - \psi_{33}\psi_{22}^2\phi_{13}^2 + \psi_{33}\psi_{22}^2\psi_{11}^2 - \psi_{33}\psi_{22}\phi_{12}^2\psi_{11} - \psi_{33}\psi_{22}\phi_{13}^2\psi_{11}$$
$$+ \phi_{23}^2\psi_{22}\phi_{13}^2 - \phi_{23}^2\psi_{22}\psi_{11}^2 + \phi_{23}^2\phi_{13}^2\psi_{11} - 2\phi_{23}\psi_{22}\phi_{12}\phi_{13}\psi_{11} - \psi_{22}^2\phi_{13}^2\psi_{11})/4\Delta$$

$$\Delta = \psi_{33}^3\psi_{22}^2\psi_{11} - \psi_{33}^3\psi_{22}\phi_{12}^2 + \psi_{33}^3\psi_{22}\psi_{11}^2 - \psi_{33}^3\phi_{12}^2\psi_{11} - 2\psi_{33}^2\phi_{23}^2\psi_{22}\psi_{11} + \psi_{33}^2\phi_{23}^2\phi_{12}^2 - \psi_{33}^2\phi_{23}^2\psi_{11}^2$$
$$- 2\psi_{33}^2\phi_{23}\psi_{22}\phi_{12}\phi_{13} - 2\psi_{33}^2\phi_{23}\phi_{12}\phi_{13}\psi_{11} + \psi_{33}^2\psi_{22}^3\psi_{11} - \psi_{33}^2\psi_{22}^2\phi_{12}^2 - \psi_{33}^2\psi_{22}^2\phi_{13}^2 + 2\psi_{33}^2\psi_{22}^2\psi_{11}^2$$
$$- 2\psi_{33}^2\psi_{22}\phi_{12}^2\psi_{11} - 2\psi_{33}^2\psi_{22}\phi_{13}^2\psi_{11} + \psi_{33}^2\psi_{22}\psi_{11}^3 + \psi_{33}^2\phi_{12}^2\phi_{13}^2 - \psi_{33}^2\phi_{12}^2\psi_{11}^2 + \psi_{33}\phi_{23}^4\psi_{11} + 2\psi_{33}\phi_{23}^3\phi_{12}\phi_{13}$$
$$- 2\psi_{33}\phi_{23}^2\psi_{22}^2\psi_{11} + \psi_{33}\phi_{23}^2\psi_{22}\phi_{12}^2 + \psi_{33}\phi_{23}^2\psi_{22}\phi_{13}^2 - 2\psi_{33}\phi_{23}^2\psi_{22}\psi_{11}^2 + \psi_{33}\phi_{23}^2\phi_{13}^2\psi_{11} - \psi_{33}\phi_{23}^2\psi_{11}^3$$
$$- 2\psi_{33}\phi_{23}\psi_{22}^2\phi_{12}\phi_{13} - 2\psi_{33}\phi_{23}\psi_{22}\phi_{12}\phi_{13}\psi_{11} - 2\psi_{33}\phi_{23}\phi_{12}^3\phi_{13} + 2\psi_{33}\phi_{23}\phi_{12}\phi_{13}^3 - 2\psi_{33}\phi_{23}\phi_{12}\phi_{13}\psi_{11}^2$$
$$- \psi_{33}\psi_{22}^3\phi_{13}^2 + \psi_{33}\psi_{22}^3\psi_{11}^2 - 2\psi_{33}\psi_{22}^2\phi_{12}^2\psi_{11} - 2\psi_{33}\psi_{22}^2\phi_{13}^2\psi_{11} + \psi_{33}\psi_{22}^2\psi_{11}^3 + \psi_{33}\psi_{22}\phi_{12}^4$$
$$- 2\psi_{33}\psi_{22}\phi_{12}^2\psi_{11}^2 + \psi_{33}\psi_{22}\phi_{13}^4 - 2\psi_{33}\psi_{22}\phi_{13}^2\psi_{11}^2 + \psi_{33}\phi_{12}^4\psi_{11} + \psi_{33}\phi_{12}^2\phi_{13}^2\psi_{11} + \phi_{23}^4\psi_{22}\psi_{11}$$
$$+ 2\phi_{23}^3\psi_{22}\phi_{12}\phi_{13} - 2\phi_{23}^3\phi_{12}\phi_{13}\psi_{11} + \phi_{23}^2\psi_{22}^2\phi_{13}^2 - \phi_{23}^2\psi_{22}^2\psi_{11}^2 + \phi_{23}^2\psi_{22}\phi_{12}^2\psi_{11} - \phi_{23}^2\psi_{22}\psi_{11}^3 - 4\phi_{23}^2\phi_{12}^2\phi_{13}^2$$
$$+ \phi_{23}^2\phi_{12}^2\psi_{11}^2 + \phi_{23}^2\phi_{13}^2\psi_{11}^2 - 2\phi_{23}\psi_{22}^2\phi_{12}\phi_{13}\psi_{11} + 2\phi_{23}\psi_{22}\phi_{12}^3\phi_{13} - 2\phi_{23}\psi_{22}\phi_{12}\phi_{13}^3 - 2\phi_{23}\psi_{22}\phi_{12}\phi_{13}\psi_{11}^2$$
$$+ 2\phi_{23}\phi_{12}^3\phi_{13}\psi_{11} + 2\phi_{23}\phi_{12}\phi_{13}^3\psi_{11} - \psi_{22}^3\phi_{13}^2\psi_{11} + \psi_{22}^2\phi_{12}^2\phi_{13}^2 - \psi_{22}^2\phi_{13}^2\psi_{11}^2 + \psi_{22}\phi_{12}^2\phi_{13}^2\psi_{11} + \psi_{22}\phi_{13}^4\psi_{11}$$

The equations in this part of the Appendix are not numbered (Voyiadjis and Kattan,[231]).

Appendix 4
Formulas for Chapter 12

In this Appendix, the expressions for the elastic and plastic stress and strain concentration tensors are given based on the Mori-Tanaka method. In this method, the equivalence principle of Eshelby is used to find the stress in the fibers. According to this principle, the stress field in the fiber material (inhomogeneity) is the same as the stress field of an equivalent inclusion that has the same material properties as the matrix. The reader is referred to Weng [137], Mori and Tanaka [106]. Gavazzi and Lagoudas [107], and Lagoudas et al.[108] for more details. The final expressions for the elastic stress and strain concentration tensors $\overline{B}^F$ and $\overline{A}^F$, respectively, for the fibers are given by:

$$\overline{A}^F = [I_4 + \overline{c}^M S : \overline{E}^{M-1} : (\overline{E}^F - \overline{E}^M)]^{-1} \tag{A22}$$

$$\overline{B}^F = [I_4 + \overline{c}^M \overline{E}^M (I_4 - S) : (\overline{E}^F - \overline{E}^M)]^{-1} \tag{A23}$$

Using equations (A22) and (A23) in the constraint equations:

$$\overline{c}^M \overline{A}^M + \overline{c}^F \overline{A}^F = I_4 \tag{A24}$$

$$\overline{c}^M \overline{B}^M + \overline{c}^F \overline{B}^F = I_4 \tag{A25}$$

We can easily obtain the elastic stress and strain concentration tensors $\overline{B}^M$ and $\overline{A}^M$, respectively, for the matrix. In equation (A22) and (A23), the fourth-rank tensor S is Eshelby's tensor for elasticity. Its nonzero components for cylindrical fibers with circular cross-section are given by [150]:

$$S_{1111} = \frac{5 - \nu^M}{8(1 - \nu^M)}$$

$$S_{2222} = S_{1111}$$

$$S_{1122} = \frac{4\nu^M - 1}{8(1 - \dot{\nu}^M)}$$

$$S_{2211} = S_{1122}$$

$$S_{2323} = \frac{1}{4}$$

534

$$S_{2233} = \frac{\nu^M}{2(1-\nu^M)}$$

$$S_{3311} = 0$$

$$S_{1133} = S_{2233}$$

$$S_{3322} = 0$$

$$S_{1212} = \frac{3-4\nu^M}{8(1-\nu^M)}$$

$$S_{3131} = \frac{1}{4}$$

The plastic strain concentration tensor $\overline{A}^{MP}$ for the matrix is obtained based on the Mori-Tanaka method using euqation (A22), except that the tensors $\overline{E}^M$ and S are no longer constant but vary with the load increments. In fact, $\overline{E}^M$ is replaced by $\overline{D}^M$ as follows:

$$\overline{A}^F = [I_4 + \overline{c}^M S^P : \overline{D}^{M-1} : (\overline{E}^F - \overline{D}^M)]^{-1} \tag{A26}$$

$$\overline{B}^F = [I_4 + \overline{c}^M : \overline{D}^{M-1} : (I_4 - S^P) : (\overline{E}^{F-1} - \overline{D}^{M-1})]^{-1} \tag{A27}$$

Substituting equations (A26) and (A27) into the constraint equations (A24) and (A25), respectively, one obtains the plastic strain and stress concentration tensors for the matrix. In equations (A26) and (A27), the fourth-rank tensor S^P is Shelby's tensor for plastic deformation. No closed-form expression exists for this tensor. Its evaluation will be performed numerically at each load increment using the procedure proposed by Gavazzi and Lagoudas [107]. Its components are given by the double integral:

$$S_{ijkl}^P = \frac{1}{8\pi} E_{mnkl}^{-M} \int_{-1}^{+1} d\xi \int_{0}^{2\pi} [G_{imjn}(\overline{\xi}) + G_{jmin}(\overline{\xi})] d\omega \tag{A28}$$

where

$$G_{ijkl}(\overline{\xi}) = \frac{\overline{\xi}_k \overline{\xi}_l \overline{N}_{ij}(\overline{\xi})}{D(\overline{\xi})} \tag{A29}$$

$$\overline{\xi}_i = \frac{\xi_i}{a_i} \tag{A30}$$

$$\xi_1 = \sqrt{1-\xi_3^2}\,.\cos\omega \tag{A31}$$

$$\xi_2 = \sqrt{1-\xi_3^2}\,.\sin\omega \tag{A32}$$

$$\xi_3 = \xi_3 \tag{A33}$$

$$D(\overline{\xi}) = \varepsilon_{mnl}K_{m1}K_{n2}K_{l3} \tag{A34}$$

$$N_{ij}(\overline{\xi}) = \frac{1}{2}\varepsilon_{ikl}\varepsilon_{jmn}K_{km}K_{ln} \tag{A35}$$

$$K_{ik} = E_{ijkl}^{-M}\overline{\xi}_i\overline{\xi}_l \tag{A36}$$

In equation (A28) and (A29), the indicial summation convention for representing tensor components is used. The double integral of equation (A28) is integrated numerically using Gauss-Legendre quadrature.

SUBJECT INDEX